Introduction to Medical Physics

Series in Medical Physics and Biomedical Engineering

Series Editors: Kwan-Hoong Ng, E. Russell Ritenour, and Slavik Tabakov
Recent books in the series:

Advanced Radiation Protection Dosimetry
Shaheen Dewji, Nolan E. Hertel

On-Treatment Verification Imaging A Study Guide for IGRT
Mike Kirby, Kerrie-Anne Calder

Modelling Radiotherapy Side Effects Practical Applications for Planning Optimisation
Tiziana Rancati, Claudio Fiorino

Proton Therapy Physics, Second Edition
Harald Paganetti (Ed)

e-Learning in Medical Physics and Engineering: Building Educational Modules with Moodle
Vassilka Tabakova

Diagnostic Radiology Physics with MATLAB®: A Problem-Solving Approach
Johan Helmenkamp, Robert Bujila, Gavin Poludniowski (Eds)

Auto-Segmentation for Radiation Oncology: State of the Art
Jinzhong Yang, Gregory C. Sharp, Mark Gooding

Clinical Nuclear Medicine Physics with MATLAB: A Problem Solving Approach
Maria Lyra Georgosopoulou (Ed)

Handbook of Nuclear Medicine and Molecular Imaging for Physicists – Three Volume Set
Volume I: Instrumentation and Imaging Procedures
Michael Ljungberg (Ed)

Practical Biomedical Signal Analysis Using MATLAB®
Katarzyna J. Blinowska, Jaroslaw Zygierewicz

Handbook of Nuclear Medicine and Molecular Imaging for Physicists – Three Volume Set
Volume II: Modelling, Dosimetry and Radiation Protection
Michael Ljungberg (Ed)

Handbook of Nuclear Medicine and Molecular Imaging for Physicists – Three Volume Set
Volume III: Radiopharmaceuticals and Clinical Applications
Michael Ljungberg (Ed)

Electrical Impedance Tomography: Methods, History and Applications, Second Edition
David Holder and Andy Adler (Eds)

Introduction to Medical Physics
Stephen Keevil, Renato Padovani, Slavik Tabakov, Tony Greener, Cornelius Lewis (Eds)

For more information about this series, please visit: https://www.routledge.com/Series-in-Medical-Physics-and-Biomedical-Engineering/book-series/CHMEPHBIOENG

Introduction to Medical Physics

Edited by
Stephen Keevil
Renato Padovani
Slavik Tabakov
Tony Greener
Cornelius Lewis

CRC Press
Taylor & Francis Group
Boca Raton London New York

CRC Press is an imprint of the
Taylor & Francis Group, an **informa** business

First edition published 2022
by CRC Press
6000 Broken Sound Parkway NW, Suite 300, Boca Raton, FL 33487-2742

and by CRC Press
2 Park Square, Milton Park, Abingdon, Oxon, OX14 4RN

CRC Press is an imprint of Taylor & Francis Group, LLC

© 2022 Taylor & Francis Group, LLC

Library of Congress Cataloging-in-Publication Data
Names: Keevil, Stephen F. (Stephen Frederick), editor. I Padovani, Renato,
editor. I Tabakov, Slavik, editor. I Greener, Tony, editor. I Lewis,
Cornelius, editor.
Title: Introduction to medical physics / edited by Stephen Keevil, Renato
Padovani, Slavik Tabakov, Tony Greener and Cornelius Lewis.
Description: First edition. I Boca Raton : CRC Press, 2022. I Series:
Series in medical physics and biomedical engineering I Includes
bibliographical references and index.
Identifiers: LCCN 2021033606 I ISBN 9781498744799 (hardback) I ISBN
9781138742833 (paperback) I ISBN 9780429155758 (ebook)
Subjects: LCSH: Medical physics.
Classification: LCC R895 .I68 2022 I DDC 610.1/53--dc23
LC record available at https://lccn.loc.gov/2021033606

ISBN: 978-1-498-74479-9 (hbk)
ISBN: 978-1-138-74283-3 (pbk)
ISBN: 978-0-429-15575-8 (ebk)

DOI: 10.1201/9780429155758

Typeset in Times
by SPi Technologies India Pvt Ltd (Straive)

Contents

Preface

The relationship between physics and medicine can be traced back thousands of years, but medical physics, in the modern sense of the term, is usually said to have begun with the discovery of X-rays by Wilhelm Röntgen in 1895. With rapidly expanding use of ionising radiation in diagnosis and therapy early in the twentieth century, there was a need for physicists to be directly involved in clinical work for the first time, and a distinctive medical physics profession was born. Since then, medical physicists have contributed to numerous advances in medical technology and practice, particularly in imaging and radiotherapy. Medical physics departments in hospitals provide scientific support for a wide range of clinical services, ensuring that physics-based technology is used as safely and effectively as possible. In areas such as radiotherapy, physics staff are involved directly in the treatment of individual patients. Today, medical physics provides many fascinating examples of the beneficial application of cutting-edge science, with discoveries in areas such as particle physics, superconductivity and astronomy finding applications in the clinic. It frequently features as an option in undergraduate degree programmes and there are a growing number of specialist postgraduate programmes, both taught and research-based. Medical physics is one of the most popular career choices for physics graduates, with fulfilling career options in healthcare services, university research and industry.

Despite the growing popularity and importance of medical physics, there are very few introductory textbooks that provide comprehensive coverage of the field. This book aims to meet the needs of advanced undergraduate and postgraduate students studying medical physics, as well as students in allied areas such as biomedical engineering, health physics and radiography. It covers all aspects of modern medical physics, including ionising and non-ionising radiation physics, the physics of imaging (using x-rays, ultrasound, nuclear medicine, MRI and optical methods), radiotherapy (including brachytherapy and molecular radiotherapy), radiation protection and the principles of image display and analysis. A final chapter looks ahead at how selected emerging techniques may influence future diagnosis and treatment.

Each chapter has been written by experts in the relevant specialism. The authors between them have many years of experience in the development and delivery of academic programmes in medical physics, including programmes specifically designed to meet the needs of graduate physicists training for clinical roles in the UK National Health Service. They also have extensive experience in medical physics research and development and in the management of clinical services, and the material is designed to provide the underpinning knowledge needed for practical training in hospital-based medical physics.

Editors

Stephen Keevil is Head of Medical Physics at Guy's and St. Thomas' NHS Foundation Trust, where he leads a team of around 180 physicists, engineers and technologists working across the full range of medical physics and clinical engineering disciplines. He is also Professor of Medical Physics at King's College London and chairs the examination board for the university's MSc programmes in medical engineering and physics. Professor Keevil studied physics at Oxford University, followed by an MSc in Medical Physics from the University of Surrey and a PhD in NMR spectroscopy from the University of London. Following a broad training in medical physics in the National Health Service, he held a series of research, academic and clinical roles in magnetic resonance imaging physics over a period of almost 30 years prior to his current appointment. He has taught MR physics at undergraduate and postgraduate levels throughout that time. He is President-Elect of the British Institute of Radiology, a member of the Administrative Council of the International Union for Physical and Engineering Sciences in Medicine and a past President of the Institute of Physics and Engineering in Medicine.

Renato Padovani, medical physicist consultant at the International Centre for Theoretical Physics (Trieste, Italy), is the Coordinator of the Master of Advanced Studies in Medical Physics, a Joint ICTP and Trieste University programme aiming to support the development of medical physics in low-middle income countries. He is also a teacher of radiation dosimetry and radiation protection and contributes to the development of medical physics activities of the ICTP. He has been Head of the Medical Physics Department at the S. Maria della Misericordia University Hospital of Udine (Italy) for 30 years developing clinical and research projects in the context of Italian and European programmes in diagnostic and interventional radiology, radiation therapy, nuclear medicine and radiation protection. He has served as an expert in IAEA missions and task groups for more than 20 years. He is co-founder and an honorary member of the Italian Association of Medical Physics and he served for 6 years as secretary-general of the European Federation of Organisations for Medical Physics (EFOMP).

Slavik Tabakov is Vice-President of the International Union for Physical and Engineering Sciences in Medicine (IUPESM, 2018-2022) and Coordinating Director of the International College on Medical Physics, ICTP, Trieste, Italy. He was the Director of three MSc programmes in Medical Physics and Engineering at King's College London, UK from 2000 to 2018. He has been an officer of the International Organization for Medical Physics (IOMP) since 2000 and President in 2015-2018. Prof. Tabakov graduated in Sofia, Bulgaria, followed by industry training in the USA and France, and a PhD in Computed Tomography densitometry. He has contributed to medical physics development in low- and middle-income countries (LMIC) and has advised the development of 16 MSc programmes in LMIC. He was a lecturer and expert in many IAEA projects in the fields of X-ray Diagnostic Radiology and education, including the development of TCS 56 "Postgraduate Medical Physics Academic Programmes". He developed and coordinated 7 international pilot projects, which produced the first e-learning and the first educational website in medical physics, the first e-Encyclopaedia of Medical Physics and related Scientific Dictionary (now in 32 languages). He is the Founding Co-Editor in Chief of the IOMP Journal Medical Physics International and is a Fellow of IUPESM, IOMP and IPEM.

Tony Greener read physics at York University graduating with a BSc in 1982. He began his career in the UK National Health Service as the North East Thames Regional Medical Physics trainee from 1982 to 1984 gaining an MSc in radiation physics as part of this early training scheme. He joined the Medical Physics department at Guy's and St. Thomas' Hospital, London in 1988 remaining until retirement in 2021. In 2013, he was appointed head of the radiotherapy physics team managing a large group supporting clinical radiotherapy services to a population of 1.9 million people in the south of London and beyond. Research areas developed in the group over this period have centred around the application of machine learning within radiotherapy. In 2016 Tony oversaw physics aspects relating to the expansion and transfer of the radiotherapy department into a purpose-built integrated cancer centre on the Guy's campus along with the opening of the new Guy's 'satellite' cancer unit at Queen Mary's Hospital, Kent. Throughout his career, Tony has taught a wide range of radiotherapy physics topics to medical physics trainees, clinical oncologists, radiotherapy radiographers and medical students. As a member of several examination boards, this has included the design of teaching modules, examination papers and the supervision of numerous MSc projects.

Cornelius Lewis graduated with a PhD in Medical Physics from Leeds University in 1979. After completing a research post between Surrey University and the University of British Columbia, he pursued a career in the UK NHS. Dr Lewis has worked in a number of NHS Trusts across London in nuclear medicine, diagnostic radiology and radiation protection. He joined the Medical Engineering and Physics Department of King's College Hospital, London in 1991 as Radiation Protection Adviser prior to being appointed as Head of Radiation Physics and from 2004 until his retirement in 2016 was Director of Medical Engineering and Physics.

Contributors

Michele Avanzo
IRCCS Centro di Riferimento Oncologico
Aviano, Italy

Elizabeth Benson
King's College Hospital NHS Foundation Trust
London, UK

Luciano Bertocchi
Abdus Salam International Centre for
    Theoretical Physics
Trieste, Italy

Paola Bregant
ASU GI
Trieste, Italy

Mauro Carrara
Fondazione IRCCS Istituto Nazionale dei
    Tumori
Milano, Italy

Elena De Ponti
ASST Monza
Monza, Italy

Charles Deehan
King's College London
London, UK

Fiammetta Fedele
Guy's and St Thomas' NHS Foundation Trust
London, UK

Tony Greener
Guy's and St Thomas' NHS Foundation Trust
London, UK (retired)

Emma Jones
Guy's and St Thomas' NHS Foundation Trust
London, UK

Stephen Keevil
Guy's and St Thomas' NHS Foundation Trust
London, UK

Andrew King
King's College London
London, UK

Cornelius Lewis
King's College Hospital NHS Foundation Trust
London, UK (retired)

Renata Longo
Università degli studi di Trieste
Trieste, Italy

Raffaele Novario
Università degli Studi dell'Insubria
Varese, Italy

Renato Padovani
Abdus Salam International Centre for
    Theoretical Physics
Trieste, Italy

Luigi Rigon
Università degli studi di Trieste
Trieste, Italy

Perry Sprawls
Emory University
Atlanta, USA

Lidia Strigari
IRCCS Azienda Ospedaliero-Universitaria di
    Bologna
Bologna, Italy

Sabina Strocchi
ASST dei Sette Laghi
Varese, Italy

Slavik Tabakov
King's College London
London, UK

Christopher Thomas
Guy's and St Thomas' NHS Foundation Trust
London, UK

Jim Thurston
Dorset County Hospital NHS Foundation Trust
Dorchester, UK

Francesco Ziglio
Ospedale Santa Chiara
Trento, Italy

1 Medical Physics, an Introduction

Perry Sprawls
Emory University, Atlanta, USA

CONTENTS

1.1 INTRODUCTION

Physics, along with biology, chemistry and psychology, is one of the basic sciences that are the foundation of medicine. Physics is especially significant because the human body is a physical environment and system. It is within the physical body that all of the other scientifically based functions take place producing and supporting life.

Medicine is the comprehensive science and practice of diagnosing, treating or preventing disease and other damage to the human body and mental system. This is usually achieved by interacting with the different scientifically based functions within the body. For example, infections are biological events and poison is a chemical condition. These would generally be diagnosed and treated based on those sciences.

Because the structure, composition and many functions of the human body are physical, physics is the basic science for the diagnosis and treatment of many conditions and is the foundation of the profession of *medical physics*.

DOI: 10.1201/9780429155758-1

1.2 MEDICAL PHYSICISTS

Medical physicists are professionals with a strong academic background in general physics, medical physics topics and other medically related subjects including anatomy, physiology and pathology. They work in research and development, education and clinical medical physics. Clinical medical physicists generally have academic degrees in medical physics, supervised work experience (for example a residency programme in the US or the UK Scientist Training Programme (STP)), and are certified by regulatory bodies or professional organisations.

In addition to physicists, there are many other medical professionals, especially radiologists, radiographers and technologists, who apply physics in their clinical activities.

1.3 PHYSICS AND MEDICINE

An overview of the relation of physics to medicine is shown in Figure 1.1.

The two types of medical procedures that are based on physics are *diagnosis* and *therapy* (treatment).

1.4 THE DIAGNOSTIC PROCESS

Diagnosis of diseases or injuries is usually a two-step process. The first step is obtaining information from the human body in the form of images that are produced by physical interactions within the body. The second step is the 'reading of the image' and interpreting or translating that information into a medical diagnosis. This last step is usually performed by qualified medical doctors, generally radiologists.

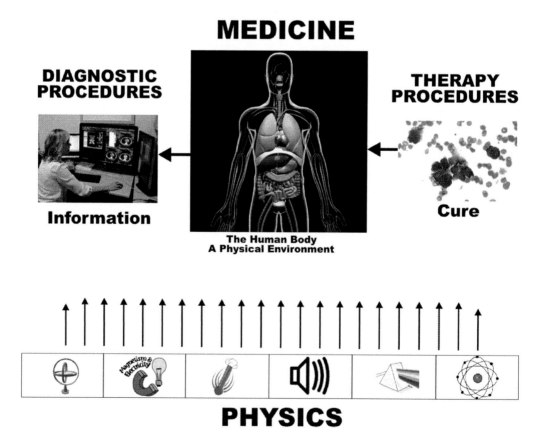

FIGURE 1.1 Overall relationship of physics to medicine.

Physicists play a major role in the first step, designing and optimising imaging methods and procedures to capture the medically significant information from within the body while managing any potential risks to patients. This can be a complex process because images are physical objects with a number of important characteristics that can affect the visibility of objects and conditions within the body. It is the physicist who has the knowledge and experience required to analyse and adjust or optimise these factors for specific clinical procedures.

Clinical medical physicists are high-level professionals on the staff of hospitals and clinics. Their role is in assuring the diagnostic quality of imaging procedures and optimising procedures with respect to image quality and potential risks to patients. This is through several activities. One activity that is often required by government and accrediting regulations is the periodic evaluation of imaging equipment performance and image quality with specific testing procedures. Physicists use their knowledge and experience to consult and collaborate with the other medical imaging professionals, especially radiologists and technologists, in developing and optimising clinical procedures. Many medical physicists are educators, teaching medical physics students and physics residents, and also trainee radiologists who require knowledge of physics to qualify as radiologists.

1.5 THE THERAPEUTIC PROCESS

There are several forms of therapy based on the application of various forms of physical energy to the body to treat and hopefully cure various diseases, injuries or other abnormal conditions. The use of ionising radiation is the medical specialty of *radiation oncology* or *radiotherapy* and is done by or under the direction of qualified medical professionals, generally physicians certified in that field.

Medical physicists play a significant role in therapeutic procedures using ionising radiation to treat cancer. The challenge and goal for each procedure is to maximise the radiation dose to the cancer tissues while minimising the radiation to the surrounding healthy tissues. This is achieved by the often complex process of treatment planning conducted by medical physics staff.

1.6 AREAS OF PHYSICS

All areas of physics, ranging from mechanical to atomic and nuclear, have applications in both diagnostic and therapeutic procedures. Some are much more significant in modern medicine as will be described.

Atomic and nuclear physics is by far the foundation of most physics applications in medicine because this is both the source of several types of radiation and the basis of the interactions of radiation within the human body.

1.7 IONISING RADIATION

Of special significance are ionising radiations, x-rays and radioactive substances, which can penetrate the human body and also produce biological effects when absorbed in tissue. Biological effects are the basis for radiation therapy to treat cancer but are generally undesirable in imaging applications. Medical physicists and the related profession of health physicists are scientists with knowledge and experience related to the exposure of humans to ionising radiation. Much of their work is in controlling and using radiation for the maximum benefit of humans.

Medical physics and the application of physics in medicine can be divided into four categories:

- medical imaging for medical diagnosis
- radiation therapy for treatment of cancer
- other physics-based medical applications
- radiation safety and risk management

The first two areas are where most medical physicists work and are often required by professional standards and legal regulations in clinical medicine activities.

1.8   MEDICAL PHYSICS AND INNOVATIONS IN TECHNOLOGY

The science of medical physics is closely related to developments and innovations in technology. Most physics activities involve equipment, and this is especially true for medical physics. Many of the advances in medical physics over the years have occurred and been made possible by developments in technology, especially computing and digital electronics. Most of the diagnostic imaging and treatment methods discussed later are based on physics interactions with the human body but are very much 'high technology'. Therefore, medical physicists need up-to-date knowledge in technology along with physics.

1.9   MEDICAL IMAGING FOR DIAGNOSIS

One of the greatest challenges in medical diagnosis, and where physicists have made major contributions, is in viewing the interior of the human body. Normal vision with light is generally limited to the surface of the body; so extending vision into the body requires radiation that can penetrate into and through the body. This is possible with radiations both below and above visible light in the electromagnetic spectrum, radio-frequency (RF) below and the ionising radiations above. It is these radiations and the associated physics that are the foundation of most modern imaging methods. The development of medical imaging has been an ongoing process for well over a century driven by physics discoveries and developments in technology. This includes digital and computer technology that is now a major component of all medical imaging methods.

1.9.1   The Discovery and the Beginning

Medical imaging, and to a great extent medical physics, had its origin in a physics laboratory at the University of Würzburg, Germany in 1895 when Professor Wilhelm Roentgen discovered a 'new kind of rays' to become known as x-radiation or Roentgen radiation. Following the discovery, he conducted extensive research to determine the properties of this radiation. This included the ability to penetrate objects and cast shadow images of internal structures and to display the images on both fluorescent screens and photographic plates. In early 1896, Dr. Roentgen gave a presentation in which he described the results of his research on the properties of the new radiation and demonstrated the production of an image showing the bones in a human hand.

It was quickly realised that this was not only a major discovery in physics but also an answer to a great need in medicine. The news quickly spread around the world and a new era combining physics and medicine began.

In many countries, the first medical x-ray images were not in hospitals but in physics laboratories and conducted by physicists. This was because the early x-ray machines consisted of partially evacuated glass tubes and high-voltage sources, already being used in many physics laboratories. Roentgen's discovery was that this type of equipment also produced x-radiation.

These early x-ray images were a major breakthrough and contribution to diagnostic medicine, but it was only the beginning for the developments of the many different and highly advanced medical imaging methods used today. All of these are based on physics principles and are a major component of the profession of medical physics.

1.9.2   The Quest for Extended Visibility

The purpose of medical imaging procedures is to provide visibility into the human body so that medical professionals can see anatomical structures, tissue characteristics, biological functions, and other signs of disease and injury. Medical imaging equipment and technology can be compared to other scientific instruments such as microscopes and telescopes used to extend visibility. All medical imaging methods use technology or equipment to provide visibility. A question might be, 'Why are there many different types of imaging methods or modalities used today?' The answer is that there is no medical imaging method that provides visibility of everything within the body. Every method has its limits. The research for more than a century has been to apply physics and develop technology that can extend visibility to a larger range of objects within the human body. Visibility of a specific object or structure, such as a cancer, within the human body is determined by the combined physical characteristics of the objects and the image as will be described later after introducing the imaging methods that are now used.

1.9.3 IMAGING METHODS OR MODALITIES

The different imaging modalities used today are based on various physics principles and technological developments. An overview is shown in Figure 1.2.

While the imaging modalities are very different in how they produce images and the visibility they provide, there are some common factors that are of special interest with respect to physics. Images are physical objects with specific physical characteristics. The application of physics in medicine is often focused on images. This includes the analysis of images from a physics perspective, matching image characteristics to specific clinical medicine requirements, selecting and adjusting imaging equipment to produce images with the appropriate characteristics. Two major functions of medical physicists are analysing image characteristics with respect to equipment performance in the context of quality control/assurance activities and providing physics consultation and knowledge to other members of the medical imaging team, physicians and technologists.

The various imaging methods and modalities will be described in detail in other chapters. Here is a brief introduction of each.

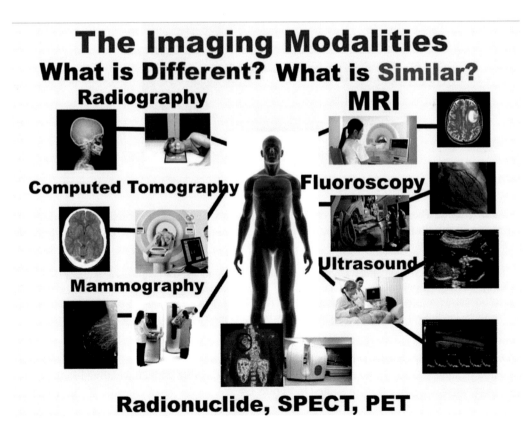

FIGURE 1.2 Imaging modalities used in medicine for diagnostic purposes.

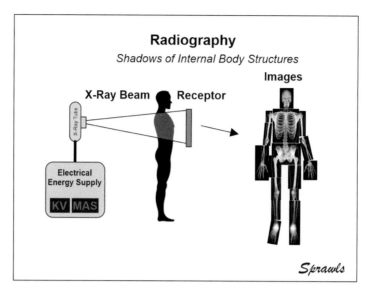

FIGURE 1.3 Radiographic process.

Mammography

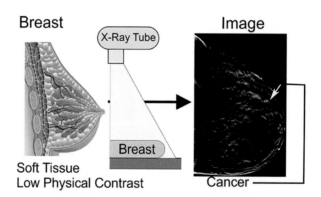

FIGURE 1.4 X-ray imaging procedure specifically for examining the breast and detecting cancer.

Radiography: The formation of an image by passing an x-ray beam through the body and recording the image on film or digital media. It is what Roentgen demonstrated. The process is illustrated in Figure 1.3.

Radiography, the creation of permanent images on film or digital media, continues as a major form of x-ray imaging. It is a relatively quick and easy process, especially with digital procedures. It has the advantage of producing images with high visibility of detail, especially bone composition and fractures, but limited to imaging structures in the body that have relatively high differences in density (e.g. bones, lungs).

Mammography: A specific type of radiography developed for breast imaging. It provides for visualisation of signs of breast cancer and other conditions not visible with conventional radiography. The mammography process is illustrated in Figure 1.4.

The female breast is a significant area for the development of cancer, and x-ray imaging and mammography has been an effective method for detecting cancers in early stages when most curable. There is a

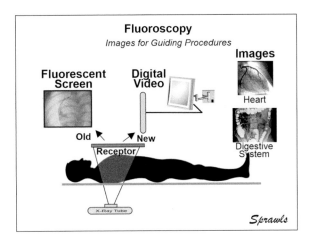

FIGURE 1.5   Basic components of a fluoroscopic system and typical images.

major challenge in that the breasts consist of soft tissues with very low differences in physical density between cancers and normal tissue to produce visible contrast in images. Also, a valuable sign of some cancers is small calcifications or 'micro-calcifications' that require imaging procedures with very low blurring and high visibility of detail. The development of mammography equipment and methods to increase visibility of cancers continues with much advancement.

Fluoroscopy: In fluoroscopy, the same principle as in radiography is applied to obtain real-time and continuous images especially useful for observing motion. It was first performed by viewing images on fluorescent screens, which led to the origin of the name fluoroscopy. Fluorescent screens have been replaced with either electronic image intensifier tubes or digital devices that generally require less radiation exposure. The fluoroscopy process is illustrated in Figure 1.5.

Fluoroscopy provides immediate and continuous viewing into the body by physicians. A major application is guiding procedures in which substances, contrast media, are injected into the body cavities or vessels to produce visible contrast. Examples are iodine compounds into the vessels of the heart and barium into the digestive system.

Computed tomography (CT): CT is an x-ray imaging method that has two major advantages over radiography. Tomography is the process of 'slicing' and producing images of individual slices of the body so they can be viewed directly. It also uses digital radiation detectors and computer technology for the process of 'windowing' that enables the visualisation of the soft tissues within the body, especially the head. The concept of CT is illustrated in Figure 1.6.

CT combines two features to increase visibility of soft tissues and especially the brain within the skull. One is producing direct images of thin slices without overlying anatomy and the other is very high contrast sensitivity using the concept of 'windowing' to enhance the visibility of low-contrast regions such as diseased tissue.

Magnetic resonance imaging (MRI): This technique produces images that often look like CT images but is based on different tissue characteristics. Images are produced by placing the body in a strong magnetic field which produces the condition of RF resonance in the tissues as illustrated in Figure 1.7.
  The resulting images are displays of RF signal intensities collected from the body. The signal intensities are determined by a combination of magnetic characteristics of the tissues.

MRI provides many advantages compared to other imaging methods. Images can be produced in any plane through the body, especially valuable for imaging along the spine. It can be adjusted to visualise

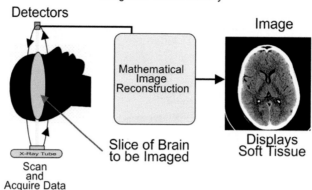

FIGURE 1.6   Concept of computed tomography and typical image.

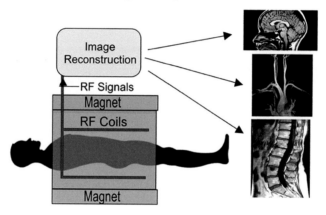

FIGURE 1.7   Magnetic resonance imaging (MRI) process.

several different tissue characteristics increasing the possibility of visualising abnormalities. It has the capability of imaging blood flow without the injection of contrast media. Associated with this imaging procedure is the possibility of spectroscopic analysis of tissues.

Ultrasound imaging: High-frequency sound, or ultrasound, is used for medical imaging in two ways. First, pulses of ultrasound are transmitted into the body, resulting in reflections or echoes which display the tissue structure characteristics. The other is using the Doppler physics principle to produce images of flowing blood. The general process and typical images are shown in Figure 1.8.

Ultrasound imaging has the advantages of not using ionising radiation and is relatively easy to perform. It has several valuable applications, as illustrated, which makes it a useful method to be used along with the other imaging methods.

Nuclear medicine or radionuclide imaging: Radioactive substances (radiopharmaceuticals) are administered to patients and images are produced showing their distribution and selective uptake in various organs and tissue conditions as illustrated in Figure 1.9.

Ultrasound Imaging

Images of Reflected Sound Echos

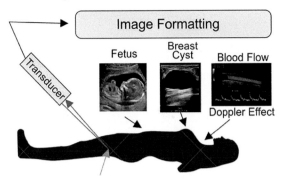

FIGURE 1.8 Images created from ultrasound reflections with the human body.

Nuclear Medicine

Images of Radioactivity in the Body

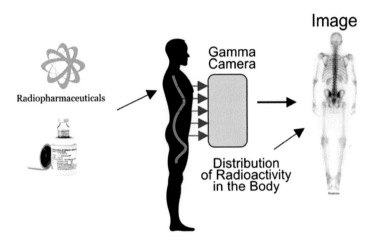

FIGURE 1.9 Using a gamma camera to image radioactive substances within the human body.

Many different radioactive isotopes and compounds are used to image various medical conditions including cancer. The gamma camera is the device used for general radionuclide imaging. It produces images of body sections. There are two additional radionuclide methods that produce tomographic images and have specific clinical applications.

The great value in nuclear medicine imaging is in the range of agents, radiopharmaceuticals, that have been developed and are available. These are various radioactive isotopes and compounds that selectively concentrate in specific organs and tissues in relation to functions and conditions including cancers.

Single-photon emission computed tomography (SPECT): This is performed using a gamma camera to produce CT images of slices of tissue as illustrated in Figure 1.10.

The term 'single photon' does not mean that just one photon is used. It distinguishes the method from positron emission tomography (PET) where each nuclear event produces a pair of photons.

Single Photon Emission Computed Tomography

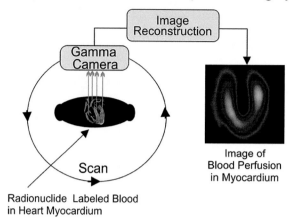

FIGURE 1.10   Tomographic images produced by scanning a gamma camera around a patient's body and reconstructing an image.

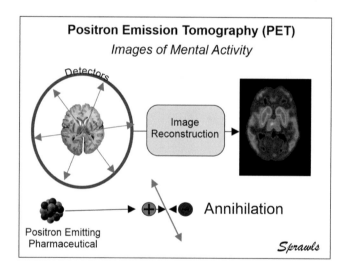

FIGURE 1.11   Concept of positron emission tomography (PET) producing images of mental activity.

SPECT is performed using a gamma camera along with the process of CT to produce tomographic slice images showing the distribution of radiopharmaceuticals in specific organs and tissues. This can be used to visualise conditions not possible with other imaging methods.

Positron emission tomography (PET): Radioactive elements that produce positrons have a very valuable role in medical imaging. They are naturally found in living biological systems including the human body and are actively involved in biological functions like metabolism. PET provides images of biological function, not just the static characteristics of tissues provided by most other imaging methods. The PET process is illustrated in Figure 1.11.

The great value of PET is in the radioisotopes that are positron emitters. They are low atomic number isotopes that are naturally active in biological functions. This provides the possibility of functional imaging including activity in various areas of the brain.

Complementarity of imaging modalities: A modern and well-equipped hospital or clinic will often have each of these imaging methods. Each one provides visibility of some structure, object, condition or function that cannot be imaged by the other methods. One of the distinguishing characteristics of each imaging method is the physical condition (density, radioactivity, etc.) that forms the image. This is the general characteristic of contrast sensitivity which can also depend on how the equipment is operated.

An important characteristic of an imaging process is visibility of detail, or the smallest objects that can be imaged. This is determined by the inherent blurring associated with each imaging method. The blur and resulting visibility of detail with a specific imaging modality is determined by the combination of equipment design characteristics and adjustments of the technique or protocol factors for each imaging procedure. This will be described for each imaging modality in later chapters.

1.10   RADIATION THERAPY

Medical physics provides one of the most effective methods for treating cancer used today – through the process of applying ionising radiation directly to cancer cells within the human body to stop them multiplying and spreading. It is the medical specialty of radiation therapy (radiotherapy) or radiation oncology. There are different methods used to deliver radiation to cancers as shown in Figure 1.12.

There are two major types of radiation therapy, depending on the location of the radiation source relative to the patient's body – external or internal (brachytherapy and molecular radiotherapy).

There are several types of radiation sources used for external radiation therapy and two major requirements – the ability to produce radiation that can penetrate well into the patient's body at a high intensity or dose rate to minimise the time required for treatment procedures.

Linear accelerators: The linear accelerator, or Linac, is widely used for this purpose. The output is a high photon energy x-ray beam that can be controlled to provide the desired distribution of radiation dose within the cancer area. One such innovation is the process of adjusting and varying the intensity of the radiation within the treatment area known as intensity-modulated radiation therapy (IMRT).

Proton therapy: Protons, which are charged particles, have some advantages in radiation therapy because they penetrate into the body and then deposit more energy at the location of the cancer and not near the surface.

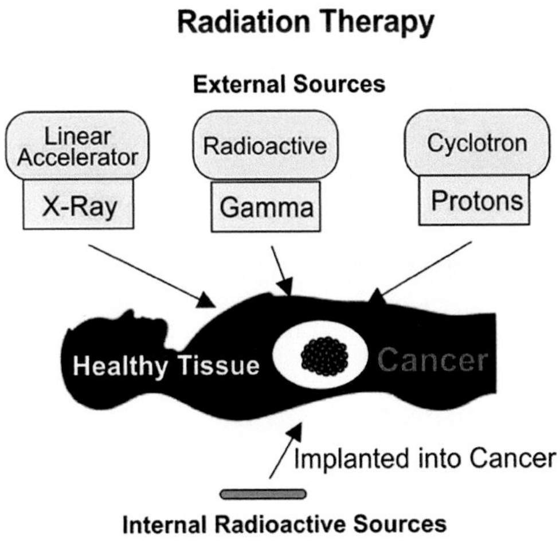

FIGURE 1.12   Different methods and types of equipment used for radiation therapy.

Cobalt-60 therapy: The radioactive isotope cobalt-60 has been a widely used source for external radiation therapy for many years. Cobalt-60 as a source for radiation therapy had many values and advantages. It was available long before the development of appropriate accelerators, had a stable and constant output over time, required relatively low maintenance, and was general economically feasible. However, with the development and many innovations in accelerator technology providing higher and more controlled dose deliveries, cobalt-60 therapy is being replaced in many facilities. The 'gamma knife' is a machine that uses many cobalt-60 sources arranged to produce small beams of radiation converging on a point within the patient's body, primarily to treat brain lesions.

1.10.1   TREATMENT PLANNING AND REQUIREMENT FOR PRECISION

The great challenge in radiation therapy is delivering the necessary radiation dose to the cancer cells to kill them and minimising, and hopefully eliminating, damage to the surrounding normal tissues and organs. This has been the focus in the development of radiation therapy methods and technology for many years. With external therapy, this is usually achieved by using multiple radiation beam directions as described and illustrated below.

For each method and each individual patient a plan is developed that will guide the actual treatment procedure. Treatment planning is performed by medical physicists and by technologists or dosimetrists.

Depending on the treatment method, the planning process can be quite complex. It is often done with a computerised treatment planning system. Using images of the cancer area and prescribed radiation dose values from the oncologist, the physicist develops a plan for the procedure. It can be a complex combination of radiation beam sizes, shapes, intensities, and times.This is illustrated in Figure 1.13.

1.10.2   RADIUM AND THE ORIGIN OF RADIATION THERAPY

The discovery of radium, and separating it from other minerals in 1898 by Marie Curie, was the foundation of radiation therapy and a major contribution to the treatment of cancer. For many years, large radium sources were used for external therapy and small sources implanted for brachytherapy. Because of its long half-life, radium was removed after the treatment time and used for other patients.

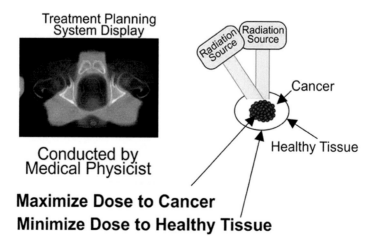

FIGURE 1.13   Display of a radiation therapy treatment plan showing characteristics of multiple radiation beams to deliver the required radiation dose to a cancer area.

1.11 OTHER PHYSICS-BASED MEDICAL APPLICATIONS

Medical physicists are significant participants in clinical medical procedures that use forms of ionising radiation as described above, both diagnostic radiology and radiation therapy. There are other medical practices that involve physical principles, including ophthalmology (sight) and audiology (hearing). Physicists might conduct research and development in these areas but are not usually involved in clinical procedures.

1.12 RADIATION SAFETY AND RISK MANAGEMENT

Because of the potential detrimental effects of ionising radiation to humans, both patients and medical staff, physicists are closely involved in risk minimisation. This specialisation is often known as health physics.

1.13 CONCLUSION

The human body is a *physical universe* in which a combination of chemical and biological interactions and processes occur to sustain life. All methods for diagnosing or treating diseases are based on one of the sciences: physics, chemistry or biology. While chemical- and biological-based diagnostic tests and medications for treatment are of great value, it is the physics-based medical imaging methods that predominate in modern clinical medicine and radiation therapy is a major method for treating cancer.

The development and effective application of these methods require physicists throughout the process and establishes the professional specialisation of Medical Physics. Physicists are highly valued and respected healthcare professionals, serving along with physicians to provide high-quality medical care around the world.

2 Radiation Interaction and Dosimetry

Renato Padovani
Abdus Salam International Centre for Theoretical Physics, Trieste, Italy

Charles Deehan
King's College London, London, UK

CONTENTS

DOI: 10.1201/9780429155758-2

2.1   INTRODUCTION

In medicine, ionising radiation plays an important role in imaging internal anatomy as well as in therapeutic applications. This is because of the attenuation properties of tissues and the capacity of ionising radiation to produce damage in biological structures, respectively.

The energy ranges that are used in medical applications are 10 keV to 25 MeV for photons and electrons, for neutrons up to 100 MeV, for protons up to 300 MeV, and for heavier charged particles up to 400 MeV/m_u (m_u – atomic mass unit).

As an understanding of radiation interactions with tissues developed, it has become possible to harness the benefits of radiation to produce highly useful diagnostic images and to successfully treat malignant and non-malignant conditions.

This chapter will describe different types of radiation used clinically and the ways in which their interaction with tissues is understood. It will also introduce elements of radiation dosimetry.

2.2   INTERACTION OF PHOTONS WITH MATTER

Radiation can be either directly or indirectly ionising. Charged particles such as electrons or protons are known as directly ionising radiation because they carry charge. In contrast, photons and neutrons are indirectly ionising through the secondary charged particles produced following their interactions.

Photons with a wide range of energies have been used for many years in diagnostic radiology (DR), external beam radiotherapy (EBRT) and brachytherapy (BT). Energies used in DR are typically in the range 10 to 150 keV and in EBRT 10 keV to 25 MeV although in practice the use of energies above 15 MeV is less common. The useful energy range in BT is from 20 keV to 3.5 MeV depending on the radionuclide used.

2.2.1   Absorption and Scatter

The energy attenuation from a photon beam as it passes through tissue can be understood in terms of the amount of resulting radiation absorption and scatter involved.

$$\text{Attenuation} = \text{Absorption} + \text{Scatter} \tag{2.1}$$

As we will see, absorption and scatter processes are energy dependent and an understanding of each of these in isolation is important.

In the energy range of interest here, there are three main processes by which photons are removed from a beam. These are the photoelectric effect, Compton effect and pair production. Photoelectric effect (PE) and pair production are essentially absorption processes without scatter. However, Compton interactions involve both absorption and scatter.

When a photon enters tissue, it can interact in a number of ways. At lower energies, it can take part in scattering interactions with single electrons or with atoms as a whole. At these energies, coherent scattering can take place where the photon does not lose energy but is deflected at an angle with respect its approach. Thomson scattering (TS) and Rayleigh scattering (RS) are examples of coherent scattering.

At higher energies, incoherent scattering can take place where the photon loses energy in the process. An example of this is Compton scattering where some photon energy is imparted to an electron which can in turn cause other ionising events. Another form of scattering can take place where a highly energetic electron produced by a photon interaction is deflected by a coulomb interaction with the nucleus producing a bremsstrahlung photon.

The processes above are illustrated in Figure 2.1 and will be discussed in greater detail later.

Absorbed radiation is the amount of energy that is removed from a beam and is deposited in matter. Measurement of this is important when determining absorbed radiation dose.

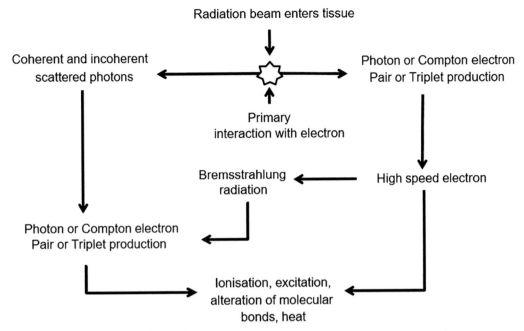

FIGURE 2.1 Photon attenuation in tissue.

Understanding absorption and scatter is important in the measurement of the half-value layer (HVL) thickness of material since this is used to determine the radiation quality of x-ray beams. For an HVL measurement to be reproducible, it must be measured under ***narrow-beam*** conditions to minimise the contribution of scattered radiation. This is illustrated in Figure 2.2 where narrow-beam conditions are achieved by careful beam collimation and experimental set-up. Measurements made under ***broad-beam*** conditions are very much dependent on the experimental set-up because of the scatter contribution.

2.2.2 ATTENUATION COEFFICIENTS

2.2.2.1 Linear Attenuation Coefficient

The cross-section for an interaction is the ratio of the number of interactions per target entity over the number of incident particles crossing per unit of area. This is measured in m^2 or barns (1 barn $= 10^{-28}$ m^2).

An important parameter used to characterise photon penetration in absorbing media is the linear attenuation coefficient, μ, which relates to the total interaction cross-section of the photon with matter for different types of interactions.

For a thin layer dx (m) of material of total atomic cross-section σ, the probability of interaction of a photon is $N_a \sigma\, dx$, where N_a is the number of atoms per unit of volume. The quantity $N_a \sigma$ is known as the linear attenuation coefficient, μ. N_a can be calculated from the Avogadro number N_A so that:

$$\mu = N_a \sigma = \frac{1000\ N_A \rho}{A} \sigma \quad \left[\text{unit: m}^{-1} \right] \tag{2.2}$$

where ρ = density.

Broad & Narrow Beam Attenuation
Half value layer (HVL)

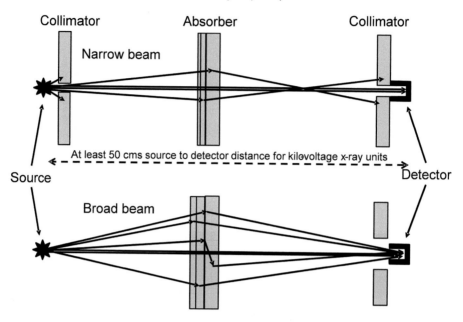

FIGURE 2.2   Broad- and narrow-photon beam attenuation.

The linear attenuation coefficient μ is the fraction of photons removed from the beam per unit thickness of absorber. For N_0 photons crossing a slab of thickness x, the expected change dN after crossing a thickness dx is given by

$$dN = -N\mu dx \quad \text{which integrates to} \quad N = N_0 e^{-\mu x} \tag{2.3}$$

This gives the well-known exponential attenuation law of photons in matter.

The mass attenuation coefficient μ/ρ is a frequently used quantity as it is independent of the material density, measured in units of m^2kg^{-1}, and the exponential law can be re-written as

$$N = N_0 e^{-\frac{\mu}{\rho}x} \tag{2.4}$$

where the units of x are now kg^1m^{-2}.

Equation 2.4 will only hold under narrow-beam conditions. Under broad-beam conditions, the relationship between incident and transmitted photons is not exponential. Narrow-beam conditions allow the reproducible measurement of HVL for keV energies and are therefore an important check of beam quality. Figure 2.3 shows the result of HVL and tenth-value layer (TVL) measurements for an absorber.

2.2.3   BEAM ATTENUATION PROCESSES

Over the range of energies used for clinical work, there are several processes that are responsible for removing energy from a photon beam. The three main processes are as follows:

- Photoelectric effect
- Compton effect
- Pair production

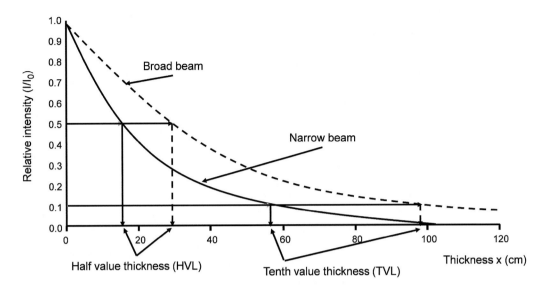

FIGURE 2.3 Broad- and narrow-photon beam attenuation.

Two minor processes are also involved these are as follows:

- Coherent scattering – Thomson and Rayleigh scattering
- Photonuclear reactions

The main energy removal processes all have their individual total mass attenuation coefficient: τ/ρ for photoelectric effect, σ/ρ for Compton effect and π/ρ for pair production.

Unlike PE and pair production, with mass absorption coefficients τ/ρ and π/ρ respectively, Compton interaction involves a combination of both absorption and scatter. Therefore, the Compton (σ/ρ) total mass attenuation coefficient is expressed as the sum of the Compton mass absorption coefficient (σ_a/ρ) plus a Compton mass scattering coefficient (σ_s/ρ) (see Figure 2.6 and discussion later).

The total mass attenuation coefficient μ/ρ is therefore:

$$\mu/\rho = \tau/\rho + \sigma_a/\rho + \sigma_s/\rho + \pi/\rho$$

As we shall see that the relative contribution of the individual coefficients to the total mass attenuation coefficient is energy dependent.

For dosimetry purposes, it is central to know the energy transferred by photons to charged particles (primarily electrons and positrons) and, finally, the energy transferred by these charged particles to matter. The mass energy transfer coefficient μ_{tr}/ρ is defined taking into account the mean energy transferred $\bar{T}$ to charged particles by photons of energy $h\upsilon$

$$\mu_{tr}/\rho = \mu \frac{\bar{T}}{h\upsilon} \tag{2.5}$$

Also the total mass energy transfer coefficient, μ_{tr}/ρ, is the sum of the mass energy transfer coefficients for each type of interaction

$$\mu_{tr}/\rho = \tau_{tr}/\rho + \sigma_{tr}/\rho + \pi_{tr}/\rho$$

where τ_{tr}/ρ, σ_{tr}/ρ and π_{tr}/ρ are the mass energy transfer coefficients for photoelectric effect, Compton effect and pair production, respectively.

Note that the TS and RS processes do not contribute to the total mass energy transfer coefficient, as these elastic interactions do not transfer energy to charged particles.

Because part of the energy transferred to charged particles is lost in radiative processes (mainly bremsstrahlung), the energy absorbed by matter (or imparted to matter) will be a fraction of the energy transferred. If we define $\bar{g}$ as the mean fraction of energy lost in radiative processes, the mass energy absorption coefficient, μ_{en}/ρ, becomes

$$\mu_{en}/\rho = \mu_{tr}/\rho\left(1-\bar{g}\right)$$

The radiative fraction $\bar{g}$ is negligible at keV photon energies used in diagnostic imaging with low Z materials. However, it becomes significant at MeV photon energies used in therapy and increases with energy especially where high Z materials are involved.

2.2.3.1 Photoelectric Effect (PE)

This process dominates up to 30 keV in the case of low Z materials or tissue. With PE the entire primary photon energy is absorbed when interacting with the target atom, i.e. the photon interacts with a bound electron giving it all its energy.

One electron, generally from the K or L shell of the atom, is then immediately ejected. If the total energy imparted by the photon ($h\upsilon$) is greater than the electron binding energy (B_e) the electron will eject with kinetic energy equal to T where:

$$T = h\upsilon - B_e \tag{2.6}$$

The mass attenuation coefficient for the PE (τ/ρ) is strongly dependent on atomic number, Z, and beam energy, E, especially at lower energies

$$\tau/\rho \propto Z^3/E^3$$

This is true up to 200 keV. At higher energies $\tau/\rho \propto Z^3/E^2$ and at even higher energies $\tau/\rho \propto Z^3/E$.

As electrons from higher levels fall to occupy gaps in the vacant lower energy levels, characteristic emission edges can be clearly detected in materials such as lead with high atomic number, Z. At low energy and low atomic number material, instead of characteristic photon energies, Auger electrons carry away extra energy from the excited atom (see Figure 2.4). Consequently, the mean energy transferred to charged particles is generally higher than T (Equation 2.5) because of the emission of Auger electrons.

Dependence of PE absorption on atomic number of the medium is the major reason for the clear contrast between bones and other tissues in diagnostic radiographs since the calcium and phosphates in the bones result in a higher effective atomic number, Z_{eff}, than muscle and fat which are mainly water. This is why lower (keV) energies are used to acquire radiographic and CT images since the contrast between soft tissue and bones is good.

At megavoltage energies contrast between soft tissue and bone is poor and images acquired for treatment planning systems often use diagnostic energy acquisitions (see Figure 2.10 and discussion below).

Radiotherapy treatment at low energies needs to be carefully considered especially with permanently implanted sources. BT can often be delivered at very low keV energies, for example, ^{125}I prostate implants (28 keV γ emission). If implanted seeds are adjacent to bony structures in the pelvis, then bone may absorb significantly more energy than the tumour with the risk of possible necrosis. Similarly, care is needed with low energy external beam delivery such as when intraoperative breast treatments are used in which the peak energy is around 50 keV near the rib cage.

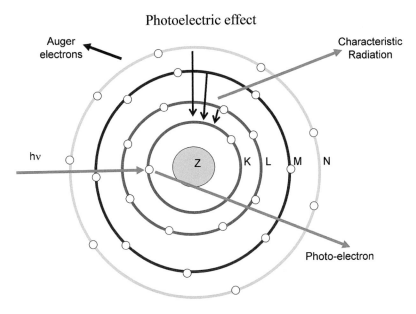

FIGURE 2.4 Photoelectric effect.

2.2.3.2 Compton Effect (CE)

For soft tissues, the CE is the most important beam attenuation process in the range of 30 keV to 10 MeV. In this case, the incident photon interacts with a bound electron. However, the electron can be regarded as being at rest and the effect of binding energy negligible.

The electron is known as the recoil or Compton electron and the photon is scattered at an angle relative to the direction of incidence (see Figure 2.5). All angles of scatter are possible and the photon is scattered with reduced energy equal to that amount imparted to the Compton electron.

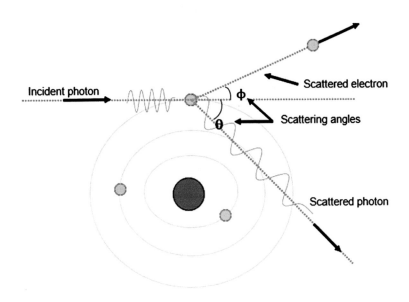

FIGURE 2.5 Compton effect.

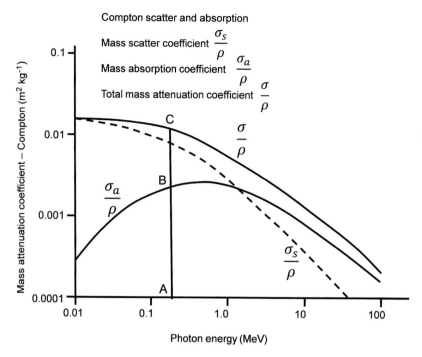

FIGURE 2.6 Compton scatter and absorption.

The mass attenuation coefficient for the CE is related to the electron density of the material and the beam energy as follows:

$$\sigma/\rho \propto \frac{\text{Electron density}}{\text{Beam energy}} \qquad (2.7)$$

Compton mass attenuation coefficients are nearly independent of Z (atomic number) because the number of electrons/g in a material varies only slightly across all elements other than hydrogen.

Because σ/ρ is proportional to electron density, CT imaging is important since it provides electron density information essential for absorbed dose calculation in radiotherapy.

Figure 2.6 is a graph of the total mass attenuation σ/ρ, mass absorption coefficient σ_a/ρ and the mass scattering coefficient σ_s/ρ for CE plotted against photon energy showing the total mass attenuation coefficient decreasing with photon energy. The mass absorption coefficient increases with energy and then decreases after about 1 MeV. However, the fraction of absorbed energy increases with energy. Thus, AB/AC increases so that the σ_a/ρ and σ/ρ curves approach each other at high energies reflecting the greater fraction of absorption by the medium as the energy increases. The mass scattering coefficient σ_s/ρ represents the average fraction of the total beam energy left to the photons after Compton scatter. The point at which the scattered photon and recoil electron share equal amounts of energy is where the curves cross at 1.5 MeV. Below this energy, on average, the scattered photon carries away more energy and above this energy the electron carries away more energy after the interaction.

2.2.3.3   Pair (PP) and Triplet Production

When the energy of the incident photon exceeds 1.022 MeV an absorption process known as pair productions may occur. This happens when the photon comes under the influence of the strong coulomb field of the nucleus of an atom where it can convert into a positron and an electron. This threshold value of 1.022 MeV is equal to the combined rest mass energy of the two particles (rest mass energy of a positron = rest mass of an electron = 0.511 MeV). Note that no net electronic

Pair and Triplet production

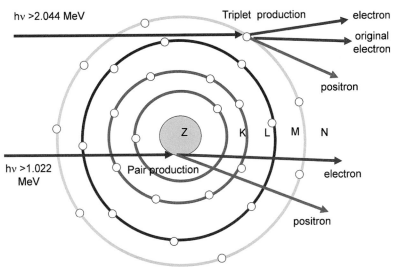

FIGURE 2.7 Pair and triplet production.

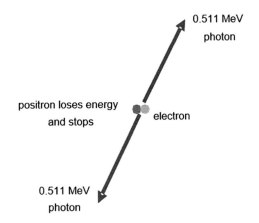

FIGURE 2.8 Electron–positron pair annihilation.

charge is created since the two particles have equal and opposite charge. Figure 2.7 illustrates pair and triplet production.

If the photon has an energy in excess of 1.022 MeV, the extra energy is shared between the positron and the electron in the form of kinetic energy.

$$h\upsilon_{photon} - 1.022 \text{ MeV} = \text{Energy}_{positron} + \text{Energy}_{electron} \qquad (2.8)$$

The proportion of energy shared by the electron and positron can take many values with one or other of them receiving almost all of the energy and the other very little or any combination in between, including equal shares.

This process causes the nucleus to recoil slightly and must strictly speaking be considered as a collision. Whilst some energy is transferred to the nucleus, it is very small and is neglected in Equation 2.8.

If it has some kinetic energy, the positron will travel through the material until it annihilates by combining with another electron. This process usually takes place when the positron has almost come to rest and in order to conserve momentum two photons are produced each with an energy of 0.511 MeV. If annihilation happens at rest then the two photons will travel in exactly opposite directions, if not then the kinetic energy remaining will add to the energy released and the two photons may not travel in exactly opposite directions. Figure 2.8 illustrates the annihilation process at rest.

The electron can take part in other interactions until it to comes to rest. Although pair production can mainly be regarded as an absorption process, if the photons leave the material without depositing all of their energy this is regarded as scattered radiation. In this case, the absorption coefficient π_a/ρ will be less than π/ρ

$$\pi/\rho = \pi_a/\rho + \pi_s/\rho \qquad (2.9)$$

However, in general, the scattering coefficient π_s/ρ is ignored since contributes little to the total scatter and so $\pi/\rho \sim \pi_a/\rho$.

Along with PP there is the probability of triplet production. This is similar to pair production but happens when the photon interacts with an electron instead of the nucleus. Three particles can be identified in this case: a positron, an electron and the electron with which the photon interacted. The threshold for this event is 2.04 MeV and the fate of the particles is the same as described above. Triplet production is a minor process compared to pair production (see Figure 2.7)

The probability of PP increases with Z number and photon energy.

$$\text{Mass attenuation coefficient } \pi/\rho \propto (E - 1.022) \times Z \qquad (2.10)$$

Figure 2.9 compares the attenuation coefficients for water (effective Z (Z_{eff}) = 7.4), aluminium ($Z = 13$) and lead ($Z = 82$). Here the total mass attenuation coefficient is shown (solid curves) and also the separate contributions of PE, CE and PP.

At low energies, PE dominates and as the Z number increases it remains predominant to higher energies. Note the L and K absorption edges feature in the plot for lead.

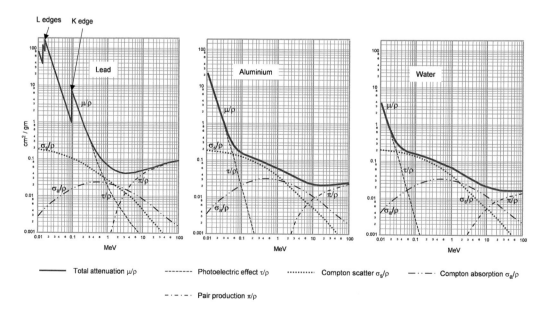

FIGURE 2.9   Mass attenuation coefficients as a function of photon energy for water, aluminium and lead.

Water and aluminium appear similar especially over the wide range of energies 0.1 to 10 MeV, i.e. the therapeutic range. The differences reflect the variation in Z values.

The higher the Z value of the material, the sooner PP begins to influence the total mass attenuation coefficient and the greater its influence becomes.

Figure 2.10 shows the absorption characteristics for some biological materials compared to air by plotting the ratio of their mass absorption coefficients μ_a/ρ. Some useful conclusions can be drawn from these plots. The good agreement between air and muscle confirms that air is a good dosimetric standard. Water follows muscle closely; this means water is a good substitute for muscle in which to perform dose measurements. Bone is only similar to air, muscle and water at higher photon energies (above about 0.1 MeV). Because of its higher Z value, μ_a/ρ increases rapidly for bone as the energy reduces (below about 0.1 MeV). At these lower energies, PE dominates and therefore high Z materials are strong absorbers at low energies, and this property results in the high contrast between tissues (e.g. muscle to bone) in DR images. At energies where CE occurs (0.1–10 MeV) bone μ_a/ρ is similar to the others (as CE is almost independent of Z). At higher energies still (> 10 MeV) pair production causes the curve for bone to rise and it would increase above the others again if the energies were high enough ($\pi/\rho \propto Z$). At higher energies, both water and muscle absorb more than air and also more than bone. This is because of electron density, bone being less than either water or muscle. CE

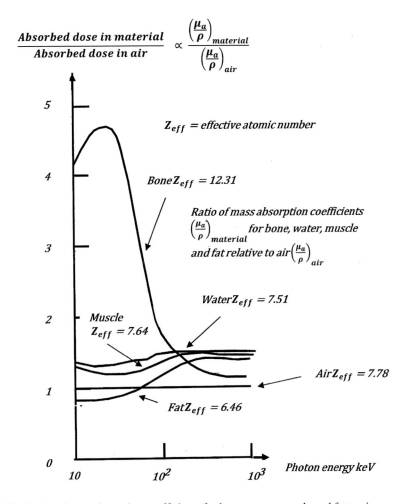

FIGURE 2.10   Ratio of mass absorption coefficients for bone, water, muscle and fat to air.

TABLE 2.1

Electron Densities in Different Materials

Material	Electron Density (Electrons/cm³)	Z
Air	3.007×10^{23}	7.64
Muscle	3.307×10^{23}	7.42
Water	3.344×10^{23}	7.42
Bone	3.193×10^{23}	13.80
Fat	3.203×10^{23}	5.92

is dependent on electron density. Table 2.1 shows the electron densities for some biological materials to demonstrate the similarity between them.

2.2.4 COHERENT OR CLASSICAL SCATTERING

CE, discussed earlier, is an example of incoherent scattering where a photon interacts with an electron, is scattered and loses energy which is transferred to the electron in the form of kinetic energy. When classical or coherent scattering occurs, a photon interacting with an electron or atom is scattered with no energy loss. The probability of this process increases with decreasing photon energy and increasing atomic number of the scattering atom.

The two types of coherent scattering that will be discussed here are Thomson or classical scattering (TS) and Rayleigh scattering (RS).

Before discussing coherent scattering, it is useful to introduce the concept of scattering cross-section. If we consider a photon that is not absorbed but deflected by a scattering centre of total cross-section σ then the degree to which it will be scattered will depend on its distance from the interaction centre. See Figure 2.11.

The area $d\sigma$ inside the total cross-section σ in which the deflection will be within the solid angle $d\Omega$ centred on the deflection angle θ is the differential cross-section $d\sigma/d\Omega$.

The relationship between the increment of the deflection angle θ and the increment of $d\Omega$ is

$$d\Omega = 2\pi \, \sin\theta \, d\theta \tag{2.11}$$

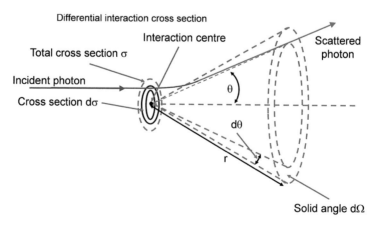

FIGURE 2.11 Differential interaction cross-section.

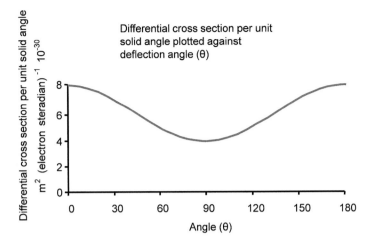

FIGURE 2.12   Differential cross-section per unit solid angle plotted against deflection angle θ.

$$\frac{d\sigma}{d\Omega} = \frac{1}{2\pi \sin\theta} \frac{d\sigma}{d\theta} \tag{2.12}$$

2.2.4.1   Thomson Scattering

This involves the interaction of an electron with an incident photon and this form of scattering is seen at low energies below 100 keV in water. Here the electron is accelerated for a short time in the electromagnetic field of the photon and as a result radiates energy. In this case, classically the differential cross-section can be expressed as follows:

$$\frac{d\sigma_0}{d\Omega} = \frac{r_0^2}{2}\left(1+\cos^2\theta\right) \tag{2.13}$$

where $\dfrac{d\sigma_0}{d\Omega}$ is the classical differential scattering coefficient per electron and per unit solid angle. Equation 2.13 is associated with TS and was named after the scientist who first derived it and r_0 is the classical electron radius. This was the classical result for low energy (referred to as zero energy) photons. The relationship plotted for different values of θ is shown in Figure 2.12.

Note this means that at low energies twice as much energy will be forward scattered (0°) and backward scattered (180°) compared to 90°.

2.2.4.2   Klein and Nishina

It was soon realised that this result did not hold for higher energies. In the 1920s using a more complex quantum mechanical treatment Klein and Nishina showed that better agreement with experimental observations was achieved by modifying Thomson's original result using the addition of a factor F_{KN}

$$\frac{d\sigma}{d\Omega} = \frac{d\sigma_0}{d\Omega} F_{KN} = \frac{r_0^2}{2}\left(1+\cos^2\theta\right) F_{KN} \tag{2.14}$$

where

$$F_{KN} = \left(\frac{1}{1+\alpha\left(1-\cos\theta\right)}\right)^2 \left(1+\frac{\alpha^2\left(1-\cos\theta\right)^2}{\left[1+\alpha\left(1-\cos\theta\right)\right]\left(1+\cos^2\theta\right)}\right) \tag{2.15}$$

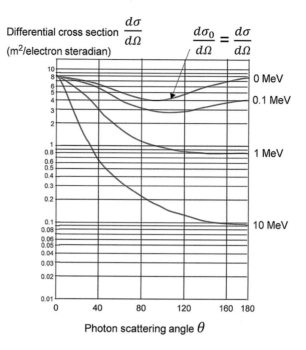

Differential cross section $\dfrac{d\sigma}{d\Omega}$

(m²/electron steradian)

$\dfrac{d\sigma_0}{d\Omega} = \dfrac{d\sigma}{d\Omega}$

0 MeV

0.1 MeV

1 MeV

10 MeV

Photon scattering angle θ

FIGURE 2.13 Differential cross-section versus angle of photon scatter for different energies using the Klein–Nishina correction.

where α is the ratio of energy of the photon to the rest mass energy of the electron ($m_o c^2$).

Using the Klein–Nishina relationship, plots for different energies, shown in Figure 2.13, agree with experimental results. It can also be seen that when α equals zero F_{KN} equals 1 and Equation 2.14 reduces to Thomson's original result.

2.2.4.3 Rayleigh Scattering

Another form of coherent scatter is that described by Rayleigh. This is a process that occurs at low energies but involves materials with high Z values and tightly bound electrons. As a result of the tight bonding, this form of scattering involves the whole atom with the electrons of the atom behaving in a cooperative manner. The atom is neither excited nor ionised instead different parts of the atomic cloud combine to produce the scattering effect.

Here a factor is also added to the original Thomson expression:

$$\frac{d\sigma_{coh}}{d\Omega} = \frac{r_0^2}{2}\left(1+\cos^2\theta\right)\left[F\left(x,Z\right)\right]^2 \tag{2.16}$$

Recalling Equation 2.11 $d\Omega = 2\pi\sin\theta\,d\theta$ and substituting this into Equation 2.16 gives

$$\frac{d\sigma_{coh}}{d\theta} = \frac{r_0^2}{2}\left(1+\cos^2\theta\right)\left[F\left(x,Z\right)\right]^2 2\pi\sin\theta \tag{2.17}$$

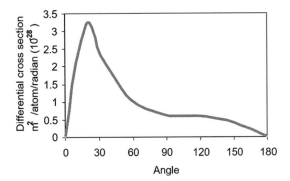

FIGURE 2.14   Differential cross-section plotted against scattering angle θ for Rayleigh scattering.

where $F(x, Z)$ is called an atomic form factor which for small θ approaches Z and for large θ

approaches zero. $x = \dfrac{\left(\sin\dfrac{\theta}{2}\right)}{\lambda}$ in Equations 2.16 and 2.17 where λ is the wavelength of the photon.

Equation 2.17 is the differential cross-section per unit angle and gives the fraction of the incident energy that is scattered into the cone containing θ and $\theta + d\theta$ (see Figure 2.11). Figure 2.14 shows this differential cross-section plotted against scatter angle. The scatter is strongly in forward direction, increases with energy and varies as Z^2.

2.2.5   PHOTO NUCLEAR INTERACTIONS

Photon and electron interactions with the atomic nucleus can lead to energy emission in the form of neutrons or protons. This has implications for room and equipment design in radiotherapy. For stable nuclei heavier than carbon the minimum beam energy for emissions lies in the range 6 to 16 MeV but, in general, no specific bunker protection is needed below 10 MeV. Where significant emission is present, lightweight materials such as polythene are used in addition to standard bunker shielding as these are most effective in reducing the neutron flux. However, where neutrons are absorbed in heavy-metal materials in the linear accelerator head, residual activity may be induced.

The relative biological effectiveness of neutrons varies with energy and where activation is a factor this has to be taken into account when designing equipment. This is particularly important in the treatment of paediatric tumours, where life expectancy is longer, in an effort to avoid induced cancers in later years.

2.3   INTERACTION OF CHARGED PARTICLES WITH MATTER

As already mentioned charged particles differ from photons in that they are directly ionising. This means that they deposit energy through direct coulomb interactions with orbital electrons. Through these interactions, the electrons may: i) lose kinetic energy through collisions and radiative losses or ii) scatter with no energy loss (elastic scattering).

This is in contrast to photons which give rise to charged particles which then deposit energy in the medium.

Treatments with charged particles most commonly use electrons. Typically, these range in energy from 10 keV to 2 MeV in BT and from 4 to 25 MeV in EBRT. Other charged particles used are protons in the range 60–300 MeV and short-range alpha particles from 1 to 8 MeV, for example, in targeted radiotherapy. The use of light and heavy ions such as fully stripped carbon, neon and argon is of growing interest.

Charged particles lose energy in quite a different way from photons (and neutrons) and as a result their interaction with tissues is modelled differently. Photons interact much less frequently than charged particles especially at high energies and as many as two out of three photons traverse irradiated material completely without interaction. Charged particles on the other hand lose energy by a gradual friction-like process involving many interactions and so a continuous slowing down approximation (CSDA) is more appropriate when describing their energy loss. The amount of energy deposition in matter depends on the details of the interactions. In low Z materials, electrons have a total cross-section about 5 orders of magnitude higher than photons in the kinetic energy range 0.01–10 MeV.

In terms of range, photons and charged particles are clearly different. For photons, range or path length is not relevant and instead mean free distance to first interaction is a more useful measure. Having said this, mean free path without interaction can be used to compare photon and charged particle interactions (see Section 3.2). For charged particles, individual range or path length is relevant and can be estimated from the statistical average over many collisions.

Fast electrons follow a tortuous track and, compared with other charged particles, produce sparse ionisation and excitation events with an almost random distribution of energy within cell-sized volumes. Other types of charged particles produce concentrated columns of ionisation and excitation along their track.

Figure 2.15 shows examples of characteristic charged particle tracks in tissue with the dots representing interactions. As can be seen, particle tracks differ in appearance with some much denser than others. Delta ray tracks are also shown. A delta ray is an electron that has acquired enough kinetic energy to deviate from the track of the particle with which it interacted causing further ionisations along a separate track.

A proton can create a track with a core diameter measuring in the region of 10 to 15 nm. A fast electron produced by such a track could travel radially outwards by as much as 100 μm, a distance of several mammalian cell diameters. Particles with a higher density of interactions will deposit more energy per unit length of track and therefore have a greater probability of causing cell damage.

For any radiation type, the energy deposition (dE) per unit length of track (dx) is known as the linear energy transfer (LET).

$$LET = dE/dx$$

Figure 2.16 shows examples of high and low LET radiation. As can be seen the types of radiation used in therapy today fall mainly in the low LET category. All of the high LET radiations interact with orbital electrons except for neutrons which interact with atomic nuclei resulting in the ejection of slow, densely ionising protons with high LET.

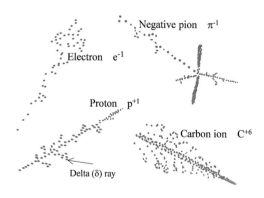

FIGURE 2.15 Charged particle tracks in tissue.

High and Low LET Radiations

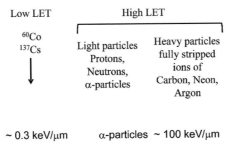

FIGURE 2.16   High and low LET radiations.

2.3.1 INTERACTIONS

The major interaction of charged particles is split into three broad groups.
These are:

- Interactions with atoms as a whole.
- Interactions with individual electrons of atoms and molecules.
- Interaction with nuclei.

Which of these three modes applies is determined by energy and distance of approach of the particle to the atom or nucleus with which it interacts. Electron energy losses usually take place in small increments and therefore in general an electron must undergo many collisions before it loses all of its energy. Two energy loss processes can be identified. One results from collisional losses involving interaction with atomic electrons. The other results from radiative losses resulting from interactions with the atomic nucleus. Collisional losses are regarded as "soft" if the distance of approach is large and "hard" if the distance is small. Radiative losses arise mainly from coulomb interactions with the external nuclear field and inelastic nuclear interactions.

For collisional losses Figure 2.17 shows the distance b known as the impact parameter which is the distance from the track of the particle and the atomic nucleus of the atom. This is large for soft collisions and small for hard collisions compared with the atomic radius. For soft collisions, the coulomb field of charged particles can interact with atoms causing excitation. Large b, soft collisional interactions have a high probability of occurrence and account for approximately 50% of collisional

Distant and close collisions between
charged particles and atoms

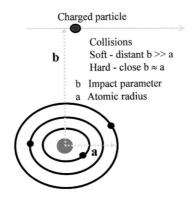

FIGURE 2.17   Distant (soft) and close (hard) collisions between charged particles and atoms.

energy losses in the medium. Here, energy is transferred from a charged particle to the atom as a whole involving many small energy transfers of a few electron volts. Although ionisation events can occur they are of a very low probability and where this happens the ejected electron carries a relatively low amount of energy. The energy transferred in soft collisions is eventually dissipated as light in a gas or as heat in a solid.

For hard collisions, the charged particle passes close enough to greatly increase the probability of interaction with a single electron. If the incident particle has enough energy to overcome the binding energy an orbital electron can be removed from the atom resulting in ionisation. Characteristic x-rays may be produced if the electron is from an inner shell. Small or large energy transfer and angular deflection of the charged particle is possible in this case. If the removed electron has an excess energy greater than or equal to 100 eV they are known as delta rays and secondary electrons can produce their own excitations and ionisations. The probabilities associated with these interactions differ for electrons or heavy charged particles and there are many more soft than hard collisions. In total, however, collisional losses, soft and hard collisions account for roughly equal amounts of energy loss.

At short distances of approach, charged particles can interact with the atomic nucleus, with 97 to 98% of these taking the form of elastic scattering events resulting from coulomb interactions with the external nuclear field. These interactions are important for electrons and positrons since there are no photon emissions or excitation of the nucleus but they are associated with larger angle scatter. Electrons can be scattered without significant energy loss through large angles in some materials. This is the basis for the production of scattering foils which are used in linear accelerators to produce a useful beam by broadening the narrow pencil beam of accelerated electrons as it emerges from the waveguide.

Not all of these very short approach events lead to inelastic scattering events and about 2 to 3% can result in the radiative losses mentioned earlier. These interactions lead to the production of bremsstrahlung radiation where a charged particle (mainly an electron or positron) is deflected by the coulomb field of the nucleus and loses some of its energy which is carried away in the form of a photon. Figure 2.18 illustrates the production of bremsstrahlung radiation. The cross-section for the production of bremsstrahlung (σ_{brems}) is directly proportional to the square of the atomic number Z and inversely proportional to the square of the mass M of the charged particle. Thus, the production of bremsstrahlung is far greater for electrons or positrons than for protons which are around 2000 times more massive.

$$\sigma_{brems} \propto Z^2/M^2 \tag{2.18}$$

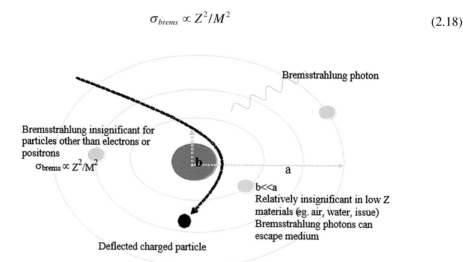

FIGURE 2.18   Radiative losses – the production of bremsstrahlung radiation.

This also means that the production of bremsstrahlung is relatively insignificant for low Z materials (e.g. air, water, tissue, etc.).

2.3.2 STOPPING POWER

For charged particles, it is necessary to have information about the manner in which they lose energy along their tracks during their passage through matter. The factor that defines this is called stopping power.

Linear stopping power (S_{dE}) is defined as:

$$S_{dE} = \frac{dE}{dx} \; MeV \, cm^{-1} \tag{2.19}$$

where dE is the fraction of energy that a particle loses during its passage through a medium along an increment of path length dx.

In a similar way, as we saw earlier with the linear attention coefficient for photons, this quantity is divided by density to obtain the mass stopping power which is independent of the physical density (solid, liquid or gas) of the absorbing material.

$$\frac{S}{\rho} = \frac{dE}{\rho dx} \; MeV \, m^2 kg^{-1} \tag{2.20}$$

Mass stopping power has two components:

(1) stopping power due to inelastic collisions with atomic electrons of the medium resulting in excitations and ionisations
(2) stopping power due to electron interactions with the electric field of the nucleus resulting in the production of bremsstrahlung.

Therefore, the total mass stopping power $\dfrac{S_{tot}}{\rho}$ can be expressed as:

$$\frac{S_{tot}}{\rho} = \frac{S_{col}}{\rho} + \frac{S_{rad}}{\rho} \tag{2.21}$$

where S_{col} is the stopping power due to collision losses and S_{rad} is the stopping power due to radiative losses. Mass stopping power relationships for different charged particles can be complex and some examples are shown below.

2.3.2.1 Mass Collisional and Radiative Stopping Power

Both distant and close collisions are accounted for in the modified Bethe relationship

$$\frac{S_{col}}{\rho} = \frac{N_A Z}{A} \frac{\pi r_e^2 2 m_e c^2}{\beta^2} \left[\ln\left(\frac{E_k}{I}\right)^2 + \ln\left(1 + \frac{\tau}{2}\right) + F^{\mp}(\tau) - \delta \right] \tag{2.22}$$

where:
N_A Avogadro's number
r_e classical electron radius
$m_e c^2$ the electron rest energy

β the incident particles velocity relative to the velocity of light (v/c)
Z, A are the atomic number and atomic weight of the target atoms
I the mean excitation energy
E_K the kinetic energy of the particle
τ equal to E_K/m_ec^2
δ accounts for effective coulomb force by atom on fast charged particles

The factor $F^{\pm}$ in Equation 2.22 can take two forms:
For electrons (Moller)

$$F^{-}(\tau) = \left(1 - \beta^2\right)\left[1 + \frac{\tau^2}{8} - (2\tau + 1)\ln 2\right] \tag{2.23}$$

and for positrons (Bhabba)

$$F^{+}(\tau) = 2\ln 2 - \left(\frac{\beta^2}{12}\right)\left[23 + \frac{14}{(\tau + 2)} + \frac{10}{(\tau + 2)^2} + \frac{4}{(\tau + 2)^3}\right] \tag{2.24}$$

From Equation 2.22 collisional losses for electrons and positrons are proportional to:

1. log of the reciprocal of the square of the excitation potential
2. reciprocal of the square of the velocity
3. Z/A of the medium

The greater binding energy of the high atomic number material makes excitation of shell electrons less likely and collision energy losses vary slowly with atomic number.
Mass radiative stopping power for electrons is given by (Bethe–Heitler):

$$\frac{S_{rad}}{\rho} = \sigma_0 \frac{N_A Z^2}{A}\left(E_K + m_ec^2\right)\bar{B}_r \tag{2.25}$$

where
$\sigma_0 = \alpha\left(e^2/(4\pi\varepsilon_o m_ec^2)\right)^2 = 5.80 \times 10 - 28$ cm²/atom
α is a constant (fine structure constant)
B_r is a function of Z and E_K (a form function varying between 5.33 to 15 in the energy range from ~ 0.5 MeV to 100 MeV)

As protons and heavy ions pass through tissue they lose energy by interactions with atomic electrons. At very low velocities the proton will capture an electron. The mass stopping power is given by:

$$\frac{S}{\rho} = \left(\frac{4\pi N_A Z}{A}\right)\left(\frac{r_e^2 m_ec^2}{\beta^2}\right)z^2\left[\ln\left(\frac{2m_ev^2}{I}\right) - \ln\left(1 - \beta^2\right) - \beta^2 - \Sigma\left(\frac{C_i}{Z}\right)\right] \tag{2.26}$$

where
z is the charge of the incident particle, Z/A the atomic no./atomic weight of the target
$\beta = v/c$, and $\Sigma C_i/Z$ are the shell correction factors.

Electrons will contribute less to the stopping power if v is comparable to the velocities in their orbits. I is the mean excitation energy (potential) of the atoms in the material.

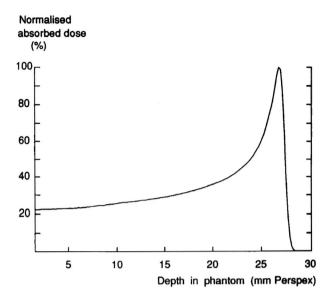

FIGURE 2.19   Bragg peak for protons of 200 MeV energy.

Note that for protons and heavy ions the rate of energy loss does not depend upon the mass of the particle only on its charge. In addition, the rate of energy loss is proportional to the square of the charge on the ion and the rate of energy loss is inversely proportional to the square of the velocity. For protons and heavy ions the energy loss is almost all due to collisional losses involving ionisation and excitation of atomic electrons. There are some losses as a result of nuclear interactions and these vary with energy (e.g. 2.5% at 100 MeV). Radiative losses are negligible since as already mentioned earlier the cross-section for the production of bremsstrahlung is inversely proportional to the square of the mass of the particle (see Equation 2.18). Unlike the case for electrons and positrons, the mass collisional stopping power for protons and heavy ions is effectively the total mass stopping power.

Characteristically, charged particles deposit most of their energy at the end of their track as they are slowing down. The energy deposition plotted against depth for protons of 200 MeV is shown in Figure 2.19. The peak of energy deposition is seen at around 26 mm depth in a perspex phantom for this energy. The peak is known as the Bragg peak. At the extreme end of their track, the average rate of energy loss decreases until the energy matches the thermal energy of the atoms in the medium.

2.3.2.2   Charge Particle Range

As a charged particle passes through tissue it will slow down, change direction and eventually lose its kinetic energy and come to rest. Photons are absorbed exponentially as they pass through tissue. This means that effectively there is no thickness that will completely stop a photon beam. Charged particles however will travel a finite distance and then come to rest in tissue. This distance is known as the range of a particle and this can be calculated from the stopping power.

If for a charged particle we know the stopping power (S_{dE} see Equation 2.19) associated with a small energy change dE then we can calculate the distance travelled by that particle.

$$\text{Distance} = dE / S_{dE} \qquad (2.27)$$

The full particle range is therefore given by:

$$Range = \frac{\int_0^E \frac{dE}{(S/\rho)}}{\rho} \qquad (2.28)$$

TABLE 2.2

Energy versus Range for Protons

Energy (MeV)	Range in Tissue (cm)
60	3.4
100	8.4
200	28.5

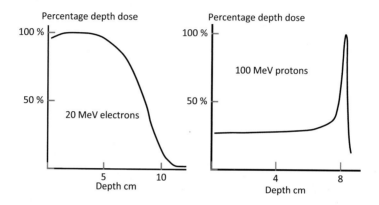

FIGURE. 2.20 Percentage depth dose plots for 20 MeV electrons and 100 MeV protons. Protons show the characteristic Bragg peak.

where ρ is the material density.

Table 2.2. shows the range of protons in tissues. Very high energies (up to 250 MeV) are needed for radiotherapy to deep-seated tumours.

Figure 2.20 shows the percentage depth dose plots for 20 MeV electrons and 100 MeV protons. The proton plot shows the Bragg peak which is characteristic of charged particles. This results from the fact that protons and heavy charged particles slow down and lose large amounts of energy at the end of their track. Because they are charged particles the plot for electrons should show a similar Bragg peak. However, they are subject to multiple changes in direction as they slow down which completely smears out the Bragg peak and it is not observed in the dose plot.

Both electrons and protons can have some advantages over photons in treatment since they have a sharp cut off in dose at energy dependant depths. Protons also have a low entry dose compared with the height of the Bragg peak as can be seen in Figure 2.20.

2.4   INTRODUCTION TO RADIATION DOSIMETRY

Just a few years after the discovery of radium and x-rays, the discussion on how to measure the effects of the interaction of radiation with matter began. In 1909 the Roentgen Society of Great Britain appointed a committee to consider how the output of an x-ray tube could be measured. The 1910 Congress of Radiology in Brussels established an international committee under Rutherford which met in Paris in 1912 and adopted an International Radium Standard prepared by Marie Curie. In 1915 Winawer and St. Sachs suggested that a beam of x-rays should be regarded as having unit energy, when by its complete absorption in air, it produces the same number of ions as the γ rays from 1 gram of radium would produce under similar conditions. In 1928 the Second International Congress of Radiology in Stockholm (Sweden) defined the roentgen.

Today radiation dosimetry is a mature physical science with a special interest in determining the energy absorbed by matter from the interaction of ionising radiation.

2.5 RADIATION FIELDS, ENERGY DEPOSITION, DOSE QUANTITIES AND UNITS

2.5.1 STOCHASTIC AND NON-STOCHASTIC QUANTITIES

Interaction of radiation with matter is a stochastic (random) process; it is not possible to predict at what position and time a given particle will interact. For most applications many particles are present, a very large number of interactions take place and radiation interactions are well approximated by non-stochastic descriptions in terms of mean or expectation values. Only in the case of few particle or few interactions, e.g. when the volume of interest is microscopic or the observation time is too short, non-stochastic descriptions are inappropriate. Figure 2.21 represents the behaviour of the energy deposition (E) per mass (m) as a function of the log of interacting mass (log m), where the stochastic nature of the phenomenon is evident with decreasing the mass.

2.5.2 RADIATION FIELDS

Central to dosimetry are the methods for a quantitative determination of energy deposited in a given medium by directly or indirectly ionising radiations and a number of physical quantities have been introduced to describe a radiation field and its absorbed dose.

Particle fluence Φ can be defined as:

$$\Phi = \frac{dN}{da}$$

with dN the differential value of the number of particles crossing an infinitesimal sphere of area da. The SI unit is m^{-2}.

Energy fluence Ψ can be defined as:

$$\Psi = \frac{dR}{da}$$

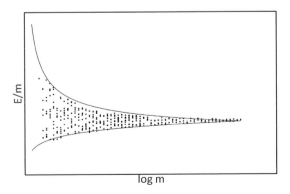

FIGURE 2.21 Energy absorbed per mass becomes a clear non-stochastic quantity when mass increases.

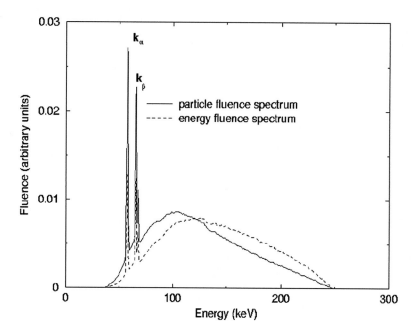

FIGURE 2.22   Fluence and Energy fluence spectrum of an x-ray beam used in diagnostic radiology.

with dR the total radiant energy crossing the infinitesimal sphere. The unit is Jm^{-2}, but is frequently expressed as $MeVcm^{-2}$. If the radiation field has particles with the same energy E then $\Psi = \Phi E$.

The energy fluence differential called the *energy fluence spectrum* is defined as:

$$\Psi_E = \frac{d\Psi(E)}{dE} = \frac{d\Phi(E)}{dE}E$$

Figure 2.22 shows an example of fluence and the energy fluence spectrum of an x-ray beam used in DR. Over the bremsstrahlung spectrum, the k_α and k_β characteristic lines from the tungsten target can be seen.

2.5.3   ENERGY DEPOSITED AND ENERGY IMPARTED

The term energy deposit ε_i refers to the energy deposited in a single interaction process and it is a stochastic quantity

$$\varepsilon_i = \varepsilon_{in} - \varepsilon_{out} + Q$$

with ε_{in} and ε_{out} the energy of incident particle and the sum of energies of all particles leaving the interaction, respectively, and Q the change in the rest energies of the nucleus and all particles involved in the interaction.

The term energy imparted ε to a small volume is the sum of all energy deposits in the volume

$$\varepsilon = \sum_i \varepsilon_i$$

2.5.4   ENERGY TRANSFERRED AND ENERGY IMPARTED

When an uncharged particle (e.g. photon, neutron) interacts with matter in a volume V, the energy transferred to charged particles ε_{tr} is the sum of all charged particle initial kinetic energies transferred by uncharged particles in the volume V

$$\varepsilon_{tr} = \left(R_{in} \right)_u - \left(R_{ou} \right)_u^{non-rad} + \Sigma Q$$

with $(R_{in})_u$ the radiant energy of uncharged particles entering the volume, $\left(R_{ou} \right)_u^{non-rad}$ the radiant energy of uncharged particles generated from non-radiative processes leaving the volume V and ΣQ the change of rest masses in V, counted as negative for an energy to mass transformation. Note that radiative processes refer to bremsstrahlung production and positron annihilation-in-flight. The latter of these is an infrequent event compared to positron annihilation where the radiative loss is the kinetic energy of the positron at the time of the annihilation.

The energy imparted in the volume V is

$$\varepsilon = \left(R_{in} \right)_u - \left(R_{ou} \right)_u + \left(R_{in} \right)_c - \left(R_{ou} \right)_c + \Sigma Q$$

where the radiant energies of all type of particles entering, $(R_{in})_u$ and $(R_{in})_c$, and of those leaving, $(R_{ou})_u$ and $(R_{ou})_c$, the volume of interest V are taken into account together with the change in the rest mass. Here $(R_{in})_c$ and $(R_{ou})_c$ correspond to the radiant energy of charged particles entering and leaving volume V respectively.

It is necessary to note that in the case of uncharged particles, e.g. photons, the transfer of energy to matter occurs in two phases or steps:

(i) the transfer of kinetic energy to charged particles
(ii) and, the loss of energy to matter by these charged particles through several collisions or radiative processes.

The first process is accounted for by the energy transfer coefficient and the second by the stopping power. In Figure 2.23, the energy imparted in the volume V by the charged particle (electron)

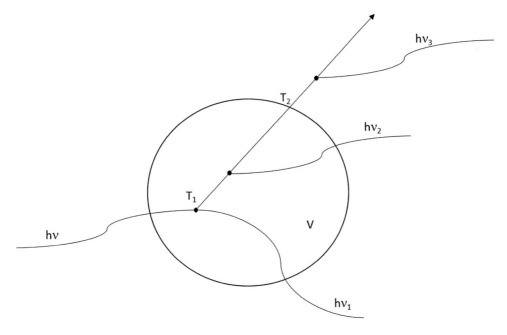

FIGURE 2.23   A photon has a Compton interaction in V and transfers the energy T_1 to an electron that undergoes two bremsstrahlung interactions.

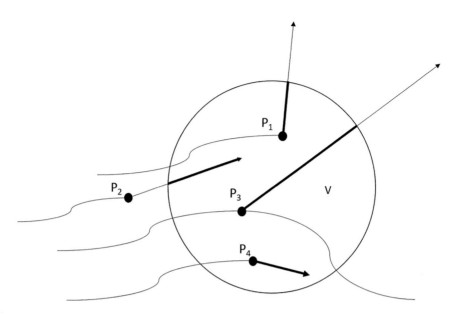

FIGURE 2.24 Photons are transferring energy to charged particles in P_1, P_2, P_3 and P_4 that impart energy along their tracks, totally or partially inside the volume V; the two processes of energy transfer and energy imparted do not take place at the same location. Note that, the photon interacting in P_2 transfers energy outside the volume V (transfer not counted for energy transfer in V) but the resulting charged particle imparts part of its energy in the volume V.

derived from the Compton interaction of the photon hv is the energy deposited T_1–T_2 with T_1 the initial kinetic energy and T_2 the kinetic energy before leaving V. Another important aspect to note is that the two processes, the energy transfer and the energy imparted, are not taking place at the same location, as better shown in the example of Figure 2.24.

2.5.5 KERMA

Kerma is the acronym for Kinetic Energy Released per unit Mass. It is defined as the average amount of energy transferred $\overline{dE_{tr}}$ in a small volume of mass dm from indirectly ionising radiation to directly ionising radiation ignoring what happens after this transfer

$$K = \frac{\overline{dE_{tr}}}{dm} \quad \text{Unit: } Jkg^{-1} \text{ with the special name of gray. 1 Gy} = 1 \ Jkg^{-1}$$

The quantity is obviously applicable only to uncharged particles or indirectly ionising radiation, e.g. photons and neutrons.

Because charged particles can experience collisions or radiative interactions, it is possible to divide kerma in two components, collision and radiative kerma

$$K = K_{col} + K_{rad}$$

The fraction of the energy lost through radiative processes can be expressed by the radiative factor $g = K_{rad}/K$ and, the relationship between K and K_{col} can be written as

$$K_{col} = K(1-g)$$

The difference between energy transferred to charged particles by uncharged particles (Kerma) and energy imparted to matter is explained in the example given in Figure 2.24. An electron receiving kinetic energy from a photon interaction outside the volume V (not accounted for in the energy transfer) transfers part of its kinetic energy in the volume of interest V (accounted for in the energy imparted to V).

From the definitions, it is easy to demonstrate that kerma can be expressed in terms of energy fluence Ψ and the mass energy transfer coefficient μ_{tr}/ρ

$$K = \Psi\,\mu_{tr}/\rho$$

and K_{col} in terms of Ψ and the mass energy absorption coefficient μ_{en}/ρ

$$K_{col} = K\left(1-g\right) = \Psi\,\mu_{tr}/\rho\left(1-g\right) = \Psi\,\mu_{en}/\rho$$

When comparing the collision kerma between medium 1 and medium 2, assuming the same energy fluence Ψ, an important and frequently used relationship is obtained

$$\frac{K_{col,2}}{K_{col,1}} = \frac{\Psi\left(\mu_{en}/\rho\right)_2}{\Psi\left(\mu_{en}/\rho\right)_1} = \left(\mu_{en}/\rho\right)_{2,1}$$

where $\left(\mu_{en}/\rho\right)_{2,1}$ is the ratio of the mass energy absorption coefficients of the two media.

2.5.6 CEMA AND ABSORBED DOSE

Similar to kerma, cema is the acronym for Converted Energy per unit Mass. It quantifies the average amount of energy converted in a small volume from directly ionising radiations, such as electrons and protons, in collisions with atomic electrons, ignoring what happens after this transfer,

$$C = \frac{\overline{dE_c}}{dm} \quad \text{Unit: } Jkg^{-1} \text{ with the special name of gray. } 1\ Gy = 1\ Jkg^{-1}$$

with $\overline{dEc}$ the mean energy lost in electronic interactions by the primary charged particles.

Absorbed dose is a quantity applicable to both indirectly and directly ionising radiations. As with kerma and cema, the absorbed dose is a non-stochastic quantity related to the stochastic quantity energy imparted ε

$$D = \frac{\overline{d\epsilon}}{dm}$$

with $\overline{d\epsilon}$ being the average energy imparted in a volume of interest of mass dm.

Because the energy imparted is due to the interaction of charged particles (e.g. electrons and positrons) with matter and the amount of energy imparted is described by the mass stopping power, the collision component in particular, the absorbed dose D in a medium m can be expressed as

$$D_m = \phi\left(S_{col}/\rho\right)_m$$

with Φ the charged particle fluence. Because charged particles are losing energy along their tracks in multiple interactions, the charged particle fluence has a spectrum of energies $\phi_m(E)$ and the absorbed dose D_m can be better expressed as

$$D_m = \int_0^{E_{max}} \phi_m(E)\left(S_{col}(E)/\rho\right)_m dE$$

One may again use a useful notation for the mass collision stopping power averaged over the fluence spectrum

$$\left(\overline{S_{col}/\rho}\right)_m = \frac{1}{\phi_m} \int_0^{E_{max}} \phi_m(E)\left(S_{col}(E)/\rho\right)_m dE$$

and D_m becomes

$$D_m = \phi_m \left(\overline{S_{col}/\rho}\right)_m$$

Comparing the absorbed dose in a medium 1 and a medium 2, for the same fluence $\Phi_{m1} = \Phi_{m2}$, then an important and useful relationship is

$$\frac{D_{m2}}{D_{m1}} = \frac{\phi_{m2}\left(\overline{S_{col}/\rho}\right)_{m2}}{\phi_{m1}\left(\overline{S_{col}/\rho}\right)_{m1}} = \left(\overline{S_{col}/\rho}\right)_{m2.m1}$$

with $\left(\overline{S_{col}/\rho}\right)_{m2.m1}$ the ratio of the average mass collision stopping power of the two materials.

2.6   RADIATION EQUILIBRIUM AND CHARGED PARTICLE EQUILIBRIUM

A uniform density volume V in Figure 2.25 contains a uniformly distributed radioactive material and it is larger than any distance of penetration of rays emitted (excluding the neutrinos). Consider a small volume s around a point P, and the plane S tangential to the volume s at P'. There is perfect reciprocity (in the non-stochastic limit) of each type and energy of particle crossing the plane at P' from both directions (as much in as out). Because of the spherical symmetry of v and the uniform distribution of the radioactivity in V, this will be true for all possible orientation of S. Then, in the non-stochastic limit, for each type and energy of ray entering v, another identical ray leaves. This condition is called radiation equilibrium (RE), that can be written for directly and indirectly ionising radiation

$\left(\bar{R}_{in}\right)_c = \left(\bar{R}_{out}\right)_c$ and $\left(\bar{R}_{in}\right)_u = \left(\bar{R}_{out}\right)_u$

The average energy imparted will be

$$\bar{\epsilon} = \left(\bar{R}_{in}\right)_c - \left(\bar{R}_{out}\right)_c + \left(\bar{R}_{in}\right)_u - \left(\bar{R}_{out}\right)_u + \Sigma Q = \Sigma Q$$

and consequently the absorbed dose D

$$D = \frac{\overline{d\epsilon}}{dm} = \frac{\Sigma Q}{dm}$$

Then, if RE exists at a point in a radioactive material, the absorbed dose is equal to the expectation value of the energy released by the radioactive material per unit of mass at that point, ignoring neutrinos.

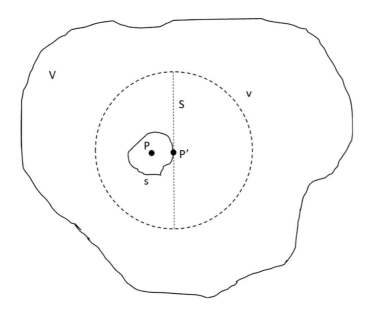

FIGURE 2.25   Radiation equilibrium (RE) in any point P of a uniformly distributed radioactive material in a volume V.

If particle equilibrium exists only for charged particles, it is called Charged Particle Equilibrium (CPE) that can be written

$$\left(\bar{R}_{in} \right)_c = \left(\bar{R}_{out} \right)_c$$

and the average energy imparted becomes

$$\bar{\epsilon} = \left(\bar{R}_{in} \right)_u - \left(\bar{R}_{out} \right)_u + \Sigma Q$$

This means that if the volume of interest is small enough to allow radiative photons (e.g. bremsstrahlung photons – part of $\left(\bar{R}_{out} \right)_u$) to escape from the volume: in CPE condition, the absorbed dose in the medium m is equal, in photon or uncharged particle fluence, to the collision kerma K_{col} or, in a charged particle fluence, to cema

$$D_m \overset{CPE}{\Longleftrightarrow} \left[K_{col} \right]_m \quad D_m \overset{CPE}{\Longleftrightarrow} \left[C \right]_m$$

This is an important relationship, as it equates the measurable quantity D with the calculable quantity K_{col} or Cema.

2.7 CAVITY THEORY

To derive the absorbed dose, D, to a medium from the absorbed dose measured with a detector of different material is a special case of a general problem to compute the absorbed dose to an arbitrary volume element in an arbitrary radiation field. There is no analytical method to calculate D in a medium or detector and this is usually done using Monte Carlo methods.

In practice, the detectors are calibrated in terms of kerma or absorbed dose to the medium in reference irradiation conditions and with a reference radiation quality Q_0. If the calibrated detector is used with a different radiation quality Q, cavity theory is required to identify the appropriate correction factor, f_Q, to apply.

For a radiation detector to be used as a dosimeter, the signal must be proportional to the mean absorbed dose D_{det} in the sensitive material of the detector (Figure 2.26). The detector can be considered a cavity (a word derived from the gas-filled ionisation chamber) in the medium m and cavity theory aims to determine the factor f_Q

$$f_Q = \left(\frac{D_m}{D_{\text{det}}} \right)_Q$$

necessary to calculate the absorbed dose D_m at the point P (Figure 2.26).

Cavity sizes are referred as 'small', 'intermediate' or 'large' in comparison with the ranges of secondary charged particles produced by primary radiation in the cavity medium (scheme in Figure 2.27). In particular, a typically used detector, e.g. ionisation chamber or solid-state detector, will be of a 'large' type at the keV x-ray energies used in DR for the very short range secondary electrons in solid medium of a few microns. Of special interest are the ionisation chambers used in radiotherapy at photon, electron or proton MeV energies that are detectors of the 'small' type where the secondary charged particles have track lengths of some centimetres.

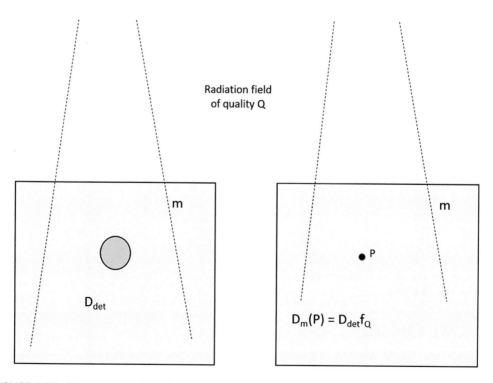

FIGURE 2.26 Measurement of the absorbed dose D in a medium m with a detector (the cavity) of different medium.

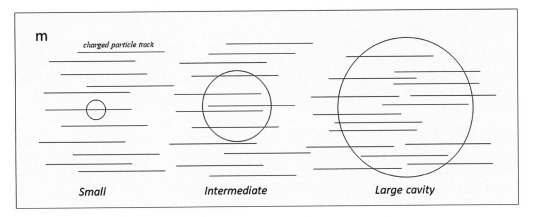

FIGURE 2.27 Small, intermediate and large cavities, classified in comparison with the track length of the secondary charged particles in the cavity.

2.7.1 SMALL CAVITY THEORY

The Bragg–Grey cavity theory (B-G) for small cavity sizes is based on the following assumptions:

1. The cavity must negligibly perturb the fluence of charged particles in the medium.
2. The absorbed dose in the cavity is deposited entirely by the charged particle entering it (crossers).

Consequently, the energy depositions within the cavity (usually air for ionisation chambers) are entirely due to the charged particles crossing the cavity and loosing energy.

The absorbed dose in the cavity (e.g. air) will be:

$$D_{cavity} = \int \phi_E \left(\frac{S_{col}}{\rho} \right)_{cavity} dE$$

with the integral over the energy spectrum of the charged particle, while the absorbed dose in the medium m (e.g. water) will be:

$$D_m = \int \phi_E \left(\frac{S_{col}}{\rho} \right)_m dE$$

With the B-G hypothesis of non-perturbation of the charged particle fluence crossing the cavity, the correction factor becomes

$$f_Q = \left(\frac{D_m}{D_{cavity}} \right)_Q = \frac{\int \phi_E \left(\frac{S_{col}}{\rho} \right)_m dE}{\int \phi_E \left(\frac{S_{col}}{\rho} \right)_{cavity} dE} = S_{col,m,cavity}$$

with $S_{col, m, cavity}$ being the ratio of the average mass collision stopping power for the two materials.

In practice, the presence of a cavity always causes some degree of fluence perturbation that requires the introduction of a fluence perturbation correction factor.

An improvement of the B-G theory is the Spencer–Attix cavity theory, not described here, that takes into account the presence of secondary electrons (δ-electrons deriving from hard-collisions of charged particles) in the fluence of the charged particles and those charged particles that end their tracks in the cavity (stoppers).

2.7.2 Large Cavity Theory in Uncharged Particle Fields

For a large cavity, where the range of the secondary charged particle is negligible compared to the size of the cavity, the secondary charged particles depositing energy in the cavity arise from interactions of photons or uncharged particles inside the cavity. Then, the ratio of absorbed dose in the medium to the cavity will be the ratio of the medium to the cavity collision kerma, equal to the ratio of the average mass energy absorption coefficients, medium to cavity

$$ f_Q = \left(\frac{D_m}{D_{cavity}} \right)_Q = \frac{K_{col,m}}{K_{col,cavity}} = \frac{\Psi \left(\mu_{en}/\rho \right)_m}{\Psi \left(\mu_{en}/\rho \right)_{cavity}} \left(\mu_{en}/\rho \right)_{m,cavity} $$

2.7.3 Intermediate Cavity: Burlin Cavity Theory for Uncharged Fields

For intermediate cavity sizes, Burlin uses a weighting technique combining B-G and large cavity theories

$$ f_Q = \left(\frac{D_m}{D_{cavity}} \right)_Q = d S_{col,m,cavity} + \left(1 - d \right) \left(\mu_{en}/\rho \right)_{m,cavity} $$

where d is a weighting factor, parameter related to cavity size approaching unity for small cavities and zero for large ones. This theory introduced on a purely phenomenological basis had relative success in calculating ratios of absorbed dose for some types of intermediate cavities.

More generally, however, Monte Carlo calculations show that, when studying ratios of directly calculated absorbed doses in the cavity to absorbed dose in the medium as a function of cavity size, the weighting method is too simplistic and additional terms are necessary to calculate dose ratios for intermediate cavity sizes. For these and other reasons, the Burlin cavity theory is no longer used in practice.

2.8 PRIMARY DOSIMETRY STANDARDS

Primary dosimetry standards are instruments of the highest metrological quality that permits determination of dose.

A worldwide network of Primary Standard Dosimetry Laboratories (PSDLs) with the Bureau International de Poids and Measures (BIMP) organise periodic comparisons to ensure consistency of the dosimetry standards.

The primary standards are not used for routine calibrations or measurements. Instead, the PSDLs calibrate secondary standard dosimeters for the network of Secondary Standards Dosimetry Laboratories (SSDLs) that in turn are used for calibrating the reference instruments used in hospitals. This assures that dosimeters used in hospitals for calibration of diagnostic and radiotherapy beams have a calibration coefficient traceable to a primary standard.

At present, there are three basic methods used for the determination of kerma or absorbed dose at the primary standard level:

1. ionometric,
2. total absorption method based on chemical dosimetry,
3. calorimetry.

Ionometric method is used to develop standards of kerma in air with free-air ionisation chambers for x-rays up to 300 kV and for kerma in air and absorbed dose in water with cavity ionisation chambers for photon and electron beams used in radiotherapy for energies over 300 keV and up to some MeV.

In chemical dosimetry, the absorbed dose is determined by measuring the chemical change produced by radiation in the sensitive volume of the dosimeter. The most widely used chemical dosimeter is the ferrous sulphate Fricke dosimeter, a water solution of $FeSO_4$, H_2SO_4 and $NaCl$. The irradiation oxidises ferrous ions Fe^{2+} into ferric ions Fe^{3+} that exhibit a strong absorption peak at a wave length of 304 nm, whereas ferrous ions Fe^{2+} do not show any absorption at this wavelength. The response is expressed in terms of its sensitivity, known as the radiation chemical yield G, the number of moles of ferric ions produced per joule of the energy absorbed in the solution. The Fricke dosimeter has a low sensitivity and the typical dynamic range is from a few Gy to about 400 Gy.

Calorimetry is the most fundamental of the three reference dosimetry techniques, because it is based on the definition of either energy or temperature. In principle, calorimetric dosimetry is simple; in practice, it is very complex because of the need for measuring extremely small temperature differences and possible in sophisticated standards laboratories. In a calorimeter, the energy imparted to matter by radiation causes an increase in temperature ΔT and the absorbed dose D_m in the matter m is

$$D_m = c_m \Delta T$$

with c_m the thermal capacity of the material of the sensitive volume. Calorimetric dosimetry, performed with graphite or water calorimeters, is the most accurate of all absolute dosimetry techniques reaching a standard uncertainty of less than 0.5%.

BIBLIOGRAPHY

Attix, F. H., 2004, *Introduction to Radiological Physics and Radiation Dosimetry*, Wiley-VCH Verlag GmbH & Co.

Hendy, P. P. and B. Heaton, 1999, *Physics for Diagnostic Radiology*, 2nd Edition, Taylor & Francis Group.

International Commission on Radiation Units and Measurements, 1993, *Fundamental Quantities and Units for Ionizing Radiation (revised)*, Report No. 51, Journal of the ICRU Vol. 11 No 1, 2011.

3 Ionising Radiation Detectors

Elizabeth Benson

King's College Hospital NHS Foundation Trust, London, UK

CONTENTS

Radiation detectors are designed to detect the quantity of radiation present in, or incident on a medium. To do this, they detect the interactions that take place between the radiation and that medium. This is the premise of the first detectors: x-ray film measures changes in the film structure brought about by ionisation, scintillating detectors detect the release of light caused by the interaction of radiation within a medium and ionisation chambers detect the charge produced when radiation interacts with particles within the chamber. This premise remains the same today in the development of similar and more advanced radiation detectors.

DOI: 10.1201/9780429155758-3

3.1 MODES OF OPERATION

In all detectors, interaction between radiation and a medium leads to the production of a charge. It is this charge that is used to indicate firstly the presence of radiation and secondly the amount of radiation present. The first basic difference between detectors is based upon how this charge is collected, or the mode in which the detector operates. Detectors may either operate in pulse mode, current mode or mean square voltage mode. The difference between these three modes is how events that occur within the detector are recorded and conveyed to the user, and the information that the user can gain from the detector as a result.

3.1.1 PULSE MODE

Pulse mode is the most commonly applied of these three operational modes. Detectors operating in pulse mode are able to detect the charge associated with individual events. In this case, the charge associated with each event is collected and the current converted to a voltage by the detector circuitry. These voltage changes are easier to measure than the current changes with which they are associated. The amplitude of the voltage measured is directly proportional to the energy deposited within the detector.

3.1.2 CURRENT MODE

When using current mode, the creation of charge within the detector gives rise to an associated current. The average output current is recorded and therefore no information on individual interactions is available to the user. Fluctuations of the energy deposited in the detector will lead to fluctuations in the output. However, increasing the length of time over which the current is averaged may mean the loss of information about the rate or nature of the interactions taking place.

3.1.3 MEAN SQUARE VOLTAGE (MSV) MODE OR 'CAMPBELLING' MODE

When operating in MSV mode the current flowing as a result of the charge produced by interactions is squared and averaged to give the detector output. Thus the amplitude of the output signal is proportional to the square of the charge produced. As a result, there is a higher differential between the outputs generated by interactions associated with different energy radiations. Detectors operating in this mode are most effective in environments where several different radiation types may be present.

3.2 DETECTOR PROPERTIES

The properties of detectors indicate the application for which they are most suitable. These may change given the circumstances of use. Of course, the application to which a detector is suited also depends upon the radioactive source and hence the radiation type to be detected. These factors must be considered together when selecting a detector for a particular application.

3.2.1 ENERGY RESOLUTION

Detectors respond differently to different types of radiation. The investigation of the energy distribution of the radiation incident on a detector is termed radiation spectroscopy. To determine a detector's energy resolution for a certain radiation we use radiation spectroscopy and consider the

TABLE 3.1
Detector Operational Modes

	Pulse Mode	Current Mode	MSV Mode
Output	Voltage pulse associated with each event and directly proportional to the energy deposited.	Average current associated with a series of events.	Average of the square of the voltage associated with a series of events.
Advantages	Higher sensitivity – information regarding the rate and amplitude of individual events detected. Lower detection limit set by background interaction rate.	Indication of the average energy associated with a series of events. Avoid dead-time losses	Higher amplitude differential between events associated with different radiation types. Avoid dead-time losses.
Disadvantages	May be affected by dead-time losses at high interaction rates.	Lower detection limit may be set by an average interaction rate greater than background. No information about individual pulses.	No information about individual pulses
Example detectors	Counters, e.g. Geiger–Müller, proportional, scintillation Spectrometers	Ionisation chamber	Neutron detectors

detector's response function for a monoenergetic source of that radiation (Figure 3.1a). A narrow spectrum indicates that there is little fluctuation of pulse height, H, about the average pulse height, H_0, for that single energy, and therefore a better energy resolution compared to a greater spread of pulse heights around H_0. That is, a lower distribution of pulse heights indicates a more uniform detector response to the incident monoenergetic radiation beam.

Following from this, for a Gaussian detector response, the energy resolution for a particular radiation may be quantified by measuring the full width at half maximum (FWHM) of the detector response (assuming that background has been removed) and applying Equation 3.1 (Figure 3.1b). Assuming that the same number, N, of interactions have taken place, the area under each of these response curves will be identical. Therefore, a lower R value indicates a better detector energy resolution.

$$Energy\ Resolution, R\left(\%\right) = \frac{FWHM}{H_0} \tag{3.1}$$

Changes in the operating characteristics of a detector, random noise within the detector and its associated instrumentation, and statistical noise in the measured signal may affect the response function of the detector. The result of a change in the response function of the detector may be a decrease in its energy resolution. It may be possible to reduce the likelihood of some of these sources of noise; however, some amount of random and statistical noise is always present.

How does energy resolution vary with radiation source type? For radiation sources that emit a spectrum of energies, detector energy resolution is very important as it allows for distinction of

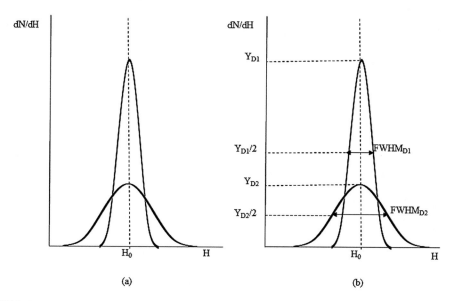

FIGURE 3.1 Measuring detector energy resolution.

energy peaks within the spectrum. For radiation sources that emit radiation at several discrete energies, energy resolution may be of less importance as the distinction between these emissions will
be easier to resolve, if the energy peaks are well separated. Applying this in medical physics, it
may be possible to identify a radionuclide from its emissions using a detector with a higher R value
(poorer energy resolution), whereas a detector with a lower R value (better energy resolution) will
be required to identify the characteristic peaks in a particular x-ray spectrum.

Where is energy resolution important? Provided that two incident radiations have energies
greater than one FWHM apart the detector should be able to separate them. Therefore, detectors
with good energy resolution are generally used in radiation spectroscopy to identify different energy
peaks within radiation spectra, and in environments where sources of different radiations are present
to distinguish between their mixed emissions.

3.2.2 DETECTION EFFICIENCY AND SENSITIVITY

To achieve perfect, or 100%, detector efficiency, every quantum of radiation entering the detector
volume would be counted. Thus, the detection efficiency of a detector is a measure of the ratio of the
number of pulses recorded in relation to the number of radiation quanta present. Two different types
of detection efficiency are defined: the absolute counting efficiency, ϵ_{abs}, and the intrinsic counting
efficiency, ϵ_{int}. The equations for these are shown (3.2 and 3.3).

$$absolute\ counting\ efficiency, \epsilon_{abs} = \frac{no.\ pulses\ recorded}{no.\ radiation\ quanta\ emitted\ by\ the\ source} \tag{3.2}$$

$$intrinsic\ counting\ efficiency, \epsilon_{int} = \frac{no.\ pulses\ recorded}{no.\ radiation\ quanta\ incident\ on\ the\ detector} \tag{3.3}$$

$$\epsilon_{int} = \epsilon_{abs} \cdot \frac{4\pi}{\Omega} \qquad (3.4)$$

When using absolute counting efficiency, counting geometry such as the distance from the source to the detector and the solid angle of the detector from the source position (Ω) must be taken into account. Therefore, it is more straightforward to use the intrinsic counting efficiency although the two may be related as in Equation 3.4. The intrinsic counting efficiency of a detector depends mainly upon the material from which the detector is made, the thickness of the detector material, and the energy of the incident radiation. There is also a slight dependence on the distance between the source and the detector – for larger source to detector distances, the energy the radiation has when it reaches the detector may be lower, and therefore the length of its path through the detector material will be shorter.

If all the interactions that occur within the detector volume are considered, then the given efficiency is defined as the total efficiency, ϵ_{total}. In practice, the concept of a peak efficiency, ϵ_{peak}, is usually used. Here, a filter is applied to remove the low-amplitude signals generally associated with noise. In this way, only the interactions that lead to complete energy deposition are counted and are not obscured by lower energy events such as noise or scattering within the detector material. Peak efficiency takes into account the number and the nature of the interactions in the detector volume. Total efficiency and peak efficiency are related by the peak-to-total ratio as shown in the following equation:

$$peak-to-total\ ratio, r = \frac{\epsilon_{peak}}{\epsilon_{total}} \qquad (3.5)$$

The sensitivity of a detector is a measure of the response it has per unit amount of energy deposited in the detector volume. Detectors with a high sensitivity have a higher amplitude response to interactions within the detector volume than detectors with lower sensitivity. Therefore, high-sensitivity detectors may be able to distinguish incident radiation events from background radiation events and give a more accurate indication of the incident radiation present. For some detectors, sensitivity may change with photon energy and therefore it is important to check a detector's sensitive range before use.

How does detection efficiency and sensitivity vary with radiation type? Charged particles such as alpha and beta are easily stopped and therefore tend to interact immediately upon entering the detector; they also tend to lose all their energy at this initial interaction giving a large signal pulse. These radiations are more likely to be counted leading to a detector efficiency approaching 100%. Uncharged radiations such as x- and gamma-rays are more difficult to stop than alpha and beta particles, and have a long path length between interactions; therefore, the chance that they will have a significant interaction within the detector volume is lower. For these radiations, detection efficiency is usually significantly less than 100%. The detector sensitivity in these cases depends upon how the response of the detector material varies with photon energy.

Where is detection efficiency and sensitivity important? Detector efficiency and sensitivity varies for different types and energies of radiation. Some detectors with 100% detection efficiency and high sensitivity for charged particles may have a very low detection efficiency and sensitivity for x- and gamma-rays as the detector material is not suitable to stop these radiations. The opposite is also true – alpha and beta particles may not be able to penetrate the detector volume of an x- or gamma-ray detector due to their short path length; as a result, they will not be detected by such a detector. Therefore, it is very important that the correct radiation detector is employed for specific radiation types.

Detector sensitivity is particularly important in nuclear medicine where low energy emissions from patients may need to be detected and distinguished from background for diagnostic purposes.

3.2.3 DEAD TIME

For a radiation detector to be able to register separate interaction events, it must be able to respond to those events individually. When a detector is waiting to receive a signal it is considered 'live', once an interaction has occurred and a signal has been received the detector is considered 'dead'. After the signal is processed the detector becomes live again and is ready to receive further signals.

A time interval or 'dead time' is defined for particular detectors that describes how long the time difference between individual events must be for them to be counted separately by the detector.

The dead time for any particular detector depends on the detector electronics and the interaction processes taking place within the detector itself. Counting losses attributed to dead time are described as non-paralysable and paralysable.

If a detector has a non-paralysable response, the detector does not respond to any additional events during the dead time period. Events that occur during the dead period are not counted. Conversely, if a detector has a paralysable response, the detector continues to respond to additional events during the dead period. These events do not register as separate counts but prolong the detector's dead time. Taking this into consideration, at high event rates, the non-paralysable detector will register more counts than a paralysable detector and this must be adjusted for to ensure counting accuracy. At low rates, counts for both detectors will be similar. It is possible to measure dead time and to calculate and correct for paralysable and non-paralysable dead time.

How does dead time vary with radiation source type? Dead time losses vary depending on the radiation source and whether emissions are pulsed or continuous. For pulsed sources, the losses will depend on the pulse frequency in comparison to the dead period of the detector. This needs to be taken into consideration when selecting a suitable detector.

Where is dead time important? At high event rates, losses due to dead time may be significant, particularly for certain detectors. Geiger–Müller (G-M) counters are particularly susceptible to dead time losses. Therefore, dead time and the radiation source under consideration should always be considered when choosing a detector.

3.2.4 ANGULAR DEPENDENCE

The angular dependence of a radiation detector describes its ability to detect incident radiation photons from different angles. An isotropic detector has no angular dependence, that is, it will collect photons from all possible directions. However, a non-isotropic detector has directional and angular dependence. Detector efficiency depends partly on angular dependence of the detector as considered in Section 3.2.2.

How does angular dependence vary with radiation type? Different radiation sources may have the same photon fluence rate in all directions (isotropic) or different photon fluence rates in different directions (non-isotropic). Radiation emitted from radionuclides is non-isotropic, as is x-ray radiation emitted from an x-ray target. Gamma-ray sources tend to be isotropic. Therefore, if the radiation emissions from these sources are to be detected the detector must have a suitable angular response that enables radiation detection from all required directions.

Where is angular dependence important? Directional dose is important when carrying out area monitoring where radiation incident on the detector is unlikely to be isotropic. For this purpose, the International Commission on Radiation Units (ICRU) has defined the directional dose equivalent, H'(d, Ω). It is important to check when using a survey meter that it is non-directional, i.e. has an isotropic response. For detectors with a non-isotropic response the positioning of the detector is very important to enable the most accurate measurement possible.

3.3   DIFFERENT TYPES OF DETECTOR

3.3.1   PHOTOGRAPHIC FILM

Photographic film is the earliest material used to detect radiation. Indeed it was through the interaction of x-rays with photographic film that the existence of x-rays was first discovered by Röntgen in 1985. Now, rather than photographic film, specialist x-ray film is used, although the principles of detection are still the same.

The construction of basic x-ray film is shown in Figure 3.2. The polyester base layer is generally around 0.15 mm thick and remains stable and rigid during exposure and development. The base laser is generally coated on both sides with an emulsion, although for some specialist applications only one side of the base layer is coated. To prevent damage, the base layer and emulsion are then coated with a protective layer.

The emulsion is the part of the film which is sensitive to radiation; it is composed of silver (Ag^+) and bromide (Br^-) ion crystals. When the emulsion is exposed to radiation, photoelectric and Compton interactions with Br^- ions cause the production of free electrons. These electrons are trapped due to impurities within the emulsion crystal lattice. The trapped electrons attract positive Ag^+ ions and combine to form neutral Ag atoms. This process continues for all interactions within the emulsion so that the crystal lattice holds a 'map' of silver ions showing where the interactions have happened. This is referred to as a latent image.

To convert the latent image to a radiographic image an alkaline agent is applied to 'reduce' the remaining Ag^+ ions from the emulsion. The film is then 'washed' with an acidic solution to remove the reduced Ag^+ ions, fix the image and harden the film. Where there have been many interactions, there are many Ag atoms or grains corresponding to a dark area on the film and where there have been no interactions there are no Ag grains, corresponding to a light area on the film.

X-ray film is sensitive across radiation energies from *c.* 10–100 keV; therefore, it is not only used in imaging, but also in personal dosimetry – detecting exposure to diagnostic radiation energies. One of the main advantages of the use of film as a personal dosimeter is its change in response to photons of different energies. To make use of this property, the film badge holder contains a number of filters. An example of a typical film dosimeter holder is shown in Figure 3.3.

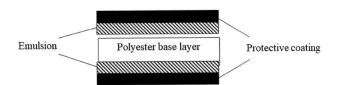

FIGURE 3.2   Structure of x-ray film.

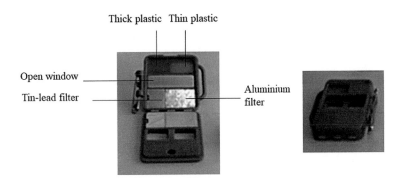

FIGURE 3.3   Typical film dosimeter holder and filters.

TABLE 3.2
Relative Properties of Film Dosimetry Badges and TLDs

	Film Badge	TLD
Dose response	0.2 mGy – 6 Gy	0.1 mGy – 10^4 Gy
Response linear with dose?	No	Yes
Response dependent on radiation energy?	Yes	No (except at low kV)
Sensitive to temperature and humidity?	Yes	No
Maximum use period	2 months	12 months
Size	Small (c. 5 cm)	Very small
Permanent, visual record?	Yes – film	No
Type of radiation indicated?	Yes	No

Each filter has a different purpose; the open window allows all radiation to pass; therefore, overall exposure can be seen, the plastic filters allow beta radiation of different energies to pass, the aluminium filters remove beta radiation but allow x- and gamma radiation to pass and the tin-lead filter extends the useful range of the film's response so that photons of energy higher than 100 keV can be detected. Different filters may be used in different areas, for example, a cadmium filter allows neutron exposure to be detected and this may be used in areas where exposure to neutrons is possible. Due to these filters, when the film is processed it is possible to tell from the pattern on the film the type of radiation that the wearer has been exposed to.

Film badges in personal dose monitoring have been largely replaced by thermoluminescent detectors (TLDs – Section 3.3.2.5). The relative properties of film badges and TLDs are shown in Table 3.2. One of the main advantages of TLDs is their size and range of dose response, although they do tend to be more expensive than film badges.

3.3.2   Gas-Filled Detectors

The operation of gas-filled detectors is based upon the interactions resulting from a charged particle or radiation photon passing through a gas. These interactions cause ionisation of the gas molecules producing a free electron and a positive ion (an ion pair). It is the detection of these ion pairs that is used to indicate the presence of radiation and in some cases the amount and type of radiation that is present.

Here the function of three different types of gas-filled radiation detector is discussed: ionisation chambers, proportional counters and G-M counters. In all of these detectors, the application of an electric field is used to collect the charge created by the interaction of radiation with the gas volume. Indeed, the electrical field applied to gas-filled detectors determines the interactions that occur within the gas and the operation of the detector. Figure 3.4 shows the different regions of operation for these detectors.

Charge collection is achieved slightly differently for each detector and there are two concepts that must be understood in relation to gas-filled detectors. The first of these, cavity theory, has been discussed previously and describes how the absorbed dose to a particular material, in this case a gas, can be inferred based on the properties of that gas. This has particular application for ionisation chambers.

The second concept, fundamental to the operation of proportional and G-M counters, is gas multiplication. When an electric field is applied across a gas volume containing ion pairs, the positive ions will drift to the cathode (negative electrode) and the negative free electrons will drift to the

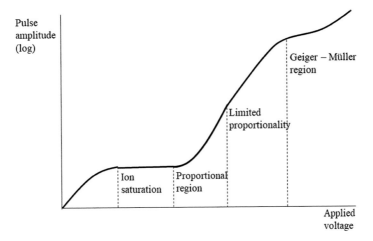

FIGURE 3.4 Gas-filled detectors; regions of operation.

anode (positive electrode). As they drift towards their respective electrodes, these particles may collide with other gas molecules. The positive ions have little energy and therefore do not produce any ionisation effects; however, due to acceleration by the electric field, the electrons may have enough kinetic energy to interact and produce a second ion pair – multiplication.

Only when the electric field exceeds a certain threshold strength will the acceleration of the electrons be enough to give them the energy to interact in this way. Therefore, below the threshold electric field strength there is no gas multiplication within the gas volume.

Electrons from secondary ionisation events are also accelerated by the electric field and may therefore go on to cause further ionisation events and more ion pairs, resulting in a Townsend Avalanche. The multiplication effect is described by the gas multiplication factor, M, which indicates the factor by which the charge associated with the original number of ion pairs has been amplified.

The application of these concepts to relevant detectors, and the characteristics of the resulting outputs, is discussed in the following sections.

3.3.2.1 Ionisation Chambers

Ionisation chambers are the simplest of the gas-filled detectors. The gas chamber is part of an electric circuit across which a voltage is applied. When radiation passes through the gas ionisation occurs, creating ion pairs. The voltage across the chamber attracts the positive ions to the cathode and the negative ions to the anode creating a flow of charge – a current. This current is then measured by an ammeter in the circuit. In practice, the current measured is very small and therefore it is generally amplified. As the current measured from the ionisation chamber is directly proportional to the number of ion pairs created, according to cavity theory, the absorbed dose to a given material can be calculated.

Ionisation chambers are operated at a voltage high enough to achieve ion saturation. That is, following ionisation, all positive ions and free electrons formed are attracted towards their respective electrodes and none are lost to recombination or diffusion from the gas volume. In this case, all the ion pairs formed by the original radiation interaction are counted. Ionisation chambers are usually operated in current mode so that an average rate of ion pair formation or energy deposition is measured; however, they can also be operated in pulse mode.

3.3.2.1.1 Ionisation Chamber Design

The most common gas used in ionisation chambers is air at normal atmospheric pressure. Under normal circumstances, a very large volume of air would be required to be able to measure interactions accurately, particularly for x-rays and gamma-rays. Therefore, this large volume of air is

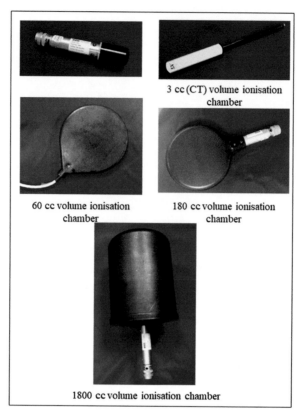

FIGURE 3.5   Common air equivalent ionisation chambers.

(Photographs courtesy of Glafkos Havariyoun.)

replaced by an air equivalent shell, generally made of plastic, which has the equivalent properties of the required volume of air, compressed.

Commonly the ionisation chamber electrical field is applied between two parallel plates. However, construction may also be cylindrical with the outer shell acting as a ground electrode and an axial conducting rod carrying the required voltage. Some common ionisation chambers are shown in Figure 3.5.

3.3.2.2   Proportional Counters

Proportional counters detect radiation events based on the principle of gas multiplication, described earlier in this section. As suggested by their name, the pulse produced as a result of radiation interactions and gas multiplication is proportional to the initial energy deposited in the gas volume and the two are related by the gas multiplication factor, M. Proportional counters are generally operated in pulse mode.

The proportional counter gas volume is usually a cylindrical tube. For each electron to undergo identical gas multiplication, the region of gas multiplication must be small in comparison to the total gas volume. If this is true, all primary free electrons are formed outside the multiplication area and drift into the area where the electric field is high enough to cause multiplication. Therefore, the multiplication process, and the associated multiplication factor, is the same for all free electrons. In practice, this is achieved by using a fine wire along the central axis of the tube as the anode and the inside surface of the tube as the cathode (Figure 3.6). Free electrons are formed in the larger gas volume and attracted towards the fine wire, which is surrounded by a volume that is small in comparison to the remaining gas volume.

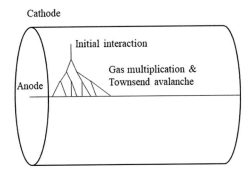

FIGURE 3.6 Common proportional counter structure, multiplication and avalanche formation.

3.3.2.2.1 *Proportional Counter Design*

Proportional counters tend to be sealed tubes with a window either at one end or along the cathode wall. Windowless counters are used for radiation sources that may lose significant energy or be completely stopped by the window. These counters are of hemispherical construction with the source placed within the detector volume.

For all proportional counters, energy resolution depends on an identical multiplication factor for all primary ion pairs. This relies on a uniform electrical field along the anode, which can be achieved by ensuring that the anode wire is as smooth and uniform as possible. The electrical field may become distorted at the ends of the anode due to the connection with the housing; this can be avoided by creating dead zones in these regions so that no multiplication can take place.

The choice of fill gas for proportional counters is particularly important to ensure the performance of the detector. Different gases may have very different electron drift and diffusion characteristics due to their atomic number and density. For ideal operation of a proportional counter, it is desirable to have a low electron diffusion coefficient, high drift velocity and a low electron attachment coefficient. Air in its natural state is not useful as a proportional counter fill gas as it has an appreciable electron attachment – therefore for sealed tube proportional counters air ingress must be prevented. This is not an issue for continuous flow proportional counters.

Generally, the noble gases are used and a combination of 90% argon with 10% methane (P-10 gas) is widely used. In proportional counters used for neutron detection and spectroscopy, gases such as hydrogen, methane, helium and other low atomic number gases are also used. When using a proportional counter for dosimetry, gases with properties approximating those of tissue are used, e.g. a combination of carbon dioxide, methane and nitrogen (64.4%, 32.4% and 3.2%, respectively).

Statistical variations in proportional counter readings arise as a result of fluctuations in the number of ion pairs formed per radiation interaction (Fano factor) and fluctuations in the number of avalanches caused by free electrons. These, combined with electrical noise, geometrical non-uniformities, such as non-uniform anode wire, and operational factors such as voltage stability, gas pressure and gas purity should be carefully controlled to maintain optimum energy resolution. Proportional counters are susceptible to age effects, for example, caused by leakage of air into the gas volume and solid depositions on the anode.

Adaptation of the proportional counter for various different applications has resulted in many design variations on the traditional proportional counter described here. These are described in Knoll [1], but are beyond the scope of this book.

3.3.2.3 Geiger–Müller Counters

The G-M counter was the earliest radiation detector to be introduced. As for proportional counters, its performance is based on gas multiplication; however, the G-M counter operates at a higher voltage than the proportional counter. As a result, unlike proportional counters, where each free electron

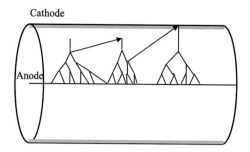

FIGURE 3.7   Common Geiger–Müller structure, gas multiplication and avalanche.

formed in an ion pair may interact to form a single Townsend avalanche, in the G-M counter each avalanche may then form another avalanche. Indeed, at the voltages used in the G-M counter, each avalanche formed by a primary free electron will give rise to at least one more avalanche.

When a primary free electron interacts to form a primary avalanche, the result is a secondary free electron, a secondary ion and excited gas molecules. These excited gas molecules lose energy quickly to return to the ground state and in doing so emit photons of energy. Some of these photons are in the visible or ultra-violet range and may be photoelectrically absorbed by other gas molecules in the chamber volume to produce a tertiary free electron. Alternatively, the photons may be absorbed in the cathode walls with the same result. These tertiary free electrons are accelerated towards the anode where they can cause further avalanches (Figure 3.7).

The fill gas in G-M counters is chosen to optimise the propagation of avalanches and their spread along the length of the anode. These avalanches originate at random points in the multiplication region and therefore there is no useful positional information associated with them in contrast to proportional counters. As the number of avalanches increases, so does the number of positive ions present as these are produced with every avalanche.

The positive ions produced move very slowly in relation to the electrons and collect in the multiplication region. This collection of positive ions is referred to as a space charge and reduces the electric field in the multiplication region. As multiplication relies on the electric field remaining above a certain threshold value, eventually this accumulation of positive ions will result in the termination of gas multiplication – the tube is referred to as 'dead'. The Geiger discharge refers to the charge produced up until the point where gas multiplication is terminated. An increase in voltage and the electric field will increase the Geiger discharge and the amplitude of the output pulse.

For a set voltage, the point at which the Geiger discharge terminates will always be the same, as the same number of positive ions will always need to be produced to reduce the electrical field below the multiplication threshold. As a result, all G-M pulses have the same amplitude, which is unrelated to the number of primary ion pairs formed. Therefore, G-M pulses do not provide any useful information on the amount of energy deposited in the gas volume and the properties of the incident radiation pulse.

For further multiplication to occur when the tube is dead the positive ions formed must drift towards the cathode, reducing the space charge and allowing the electrical field across the multiplication region to recover to a level above the multiplication threshold. As the electric field recovers, multiplication can take place; however, there are still some positive ions present and so the next Geiger discharge takes less time to terminate, has a lower amplitude and may not be registered. Only when all the positive ions have returned to the cathode can a full Geiger discharge occur and a full amplitude pulse be registered.

The dead time of the G-M counter is defined as the time between the original pulse and the time at which a second Geiger discharge can be registered – regardless of its amplitude. This can also be referred to as the resolving time and is generally about 50–100 µs, a long dead time when compared to other detectors. The time taken for the tube to be able to respond with a full Geiger discharge is referred to as the recovery time.

3.3.2.3.1 Geiger–Müller Counter Design

The most important consideration in the design of a G-M counter is the fill gas. G-M counters are the only detectors that must be quenched in order to ensure that interactions causing avalanches are stopped. The requirement for quenching arises when using a single fill gas. In this case, all the positive ions formed during primary and subsequent interactions are ions of that gas. When these ions reach the cathode, they are neutralised by recombination with an electron at the cathode surface. This causes a release of energy, the work function, equal to the difference between the ionisation energy of the gas and the energy required to free an electron from the cathode surface.

If the energy released is greater than the work function of the cathode surface, then a free electron may be produced from the cathode. The more ions there are arriving at the cathode surface, the higher the probability that this will occur. Any free electrons produced will be accelerated towards the anode causing further gas multiplication, avalanches and output pulses. This may also occur in proportional counters, although the probability is lower. In proportional counters the output produced is lower as only one avalanche is produced per free electron; therefore, a spurious pulse rather than continuous output will be the result.

The process of quenching is used to prevent the continuous output associated with this process. Quenching may be internal or external although internal quenching is more common. In internal quenching a quench gas is added to the fill gas (around 5–10%). The quench gas must have lower ionisation potential and more complex molecular structure than the fill gas. As positive ions are formed and drift back to the cathode they interact with the quench gas molecules. As a result of the lower ionisation potential of the quench gas, the positive charge of the ions is transferred to the quench gas molecules. The product of this interaction is a neutral atom of the fill gas and a positive ion of the quench gas.

The positive ion of the quench gas continues to drift towards the cathode. At the cathode the excess energy produced is absorbed in dissociation of the more complex quench gas molecule, rather than in freeing an electron from the cathode surface. Provided the concentration of the quench gas is high enough, the majority of positive ions arriving at the cathode will be quench gas ions. Therefore, the probability of the formation of a free electron as a result of energy release from a positive ion of the fill gas is very much lower.

Throughout the G-M counter's lifetime the molecules of the quench gas are dissociated and so the quench gas concentration gradually decreases. Unless a quench gas that can be replenished is used, i.e. halogens such as chlorine or bromine, the G-M counter has a definite lifetime, defined by quench gas and the number of interactions that can take place before it is no longer effective. The lifetime of the G-M counter is also reduced by contamination of the gas with the products of dissociation and deposition of these dissociation products on the anode surface. Quench gases may be added to the fill gas of proportional counters. In this case it is to absorb any of the visible and ultraviolet photons released during gas multiplication which may cause extra avalanches, adding to detector noise.

In external quenching, the high voltage applied to the G-M counter is reduced for a finite period following each output pulse – generally a few hundred microseconds. This ensures that the electric field across the gas is not high enough to produce gas multiplication and avalanching. The disadvantage of this approach is the time taken for the electric field to regain its full strength and therefore produce full Geiger discharge. This approach is only really effective at low count rates.

Due to the requirement for quenching, the choice of fill gas for a G-M counter is one of the most important aspects of its design. In some respects, the properties of fill gases used for G-M counters are very similar to those chosen for proportional counters. As for proportional counters, gases that produce negative ions, such as oxygen, cannot be used. As a result, in a similar way to proportional counters, noble gases argon and helium are often used. Unlike proportional counters, in order to achieve the high electric field required for multiple avalanches, the gas is often pressurised to below atmospheric pressure.

In this configuration, the tube window must be thin enough to allow short-range particles such as alpha particles to pass but rigid enough to support the low pressure within the tube. For longer range radiations such as beta particles and low energy x-rays and gamma rays the window may be thicker. Continuous flow G-M counters are available for very soft radiations. As the output pulses from the G-M counter spread all along the anode wire, and do not provide quantitative information, uniformity of the electric field along the anode wire is of less importance than in proportional counters.

In theory, a detection efficiency of 100% for charged particles can be achieved using the G-M counter. However, this relies on all the particles passing through the tube window. G-M detection efficiency for gamma and x-rays is low, generally a few per cent.

For these radiations, the particle interaction occurs not with the fill gas, but with the tube walls. Therefore, detection efficiency is reliant upon the probability that a photon will interact with the wall, and that those interactions occur close enough to the wall edge for any free electrons produced to travel through the tube wall to the cavity where they can produce ions. The possibility of photon interaction with the tube wall increases with increasing atomic number and so Bismuth ($Z = 83$) is often used as the tube wall material. Similarly, the probability of interaction with the fill gas may be increased by using a gas with high atomic number, such as xenon or krypton. These methods are effective for low energy x-ray and gamma radiations (< 10 keV); however, for high energy x- and gamma-ray detection efficiency using a G-M counter remains low.

3.3.3 SOLID-STATE DETECTORS

Due to the dependence of radiation interactions on material properties, such as a high atomic number and high electron density, solid materials often make more efficient radiation detectors than gases. Solid-state detectors include scintillation detectors (Section 3.3.3.1), semi-conductor detectors (Section 3.3.3.2) and thermoluminescent detectors (Section 3.3.3.3).

3.3.3.1 Scintillation Detectors

The molecules of certain materials release light when radiation interacts with them. Scintillation detectors detect the light that is released and convert it to information about the incident radiation. The materials most commonly used in scintillation detectors are inorganic alkali halide crystals, of which sodium iodide (NaI) is the most popular, and organic based liquids and plastics. The following characteristics are desirable in a scintillation material, although in practice no single material possesses all of these criteria:

- High scintillation efficiency – efficient conversion of radiation kinetic energy to light.
- Linear conversion of deposited energy to light.
- Transparent to the wavelength of light that the material itself emits for optimal light collection.
- Fast decay time of light produced so that the material can respond at high count rates (low 'after-glow')
- Possible to manufacture as a practical detector.
- Refractive index that allows efficient coupling to a photomultiplier (generally close to that of glass, 1.5).

3.3.3.1.1 Inorganic Scintillants

The use of inorganic scintillant detectors has developed rapidly in recent years, largely due to the discovery of new materials. These materials are solid crystals and produce light dependent on the energy bands within the crystal structure. Three bands are defined within the crystal lattice, the valence band, the conduction band and the forbidden band.

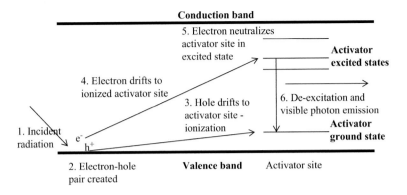

FIGURE 3.8   Scintillation in inorganic crystals.

Electrons in the valence band are bound to sites within the lattice. Electrons in the conduction band are free and have enough energy to move around the crystal. The forbidden band lies between the valence band and the conduction band and has intermediate energy states (activator states), created by intentionally introduced impurities (activators) in the crystal. These activator states are referred to as luminescence or recombination centres and electrons can de-excite through these states to return to the valence band. The energy lost in these transitions is less than that between the conduction band and the valency band and is designed to produce a photon with a visible wavelength – scintillation.

A charged particle or photon passing through the crystal will transfer energy to several electrons in the valency band, allowing them to migrate to the conduction band. The elevation of these electrons leaves a corresponding number of positive 'holes' in the valency band. This process of electron–hole pair production in solid materials is analogous with the process of ion-pair production in gases. The positive holes drift to activators in the crystal and ionise them. The electrons move around the lattice until they can combine with an ionised activator to create a neutral configuration. If the electron combines with an excited activator site where de-excitation to the activator ground state is possible then this occurs very quickly (30–500 ns), with a high probability that a visible photon will be emitted. Alternatively, the electron–hole pair may migrate together, as an exiton, to an activation site. Similar excitation and de-excitation processes occur and a visible photon is emitted.

It may be the case that the electron neutralises an excited activator site whose transition to the activator ground site is forbidden. If this happens, more energy will be required to raise the configuration to an energy state where de-excitation to ground is possible. This energy may be provided thermally and the resulting emission is phosphorescence. Phosphorescence is a slow response that is responsible for afterglow in inorganic crystal scintillation detectors.

Alternatively, for transfer between some states, energy may be lost with no resulting visible photon transmission. This represents a loss of energy in the system and is referred to as quenching.

These inefficiencies need to be taken into account, along with others. However, the light resulting from these interactions is approximately proportional to the radiation energy deposited. The scintillation signal from inorganic scintillators tends to be more proportional to the incident radiation than that from organic scintillators.

3.3.3.1.2   Inorganic Alkali Halide Scintillators

Of the alkali halide scintillators, sodium iodide activated with thallium iodide – NaI(Th) – has proved to be one of the most successful and enduring for detecting x-ray and gamma radiation. Caesium iodide activated with thallium or sodium (CsI(Th), CsI(Na)) has also proved to be a popular scintillation detector choice in recent years. The different properties of these scintillators are compared in Table 3.3.

TABLE 3.3

Properties of NaI(Th), CsI(Th) and CsI(Na) Scintillation Crystals

Property	NaI(Th)	CsI(Th)	CsI(Na)
Light yield (Photons/MeV)	38,000	65,000	39,000
Ease of manufacture	Crystals can be formed together to create unusual shapes, although they are fragile and can be damaged thermally or mechanically	Less fragile than NaI(Th) – withstands shock and vibration. Can be cut into thin sheets	
Applications	Gamma and x-ray detection and spectroscopy.	Gamma and x-ray imaging Space instrumentation	Gamma and x-ray detection Space instrumentation
Temperature response	Reduced scintillation yield at higher temperatures. Increased decay time at lower temperatures.	Reduced scintillation yield at high and low temperatures.	Reduced scintillation yield at high and low temperatures.
Scintillation decay time for gamma radiation	0.23 µs	Variable gamma decay time 0.68 µs, 3.34 µs. Variable decay time for different particles.	Variable decay time 0.46 µs, 4.18 µs
Special properties	Hygroscopic must be protected from air to avoid water absorption and deterioration	May be grown in thin layers with a microstructure that inhibits the spread of light to achieve excellent spatial resolution for radiation imaging applications. Less hygroscopic than NaI(Th), deteriorates in water or high humidity.	Hygroscopic must be protected from air to avoid water absorption and deterioration

3.3.3.1.3 Organic Scintillants

Fluorescence in organic scintillants occurs due to energy transitions within single molecules of the material. These transitions can occur whether the material is in a solid, liquid or gaseous form. Within each molecule, there are two sets of energy levels, singlets and triplets, with each having multiple energy levels (Figure 3.9).

Kinetic energy is absorbed from passing charged particles such as free electrons, causing the singlet states in the molecule to become excited. Following excitation, the singlet states are quickly de-excited to a singlet state above ground level, in a transition that does not produce any radiation. The molecule is left in an excited state above ground as a result of excitation caused by incident radiation.

Excited molecules return to the ground level via de-excitation and emit visible light – fluorescence. Some molecules may undergo transitions referred to as inter-system crossing to a triplet energy state. When they are de-excited from this state, there is a delayed light emission – phosphorescence. If a molecule in an excited triplet state is excited back to an excited singlet state and is then de-excited this delayed transition is termed delayed fluorescence. A summary of these processes can be seen in Figure 3.9. As for inorganic scintillants, organic scintillants are subject to quenching (radiationless transitions) which decrease the scintillation efficiency of the material.

Several different materials may be used as organic scintillants (Table 3.4).

The light output produced by scintillation in a material is generally very low intensity. Therefore, care must be taken when collecting the light to preserve as much of the signal as possible. The light

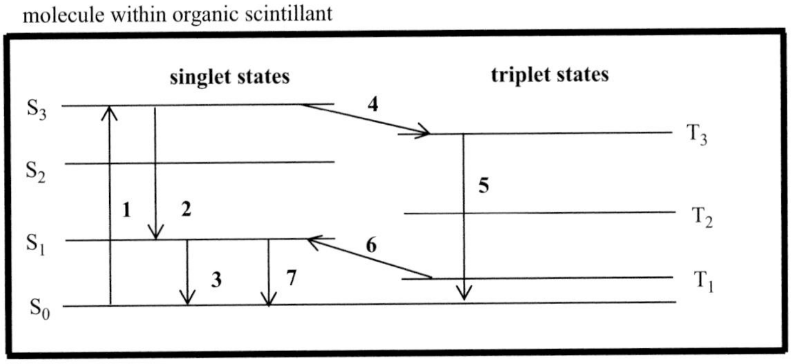

molecule within organic scintillant

1. Energy absorbed from charged particle.

2. Singlet state de-excited to a state above ground level, no energy is released.

3. **Fluorescence** - de-excitation to allow molecule to return to the ground state.

4. **Inter-system crossing** – de-excitation from singlet to triplet state.

5. **Phosphorescence** – light emission caused by de-excitation from a triplet state.

6. **Inter-system crossing** – molecule in an excited triplet state excited to an excited singlet state.

7. **Delayed Fluorescence** – de-excitation to allow molecule to return to the ground state.

FIGURE 3.9   Structure and energy transfer processes within organic scintillants.

TABLE 3.4
Organic Scintillants

Property	Pure organic Crystals, e.g. Anthracene, Stilbene	Liquid Organic Solutions	Plastic Scintillators
Structure	Fragile, not usually available in large sizes	Organic scintillator dissolved in appropriate solvent. Often in sealed containers and used as solid scintillators, therefore can be quite large volume.	Organic scintillator dissolved in a solvent and polymerised to form a plastic. Relatively inexpensive and easy to machine therefore available in several forms – rods, cylinders, flat sheets, etc. Degradation in light.
Scintillation efficiency	Dependent on orientation of ionising particle – directional variation	Lower than Anthracene	Lower than Anthracene

is collected via several methods such as light pipes or fibre scintillators optically coupled to the scintillating material. These structures are used to deliver the light signal to a photomultiplier tube where the signal is amplified to a useful level. An important consideration in the use of scintillation detectors for radiation imaging is the 'after-glow' of the material, i.e. the emission of light photons after radiation exposure has ceased. This affects the dead time of scintillation detectors.

3.3.3.1.4 Photomultiplier Tubes

For the weak light output from scintillating materials to be of any practical use the signal must be detected and converted into an electrical pulse. This is achieved efficiently and with as little additional noise as possible using photomultiplier (PM) tubes. A diagram of a photomultiplier tube is shown in Figure 3.10 and the processes that occur in the tube are discussed with relation to the diagram.

1. **Photocathode:** The photocathode converts visible light photons to electrons by the process of photoemission: incident light photons are absorbed by the photocathode, transferring their energy to an electron within the photocathode. The electron migrates to the surface of the photocathode and escapes.
2. **Electron acceleration:** Electrodes create an electric field that accelerates the electrons from photoemission towards another electrode, termed a dynode.
3. **Electron multiplication:** Dynodes are designed to allow the emission of secondary electrons via photoemission. Due to the kinetic energy gained during acceleration the electron energy deposited at the dynode is very much greater than the electron's initial energy and therefore several secondary electrons may be emitted. The number emitted depends on the energy of the incident electrons.
4. **Multiple stage multiplication:** The secondary electrons produced at the first dynode are directed and accelerated by further electric fields through a series of dynodes.

Standard photomultiplier tubes are capable of achieving high electron gains from the original electron output at the photocathode. The final gain achieved is dependent on the electric field voltages applied. Due to the different processes involved in obtaining an output from a scintillation detector they tend to have low energy resolution in comparison to other radiation detectors. Several different designs of photomultiplier tube are available and design depends upon the nature of the input and the required output.

3.3.3.2 Semi-conductor Detectors

Semi-conductor detectors have better energy resolution than scintillation detectors. They also tend to be compact and respond quickly. They are widely used in several areas in medical physics.

Semi-conductors are crystalline in structure and, like scintillators, the lattice structure contains energy bands: the valence band and the conduction band, separated by a band gap. The process of radiation detection within the lattice is comparable to that of gas ionisation and is described in Figure 3.11.

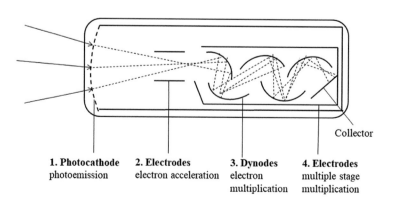

FIGURE 3.10 Photomultiplier tube processes.

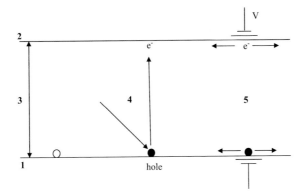

1. **The valence band:** outer shell electrons are bound to lattice sites in the valenceband.

2. **The conduction band** contains free electrons that are free to migrate around the crystal,

causing the flow of charge.

Without external excitation all electrons exist in the valence band and none in the conduction band.

3. **The band gap:** the amount of energy that is required for an electron to leave its lattice site

in the valence band and join the conduction band.

4. **Electron-hole pairs:** When excitation occurs an electron may gain enough energy to cross

the band gap and enter the conduction band, leaving its lattice site in the valence band empty

– a hole – which has a positive charge. For every electron that gains enough energy to enter

the conduction band there will be a hole in the valence band, hence an electron-hole pair.

5. **Movement of charge:** As in gases,under the influence of an electric field both the

electron and the hole will move through the crystal. The rate at which they do this governs

the conductivity of the semi-conductor.

FIGURE 3.11   Electron–hole formation in semi-conductors.

For some materials, this process may occur at room temperature via thermal excitation within the structure – these materials are conductors. In the absence of any excitation stimulus the electron–hole pairs recombine and therefore some semi-conductor materials do not conduct charge at room temperature and are insulators. This property of the material will depend on its band gap. Semi-conductor materials that allow electrons to move from the conduction band without external stimulation are intrinsic semi-conductors and are pure. Silicon and germanium are the semi-conductors with the highest purity commonly available.

3.3.3.2.1   n-Type and p-Type Semi-Conductors

Semiconductor materials tend to be doped with impurities. The most commonly used are n-type and p-type semiconductors. The 'n' refers to concentration of electrons in the conduction band and 'p' to the concentration of holes in the valence band. For an intrinsic semi-conductor $n_i = p_i$.

Considering silicon and germanium, both form four covalent bonds (tetravalent) with surrounding atoms. If a pentavalent doping atom (able to form five covalent bonds) is substituted into the semi-conductor array there will be a spare electron. This electron is only loosely bound to the array and therefore very little energy is required to free it. When the electron is freed it moves to the conduction band and has no corresponding hole pair.

As these impurities readily add electrons to the conduction band they are called donor impurities. In *n*-type semiconductors, the number of electrons in the conduction band is much greater than the number of holes in the valence band. Therefore, the flow of charge is caused mainly by the movement of electrons (majority carriers), rather than the movement of holes (minority carriers).

In *p*-type semiconductors, an impurity with a trivalent bond (able to form three covalent bonds) is introduced into the lattice. As a result, one covalent bond is left incomplete, the equivalent of a hole, but with slightly different energy characteristics because it is an impurity. These holes are not in the valence band, but are formed as electron attracting sites in the forbidden energy gap between the valence band and the conduction band – they are termed acceptor sites. As there is only a small energy gap between the acceptor sites and the valence band, electrons from the valence band are easily attracted to fill these acceptor sites, leaving a hole in the valence band.

In *p*-type semi-conductors electrons from the conduction band are attracted to the increased number of holes in the valence band. The number of holes in the valence band exceeds the number of electrons in the conduction band. Therefore the holes are the majority carriers and the electrons are the minority carriers.

When a charged particle or radiation photon interacts with any semi-conductor, energy is deposited and electron–hole pairs are formed. The number of electrons and holes formed will always be equal for all types of semiconductor, pure or doped. However, the main advantage of semi-conductors over gas-filled detectors is that the energy required to produce an electron–hole pair (ionisation energy) is very low – around 3 eV, compared to around 30 eV. As a result, the energy resolution for semi-conductor detectors is better than that for gas-filled detectors.

When *p*-type and *n*-type semi-conductors are joined with good thermodynamic contact, electrons and holes are able to migrate across the junction. This is because of the diffusion gradient between the two materials. If an *n*-type crystal is next to a *p*-type crystal the *n*-type crystal has a higher concentration of conduction electrons than the *p*-type and a diffusion gradient exists. The electrons diffuse from the higher concentration to the lower concentration, across the junction to the *p*-type crystal where they are able to combine with holes. The opposite is true for the valence holes which diffuse from the *p*-type crystal to the *n*-type crystal.

The diffusion of electrons and holes between crystals leaves positive, immobile impurity ions in the *n*-type crystal and negative, immobile impurity charges in the *p*-type crystal. Therefore on the *n*-side of the junction there is a positive space charge and on the *p*-side of the junction a negative space charge. The concept of a space charge has been discussed previously with regard to proportional counters and refers to a region over which the electric field formed reduces the possibility for charge diffusion. At equilibrium, this electric field prevents charge diffusion in this region, called the depletion region.

There are no electrons or holes within the depleted region – all electrons are attracted to the positive *n*-type crystal and all holes are attracted to the negative *p*-type crystal. Therefore, any additional charge created here must be caused by an external event, such as the interaction of a radiation particle or photon. Any electron–hole pairs formed by such interactions are swept from the depletion region and detected as a flow of charge.

In practice, an external voltage is applied across the semi-conductor junction. This voltage acts to increase the velocity with which charges, created as a result of external interactions, move through the junction – reducing the possibility of recombination post-radiation interaction and increasing the efficiency of charge capture. The direction in which the external voltage is applied affects the behaviour of the charge carriers intrinsic to the semi-conductor crystals and therefore the properties at the semi-conductor junction.

If the external voltage applied across the junction is forward biased, so that the *p*-crystal is positive and the *n*-crystal is negative; the majority carriers (electrons), intrinsic to the *n*-crystal, will be attracted across the junction to the positive *p*-crystal, and the majority carriers (holes) from the *p*-crystal will be attracted to the negatively charged *n*-crystal. In this case the flow of current due to internal charge carriers will be greatly increased across the junction. However, if a voltage is applied in the reverse direction (reverse biased) then the intrinsic minority charge carriers will be attracted across the junction and the current due to internal charge carriers will be small in comparison. In this way, the semi-conductor functions like a diode – allowing internal charge to pass easily in one direction but resisting flow of charge in the opposite direction.

The application of a reverse bias voltage enhances the intrinsic potential difference across the semi-conductor junction, increasing the space charge either side of the junction and the size of the depletion region. This effectively expands the active area of the radiation detector to more efficiently detect charge carriers created by external events. Therefore, practically, reverse-biased voltages are applied.

3.3.3.2.2   Germanium Semi-Conductors as Gamma-Ray Detectors

The detection of gamma rays using the semi-conductor configurations described in the previous section is difficult due to their structure. The limiting factor is the depletion region that can be formed, which are usually not deep enough to compensate for the penetration depth of gamma photons. The relationship between the depletion depth (d), reverse bias voltage (V) and the net impurity concentration (N) is shown in Equation 3.5. Increasing the reverse bias voltage to too high a level may cause failure of the semi-conductor diode; therefore, in practice the impurity concentration in the semi-conductor is reduced.

$$d \propto \frac{V}{N} \tag{3.6}$$

Two methods can be used to reduce impurities. Firstly, in the manufacturing process to further refine the semi-conductor material and remove impurities. This has only been achieved for germanium to create hyperpure germanium detectors (HPGe) with depletion depths of several centimetres. Secondly, by a process called lithium-ion drifting which can be applied to both silicon Si(Li) and germanium (GeLi). In this process, the crystals are doped with lithium after they have been grown to remove either predominant holes or conduction electrons in *p*-type and *n*-type crystals.

The detection efficiency and energy resolution characteristics of GeLi and HPGe detectors are very similar. However, the major difference between these two detectors is their temperature dependence. HPGe detectors can be used at room temperatures whereas GeLi detectors must be maintained at low temperatures. Thus, HPGe detectors are more convenient to use and are also easier to manufacture. As a result, HPGe detectors are the most commonly used of the two types.

3.3.3.3   Thermoluminescent and Optically Stimulated Detectors (TLDs and OSDs)

As described earlier in this chapter, some inorganic materials can be used as scintillants for the detection of radiation. Thermoluminescent detectors (TLDs) are also inorganic materials; however, these materials are prepared to produce thermally or optically stimulated luminescence (TSL or OSL). The key importance of these materials for radiation detection is that charge deposited in the material is stored and released at a later time via a thermal or optical stimulus. As a result of this property, and their compact size, TLDs are widely used for personal dose monitoring and have several properties that make them more useful for this purpose than traditional film badges (see Table 3.2).

So that the material is able to store charge, a high concentration of energy levels is introduced into the band gap between the valence band and the conduction band-trapping centres. Traps as far as possible from the extremes of the band gap are preferred as these store charge for longer and are termed 'deep traps'. Following the production of an electron–hole pair by an incident charged particle or radiation photon, the electron and the hole are free to migrate in their respective bands. They may then be captured in a trapping centre. This process continues for cumulative radiation exposures until the traps are full; therefore, there will be a point of saturation at which the material can store no further charge.

To release the trapped charges the material is exposed either to temperature (around 300–400°C) or to light in a specialised reading unit. When heating the material the temperature must be uniform, usually achieved via a stream of heated gas or a heated contact plate. The trapped electrons are thermally excited and move back to the conduction band where they migrate until they meet a hole in a trapping centre – now termed a luminescence centre. Alternatively, the trapped holes may be thermally released to the valence band and recombine during migration with an electron in a trapping centre (luminescence centre). Recombination of the electron–hole pairs at the luminescence centres causes the emission of a visible photon. During optically stimulated luminescence the charges are released using an appropriate wavelength of laser light. Phosphor storage plates used in radiation imaging are an example of OSDs.

The visible photons emitted during read-out are detected and amplified with a photomultiplier tube (Section 3.3.2.4.4). The intensity of the emission from the TLD is measured and when correctly calibrated for a known radiation field can be used to calculate a radiation dose. The TLD processing procedure removes all charges from the trapping centres within the material allowing the TLD material to be re-used.

3.3.3.3.1 Thermoluminescent Materials and Properties

Due to the use of TLDs for personal dose monitoring, there is a requirement for the materials from which they are manufactured to be tissue-equivalent. Therefore, although there are many different materials that have been used historically as TLDs, there are two that have prevailed due to their x-ray and gamma-ray response and physical properties: LiF:Mg,Ti and LiF:Mg,Cu,P. The basic characteristics of these two materials are summarised in Table 3.5.

3.4 DETECTOR APPLICATIONS

The applications of radiation detectors in medical physics are numerous and diverse. Some of these applications are discussed in this section; however, this is by no means exhaustive. Instead, this section should be used as an indicator of the kind of detectors that are suited to different applications and the reasons why. It should also be noted that new types of materials and detectors are always under development, particularly for the optimisation of radiation image receptors.

3.4.1 CONTAMINATION AND ENVIRONMENTAL MONITORING

All types of radiation detectors discussed in the previous sections may be used in contamination and environmental monitoring. Table 3.6 discusses the properties of different detector types in relation to environmental and contamination monitoring and the situations they are most commonly used in.

TABLE 3.5
Characteristics of LiF:Mg,Ti and LiF:Mg,Cu,P

	LiF:Mg,Ti	LiF:Mg,Cu,P
Commercial name	TLD-100	GR-200
Sensitivity	Low limited to dose measurements > 100 μGy	High c. 30 x greater than TLD-100 Can measure doses << 100 μGy
Applications	Personnel dose monitoring, environmental dose monitoring	Personnel dose monitoring, environmental dose monitoring
Disadvantages	Non-linear response at doses > 1 Gy therefore must be carefully used and calibrated in radiotherapy applications	Loss of thermoluminescent efficiency at doses > 1 Gy Loss of sensitivity if heated to > 240 °C
Advantages	Detects low energy neutrons and can be adapted to other forms for detection of fast neutrons.	May be adapted to detect neutrons

3.4.2 Detectors in Imaging

Detectors are used in all areas of radiation imaging, generally for similar purposes. In the following sections, the use of radiation detectors in diagnostic radiology, nuclear medicine and radiotherapy is considered with regard to radiation imaging, quality assurance and quality control testing, patient dosimetry and personnel dosimetry.

TABLE 3.6
Radiation Detectors for Contamination and Environmental Monitoring

Detector	Properties	Application
Geiger–Müller tube	Cylindrical gas-filled tube with thin window detects α and β particles. x-ray and gamma radiation is stopped in the chamber walls and can also be detected. Large amplification therefore useful at very low radiation levels. Gives an indication of whether radiation is present but no indication of the level of contamination. Long dead times, therefore not useful at high count rates (over a few hundred counts per second).	Contamination monitoring for α and β emitting radionuclides and x-ray and gamma radiation, e.g. in Nuclear medicine and radiotherapy departments or laboratories. Environmental monitoring to indicate presence of radiation but not for accurate measurement due to energy dependence and long dead times.
Proportional counter	Cylindrical gas-filled tube with thin window for detecting α and β particles; x-ray and gamma radiation is stopped in the chamber walls and can also be detected. More sensitive than ionisation chambers. Output is proportional to the amount of energy deposited in the chamber. Neutrons may be detected by coating the gas chamber wall with boron or by using BF_3 gas. A moderator may be added to thermalise and detect fast neutrons.	Contamination monitoring for higher energy α, β, x, and γ radiation, and neutrons. Environmental and shielding measurements for α, β, x, γ and neutrons.
Ionisation chamber	Tend to be larger volume for environmental monitoring, e.g. 1800 cc, to ensure higher sensitivity for low levels of radiation. Output is proportional to the amount of energy deposited in the chamber.	Capable of measuring diagnostic x-ray and gamma radiation. Measurement of the amount of radiation present therefore useful in shielding assessment.
Scintillation counter	Thick crystal cylinder used for contamination and environmental monitoring, increases sensitivity to higher energy radiations such as x-ray and gamma radiation in comparison to G-M and proportional tubes.	Contamination and environmental monitoring for x-ray and gamma radiation.
Semi-conductor detectors	Higher sensitivity than gas-filled detectors (around a factor of 10^4) Typically very small	Semi-conductor detectors are beginning to replace ionisation chambers for environmental monitoring applications.
TLDs	Detect x-ray and gamma radiation. High threshold for β (> 50 keV) due to casing. Can be adapted for neutron monitoring. Must be calibrated to produce a reading that relates to tissue-equivalent dose. Typically very small	Environmental monitoring for x-ray and gamma radiation Where adapted, environmental monitoring for neutrons. Capable of accumulating charge over time therefore may be used to monitor dose to an area over time.

3.4.2.1 Radiation Imaging

The use of radiation detectors for imaging is determined by the type of radiation to be detected and the energy range over which the detector needs to be sensitive. Therefore for diagnostic radiology, devices must detect x-ray radiation generally ranging from around 25 to 140 keV. In nuclear medicine, imaging devices are used to detect gamma rays and annihilation photons (40–700 keV), and x-rays (80–140 keV). In radiotherapy, devices are used to detect x-rays (40–140 keV) and annihilation photons (40–700 keV) for imaging applications. Some of the common imaging applications in each of these modalities, and the type of detectors they use, are shown in Table 3.7.

TABLE 3.7
Imaging Detectors by Modality and Application

Detector	Detectors
Diagnostic radiology	
Mammography	Resolution in mammography images is very important therefore until recently film was still widely used in mammography imaging due to its superior resolution. Film has been replaced by phosphor flat-panel digital detectors which are based on scintillation detectors such as caesium iodide (CsI), barium fluoride (BaF) and semi-conductor arrays such as amorphous selenium (α-Se).
Dental x-ray	Film is still used for dental applications, along with photostimuable phosphors (scintillation detectors) such as those used in CR and semi-conductor arrays.
Computed radiography (CR)	Photostimuable storage phosphors (scintillation detector) are used in CR. An optically stimulated detector such as caesium iodide (CsI:Na, CsI:Tl), barium fluoride (BaF:Cl:Eu^{2+}) and sodium iodide (NaI:Tl). The signal is read out using a laser.
Direct Digital radiography (DR)	Scintillation detectors such as those described for mammography: caesium iodide (CsI), barium fluoride (BaF) for indirect DR. Semi-conductor arrays such as amorphous selenium (α-Se) for direct DR.
Fluoroscopy and Interventional radiology	Traditionally image intensifiers that used scintillation detectors were used. These are now being replaced by digital detectors with scintillation detectors and semi-conductor detectors such as those described for mammography and DR.
Computed Tomography (CT)	Scintillation detectors generally used, the most common are gadolinium oxysulfide (Gadox), cadmium tungstate (CdWO$_4$), and CsI although there is much research into the development of new detectors for improved image quality and system performance, such as doped rare-earth ceramic scintillation materials. Photomultiplier tubes are also being replaced by high-purity silicon photodiodes.
	Gas-filled detectors may also be used, high atomic number gases such as xenon and krypton are the most popular in the form of an ionisation chamber.
Nuclear Medicine	
Gamma camera	Scintillation detectors such as NaI:TI for detection of gamma rays emitted from patients.
Positron Emission Tomography (PET CT)	Requires the detection of high energy (0.51 MeV) annihilation photons therefore specialist detection materials are required such as bismuth germinate (BGO), a scintillation detector.
Single Photon Emission Computed Tomography (SPECT CT)	Scintillation detectors as those used for gamma cameras are used to detect photons emitted from the patient. CT detectors as described above are used in the CT aspect of the modality.
Radiotherapy	
CT simulator	As for diagnostic CT scanners.
PET CT	As for Nuclear medicine PET scanners

TABLE 3.8
QA and QC Radiation Detectors by Application

Detector	Detectors
Diagnostic radiology	
Output and kilovoltage measurements on all diagnostic radiology x-ray sources.	Ionisation chambers and solid-state semiconductor detectors.
Nuclear Medicine	
Gamma camera & SPECT tests (*)	All gamma camera and SPECT QC tests are carried out using the scintillation counters of the gamma camera as detectors. In some cases, a specialised source known as a flood source is used.
PET tests (*)	As for gamma cameras, the scintillation detectors of the unit itself are used as detectors for QA and QC tests.
Radiotherapy	
Linear Accelerator (LINAC) tests	Ionisation chambers (Farmer chamber, plane-parallel chamber) are the main detectors used to measure radiation and electron outputs. Solid-state semiconductors are used, these are silicon or diamond based. Thermoluminescent detectors are also used extensively for calibration purposes.
CT simulator	Ionisation chambers and solid-state semi-conductor detectors (as for CT scanners in diagnostic radiology)

* NB. Where a CT scanner is used in conjunction with a gamma camera such as in SPECT-CT, x-ray radiation output tests are carried out using an ionisation chamber or solid-sate semi-conductor detector as for diagnostic radiology CT scanners.

3.4.2.2 Quality Assurance and Quality Control

The use of detectors for quality assurance (QA) and quality control (QC) of radiation equipment also depends upon the type of radiation to be detected and the energy range. As QA and QC are carried out to assess the performance of imaging equipment, these properties are the same as those seen in the previous section although without the additional capability to form images. The most common types of detector and their application in different areas are shown in Table 3.8.

3.4.2.3 Patient Dosimetry

Dose measurement is an essential requirement for medical physics equipment. There is a legal obligation under the EU Basic Safety Standards (Directive 2013/59/EURATOM) that all interventional radiology equipment, and CT equipment, as well as any new medical radiodiagnostic equipment that produces ionising radiation, is equipped with a device that either reports the radiation dose, or parameters that allow calculation of the radiation dose, post procedure. Therefore all modern medical equipment utilising ionising radiation is equipped with such a device and examples of patient dose measurement in different modalities are summarised in Table 3.9.

Patient dose in nuclear medicine arises from the administration of a radiopharmaceutical and therefore is not measured using a detector but instead is tabulated by the Administration of Radioactive Substances Advisory Committee (ARSAC) for given radiopharmaceuticals. Where CT is used as an imaging component the patient dose is assessed as for diagnostic radiology CT scanners.

3.4.2.4 Personnel Dosimetry

Personnel dosimetry is also required in law in order to ensure that constraints and limits for staff exposure to ionising radiation are not exceeded. Previously, film badges were used in all modalities for this purpose; however, these have been widely replaced by thermoluminescent detectors (TLDs).

TABLE 3.9
Patient Dosimetry Detectors

Detector	Detectors
Diagnostic radiology	**Dose area product (DAP) meter** – Ionisation chamber used to measure the dose in a beam of radiation. Applied in many diagnostic radiology applications – fluoroscopy, CR and DR.
CT scanners	**Dose length product (DLP)** – radiation dose in a phantom is measured and adjusted according to certain factors to give a dose over the full length of any CT scan. This parameter is calibrated at installation using an ionisation or solid-state detector.
Radiotherapy	**In vivo dosimetry** – Patient radiation dose tends to be assessed during treatment via TLDs or semiconductors such as MOS-FETs (Metal Oxide Semiconductor Field Effect Transistor) **Electronic Portal Imaging Devices (EPIDs)** – scintillation or ionisation detector used for treatment localisation and verification.

The relative characteristics of these are discussed in Section 3.3.1. TLDs must be calibrated for use with different radiations, and specifically for doses greater than 1 Gy in radiotherapy. One of the main advantages of TLDs is their compact size. Electronic, semiconductor-based personal dosimeters may also be used to obtain a real-time indication of personnel dose.

BIBLIOGRAPHY

AAPM, 2008, The Measurement, Reporting, and Management of Radiation Dose in CT, AAPM Report No. 96.

Attix, F. H. 2004. *Introduction to Radiological Physics and Radiation Dosimetry*. Wiley-VCH Verlag GmbH & Co.

Bilski, P. 2002. Lithium Fluoride: From LiF:Mg,Ti to LiF:MG,Cu,P. *Radiation Protection Dosimetry* 100(1–4), 199–206.

Busemann Sokole, E., Plachcinska, A., and Britten, A. 2010. Routine Quality Control Recommendations for Nuclear Medicine Instrumentation. *European Journal of Nuclear Medicine and Molecula Imaging* 37, 662–671.

Calibration of Radiation Protection Monitoring Instruments, 2000, IAEA Safety Report Series No. 16, IAEA.

Council Directive 2013/59/EURATOM laying down basic safety standards for protection against the dangers arising from exposure to ionising radiation, and repealing Directive 89/618/Euratom, 90/641/Euratom, 96/29/Euratom, 97/43/Euratom and 2003/122/Euratom, 2013, OJ L 13.

Hendy, P. P. and Heaton, B. 1999. *Physics for Diagnostic Radiology*, 2nd Edition. Taylor & Francis Group.

International Commission on Radiation Units and Measurements, Quantities and Units in Radiation Protection Dosimetry, 1993, Report No. 51, ICRU, Bethesda, MD.

Kim, Siyong and Suh, Tae-Suk. 2006. Imaging in Radiation Therapy. *Nuclear Engineering and Technology* 38(4), 327–342.

Knoll, Glenn F. 2010. *Radiation Detection and Measurement*, 4th Edition. John Wiley & Sons Inc.

Lanca, L. and Silva, A. 2013. *Digital Imaging Systems for Plain Radiography Detectors*. Springer.

Low, D. A., Moran, J. M., Dong, L., and Oldham, M. 2011. Dosimetry Tools and Techniques for IMRT. *Medical Physics* 38(3), 1313–1338.

Prekeges, Jennifer. 2013. *Nuclear Medicine Instrumentation*. Jones and Bartlett Learning LLC.

Rajan, G. and Izewska, J. 2005. Radiation Monitoring Instruments. In *Radiation Oncology Physics: A handbook for Teachers and Students*. Ed. Podgorsak, E. B. 101–121. IAEA.

Thwaites, D. I., Mijnheer, B. J., and Mills, J. A. 2005. Quality Assurance of External Beam Radiotherapy. In *Radiation Oncology Physics: A handbook for Teachers and Students*. Ed. Podgorsak, E. B. 101–121. IAEA.

Yaffe, M. J. 2010. Detectors for Digital Mammography. In *Digital Mammography*. Ed. Bick, U. and Diekmann, F. 13–31. Springer.

Yagi, M., Ueguchi, T., Koizumi, M. et al. 2013. Gemstone Spectral Imaging: Determination of CT to ED Conversion Curves for Radiotherapy Treatment Planning. *Journal of Applied Clinical Medical Physics* 14(5), 173–186.

4 Biological Effects of Ionising Radiation

Michele Avanzo
IRCCS Centro di Riferimento Oncologico, Aviano, Italy

Cornelius Lewis
King's College Hospital NHS Foundation Trust, London, UK

CONTENTS

DOI: 10.1201/9780429155758-4

4.1 RADIATION DAMAGE AT A CELLULAR LEVEL

In very simple terms, human cells have two components: the cytoplasm, which supports all the metabolic activities of the cell, and the nucleus, which contains the genetic information coded onto DNA (Figure 4.1).

Damage to DNA in the cell nucleus leads to deleterious effects on the whole organism, and this may be caused by ionising radiation through one of two mechanisms: direct action and indirect action.

4.1.1 DIRECT ACTION

Deposition of energy through ionisation occurs along the paths of photons or particles as they pass through tissue. Direct action is when an ionising event occurs within the cell nucleus causing disruption to the DNA in either the sugar phosphate backbone of the DNA structure or the base pairs (composed of adenine, thymine, guanine and cytosine) (Figure 4.2).

4.1.2 INDIRECT ACTION

Ionisation removes electrons from atoms or molecules. This may affect the valency of the atom or molecule creating a 'radical' which is highly chemically reactive. Radicals readily recombine and in doing so may change the chemistry of the molecules.

The water molecule, H_2O, is abundant in human cells. Ionisation of water molecules can produce a hydroxyl radical, $HO^{\bullet}$, which upon recombination can create a chemically stable peroxide molecule, H_2O_2; that is,

$$HO^{\bullet} + HO^{\bullet} = H_2O_2$$

Peroxide molecules within the cell may lead to biological changes which can ultimately damage the cell and, specifically, the DNA.

4.1.3 OUTCOMES OF RADIATION DAMAGE

Human cells have the capacity to repair themselves. In the majority of cases, following radiation damage, the cells will repair and their function will not alter. Of those cells that do not repair successfully, the resulting damage will not adversely affect cell function in the majority of cases.

Cells that do not repair successfully may have suffered such significant damage that they are unable to function and will die either when they attempt to divide or through a form of stress response referred to as apoptosis.

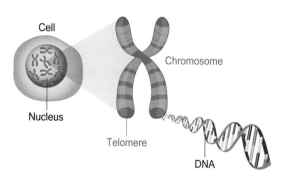

FIGURE 4.1   DNA in cells.

DNA double helix

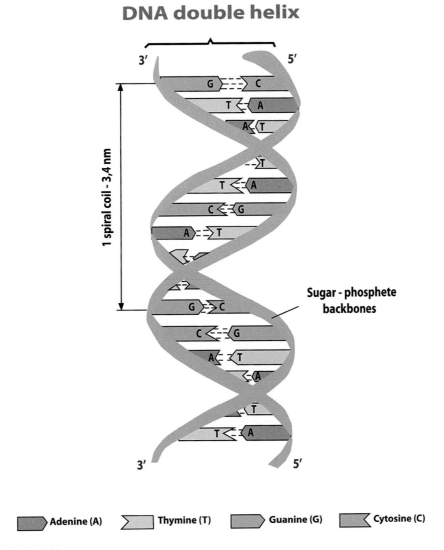

FIGURE 4.2 DNA structure.

Some cells that have suffered radiation damage may survive and continue to reproduce containing some form of genetic mutation. This is sometimes suggested as a possible mechanism for evolution. Alternatively, cells may survive but have been modified in such a way that the mechanism controlling cell proliferation is damaged which may subsequently manifest as a cancer.

4.1.4 DAMAGE MANIFESTATION

Cell damage following radiation exposure usually affects the individual who received the exposure. This is referred to as somatic damage.

If the cells damaged by radiation exposure are germ cells, there is a potential for future generations to be affected. This is referred to as genetic damage.

The foetus is particularly susceptible to radiation damage because the cells of the foetus divide more rapidly and are thus more radiosensitive. Damage to the foetus is referred to as teratogenic damage.

4.2 DETERMINISTIC EFFECTS

Radiation damage leading to cell death is referred to as deterministic damage. The body can suffer cell loss without manifestation of that damage. However, if sufficient cells are lost, then the damage will become apparent, and the greater the number of cells lost, the more significant the damage will be.

Deterministic effects are characterised by a threshold below which the damage will not be apparent. Above the damage threshold, the severity of any effect increases with the radiation dose received (Figure 4.3).

4.2.1 SKIN DAMAGE

Skin is a common site for deterministic damage to occur. In a similar manner to damage caused by too much exposure to the sun, ionising radiation damage to the skin initially appears as erythema, a slight reddening of the skin at radiation doses in excess of approximately 2Sv. Higher doses of ionising radiation may cause the damage progress to blistering and, following large exposures, ulceration.

Patients who have received radiotherapy treatment for cancer can sometimes exhibit skin damage, usually limited to mild erythema. Less commonly, patients undergoing complex interventional radiological procedures such as angioplasty (unblocking arteries in the heart) can exhibit skin damage.

4.2.2 EYE DAMAGE

The eye is also subject to deterministic damage, principally resulting in the formation of lens opacities and cataracts. Recent research on radiation damage to the eye [1] has shown that the threshold for cataract formation to occur is lower than originally thought to be the case with threshold doses in the region of 500 mSv (equivalent dose).

This is of particular concern to radiologists and cardiologists performing interventional procedures and has led to a significant reduction in the occupational doses they are permitted to receive.

4.2.3 OTHER ORGANS

The reproductive organs, testes and ovaries, which undergo relatively rapid cell turnover are also sensitive to deterministic effects. Effects may range from temporary to permanent sterility in both

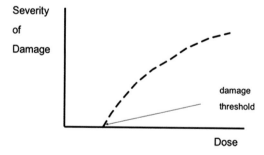

FIGURE 4.3 Graphical representation of deterministic effect.

sexes. The lowest threshold equivalent dose, for temporary sterility in the male, is believed to be approximately 150 mSv. Permanent sterility can occur at threshold equivalent doses above 2.5 Sv.

4.2.4 WHOLE BODY EXPOSURES

Radiation exposures to the whole body such as those received as the result of a nuclear accident (i.e. Chernobyl) or experienced by casualties following the atomic bombs dropped in Japan at the end of the Second World War can have lethal effects although the threshold doses for such effects are relatively high.

The effects of whole body exposure are generally divided into three categories related to the level of exposure (Table 4.1)

Absorbed doses of 1–2 Gy may initiate haematopoietic syndrome in which damage to bone marrow prevents the production of new blood cells. A reduction in the number of red blood cells will lead to poor oxygenation of tissue causing fatigue. More seriously, a reduction in the number of white blood cells will reduce the ability of the immune system to fight infection. Individuals who have received these levels of whole body exposure will most probably have symptoms of nausea and vomiting and will develop infections. However, they should survive if given prompt medical care.

At absorbed doses above approximately 5 Gy the cells which line the gut are affected. This causes a variety of symptoms including intestinal bleeding, vomiting and diarrhoea collectively referred to as gastrointestinal syndrome. Individuals who have received doses at this level and above will be weakened by the loss of blood, infections and nutritional problems. The risk of death for these individuals is high despite good medical care. It is estimated that the $LD_{50/30}$ (the lethal dose from which 50% of the affected population would die in 30 days) is in the region of 2.5–4 Gy.

Absorbed doses above approximately 30 Gy result in neurovascular syndrome. At such high doses, individuals suffer a variety of symptoms associated with the destruction of the nervous system including seizures, tremors, ataxia (lack of muscle co-ordination) in addition to the haematopoietic and gastrointestinal symptoms. Whole body doses at this level are invariably fatal with death occurring within days of exposure.

Other than the casualties following the atomic bombing of Hiroshima and Nagasaki at the end of the Second World War, there are relatively few incidents involving whole body exposure in humans. The first time that whole body radiation effects were observed was the result of accidents during the Manhattan Project. More recently, a number of the 'liquidators' who brought the Chernobyl reactor accident under control in 1986 suffered radiation sickness with approximately 30 fatalities. A well-publicised case of whole body radiation damage was the 2006 murder of Alexander Litvinenko in London. The former Soviet spy was poisoned with an estimated 4.4 GBq of the alpha-emitting isotope polonium-210.

TABLE 4.1
Effects of Whole Body Exposures

Absorbed Dose (Gy – Equivalent to Sv for Whole Body Exposure)	Syndrome or Tissue Involved	Symptoms
1–10	Bone marrow syndrome	Leucopenia, thrombopenia, haemorrhage, infections
10–50	Gastrointestinal	Diarrhoea, fever, electrolytic imbalance
> 50	Neurovascular syndrome	Cramps, tremor, ataxia, lethargy, impaired vision, coma

4.2.5 ACUTE AND PROTRACTED EXPOSURES

Most of the data on deterministic effects is derived from single, acute exposures. Damage thresholds for exposures received over an extended period are believed to be higher because of the ability of the body to repair. There is very little data on damage thresholds for protracted irradiation and, as a result, the data on which radiation protection is based is derived from the evidence from acute exposures. The exception to this is deterministic effects to the eye.

4.3 STOCHASTIC EFFECTS

Radiation damage leading to cell modification is termed a stochastic effect, sometimes referred to as a probabilistic effect. The characteristics of a stochastic effect are that the probability (risk) of an effect occurring is related to the radiation dose received but also, and importantly, there is no threshold below which an effect will not occur (Figure 4.4). Stochastic effects are those which lead to cancer.

4.3.1 DETERMINING RISK

Radiation-induced carcinogenesis is difficult to quantify because, unlike deterministic effects, the outcome of stochastic damage may not become evident for years after the original exposure. Latent periods vary with the type of cancer and may be up to 50 years for some solid tumours (e.g. lung cancer).

Stochastic risks are determined on the basis of epidemiological evidence in which two populations are compared that are identical in every respect other than one population was exposed to ionising radiation.

There are three broad groups in which radiation epidemiological studies have been performed. These are individuals who have been occupationally exposed to ionising radiation, patients treated with radiation and populations affected by nuclear explosions.

4.3.1.1 Occupationally Exposed Groups

X-rays and radioactivity were discovered in 1895 and 1896, respectively. Prior to that, there was no non-invasive method of investigating internal structures of the human body and a variety of diseases, including cancer, that could not be treated. Physicians were quick to realise the enormous benefits that could be derived from the diagnostic and therapeutic uses of ionising radiations. It was not until much later that the potential detriments were understood and many of those pioneering the use of radiation suffered and sometimes died as a result. Madame Curie, who developed the technique to separate radium from pitchblende to use for therapy, died from aplastic anaemia induced by her work with ionising radiation.

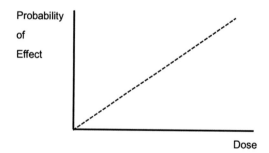

FIGURE 4.4 Graphical representation of stochastic effect.

Whilst the risks of using radiation became evident in the early pioneers, there was no way to quantify the dose/risk relationship. Indeed, there was no measure of radiation dose until the third decade of the twentieth century when the International Commission on Radiation Units (originally known as the International X-Ray Unit Committee) proposed a definition of the roentgen.

Another occupationally exposed population who developed cancers as a result of their work were the, so-called, dial painters. These were a group, predominantly female, who worked in a factory in New Jersey, USA, to apply luminous paint to clocks, watches and aircraft instruments. The active ingredient in the paint was radium which was applied with a fine paintbrush. To maintain a fine tip on the paintbrush, the workers licked the bristles and thus ingested quantities of radium leading to cancers. The practice was stopped in about 1930 when the link between the radium paint and the fate of the workers was established. As with the medical pioneers, there was no method to establish the doses received by these workers.

Mine workers, and particularly those in the uranium mining industry, are also subject to potentially high levels of radiation exposure. There have been a number of epidemiological studies of mine workers demonstrating increased lung cancer risks in the study populations, but few studies have attempted to quantify the doses received because of the very significant complexities in determining the quantities of uranium dust or radon gas inhaled.

4.3.1.2 Medically Exposed Groups

Photon radiations (x-rays and gamma rays) are used extensively in medicine for diagnosis and therapy and have been since their discovery at the end of the nineteenth century. With the knowledge we now possess on the risks of exposure to ionising radiation, current use is tightly controlled to minimise risk. However, during the development of diagnoses and therapies, there are examples which have led to detriment rather than benefit.

Treatment for cancer using radiotherapy is now well established and, with advances in treatment delivery, highly successful. However, during the development of radiotherapy techniques, treatments were given for conditions other than cancer. Non-malignant conditions treated with radiotherapy included ringworm of the scalp in children, menorrhagia (heavy bleeding) during the menopause and ankylosing spondylitis (AS). Studies of patient groups treated for these conditions demonstrated that whilst the treatments were successful, significant numbers of patients subsequently developed cancers. In the case of children with ringworm, it was cancers of the head and neck and for menorrhagia, cervical cancer. The treatment that has seen the most comprehensive studies for subsequent cancer development is AS [2].

AS is a rheumatic condition which causes the joints between the spinal vertebrae to become inflamed. The joints between the pelvis and the spine are similarly affected. As the inflammation increases, the spine becomes bent and inflexible. AS is incurable and can cause significant suffering. For example, stiffness in the spine prevents the chest from expanding fully so sufferers can have difficulty breathing and become prone to chest infections. Progression of the disease is now controlled through the use of steroids and other anti-inflammatory drugs.

In the 1930s–1950s, some patients with AS were treated using radiotherapy because ionising radiations will kill actively dividing cells such as those responsible for inflammatory processes. These patients typically received irradiation of the whole spine and pelvis. During follow-up of these patients, it was observed that more patients than expected appeared to be developing cancers, specifically leukaemias, and a report was commissioned to investigate more fully. Court-Brown and Doll [2] examined the medical records of over 14,000 patients treated for AS between 1935 and 1954. They found a significant increase in a number of cancers observed in the treated group compared to the numbers that would be expected from a comparative, untreated group. The largest increase was for leukaemia with an incidence of up to five times greater in the treated population, but there were a variety of other cancers including stomach, colon and lung. Data on the treatment doses given to these patients has also been extracted and used to develop dose/risk relationships [3].

4.3.1.3 Nuclear Explosions

By far, the largest groups of people from whom data on stochastic radiation effects has been gathered are those irradiated as a result of nuclear explosions. This category includes populations affected by nuclear testing, nuclear reactor accidents and, of course, the atomic bombs dropped on Japan at the end of the Second World War.

Following the development of the atomic bomb by the USA other major powers decided they needed to develop nuclear weapons. These were tested at remote sites around the world. The UK conducted testing in the Australian outback (Woomera) and also in the Pacific Ocean near Christmas Island. France tested weapons at Mururoa Atoll, part of the Polynesian archipelago, and the former USSR at Semipalatinsk in the north-eastern steppes of Kazakhstan. Subsequent research has shown that the indigenous peoples close to these sites were affected, principally through fall-out but also, in the case of the French nuclear tests, through radioactive materials entering the food chain [4].

The use of nuclear energy for power generation was once considered to be a replacement for fossil fuels. However, following a number of well-publicised accidents, there is considerable opposition in many countries to utilise this method of power generation. Perhaps the most widely reported accidents were at Windscale (now Sellafield) in 1957, Three Mile Island (USA) in 1979, Chernobyl (Ukraine) in 1986 and, most recently, Fukushima (Japan) in 2011. Studies of the Chernobyl accident have shown excess thyroid cancers in children [5], presumed to be due to the release of iodine-131 in the accident. However, studies have not shown the levels of other cancers that, based on existing dose/effect models, would be expected in the exposed populations. Predictions of excess cancers from the Fukushima incident have been made [6], but it will be some years before the full radiological impact of the accident is understood.

By far, the largest populations to have received significant exposures from nuclear explosions were the residents of Hiroshima and Nagasaki. The atomic bombs dropped on these two Japanese cities in August 1945 brought an end to the Second World War in the Pacific region but with devastating consequences. At the time of the bombing, the population of Hiroshima was approximately 350,000 and that of Nagasaki, approximately 250,000. Acute deaths in Hiroshima due to the effects of the explosion, radiation burns or very high exposure have been estimated as between 90,000 and 165,000. Those at Nagasaki were estimated as between 60,000 and 80,000. Within months of the bombings, a joint Japanese and American commission was established to study long-term effects in the exposed populations. This was originally called the Atomic Bomb Casualty Commission (ABCC) but was subsequently renamed the Radiation Effects Research Foundation (RERF).

Several studies of bomb survivors were established by the ABCC, the most significant of which was the Lifespan Study (LSS) [7]. This cohort of approximately 120,000 survivors who were resident in either Hiroshima or Nagasaki at the time of the bombings have been studied extensively throughout their life. The exposures they received have been determined, principally through mathematical modelling, and their medical histories were recorded. The LSS is the principal source of data that has been used to determine radiation risk.

4.4 DETERMINING STOCHASTIC RISK

4.4.1 Dose–Response Relationship

Evidence from the epidemiological studies discussed above, but principally the LSS, has enabled the relationship between radiation effects (excess cancers), and the doses received by the individuals who experienced an effect to be examined. There are, of course, a number of issues with the data. The doses received by affected individuals and groups are estimated rather than measured which introduces considerable uncertainties and the medical histories are, in some cases, of questionable reliability. However, the most significant issue is that the doses to affected populations are in a relatively high and restricted range as illustrated in Figure 4.5.

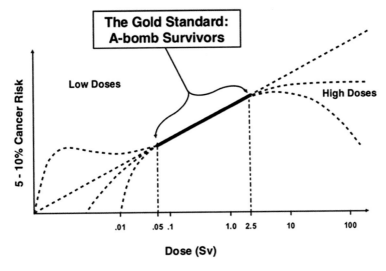

FIGURE 4.5   Radiation risk.

It is assumed that at zero dose there will be zero risk which adds one further data point to assist in developing a risk model. A number of theories have been proposed to describe the form of the relationship between dose and risk. The most basic of these is the linear no-threshold model which assumes a linear relationship between dose and risk at all dose levels. The principal alternative is the linear quadratic model which proposes that as the dose increases the effect also increases. The lack of risk data at low radiation doses, which is the main area of interest, has led to the development of many competing models [8].

4.4.2   Influence of Dose and Dose Rate

A principal reason for determining stochastic risk is to inform guidance on the safe use of ionising radiation whether in medicine, industry or to the general population from natural radiation sources. The majority of individuals and populations exposed will receive relatively small radiation doses delivered at low dose rates. This questions the validity of developing risk criteria based on populations receiving high doses at very high dose rates because of the perceived ability of the cell to undergo repair.

Experiments performed both *in vivo* and *in vitro* [9] have demonstrated that high doses delivered at high dose rates will produce greater damage per unit dose received than low doses delivered at low dose rates. One reason suggested for this effect is that at low dose/low dose rates, time is available for cell recovery. Radiobiological models account for this effect by introducing a modifying factor. This is referred to as the Dose and Dose Rate Effectiveness Factor (DDREF). Although there is evidence on DDREFs from a variety of situations, the value of this factor varies considerably depending on the particular biological endpoint under investigation. International bodies involved in making recommendations on radiation risk have proposed a DDREF of 2. This implies that the radiation risk determined from the LSS and similar data should be divided by a factor of 2 to make it applicable to low dose/low dose rate exposure situations.

4.4.3   Influence of Radiation Type

The energy deposited by ionising radiations is influenced by the charge carried by the radiation and hence the density of ionisation produced. Protons and heavy ions are densely ionising whilst photons are lightly ionising. Experimental evidence shows that densely ionising radiations are more effective in producing damage than lightly ionising radiations for the same total amount of energy deposited (absorbed dose). This is believed to be because there is a greater probability that a densely

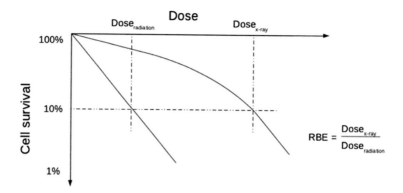

FIGURE 4.6   RBE definition.

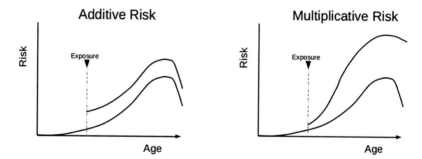

FIGURE 4.7   Additive and multiplicative risk.

ionising track will interact with a critical part of the cell. The greater effectiveness of densely ionising radiation is accounted for by using a further modifying factor, Relative Biological Effectiveness (RBE). This factor is calculated by comparing the dose at which a specific level of damage occurs for a particular radiation compared to the level at which it occurs for x-rays (usually of 250 kVp energy). For example, in Figure 4.6, the dose to reduce cell survival to 10% is compared for the two radiations. RBE is the quotient of the dose required to produce the damage using x-rays over the dose required using the other radiation type.

RBE values for densely ionising radiations (protons, alpha particles, heavy ions) vary depending on their energy and the biological system being irradiated but are generally in the region of between 2 and 20.

4.4.4 LATENCY

The initial damage to a cell produced by ionising radiation does not produce an immediate stochastic effect. Changes to the DNA of the cell may take many years to develop into a cancer. The time between the initial causal event and the appearance of a cancer is termed the latent period.

There is evidence to suggest that the subsequent risk of cancer following the initiating event may be either additive or multiplicative. In the additive model, as the name implies, the risk of cancer induction is simply added to the underlying natural risk which normally increases with age. However, in the multiplicative model, the radiation risk acts as a multiplier to the underlying natural risk.

Radiation risk estimates for specific cancers generally assume a multiplicative risk model, principally because there is a lack of evidence to suggest otherwise and this is the most conservative assumption (i.e. producing the higher risk estimate). For some organs, notably breast and bone marrow, there is sufficient evidence to suggest that an additive model produces the best estimates of subsequent cancer risk [10].

4.4.5 RADIOSENSITIVITY

Sensitivity of tissues to damage caused by radiation exposure varies and this variation is termed radiosensitivity. In 1906, Bergonie and Tribondeau reasoned why some cells and tissues are more radiosensitive than others. Their 'laws of radiosensitivity' proposed that the radiosensitivity of cells is greater if

- they are highly mitotic (divide rapidly),
- they are undifferentiated (i.e. they are stem cells which can develop into any cell type),
- they continue to divide for a long time.

Tissues that are particularly radiosensitive include blood cells formed in the bone marrow, male and female reproductive tissues and breast tissues. As a whole, the tissues of young people will also be more radiosensitive.

4.5 RISK QUANTITATION

Understanding the relationship between radiation dose and stochastic risk is essential to ensure that the benefits of using ionising radiations can be realised without significant detriment. It is a complex relationship dependent on many factors including radiation type, tissue radiosensitivity, age, etc. To facilitate estimation of radiation detriment, assumptions and simplifications are made, which allow the effects of different radiations and different tissue sensitivities to be taken into account when estimating detriment. Two quantities widely used in radiation protection have been developed to evaluate the effects of exposure to ionising radiation. The first of these is the quantity *equivalent dose, H_T,* which accounts for exposure to different radiation types and the second is *equivalent dose, E,* which accounts for differing tissue radiosensitivities when determining effects to the whole individual.

4.5.1 EQUIVALENT DOSE

As discussed in Section 4.3, tissue damage caused by densely ionising radiations will be greater than that caused by sparsely ionising radiations. This is characterised by the RBE of the radiation. Because RBE values vary depending on the particular biological system being investigated, simplifying assumptions are made for radiation protection purposes with a *radiation weighting factor w_R* assigned for each radiation type.

As the comparator for all measurements of RBE is x-rays, the weighting factor for x-rays, gamma rays and electrons is set to a value of 1. Protons are assigned a weighting factor of 2 and alpha particles and heavy ions a weighting factor of 20. The RBE of neutrons varies according to energy because this will dictate the interactions neutrons undergo and hence the density of ionisation produced. Table 4.2 below gives the weighting factors assigned to different radiation types.

TABLE 4.2
Radiation Weighting Factors

Radiation Type	Radiation Weighting Factor, W_R
Photons	1
Electrons and muons	1
Protons and charged pions	2
Alpha particles, fission fragments, heavy ions	20
Neutrons	Energy dependent

Equivalent dose is calculated from the physical absorbed dose by multiplying by the radiation weighting factor. If the tissue is irradiated by a heterogeneous radiation field, the equivalent doses for each radiation are summed.

$$H_T = \sum_R w_R . D_{T,R}$$

where

w_R is the weighting factor for radiation type R

$D_{T,R}$ is the absorbed dose to the tissue T from radiation R

Although the units of equivalent dose are dimensionally equivalent to absorbed dose, the unit is given the special name sievert (Sv).

4.5.2 EFFECTIVE DOSE

In most irradiation situations, more than one organ or tissue will be exposed. To enable comparison between different situations, the quantity *effective dose, E* includes a *tissue weighting factor, w_T,* to take account of different tissue radiosensitivities. These are determined on the basis of individual organ sensitivities assessed using published evidence, although principally from the Life Span Study [11]. Individual organs are grouped into four differing levels of sensitivity (Table 4.3). Organs or tissues that are not specifically mentioned are included in the 'Remainder Tissues' category in which the average equivalent dose to all the remainder organs is assigned a tissue weighting factor of 0.12.

Effective dose is calculated by summing the equivalent dose to each tissue weighted by the relevant tissue weighting factor.

$$E = \sum_T w_T . H_T$$

where

w_T is the tissue weighting factor

H_T is the equivalent dose to that tissue

sievert (Sv) is also used as the unit of effective dose.

4.5.3 WHOLE BODY RISKS

Using the quantities equivalent and effective dose, it is possible to estimate the whole body stochastic risk. In effect, this is the risk of detriment due to any cancer being induced as a result of any ionising radiation exposure. In this context, detriment does not simply mean death due to cancer, it also takes account of reduction in the quality of life due to living with a serious disease such as cancer and the overall years of life lost.

TABLE 4.3

Tissue Weighting Factors

Tissue	W_T	ΣW_T
Bone marrow, colon, lung, stomach, breast, remainder (14)	0.12	0.72
Gonads	0.08	0.08
Bladder, oesophagus, liver, thyroid	0.04	0.16
Bone surface, brain, salivary glands, skin	0.01	0.04

Lifetime Risk

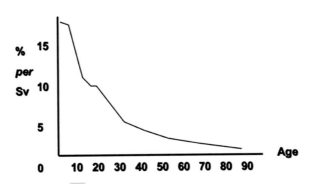

FIGURE 4.8 Approximate relationship between age of exposure and radiation risk.

The majority of the risk is somatic (to the individual who received the exposure) but a small component is thought to be genetic (affecting future generations) although there is no direct evidence for genetic effects in human populations.

Taking into account the assumptions and simplifications inherent in developing the risk data, a value of 0.05 Sv^{-1} is often used as the headline figure for somatic risk. Another way of expressing this is that for every sievert of effective dose received, there is a 5% chance (1 in 20) of detriment due to cancer induction. For every 1 mSv (a typical radiation dose received in an x-ray diagnostic test), the resulting probability of cancer detriment will be 1 in 20,000.

The 0.05 Sv^{-1} headline figure specifically relates to a young adult population. Children are at much greater risk, whereas the elderly have a lower risk. This is due to the combined effects of lower rates of cell division as the body ages and the lack of time (latent period) for a cancer to be expressed after the initial damage has occurred. The approximate relationship between risk and age at which the exposure occurs is illustrated in Figure 4.8.

4.6 THE RADIOBIOLOGICAL BASIS OF RADIOTHERAPY

4.6.1 Survival Curve

Radiation can kill cancer cells, a property that is used in radiation therapy to cure cancer patients. Clonogenic cells are defined as those neoplastic cells within the tumor that have the capacity to produce an expanding colony of descendants and, therefore, the capacity to regenerate a tumor if left intact at the end of radiation treatment [12]. In order to achieve local control of the tumor, all the clonogenic cells must be killed, that is, the fraction of cells surviving radiation (surviving fraction, SF) must be reduced to zero. Cell survival can be represented by plotting the surviving fraction on a logarithmic scale against dose on a linear scale. The dose–response curve has an initial linear slope, followed by a shoulder; at higher doses, the curve tends to become straight again.

The mechanisms that regulate killing of clonogenic cells are mainly governed by five processes. The first is the *repair* of sub-lethal radiation damage. The second involves differences in radiosensitivity during the cell cycle. There are significant variations in radiosensitivity during the cell cycle. A greater proportion of cells are killed in the sensitive portions of the cell cycle, such as mitosis, a smaller proportion in other phases. Resistance is usually greatest in the latter part of S phase. The overall effect is that radiation doses tend to synchronise the cycling among the cells, an effect that is called *reassortment*. Third, there is the *repopulation* of cells resulting from the division of the surviving cells. The fourth 'R', *reoxygenation*, is discussed in the next section.

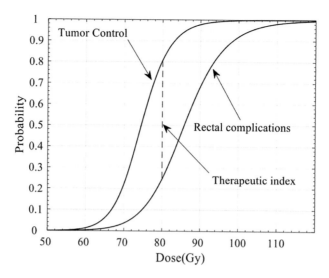

FIGURE 4.9 Dose–response curves for tumor control of a high-risk cancer of the prostate of intermediate risk and rectal bleeding. The curves have been calculated using the EUD-based TCP and NTCP models [15] (see Section 6.5) with parameters from Ref. [14]. It is assuming that 30% of the rectum receives the prescribed dose.

The final "R" is the *radiosensitivity* which is intrinsic to the cell line of the tumor. For example, dysgerminomas are relatively radiosensitive and respond to low doses of radiation (20–30 Gy) compared with the more radioresistant cervical cancer tumors, which may require more than 70 Gy to obtain a cure [13].

These processes are called the four R's and their effect on cell survival after treatment can be modelled by inclusion of correction factors in the Linear Quadratic (LQ) model.

Radiotherapy (RT) may also damage healthy tissues and organs. The ratio of the tumor response for a fixed level of normal-tissue damage has variously been called the *therapeutic ratio* or *therapeutic index*. In the example in Figure 4.9, a dose–response curve for the local control of prostate cancer [14, 15] is compared to the risk for rectal bleeding following prostate radiotherapy. The most favorable therapeutic ratio is at a dose of 80 Gy, where the tumor control probability is 80% and the risk for complication is 25%.

In recent years, it has been demonstrated that radiotherapy induces an immune response that can contribute to killing cancer cells, including outside the treatment volume [16], an effect called the abscopal effect from the latin prefix ab- "position away from" and -scopos "mark or target for shooting" [17].

4.6.2 The Linear Quadratic Model

The early basis for the LQ model was due to the work of Sax and Lea, who noted that the average yield of exchange chromosomal aberrations in Tradenscantia, is well described by a linear quadratic expression [18]. In vitro experiments use colony assay to measure cell survival or DNA flow cytometry for counting cell numbers [19] (Figure 4.10).

In the LQ model, the fraction of cells surviving a fractionated irradiation, assuming complete recovery from sub-lethal damage between fractions, is given by [20]:

$$SF = e^{-\alpha d - \beta d^2}$$

(4.1)

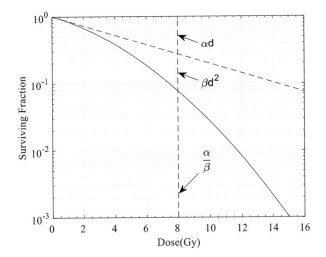

FIGURE 4.10 Survival curve calculated using the linear quadratic model with $\alpha = 0.16$ and $\beta = 0.02$.

Where d = dose in a single fraction. The linear term α and the quadratic term β are interpreted as the probability of killing the cell with a single and double hit, respectively. α describes the initial slope of the curve. The ratio α/β gives the dose at which the lethal effect of both components is equal.

For a fractionated treatment, with N fractions of dose d:

$$SF = \left(e^{-\left(\alpha d + \beta d^2\right)} \right)^N = e^{-N\left(\alpha d + \beta d^2\right)} \tag{4.2}$$

Two treatments with different doses per fraction d_1 and d_2 and number of fractions N_1 and N_2, are assumed to be equivalent when they have equal survival fractions [20, 21]:

$$SF_1 = SF_2$$

This is satisfied if:

$$SF_1 = SF_2 \Leftrightarrow N_1 d_1 \left(\frac{\alpha}{\beta} + d_1 \right) = N_2 d_2 \left(\frac{\alpha}{\beta} + d_2 \right) \tag{4.3}$$

where the ratio α/β accounts for the sensitivity of tumor or tissue to a change in fractionation. The biological effective dose (BED) is then determined as:

$$BED \equiv Nd \left(1 + \frac{d}{\frac{\alpha}{\beta}} \right) \tag{4.4}$$

It follows from Equation (4.4) that two treatments are equivalent if they have the same BED. For this reason, BED is often used to study the effect of fractionation in clinical routine, for example, for designing hypofractionated treatments [22].

Usually a value of α/β of 10 Gy is used for early complications and tumors, 3 Gy for late complications, or 2 Gy for late complications in the central nervous system or kidneys [23]. Some tumors, such as in the breast and prostate, have lower α/β, of 1.5 and 4, respectively.

Very often, in order to compare different fractionations, the equivalent dose with a 2 Gy per fraction regimen, EQD2, is used:

$$\text{EQD2} = \frac{D\left(1 + \dfrac{d}{\dfrac{\alpha}{\beta}}\right)}{D\left(1 + \dfrac{2}{\dfrac{\alpha}{\beta}}\right)} \tag{4.5}$$

For proliferating tissues and tumors, a repopulation term is added to BED:

$$BED = Nd\left(1 + \frac{d}{\dfrac{\alpha}{\beta}}\right) - \frac{\ln 2}{\alpha}\max\left(0, \frac{t + t_k}{t_p}\right) \tag{4.6}$$

Where t is the overall duration of the RT treatment in days, t_p is the time needed to duplicate the number of clonogenic cells, called potential doubling time, t_k is the starting time of repopulation (delay in onset of repopulation) [24]. For example, $t_k = 7$ days and $t_p = 2.5$ days for mucositis.

By using the formula of BED with the repopulation term, it is possible to compensate for interruption of radiotherapy treatments as explained in detail in [25].

4.6.3 TIME FACTORS

LQ model predictions can also be corrected for temporal effects. For low dose–rate exposures or closely-spaced fractions, incomplete *sub-lethal damage repair (SDLR)* can be significant:

$$SF = e^{-\alpha D - \beta G D^2} \tag{4.7}$$

where D is the total dose of the treatment, and G is called the dose protraction factor:

$$G = 2\frac{\left(\lambda T + e^{-\lambda T} - 1\right)}{\left(\lambda T\right)^2}, \text{where } T_{\frac{1}{2}} = \frac{\ln 2}{\lambda} \tag{4.8}$$

$T_{\frac{1}{2}}$ is the repair half-time, that is, the time required to repair half of the damaged cells.

Turesson and Thames [26] estimated repair half-times $T_{1/2}$ of 1.1–1.3 h for erythema and desquamation, and 3.5 h for telangiectasia. But repair based on a single exponent function underestimated the sparing effect in inter-fraction interval by 0.25 h.

They concluded that the results suggested a two-component SLDR, with a fast component of 20 min repair half-time for both early and late effects, and a slow component of 1.2 h repair half-time for early effects, and 3.5 h for late effects, a hypothesis that agrees well with the existence of the two repair mechanisms of DNA, one faster, nonhomologous end joining, and one slower, homologous recombination.

Another model accounting for incomplete recovery from sub-lethal damage between fractions uses a correction to the quadratic term [27].

4.6.4 OXYGENATION

After receiving radiation, hypoxic cells become oxygenated. Oxygen acts as a radiosensitiser because it participates in the chemical reactions that lead to DNA damage. Cells that are anoxic during irradiation are about three times more resistant to radiation than cells that are well oxygenated at the time of irradiation.

Very low levels of O_2 are required for radiosensitisation. Radiosensitivity increases as the oxygen tension increases from anoxia to ~10 Torr [28]. Because of rapid proliferation, many tumor cells lack blood supply. Acute hypoxia is due to the closure of capillaries or arterioles for a short time in the proximity of the tumor. Splitting the dose into fractions raises the possibility of the closed vessel being open, improving oxygenation before the next fraction.

Chronic hypoxia is due to the poor vasculature of tumours and the distance of the tumor from the capillaries. Fractionated radiotherapy kills cells that lie close to the capillary more effectively and the tumor cells may re-oxygenate.

The effect of oxygen reduction is described by the oxygen enhancement ratio (OER) [29]:

$$OER = \frac{\text{Dose with reduced oxygen}}{\text{Dose with full oxygenation}} \tag{4.9}$$

Tumour hypoxia is associated with resistance to RT and chemotherapy and is also associated with a malignant and aggressive phenotype [30]. In vitro values of OER have been measured for prostate cancer (1.75 and 3.25 for OER_a and OER_b) [31]. According to Carlson [32] et al., $OER_\alpha = OER_\beta = 1.5$.

4.6.5 INHOMOGENEOUS IRRADIATION TO ORGANS AT RISK

The dose distribution from external radiotherapy is often inhomogeneous. To account for an inhomogeneous dose distribution, the generalised equivalent uniform dose (gEUD) is used. This is the dose that, when delivered uniformly to the organ or tissue, produces the same Normal Tissue Complication Probability (NTCP) as the inhomogeneous dose distribution [33]:

$$gEUD \stackrel{\text{def}}{=} \left(\sum_i v_i (D_i)^a \right)^{\frac{1}{a}} \tag{4.10}$$

where v_i is the ith sub-volume of the organ irradiated with dose D_i in the differential dose volume histogram (DVH). The parameter a describes the volume effect of the irradiated organ or tissue. Serial organs have a high value of a, so that gEUD becomes close to the maximum dose. Parallel organs have a close to one, so that gEUD is close to the mean dose. Since for negative a the EUD becomes sensitive to low values of D_i due to cold spots, a negative value of a is used for tumors.

4.6.6 MODELS FOR NORMAL TISSUE COMPLICATION PROBABILITY (NTCP)

Models for Normal Tissue Complication Probability allow estimation of the risk of complications for a given treatment plan. In the Lyman-EUD (LEUD) NTCP model, the probability of side effects is calculated from the equivalent uniform dose, EUD (see Equation 4.2) using equations:

$$NTCP = \Phi(T) = \frac{1}{\sqrt{2\pi}} \int_{-\infty}^{T} e^{-\left(\frac{t^2}{2}\right)} dt \tag{4.11}$$

and

$$t = \frac{EUD - TD_{50}}{mTD_{50}} \qquad (4.12)$$

where TD_{50} is the dose resulting in a 50% complication probability. Parameter m describes the slope of the NTCP curve at TD_{50}. The function Φ in Equation (4.11) is called probit function. Other NTCP models include the Logit-EUD model (LogEUD) [15]:

$$T = \frac{1}{1 + \left(\dfrac{TD_{50}}{EUD}\right)^{4\gamma}} \qquad (4.13)$$

where γ_{50} describes the slope of the dose response. This formalism is also used for tumor control probability (TCP) models.

Other models include the Relative Seriality model [34], and the population averaged critical volume model (CV) [35]. Recently, voxel by voxel models, trying to establish correlation between dose at the voxel level and the occurrence of toxicity, have been introduced [36].

4.6.7 MODELS FOR TUMOR CONTROL PROBABILITY (TCP)

Tumour Control Probability (TCP) is the probability of having n = 0 surviving clonogens in the irradiated volume at the end of radiotherapy. This can be estimated using Poisson statistics [37] as:

$$\text{TCP} = e^{-N} \qquad (4.14)$$

Where N is the average number of cells which survive irradiation, and it is the product of the survival fraction (SF) from Equation (4.2) and the initial number of clonogenic cells in the tumor volume, N_0. The TCP can be then expressed as:

$$TCP = e^{-N_0 e^{-\alpha D\left(1 + \frac{\beta}{\alpha}d\right)}} \qquad (4.15)$$

In this equation, the target volume is assumed to receive a uniform dose. For inhomogeneous dose distributions, it can be assumed that the tumour volume is composed of a series of sub-volumes, each one receiving a homogeneous dose. TCP in every sub-volume is described by Equation (4.15), and the TCP of the tumour is the product of TCP in every sub-volume in the tumour [38].

Many corrections have been proposed to increase the accuracy of TCP models. For example, Webb and Nahum [39] incorporated inter-patient variability into TCP, by assuming that the radiosensitivity term α follows a Gaussian function in a population of patients. The effect of chemotherapy can be modelled in two ways [40]. In the case of a drug that acts as a radio-sensitiser, the effect associated with each fractional dose is enhanced by the drug. This is modelled by multiplying the total and fractional doses by a dose modifying factor [41].

In the case of a chemotherapeutic agent that, if used alone, can kill tumour cells, the total cell kill is a result of the summed effect of radiotherapy and chemotherapy (E_c) [40].

REFERENCES

[1] Ainsbury EA et al., Public Health England survey of lens dose in the UK medical sector, *J. Radiol. Prot.* 2014;34:15–29.

[2] Smith PG, The 1957 MRC report on leukaemia and aplastic anaemia in patients irradiated for ankylosing spondylitis, *J Radiol Prot.* 2007;27:B3–B14.

[3] Darby S C et al., Long term mortality after a single treatment course with x-rays in patients treated for ankylosing spondylitis, *Br. J. Cancer* 1987;55(2):179–190.

[4] de Vathaire et al., Thyroid cancer in French Polynesia between 1985 and 1995: Influence of atmospheric nuclear bomb tests performed at Mururoa and Fangataufa between 1966 and 1974. *Cancer Causes Cont.* 2000;11:59–63.

[5] Cardis E and Hatch M, The Chernobyl accident – an epidemiological perspective. *Clin. Oncol.* 2011;23(4):251–260.

[6] Harada KH et al., Radiation dose rates now and in the future for residents neighbouring restricted areas of the Fukushima Daiichi nuclear power plant, *PNAS*, 2014;E914–E923.

[7] Ozasa K et al., Studies of the mortality of atomic bomb survivors, Report 14, 1950 – 2003: An overview of cancer and noncancer diseases. *Radiat Res.* 2012;177:229–243.

[8] Yang T et al., A review of low-level ionising radiation and risk models of leukaemia, *J Radiol Oncol.* 2013;2:263–270.

[9] Rühm W et al., Dose-rate effects in radiation biology and radiation protection. *Ann ICRP* 2016;45(1) suppl 1:262–279.

[10] ICRP, The 2007 Recommendations of the International Committee on Radiological Protection. *Ann ICRP*, 2007;37(2–4).

[11] ICRP, The 2007 Recommendations of the International Committee on Radiological Protection, *Ann ICRP, Appendix* 2007;A4, 173–213.

[12] Minniti G, Goldsmith C, Brada M. Chapter 16 - Radiotherapy. *Handb. Clin. Neurol.* 2012;104:215–228.

[13] Yashar CM. Basic Principles in Gynecologic Radiotherapy. *Clin. Gynecol. Oncol.*, 2018;586–605.e3.

[14] Kim Y, Tome WA. Risk-adaptive optimization: selective boosting of high-risk tumor subvolumes. *Int. J. Radiat. Oncol. Biol. Phys.* 2006;66:1528–1542.

[15] Gay HA, Niemierko A. A free program for calculating EUD-based NTCP and TCP in external beam radiotherapy. Phys. Med. Eur. *J. Med. Phys.* 2007; 2007;23:115–125.

[16] Muraro E, Furlan C, Avanzo M, Martorelli D, Comaro E, Rizzo A, et al. Local high-dose radiotherapy induces systemic immunomodulating effects of potential therapeutic relevance in oligometastatic breast cancer. *Front. Immunol.* 2017;8:1476.

[17] Ozpiskin OM, Zhang L, Li JJ. Immune targets in the tumor microenvironment treated by radiotherapy. *Theranostics* 2019;9:1215–1231.

[18] Sax K. Chromosome aberrations induced by x-rays. *Genetics* 1938;23:494–516.

[19] Brenner DJ. The Linear-Quadratic Model Is an Appropriate Methodology for Determining Isoeffective Doses at Large Doses Per Fraction. *Semin. Radiat. Oncol.* 2008;18:234–239.

[20] Fowler JF. The linear-quadratic formula and progress in fractionated radiotherapy. *Br. J. Radiol.* 1989;62:679–694.

[21] Barendsen GW. Dose fractionation, dose rate and iso-effect relationships for normal tissue responses. *Int. J. Radiat. Oncol. Biol. Phys.* 1982;8:1981–1997.

[22] Avanzo M, Trovo M, Stancanello J, Jena R, Roncadin M, Toffoli G, et al. Hypofractionation of partial breast irradiation using radiobiological models. *Phys. Med.* 2015;31:1022–1028.

[23] Fowler JF. 21 years of biologically effective dose. *Br. J. Radiol.* 2010;83:554–568.

[24] Roberts SA, Hendry JH. The delay before onset of accelerated tumour cell repopulation during radiotherapy: a direct maximum-likelihood analysis of a collection of worldwide tumour-control data. *Radiother. Oncol.* 1993;29:69–74.

[25] Bese NS, Hendry J, Jeremic B. Effects of prolongation of overall treatment time due to unplanned interruptions during radiotherapy of different tumor sites and practical methods for compensation. *Int. J. Radiati. Oncol. Biol. Phys.* 2007;68:654–661.

[26] Turesson I, Thames HD. Repair capacity and kinetics of human skin during fractionated radiotherapy: erythema, desquamation, and telangiectasia after 3 and 5 year's follow-up. *Radiother. Oncol.* 1989;15:169–188.

[27] Thames Thames HD. An 'incomplete-repair' model for survival after fractionated and continuous irradiations. *Int. J. Radiat. Biol. Relat. Stud. Phys. Chem. Med.* 1985;47:319–339.

[28] Rockwell S, Dobrucki IT, Kim EY, Marrison ST, Vu VT. Hypoxia and radiation therapy: past history, ongoing research, and future promise. *Curr. Mol. Med.* 2009;9:442–458.

[29] Avanzo M, Stancanello J, Franchin G, Sartor G, Jena R, Drigo A, et al. Correlation of a hypoxia based tumor control model with observed local control rates in nasopharyngeal carcinoma treated with chemo-radiotherapy. *Med. Phys.* 2010;37:1533.

[30] Bussink J, Kaanders JH, van der Kogel AJ. Tumor hypoxia at the micro-regional level: clinical relevance and predictive value of exogenous and endogenous hypoxic cell markers. *Radiother. Oncol.* 2003;67:3–15.

[31] Nahum AE, Movsas B, Horwitz EM, Stobbe CC, Chapman JD. Incorporating clinical measurements of hypoxia into tumor local control modeling of prostate cancer: implications for the alpha/beta ratio. *Int. J. Radiat. Oncol. Biol. Phys.* 2003;57:391–401.

[32] Carlson DJ, Stewart RD, Semenenko VA. Effects of oxygen on intrinsic radiation sensitivity: A test of the relationship between aerobic and hypoxic linear-quadratic (LQ) model parameters. *Med. Phys.* 2006;33:3105–3115.

[33] Niemierko A. A generalized concept of Equivalent Uniform Dose. *Med. Phys.* 1999;26:1100.

[34] Kallman P, Agren A, Brahme A. Tumour and normal tissue responses to fractionated non-uniform dose delivery. *Int. J. Radiat. Biol.* 1992;62:249–262.

[35] Stavrev P, Niemierko A, Stavreva N, Goitein M. The application of biological models to clinical data. *Phys. Med.* 2001;17:1–12.

[36] Avanzo M, Barbiero S, Trovo M, Bissonnette JP, Jena R, Stancanello J, et al. Voxel-by-voxel correlation between radiologically radiation induced lung injury and dose after image-guided, intensity modulated radiotherapy for lung tumors. *Phys. Med.* 2017;42:150–156.

[37] Munro TR, Gilbert CW. The relation between tumour lethal doses and the radiosensitivity of tumour cells. *Br. J. Radiol.* 1961;34:246–251.

[38] Webb S, Nahum AE. A model for calculating tumour control probability in radiotherapy including the effects of inhomogeneous distributions of dose and clonogenic cell density. *Phys. Med. Biol.* 1993;38:653–666.

[39] Webb S. Optimum parameters in a model for tumour control probability including interpatient heterogeneity. *Phys. Med. Biol.* 1994;39:1895–1914.

[40] Jones B, Dale RG. The potential for mathematical modelling in the assessment of the radiation dose equivalent of cytotoxic chemotherapy given concomitantly with radiotherapy. *Br. J. Radiol.* 2005;78:939–944.

[41] Steel GG. The search for therapeutic gain in the combination of radiotherapy and chemotherapy. *Radiother. Oncol.* 1988;11:31–53.

5 Introduction to Diagnostic Radiology (X-Ray and Computed Tomography Imaging)

Slavik Tabakov
King's College London, London, UK

Paola Bregant
ASU GI, Trieste, Italy

CONTENTS

DOI: 10.1201/9780429155758-5

5.1   INTRODUCTION

The use of X-rays to obtain images of anatomical structures is widely accepted as the beginning of medical physics. Roentgen realised the potential value of his discovery of the 'new kind of rays' in medicine, and the resulting X-ray imaging was considered one of the most important innovations in medicine in the 19th century. For the first time, it was possible to see anatomical structures in the body without surgical intervention. The medical application of this discovery triggered the establishment of Nobel Prizes and, not surprisingly, the inaugural physics prize was presented to W. K. Roentgen.

Soon after this, a new branch of medicine was established – Diagnostic Radiology (called Roentgenology in some countries) – the field broadened to include other areas of applied physics (e.g. radioisotopes, ultrasound, magnetic resonance), and became a cornerstone of medical diagnostics. This branch of medicine expanded enormously, and since the 1990s is known by the more general term Medical Imaging. Medical physicists and engineers played the most important role in the invention of new imaging methods and their medical applications.

This chapter discusses medical imaging techniques using X-rays. These are still the most frequently performed medical imaging examinations. The 2008 UNSCEAR report (United Nations Scientific Committee on the Effects of Atomic Radiation) shows that annually 7.5 million radiotherapeutic procedures are performed worldwide as well as 37 million nuclear medicine procedures and 3,600 million X-ray diagnostic procedures [1].

If we take the UK as an example – with its National Health Service – we can see that within all medical imaging procedures, those using X-rays are the most frequently performed. For example, during the financial year 2012–2013, X-ray procedures (radiography, fluoroscopy and CT scanning) amounted to approximately 70% of all medical imaging procedures [2]. Radiography alone amounted to approximately 55% of all medical imaging procedures.

X-ray medical imaging procedures deliver the highest radiation dose to the population from artificial sources of radiation. As an example, the average effective dose per capita to the US population from all major sources of exposure during 2006 shows that 37% is associated with radon; 13% with other background radiation; 24% with Computed Tomography; 7% with interventional radiological procedures; and 5% with radiographic and fluoroscopic procedures, etc [3]. Although in various countries and various years these figures differ, they are of similar magnitude, hence diagnostic radiology delivers the highest percentage of effective dose from artificial sources of radiation (per capita).

This requires special attention to be given to all medical imaging examinations using X-rays. This chapter will discuss the production of X-rays – the X-ray tube and generator; the formation of the X-ray image; the main X-ray imaging methods (radiography, digital radiography, mammography, fluoroscopy, dual energy imaging and computed tomography).

The chapter will also discuss the evolution of the detectors used in X-ray medical imaging. The original detector, X-ray film, will be mentioned only briefly, as it has now been almost completely replaced by digital detectors. The assessment of the main physical parameters of an X-ray image and the patient dose will also be discussed and we shall also briefly mention the latest X-ray imaging methods.

5.2 X-RAY TUBE AND GENERATOR AS A SOURCE OF RADIATION

5.2.1 PRODUCTION OF X-RAYS AND THE BREMSSTRAHLUNG SPECTRUM

The X-ray region of the electromagnetic spectrum covers energies in the range 10^2 to 10^7 eV. X-rays used for medical imaging purposes (normally in the range 20–150 keV) are produced in an X-ray tube, where electrons, accelerated in a high voltage field (kV), bombard a positively charged material (the anode target). As a result of this bombardment, the kinetic energy of the accelerated electrons is converted into other forms of energy – heat and electromagnetic radiation (X-rays) – Figure 5.1. Most of the accelerated electrons react with orbital electrons in the atoms of the target material, thus producing heat. However, a few of the accelerated electrons react with the nuclei of the target material producing photons with a variety of energies (the X-ray spectrum). If the target is made of heavy metal (e.g. tungsten), this reaction is due to the strong Coulomb field of its relatively large nucleus and results in a change in direction of the accelerated electrons and their velocity (mainly decreasing the velocity). During this process, the accelerated electrons lose energy in the form of electromagnetic radiation (X-ray photons). The energy of each X-ray photon is equal to the difference in energy of the incoming accelerated electron and the energy with which it continues after the influence of the nuclear coulomb field.

Very few electrons directly hit the nucleus, which is a very small part of the target (e.g. approximate radii: electron – 10^{-15} m; nucleus – 10^{-14} m, atom – 10^{-10} m). These few electrons stop immediately and transfer all their kinetic energy to a single photon with maximum energy (equal to the energy of the accelerating high voltage field). Most of the electrons approach the nuclei differently and pass at different distances. This leads to the production of X-ray quanta with a range of energies. Electrons deflected by the first nuclei (on the surface of the target) continue their path inside through the target material and during this penetration are further deflected and decelerated by other nuclei, producing X-ray quanta with gradually decreasing energies. The resulting sum of X-ray quanta forms a continuous polyenergetic spectrum limited by the maximum energy of the applied high voltage field. This process of producing X-rays by 'stopping' the electrons is called 'bremsstrahlung' (in German, braking radiation), the name given by W.K. Roentgen.

In this interaction with the nuclei, a very small number of accelerated electrons directly hit some of the orbital electrons. If the energy of the bombarding electron is larger than the binding energy of the atomic electron with which it has collided, it will displace it, creating an orbital vacancy. Electrons from other atomic orbits quickly fill the vacancy – a process resulting in the creation of X-ray photons with material-specific energies called 'characteristic radiation'. The latter is superimposed on the continuous polyenergetic spectrum, thus forming the '**X-ray spectrum**'. For tungsten, the K-shell binding energy (K-edge) is 69.5 keV, and for molybdenum, 20 keV. However, the energy of the bombarding electrons needs to be significantly above the K-edge in order to produce significant levels of characteristic radiation.

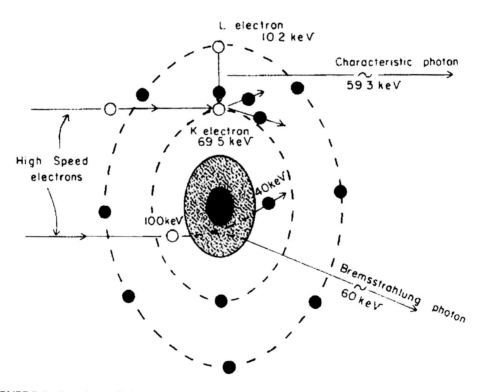

FIGURE 5.1   Stopping radiation (Bremsstrahlung) and characteristic radiation.

(Courtesy of Sprawls Resources – available as a free online resource at www.sprawls.org.)

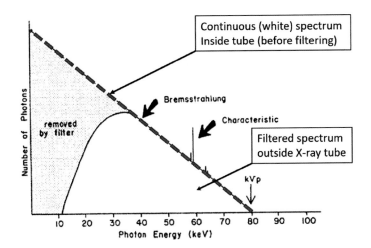

FIGURE 5.2 Typical X-ray spectrum of tungsten anode target, produced from electrons accelerated to 80 kV. The effective energy of this spectrum has been measured as approximately 30 kV.

(Courtesy of www.sprawls.org.)

Only a few of the accelerated electrons interact with the nuclei of the atoms at the surface of the target. The majority pass through these atoms and react with inner atoms. The X-rays produced in this way are absorbed inside the material of the target. The whole process of X-ray formation is very inefficient, as only about 1% of the energy of the beam of accelerated electrons transfers to useful bremsstrahlung X-rays (the remaining 99% is transformed to heat in the target – i.e. the anode).

The continuous X-ray spectrum created by an X-ray tube includes photons with a range of energies (measured inside the X-ray tube – immediately in front of the anode). However, the lowest energy photons are absorbed immediately by the material of the X-ray tube housing (a process of "energy filtration"). The remaining low energy photons are further filtered by added plates of various materials (e.g. aluminium), as otherwise these will be absorbed at the surface of the patient producing no useful information but adding to patient dose (such 'filter' plates are included in the X-ray tube housing). This additional filtration shapes the final X-ray spectrum – Figure 5.2.

There are two other main characteristics of the X-ray spectrum:

- the maximum energy (kVp – 80 kVp in Figure 5.2) – which is at the maximum accelerating energy of the electrons bombarding the target. The X-ray spectrum includes very few X-ray photons with maximum energy as very few nuclei on the target surface are directly hit by the accelerated electrons;
- The maximum intensity (amplitude and energy – 35 kVp in Figure 5.2) of the spectrum.

Using various kV and mA combinations, as well as metal filtration of the X-ray beam emitted from the tube, the X-ray spectrum is optimised to allow better visualisation of specific anatomical structures (e.g. bones, soft tissues, etc). The aim is to allow differential absorption of X-rays in various body tissues creating a 'latent X-ray image' (modulation of the spectrum by the object). X-rays exiting the object must also be compatible with the attenuation characteristics of the X-ray detector (e.g. X-ray film, digital detector, etc). This process results in an image with good quality, produced with low dose to the patient. However, more than 90% of the X-ray beam exiting the tube is absorbed inside the patient – that is, only a small percentage of the radiation produced by the X-ray tube reaches the detector to create a visible X-ray image. As mentioned above, filtration is important in ensuring that the patient doses are optimised.

5.2.2 X-Ray Tube Construction

As a source of radiation, the tube is the most important part of the X-ray equipment. In the early days of X-ray imaging, cold cathode gas-ion tubes were used but these were replaced by heated cathode vacuum X-ray tubes in 1913.

Most conventional X-ray tubes consist of a glass envelope, which contains one negative and one positive electrode – a typical 'vacuum tube' (Figure 5.3). The negative cathode is heated in order to emit electrons (thermal electrons), which travel in the accelerating electrical field (20–150 kV) to the positive anode target, colliding with it to produce X-radiation. There is a high vacuum inside the glass envelope to eliminate electron collisions before reaching the anode.

The cathode assembly of all contemporary tubes consists of a heated tungsten wire, which produces thermal electrons. The emission of thermal electrons depends on the temperature of the cathode – the density of the emission current is proportional to the square of the temperature (as per the Richardson-Dushman equation). In order to obtain high anode currents of the order of an ampere (current of accelerated electrons between the cathode and the anode), the cathode filament must be heated to a high temperature (around 2500°C depending on tube construction). Tungsten, with a melting point of 3410°C and ability to be drawn into a very thin wire, is very suitable as a cathode material.

Normally the cathode is a long (approx. 5–20 mm), thin (approx. 0.1–0.3 mm) tungsten wire in a spiral coil with diameter of the order of 0.2 mm. The filament is connected to the negative side of the high voltage.

Thermal electrons emitted from the heated cathode are accelerated towards the anode by the electric field between cathode and anode, forming the anode current. The area of the anode, which is bombarded by the thermal electrons and produces X-rays, is called the actual (or thermal) focal spot. The size of the actual focal spot depends on the size of the cathode wire (in fact the size of the electron beam generated by it). Most X-ray tubes have two foci – one small filament wire (up to 1 mm long) – known as the 'fine focus' and one bigger filament wire (approx. 2–3 mm long, made of thicker wire) – known as the 'broad focus' – Figure 5.4. The foci can be placed next to each other or one behind the other (there are many different designs).

The anode assembly of the X-ray tube is located opposite to the cathode at a distance of approximately 25 mm. In most tubes, the anode is angled at 10–20 degrees. The small region of the anode bombarded by the thermal electrons and producing X-rays is referred to as the **target**. Around 99% of the energy imparted to the target by the electrons is converted to heat and generation of secondary electrons (strictly speaking the energy converted directly to heat is less than 80%). Because of this, the target material is also usually tungsten – a material with a very high melting point and high

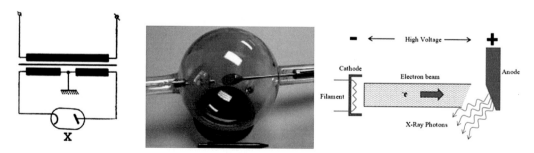

FIGURE 5.3 X-ray tube: Left: Sketch of X-ray tube with high voltage transformer; Centre: X-ray tube with glass envelope (stationary anode on right) – old type; Right: X-ray tube function sketch – the cathode creates a beam of thermal electrons, which bombard the anode (rotating anode), which produces X-ray photons.

(Courtesy of EMERALD free online resources: www.emitel2.eu.)

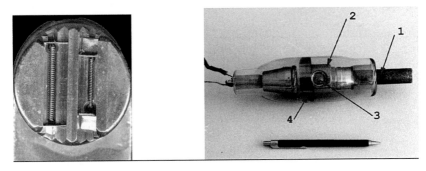

FIGURE 5.4 Left: cathode assembly with two foci (broad – left and fine – right), both in grooves to allow better focusing of the beam of thermal electrons; Right: X-ray tube with static anode (Low power): 1 – anode stem; 2 – electron capture hood; 3 – beryllium window over the anode; 4 – metallisation of the glass envelope.

(Courtesy of www.emerald2.eu.)

atomic number. The latter is important for the effective conversion of the energy of electrons to X-rays. The 'Bremsstrahlung generation efficiency' (η) is the ratio of the intensity of X-ray radiation (W) to the energy flux of the electron beam (E) - $\mathbf{E \sim I\,U}$

The intensity of X-ray radiation W (X-ray energy flux density) is:

$$\mathbf{W \sim I\,U^2\,Z,}$$

where W is the intensity of X-ray radiation; I the anode current; Z the atomic number of the anode; U the accelerating high voltage (anode tension – kV).

From the above it follows that

$$\eta \sim W\,/\,E \sim k\,U\,Z$$

The constant k has a value of approximately 1.1×10^{-9}. Thus, for tungsten (Z = 74) at 100 kV, the bremsstrahlung generation efficiency is approximately 0.8%.

X-rays generated by the target are emitted in different directions (isotropic X-ray emission) and just a small proportion of them leave the tube in the direction of the patient. The remainder are absorbed in the tube housing and in the added filtration. Less than 0.1% of the energy imparted to the anode is converted to useful radiation towards the patient. Thus, a high electron beam energy and intensity is required for the production of useful X-rays. As a result, the target of the anode heats up to very high temperatures during the exposure so anode materials must have high thermal conductivity.

Obviously, the anode cannot be made entirely from tungsten – firstly, it would be too expensive, and secondly, other metals have better thermal conductivity. The anode assembly is a composite of materials with good thermal conductivity (e.g. copper), while only its surface is coated with a thin layer of tungsten (or a specific tungsten alloy).

There are two main types of classical X-ray tubes – stationary anode and rotating anode. X-ray tubes with stationary anodes are used for low power X-ray equipment – for example, some small mobile and dental X-ray equipment. The simplest construction of a stationary anode is a large copper stem with a small tungsten plate (2–3 mm thick) embedded as the target – Figure 5.4.

X-ray tubes with rotating anodes allow for the heat of the anode to be distributed over a larger area – Figure 5.5. The rotating anode is mounted on a stem which rotates at high speed driven by an electrical motor. The rotating anode and rotor are within the glass envelope (supported by ball bearings), and the stator producing the rotating magnetic field is fitted outside the glass envelope.

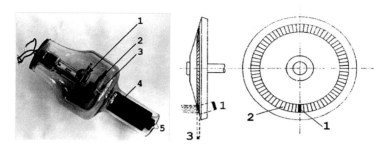

FIGURE 5.5 X-ray tube with rotating anode (high power): Left: 1 – cathode assembly; 2 – rotating anode; 3 – anode stem (molybdenum); 4 – rotor; 5 – glass envelope. Right – rotating anode: 1 – electric (momentary) focus, corresponding to the actual focus of stationary anode tube; 2 – thermal focus (thermal path or thermal track); 3 – effective (apparent or optical) focal spot.

(Courtesy of www.emerald2.eu.)

Increasing the speed of rotation (up to 9000 rpm) leads to increased tube power because the bombarded area of the track passes quickly over the fixed electron beam and has time to cool during the remaining rotation. Further demand for long sequences of powerful exposures (especially in angiography and CT scanners) has led to constructing rotating anodes with larger heat storage capabilities.

The power of an X-ray tube is measured by the ability of its anode to accumulate heat (in either Heat Units or joules). The heat capacity of a powerful X-ray tube can reach several megajoules.

The glass envelope of the X-ray tube (Figure 5.5) supports the anode and cathode assemblies. It also keeps a high vacuum in the X-ray tube (minimum 10^{-6} mbar) and provides high voltage insulation between the anode and cathode. The glass used should have low X-ray absorption and high thermal resistance.

The X-ray tube housing contains the fragile X-ray tube and provides it with mechanical strength and protection. Most housings are made of steel or aluminium alloy. The housing is earthed and filled with a special insulating oil which assists in cooling the X-ray tube and provides the necessary electrical insulation from the high voltage supplied to the anode/cathode.

The X-ray anode generates radiation in all directions. However, only a small proportion of this radiation is towards the patient (supplied through a small window in the housing). The remaining radiation generated by the X-ray tube is absorbed by a lead lining inside the housing (several millimetres). However good this shielding may be, some radiation may escape (leakage radiation) and special care is taken to minimise leakage radiation to protect staff and patients from unwanted exposure.

X-ray tubes with metal, instead of glass, envelopes were introduced during the 1980s and are currently extensively used for high power X-ray exposures (e.g. for CT scanning). These X-ray tubes have metal envelopes and a different construction to allow better heat dissipation.

5.2.3 RELATION BETWEEN FOCAL SPOT SIZE AND IMAGE RESOLUTION

Ideally the source of radiation (the focal spot) should be a point source. However, in reality, the focal spot has finite size and is rectangular with side dimensions varying from 0.3–0.5 mm (fine focal spot) to 0.8–1.5 mm (broad or large focal spot). X-rays are generated over the whole area of the focal spot but, because of the finite size of the focal spot, blurring is seen at the edges of any object that is imaged. These 'blurred' borders are known as **penumbra** – see Figure 5.12.

The larger the focal spot size, the larger the penumbra, which results in more blurred borders – hence reduced spatial resolution. For this reason, imaging methods requiring high resolution (e.g. mammography) use X-ray tubes with small (fine) effective focal spots.

5.2.4 X-Ray High Voltage Generator (XG)

All electrical circuits which supply and control high voltage to the X-ray tube (XT) are called high voltage X-ray generators (for short X-ray generators or HV generators). Initially, high voltage generators consisted of just one high voltage transformer (HVT). Later, high voltage rectifiers (now semiconductor diodes) and controlling circuitry were added to the XG. The XG controls both the high voltage (kV) and the anode current (mA). The main part of the XG is the high voltage step-up transformer with a fixed transformation ratio (of the order of 500–600). This produces the required kilovoltage which, after rectification, is applied across the cathode and anode of the X-ray tube. The XG also controls the current through the cathode filament controlling the cathode heat and thus the production of thermal electrons.

Generators for most contemporary X-ray equipment do not use mains power directly (50 or 60 Hz) but generate medium frequency electricity (2–20 kHz) from the mains supply, which allows a reduction in the size of the high voltage transformer. The principle behind this is based on the relation:

$$U/f \sim An,$$

where: U is the accelerating voltage, A is the cross-sectional area of the transformer core; n is the number of windings and f is frequency.

It is obvious that increasing the frequency, **f**, enables the use of a transformer with a smaller core **A**. Such medium frequency transformers are now produced using special ferrite materials which have high magnetic permeability. The formula above shows that at a fixed size and winding number of this transformer (**A n**), the frequency **f** is proportional to the voltage **U**. In this way, the voltage (kV) can be controlled by changing the frequency produced by the DC chopper (inverter) – Figure 5.6 (right) - allowing for very precise kV adjustment.

Figure 5.6 (left) shows a block diagram of a contemporary X-ray generator. The mains power supply is rectified and then an electronic DC inverter (chopper) is used to convert the DC voltage to a series of pulses (with frequencies of the order of 2–20 kHz). This medium frequency electricity is transformed to high voltage by the special HV ferrite transformer. It is again rectified before being delivered to the X-ray tube through high voltage cables (see Figure 5.7).

Most contemporary radiographic equipment includes a system for **Automatic Exposure Control (AEC)**. A radiation detector immediately in front of the imaging detector interrupts the exposure when a certain pre-determined dose level is reached. The system maintains the detector input dose independent of any change in X-ray parameters and patient size (thickness). The AEC system senses the exposure dose, compares it with the pre-set value, takes into consideration the sensitivity of the detector and on this basis interrupts the exposure when the dose necessary for this detector is reached. There are many different types of AEC systems. Normally the difference relates to the

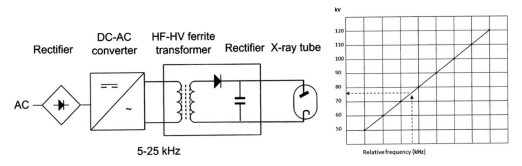

FIGURE 5.6 Left – Typical block diagram of contemporary X-ray generator. Right – control of kVp (kV peak) through change of frequency.

(Courtesy of www.emitel2.eu)

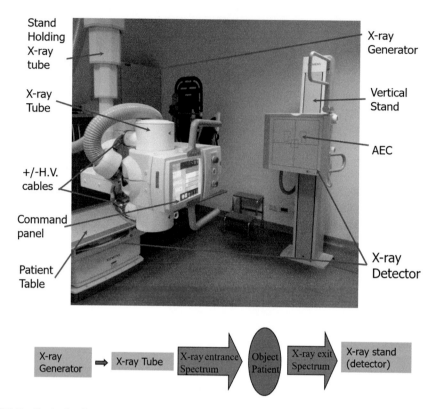

FIGURE 5.7 Typical radiographic equipment with main components shown. Under the image is an indicative diagram of the equipment components creating the X-ray spectrum and its attenuation by the patient and detection (X-ray detector).

type of detector used (ionisation chamber, photo-timer or solid-state detector). The AEC system is normally included in the X-ray equipment stand – see Figure 5.7.

The parameters of the XG are set through a control panel, which can be in a separate room, inside the X-ray room, or (more recently) directly on the X-ray tube – see Figure 5.7.

5.2.5 X-Ray Stand

The X-ray stand holds the digital imaging detector and keeps it at a fixed distance from the X-ray tube – Figure 5.7. Usually, the detector is mounted beneath the patient table or in a vertical (chest) stand. The table and vertical stand can be moved in different directions in order to allow easy positioning of the patient and magnification of the radiographic image. Most radiographic stands provide an X-ray image which is magnified by a factor of about 1.2.

5.3 X-RAY IMAGE FORMATION

5.3.1 X-Ray Attenuation and Latent Image

When a beam of X-ray photons passes through an object, part of it is attenuated or scattered by the structures within the object, so that the exiting X-ray beam is transformed or *modulated* (Figure 5.7). As an example in chest X-ray radiography (Figure 5.8), a rectangular X-ray beam with relatively homogenous intensity distribution (per unit area) passes through the chest. It will exit with variable intensity following absorption and scatter of the X-ray photons by anatomical structures – the *Latent Image*. The latent image is visualised by the detector which transforms the variation of

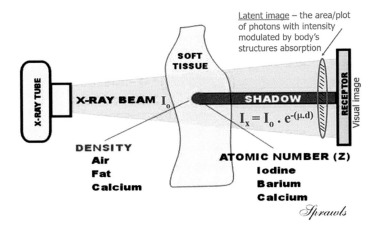

FIGURE 5.8   Formation of radiographic image.

(Courtesy of www.sprawls.com.)

X-ray intensities into variation of visible light (e.g. on a diagnostic monitor). This is the main image formation process in X-ray radiography and fluoroscopy. Image formation in X-ray Computed Tomography is different and will be discussed later.

Once the X-ray radiograph is formed it can be recorded on a digital detector (digital X-ray medical image) or on X-ray radiographic film (now largely superseded).

X-ray radiographic (photographic) film was the original X-ray detector recording X-ray intensity differences across the latent image. To enhance the photographic effect, X-ray film is placed inside a cassette incorporating phosphor plaques (intensifying screens). X-ray photons passing through these phosphors create light photons increasing the exposure of the X-ray film. This allows the radiograph to be acquired using a lower X-ray intensity, thus less radiation dose to the patient. Since the late 1980s, digital X-ray detectors have gradually replaced X-ray film.

5.3.2   X-Ray Image Contrast and Attenuation in Anatomical Structures (Tissues)

X-ray beams passing through objects (anatomical structures) are attenuated due to absorption and scattering. Scattered X-ray photons reaching the detector do not carry any information about the object, they merely degrade the resulting image. Only X-ray photons which are attenuated carry useful information about the object.

The X-ray attenuation properties of anatomical tissues modulate the intensity of the X-ray photons which penetrate them. These properties are described by their *attenuation coefficient* (μ).

The attenuation of X-ray photons is an exponential process, represented by the equation:

$$\mathbf{I} = \mathbf{I_o e^{-\mu d}}$$

where
  I_o is the initial beam intensity
  I is the final beam intensity
  d is the distance travelled by the beam (i.e. the object thickness)
  μ is the attenuation coefficient of the object (the attenuation of the tissue/material).

The *linear attenuation coefficient* represents the fraction of photons interacting per unit thickness of material (measured in m^{-1}). The mass attenuation coefficient is the rate of photon interactions per unit mass (cm^2/g). These coefficients are connected through density:

Mass Attenuation Coefficient = Linear Attenuation Coefficient / Density

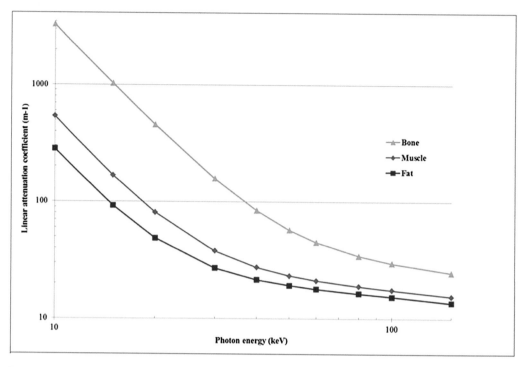

FIGURE 5.9 Dependence of the linear attenuation coefficient of various tissues on the photon energy.

The attenuation properties of different tissues result in differences in X-ray photon absorption, hence generating radiographic image contrast. Attenuation properties of different materials are energy dependent (Figure 5.9)

The figure shows that using photon energies of 100 kV will result in a significant difference between μ of bone and muscle and this will result in an image with good contrast between these two tissues. However, this photon energy will not produce good contrast between muscle and fat, and in order to have contrast (greater differential attenuation) between these two tissues, the photon energy has to be lower. X-ray photon energy is therefore an important consideration in image contrast.

As discussed in Section 5.2.1, the X-ray spectrum is polyenergetic. Most of the low energy photons produced by the X-ray tube do not reach the detector, being absorbed by the tissues. These photons increase patient dose but do not contribute to the image formation. Special metal filters are placed inside the X-ray tube housing aiming to absorb the low energy photons. For general radiography, the filters are made from aluminium. Molybdenum and rhodium are the two most common filter materials used in mammography. Figure 5.2 shows the amount of radiation removed by the filter.

An X-ray spectrum with low energy photons will produce a radiograph with high contrast. This is useful in mammography, where we are looking for contrast between tissues with very similar attenuation properties.

5.3.3 Scattered Radiation and Contrast

As discussed in Section 5.3.2, scattered photons which reach the detector do not contribute to the creation of a useful radiograph. These scattered photons (also known as "secondary radiation") are in a direction different from that of the photons from the X-ray tube ("primary radiation"). Thus the scattered photons degrade (blur) the image – Figure 5.10.

The amount of scattered radiation in the image is dependent on the primary X-ray beam energy. This is described by the Scatter-to-Primary (S/P) ratio. When the initial beam energy is of the order of 100 keV, the S/P ratio may be as high as 5. However, with lower energy beams, for example in breast imaging, the ratio is much lower, around 0.5.

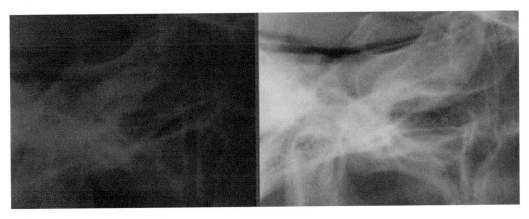

FIGURE 5.10 Radiography of bone structures inside the skull – on the left the radiograph includes both the absorbed primary radiation and the scattered secondary radiation; on the right – same with anti-scatter grid.

The amount of contrast lost due to scattered radiation can be described by the Contrast Degradation Factor (CDF), which is always less than 1:

$$CDF = 1/(1 + S/P)$$

In order to reduce scattered radiation a special device, the anti-scatter grid is placed between the patient and the detector. These grids have parallel strips of a highly attenuating material (e.g. lead) with the spaces between the strips filled with low attenuating materials. X-ray photons incident normally will pass through the grid whilst scattered photons will be absorbed in the lead strips. Different types of anti-scatter grids are produced with strips of different height and spacing. These two parameters form a descriptor of the grid known as Grid Ratio (t/D) – see Figure 5.11. The higher the grid ratio (i.e. the denser the grid), the less scattered radiation will pass through it. Using anti-scatter grids improves image quality (the contrast between the examined anatomical structures), but also increases the patient dose, as absorption in the grid itself reduces the X-ray beam incident on the detector requiring higher primary beam intensities.

Anti-scatter grids are used both in radiography and fluoroscopy.

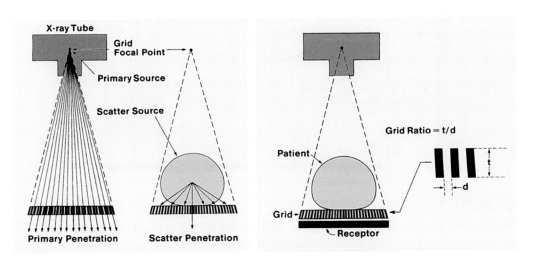

FIGURE 5.11 Anti scatter grid preventing scatter radiation reaching the detector and (right) grid ratio.

(Courtesy of www.sprawls.org.)

5.3.4 CONTRAST MEDIA

Differential attenuation in anatomical tissues creates image contrast. Visualisation can be increased using contrast media introduced into the body. Contrast media are of two main types:

1. Positive contrast media – most frequently used, increases contrast by introducing substances with high attenuation coefficients. For example, a barium meal is used to investigate the gastrointestinal tract, or iodine is injected in the blood stream to visualise blood vessels for example when performing investigations of the heart.
2. Negative contrast media – similarly increases contrast but by using substances with very low attenuation coefficients – for example, gas, air.

5.3.5 GEOMETRY OF THE X-RAY BEAM AND SPATIAL RESOLUTION

The relation between the X-ray tube focal spot size and image resolution was discussed in Section 5.2.3. Larger focal spots produce images with greater penumbra resulting in blurring – that is, the image will have geometric unsharpness.

The size of the penumbra also depends on the distances between the focal spot, the object (i.e. the patient) and the detector. As the object is always between the X-ray tube and the detector, the projection ('shadow') of the object will always be enlarged at the detector plane as will the penumbra.

The magnification of the image is defined as (see Figure 5.12):

$$m = \frac{FFD}{FFD - OFD}$$

If the object is directly over the detector its magnification is almost 1, but if it is further away the magnification will be greater, up to 1.5.

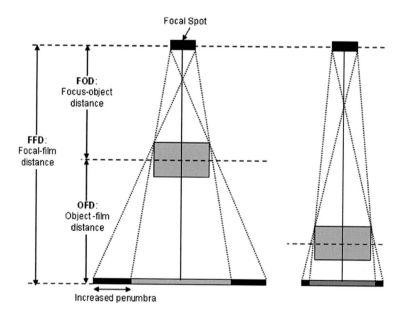

FIGURE 5.12 Magnification and geometric unsharpness of planar radiograph. Note also that the finite size of the focal spot is associated with the presence of penumbra around the image of an object. Large focal spots will have large penumbra. Large magnification (left) further increases this penumbra. The penumbra blurs the edges of the shadow/image of the object and leads to a blurred image – that is, an image with decreased spatial resolution.

(Courtesy of www.spraws.org.)

5.3.6   BASIC X-RAY IMAGE RESOLUTION

Spatial resolution (often simply called 'resolution') describes the ability of the detector (or the imaging system) to 'resolve' (i.e. to see) the separation between two small and closely spaced objects. A high-resolution detector allows smaller elements in the object to be visualised. Spatial resolution is directly linked with blurring. As in any optical system, the blurring of an X-ray imaging system is a specific characteristic of the system. An imaging system (e.g. X-ray radiographic equipment) with high blur will not be able to visualise small objects.

Image resolution can be measured in 'line pairs per millimetre (lp/mm)'. A line pair is one white object and an adjacent identical black object in a resolution bar-phantom test object (Figure 5.14). The spatial frequency (in the space domain) is the number of contrasting lines with equal size (black and white) per unit width – that is, line pairs per mm – a concept similar to positive and negative wave cycles per unit time, measured in Hz. The limiting spatial resolution of a visualising system (detector) is the smallest (in terms of width) line pair that can be seen. Line pairs below the threshold width will be so blurred as to be indistinguishable from their background.

To estimate the limiting spatial resolution of an imaging system, a test object with differing spatial frequencies (Figure 5.14) is imaged. The smallest set of line pairs observed determines the limiting value. For example, if the limiting spatial resolution of a system is 4 lp/mm, the smallest detail that could be seen is 1/[2*4] = 0.125 mm.

The best quantitative descriptor of spatial resolution is the Modulation Transfer Function (MTF) – an overarching image parameter related to the effect of the imaging system on the object being imaged, that is, how the difference of tissue attenuation will be transferred into a difference in image brightness (contrast) on the radiograph. MTF represents the ratio between the output modulation and the input modulation. MTF, which is frequency dependent, quantifies the change of signal amplitude as a result of the imaging process. By definition, the Modulation Transfer Function (MTF) is

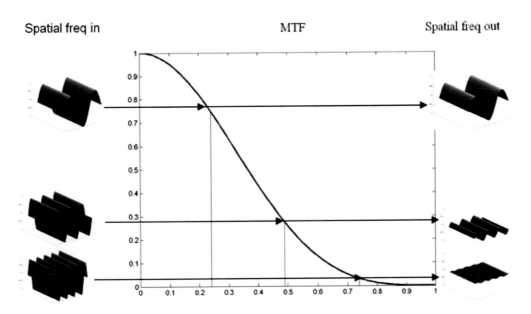

FIGURE 5.13   A real MTF showing the decrease of the input spatial frequency amplitude after this signal is visualised with the imaging system – that is, the radiographic image will preserve only low spatial frequency information, while thinner objects will be blurred.

(Courtesy of www.emitel2.eu.)

represented by the modulus of the Fourier Transformation (FT) of the Line Spread Function (LSF) or Point Spread Function (PSF) of the image of a line or point:

$$\mathrm{MTF}(f) = \left| \mathrm{FT}\left\{ \mathrm{LSF}(x) \right\} \right|$$

In other words – the image of a point will include some blur – that is, its image will not be sharp, but spread (hence PSF). Fourier transformation of the signal from an object composed of a range of frequencies will show how much the signal amplitude decreases as spatial frequency increases. At a certain high spatial frequency (limiting frequency), the signal amplitude (i.e. the visual contrast) becomes so small, that it is no longer distinguishable – Figure 5.13. A good X-ray imaging system will have a limiting spatial resolution of the order of 8–10 lp/mm.

Contrast resolution represents the ability of an imaging system to resolve small differences in object contrast – that is, the ability to resolve (visualise) objects with similar attenuation coefficients. Contrast resolution is limited mainly by noise in the imaging system. When the amplitude of the noise becomes similar to the amplitude of the signal, the observer cannot distinguish between them (Figure 5.15).

In X-ray imaging, noise is the statistically random distribution of photons received by the detector. It is observed over the radiograph as granularity in the image of a homogenous object. Sources of noise include the source of radiation; the parameters of the detector; the digital image matrix and/or image processing and electrical components in the equipment. High noise levels lead to low contrast (Figure 5.15).

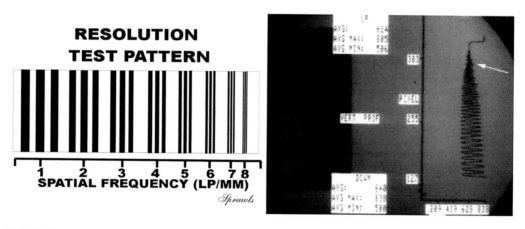

FIGURE 5.14 A test pattern with varying spatial frequency (measured in lp/mm) and radiograph of such a test object (right) showing the decreased signal (amplitude of contrast) of the test objects with increasing spatial frequency. The arrow shows the point when the contrast becomes zero and the signal (the image of the test object) cannot be visualised anymore – the so-called limiting spatial frequency.

(Courtesy of Left: www.sprawles.org; Right: www.emerald2.eu.)

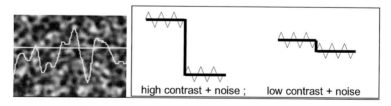

FIGURE 5.15 Left: Magnified amplitudes of noise (granularity) over a digital image of a homogeneous object (water) and Right: comparison of contrast amplitude and noise amplitude – note how in low-contrast object the noise amplitude decreases visual detection of the actual contrast.

Another general term used in imaging is contrast scale. As the X-ray radiograph is usually a black and white image, contrast scale is the number of visibly distinguishable shades of grey in the radiographic image.

Digital images are usually presented using a grey (contrast) scale with up to 4096 levels. As the human eye cannot distinguish more than approximately 200 grey levels, visualisation of image details requires 'windowing' (explained below in relation to CT images).

Contrast resolution can be described quantitatively by the SNR (Signal to Noise Ratio) or NPS (Noise Power Spectrum) of the system. These parameters are particularly suitable for measurements in digital imaging systems.

Another important parameter of spatial resolution in digital images is the matrix size of the radiograph (e.g. 1024×1024 vs. the more detailed 2048×2048 matrix).

5.4 X-RAY IMAGING METHODS AND THEIR APPLICATION IN MEDICINE

5.4.1 RADIOGRAPHY

The oldest and the most frequently used X-ray imaging method in a wide variety of applications is (plain film) radiography. The radiograph is essentially a static projected image of the attenuation pattern of anatomical structures. X-ray exposures are of the order of tens to hundreds of milliseconds at photon beam energies in the range 50 – 150 kV. Specific radiographic equipment exist for imaging specific anatomical structures.

X-ray Imaging Detectors are mainly of two types:

- Computed Radiography (CR) detectors using photostimulable phosphors;
- Digital radiography detectors – flat panel detectors (FPDs) using semiconductors. There are two types – with indirect and direct conversion of the latent image into a digital signal.

5.4.1.1 Computed Radiography

Computed Radiography (CR) was the first digital radiographic technology introduced into healthcare in the 1980s.

CR uses photostimulable phosphor (PSP) materials as the basis for image formation. PSP materials used in commercial CR readers are all barium fluoro-halide materials doped with europium (BaFHa:Eu). The relative proportions of fluorine and other halide elements (typically bromine and iodine) vary from manufacturer to manufacturer. The presence of barium and the heavier halide elements means that diagnostic energy X-ray photons will interact strongly with the PSP material, chiefly via the photo-electric effect.

Before exposure the europium is in the form of the Eu^{2+} ion. The presence of radiation leads to a free electron being released and the europium becoming Eu^{3+}. The number of free electrons created is proportional to the radiation intensity. These migrate to energy traps in the BaFHa between the conduction and valence bands thus preventing the recombination of a small proportion of the electron-hole pairs. The X-ray latent image in the detector is formed by the number of trapped electrons. If there is no additional external energy these electrons can stay trapped for a long period storing the 'electronic' latent image in the CR plate (Figure 5.16).

The release of electrons from these energy traps can be stimulated using external energy (either heat or light – e.g. laser). When electrons escape the energy traps, they recombine with holes releasing energy in the form of light photons at discrete energies characteristic of the energy difference between the valence band and the energy trap. The CR reader uses a laser to release the trapped electrons (photo-stimulated luminescence). The more electrons trapped in one region of the CR cassette, the more intense the light produced by this region. A light guide connected to a photomultiplier tube scans and samples the light released and converts it to an electrical signal forming the latent X-ray image which is digitised, displayed and stored. The rate of sampling determines pixel size in the image (Figure 5.17).

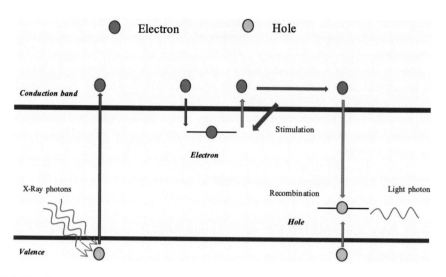

FIGURE 5.16   Energy level diagram for PSP material.

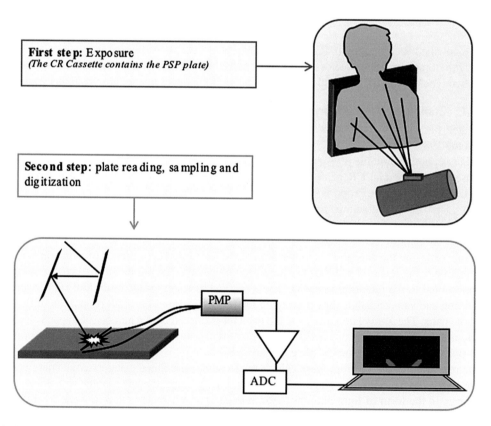

FIGURE 5.17   Main stages of the process of CR digital radiography (PMP – photomultiplier, ADC – analogue to digital converter).

Once electrons are released from the traps, the europium returns to the Eu^{2+} state although before re-use the CR plate is exposed to a strong light to ensure it has been fully 'erased'. CR plates can be put through several tens of thousands of exposure/erasure cycles before they need replacement.

Digital images can be post-processed to enhance the quality of the image, digitally recorded and visualised on a monitor.

5.4.1.2 Indirect Imaging Detectors for Radiography

Unlike CR, FPDs are typically a component part of the X-ray equipment. There are two types of FPD using indirect or direct X-ray photon detection.

Indirect FPDs use a phosphor layer which converts the energy from individual X-ray photons into visible light photons via the process of fluorescence (as described in the section on fluoroscopic systems). A matrix of photodetectors (photodiodes) is then used to capture the fluorescent light and transfer it into an electrical signal which is used to form the image matrix. The phosphors used are typically either caesium iodide doped with thallium (CsI:Tl), or gadolinium oxysulphide (Gd_2O_2S sometimes referred to as GOS). The former has the advantage that it can be grown in a columnar structured phosphor material. This structure channels the fluorescent light in a manner similar to a fibre optic cable and consequently limits the lateral spread of the light away from the point of photon interaction thus improving the resolution properties of the phosphor. The main components of an indirect X-ray imaging detector are shown in Figure 5.18.

When light is incident on the phosphor layer, it creates electron hole pairs in the photodiode. This charge is then stored on a capacitor associated with each pixel. The amount of charge stored on the capacitor will be determined by the incident X-ray flux and, in turn, the flux of light photons created through fluorescence. A TFT element controls storage and readout of the charge on the capacitor. All three elements – photodiode, capacitor and TFT – form one pixel under the phosphor.

The detector architecture is such that all the capacitive elements in a single row of the detector can be read out simultaneously along separate data lines – see Figure 5.19. Each data line is connected to a separate amplifier and analogue to digital converter. After one row of the detector is read out, the transistor switches in the line are closed and the next data line is read. Detector signals are then transferred to an image memory matrix with the same number of pixels as the detectors. The charge stored in the capacitor of each pixel is digitised and displayed on the diagnostic monitor with respective brightness.

Imaging systems with FPDs have three pixel types:

- Pixels of the FPD (the photodiodes with capacitors)
- Pixels in the image memory (containing numbers corresponding to the charge in the capacitor)
- Pixels of the digital monitor screen (displaying brightness corresponding to the image memory matrix).

Pixels in these three matrixes have identical coordinates. The brightness of pixels in the displayed image is determined by the charge stored in the individual capacitors which is determined by the

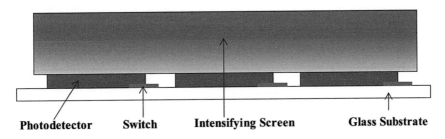

Photodetector   **Switch**   **Intensifying Screen**   **Glass Substrate**

FIGURE 5.18   The main components of an indirect DR system.

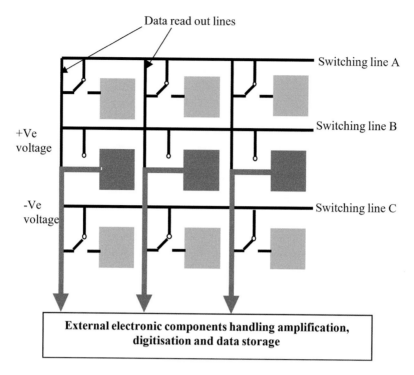

FIGURE 5.19   Separate pixels in the DR structure.

X-ray flux incident on them. Various image processing methods and a system of windowing are used to manipulate the image contrast.

After reading, the capacitors are discharged and the detector is ready to be used again. An FPD is capable of producing 30 frames in 1 second.

5.4.1.3   Direct Imaging Detectors for Radiography

Direct capture detectors do not require a phosphor to initially capture the X-ray photon and convert it to light. Instead they use an amorphous selenium photodiode/TFT (thin film transistor) array – Figure 5.20. This is a photodiode array similar to that used by indirect imaging detector systems, but with amorphous selenium replacing the phosphor. Selenium has an atomic number of 34 compared to 14 for silicon. Consequently, diagnostic energy X-ray photons interact much more strongly with selenium, creating electron hole pairs which can be stored on the pixel capacitors and read out in a similar manner to amorphous silicon photodiodes.

Initially the amorphous selenium is charged using a high voltage (of the order of thousands of volts). When the charged plate is exposed to X-rays, the electron-hole pairs produced discharge

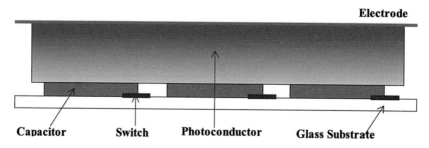

FIGURE 5.20   The main components of a Direct DR system.

the exposed area. The percentage discharge is proportional to the intensity of radiation producing a 'voltage' latent image over the amorphous selenium plate. The charge pattern (latent image) on the detector is read out using a TFT array and transferred, pixel by pixel, to the image memory and displayed on the diagnostic monitor. Again, various image processing methods and windowing are used to manipulate the contrast of the image.

After reading, the amorphous selenium plate is electrically 'erased' and charged again for the next exposure.

5.4.2 Mammography

X-ray imaging of the breast, referred to as 'mammography', is a fundamental aid in the early detection and diagnosis of breast diseases in women.

Breasts are composed of glandular and adipose tissues, in a percentage depending on the age and physical features of the woman. The presence of masses, microcalcifications, asymmetric densities and/or architectural distortions in the image may indicate disease. The capability to distinguish between very small differences in physical density (masses) and the visibility of very small high contrast objects (microcalcifications) is important in mammography.

The breast is a radiation-sensitive organ and it is also important to minimise patient dose in mammography.

5.4.2.1 Mammographic Equipment

Since 1965, dedicated X-ray equipment has been used to obtain breast images. Over time, various technical improvements have been introduced to optimise image quality and reduce dose. The essential features of a breast radiography system are described in Figure 5.21.

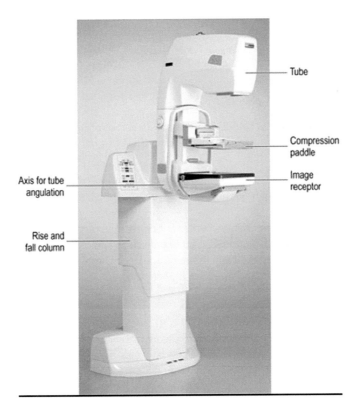

FIGURE 5.21   Typical mammographic equipment.

A mammography unit consists of an X-ray tube combined with an image receptor on a C-shaped arm. The focus-detector system distance is fixed (usually around 65 cm) and the tube assembly can be rotated and adjusted for height to suit individual patients.

Rotation allows the acquisition of oblique or lateral views. Optimal imaging requires the breast to be compressed so the system also incorporates a compression paddle.

5.4.2.2   X-Ray Tube for Mammography

To enhance the visibility of low contrast details, the X-ray tube of a mammographic unit operates at low voltages (typical range: 24–35 kV). Furthermore, X-ray spectral shaping is required to obtain high contrast sensitivity. As the technique of mammography has developed, different solutions have been proposed by various manufacturers. The first fundamental improvement was the use of molybdenum both as an X-ray tube anode material and X-ray beam filter. Due to the atomic number of this material ($Z = 42$), the anode produces characteristic X-ray peaks at 17.6 keV and 19.7 keV and the filter, with an attenuation K edge at 20 keV, absorbs most of the hard X-rays. This combination produces an X-ray spectrum with a peak around 20 keV which is the optimal energy to visualise low contrast details at reasonable patient radiation dose, particular for smaller breasts.

To optimise image quality for larger and denser breasts, equipment with two anode targets (dual track X-ray tube) and two filters, using molybdenum and rhodium, respectively, was developed. Depending on the composition and the thickness of the breast, different target/filter combinations are used:

- Mo target and 30 micron Mo filter, for fatty breasts up to 4 cm thick;
- Mo target and 25 micron Rh filter, for glandular breasts from 5 to 7 cm thick;
- Rh target and 25 micron Rh filter, for breast thicknesses above 7 cm.

With the advent of direct digital detectors, different technical solutions are used. Instead of dual track X-ray tubes, modern equipment uses a tungsten anode with rhodium or silver filter materials.

To minimise blurring and detect small calcifications, the focal spot should be as small as possible. Typically, an X-ray tube for mammography has two selectable focal spots with nominal dimensions of 0.3 mm or less for large focus (LF) and 0.15 mm or less for small focus (SF), the latter used for magnified images.

FPD mammography equipment uses the detector as an automatic exposure control (AEC) sensor. This provides some flexibility to operate with differing SNR (Signal to Noise Ratio) values.

5.4.2.3   X-Ray Beam Geometry in Mammography

In comparison to conventional radiography, only about half the beam in mammography is used to form the image.

Due to the anode heel effect, the X-ray beam is not uniform in the direction parallel to the anode-cathode axis of the X-ray tube. This effect is used in mammography to improve image homogeneity. The breast is thicker near the chest wall and thinner towards the nipple. By aligning the cathode over the chest wall and the anode over the nipple, a higher beam intensity is available to penetrate the thicker area and vice versa.

5.4.2.4   Compression Device

The presence of a motorised compression device is a key feature of mammographic equipment. With compression, breast tissue thickness is reduced and becomes more uniform. At the same time, this minimises motion and geometric unsharpness and decreases scatter and beam hardening effects. Whilst essential for optimising image quality and reducing patient dose, it is uncomfortable for the patient.

To compress the breast, a paddle is placed on the breast and a force of between 100 and 150 N is applied. Systems are generally equipped with different sized compression paddles to match the sizes of all full-field image receptors provided for the system and also to enable 'spot compression'.

5.4.2.5   Image Detector

As with other radiographic equipment, contemporary mammographic equipment mainly uses either photostimulable phosphor systems (CR) or FPDs.

The FPD, using direct or indirect conversion, is integrated into the mammography equipment and the final image can be viewed on the console within a few seconds. Detectors for mammography have higher resolution capabilities than those used in normal radiographic equipment.

5.4.2.6   Screening and Diagnostic Mammography

Screening mammography is performed on asymptomatic women. This is to detect cancers at an early stage with the aim of reducing breast cancer mortality. A screening investigation requires two exposures of each breast: a cranial-caudal view (CC) and an oblique or angled view (mediolateral-oblique, MLO). Diagnostic mammography is performed on women who present with symptoms requiring investigation. In diagnostic mammography, supplemental and special views may also be used (spot compression, magnification, etc).

5.4.3   FLUOROSCOPY

Fluoroscopy is an imaging method used to produce dynamic images of anatomical structures (e.g. of the heart, intestines, etc). This requires the use of a different type of detector, originally an image intensifier connected to a video camera. These have been replaced in contemporary systems with digital (flat panel) detectors. Fluoroscopy equipment also allows still images (spot radiographs) to be acquired during the investigation.

Compared with radiography, fluoroscopic investigations are associated with higher patient dose. Although fluoroscopic images are acquired with significantly lower tube currents than conventional (still) radiographs, the duration of the exposure is significantly longer.

Fluoroscopy is used in a wide variety of investigations and interventional procedures to diagnose or treat conditions.

Common procedures using fluoroscopy include:

- study of the gastrointestinal tract to diagnose structural or functional abnormalities;
- catheter insertion and manipulation, for example, in bile ducts, in the urinary system or in the stomach (gastrostomy);
- orthopaedic surgery (during bone fracture treatment and guiding the placement of implants);
- interventional radiology treatment (angiography and angioplasty of cerebral, lung, abdominal and peripheral vessels, cancer treatment, embolism, vertebroplasty, etc);
- haemodynamic procedures (coronary angiography and angioplasty with stent placement).

The majority of fluoroscopic procedures require the use of contrast media (see Section 5.3.4), to enhance the visibility of internal structures. Iodinated contrast is the most common choice for blood vessels and for studies of the urinary tract whilst barium sulphate is mainly used for imaging the digestive system. The effect of these substances is to increase the radiographic contrast between the area containing the agent and the surrounding tissue (see Figure 5.22 below).

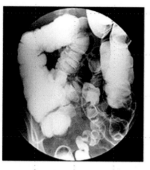
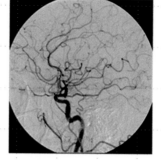
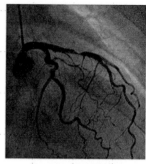

Barium enema   Cerebral Vessels   Coronary arteries

FIGURE 5.22   Some typical fluoroscopic examinations: a – gastrointestinal fluoroscopy with barium contrast; b - angiography of cerebral vessels with iodine contrast – this image is made with a digital detector using 'digital subtraction angiography' (an image processing method minimising image of bones); c – fluoroscopy of the heart with iodine contrast (angiography).

5.4.3.1   Fluoroscopic Equipment

Different fluoroscopic modalities are available with each configuration optimised for a specific clinical task. The most common (Figure 5.23 below) are;

- remote-controlled fluoroscopy systems with the X-ray tube above (A), or below the X-ray table, mainly used for gastrointestinal imaging;
- mobile C-Arm X-ray device (B), typically employed during surgery
- fixed fluoroscopy systems (C), sophisticated equipment used in interventional procedures.

Components common to all fluoroscopic devices are an X-ray generator, X-ray tube, collimation device, filtration, anti-scatter grid, image receptor system (image intensifier or digital detector), automatic exposure control and a monitor to visualise the images.

The X-ray generator and tube used in fluoroscopy systems do not differ substantially from those used in general radiography systems. However, systems dedicated to interventional radiology and haemodynamic procedures require a high heat capacity tube to enable prolonged exposures. The X-ray beam output can be either continuous or pulsed.

To reduce entrance skin dose, aluminium and/or copper filters may be added. Depending on the system, the added filtration can be automatic or user-selectable.

Except for paediatric or very thin patients, anti-scatter grids are used to minimise the negative effect of scattered radiation.

The two main image receptors used in fluoroscopy are the X-ray Image Intensifier and the Digital Detector (FPD).

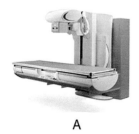
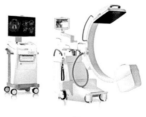
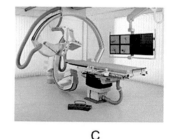

A   B   C

FIGURE 5.23   Main types of fluoroscopic equipment – see text.

5.4.3.2   X-Ray Image Intensifier

The main elements of an X-ray Image Intensifier (II) are shown in Figure 5.24.

The input phosphor, commonly made of caesium iodide (CsI) crystals, converts the X-ray beam into light photons. These photons strike a photocathode, placed very close to the input phosphor, and photoelectrons are released. A high potential difference (25–30 kV) between the anode and the cathode accelerates the electron beam through focusing electrodes. When this accelerated beam of photoelectrons reaches the output phosphor, the energy of electrons is converted into light. The light image from the output phosphor passes through a glass window and via an optical system is transferred to the TV camera tube (and from the TV tube to the diagnostic TV monitor). In modern systems, the camera image is digitised.

This process achieves significant intensification of the initial signal through the electronic gain (obtained from acceleration) and the minification gain (the signal from a large input area is concentrated on a small output area). Total intensification is of the order of several thousand times.

The parameter used to express gain is the conversion factor. This is defined as the ratio of luminance at the output phosphor to the incident X-ray air kerma rate at the input phosphor (typical range for conversion factor values: 9–27 $cd \cdot m^{-2}/\mu Gy \cdot s^{-1}$).

The electrical potential across the intensifier is approximately 30 kV. The equipotential lines of the accelerating electric field form an 'electric lens' inside the II, which focuses the beam of photoelectrons on the output phosphor. External electromagnetic fields can distort the 'electric lens' and hence the image thus the intensifier must be shielded (conventionally with mu-metal).

The input surface of image intensifiers is always circular, with diameters ranging from about 10–15 cm up to 40 cm. The optimal II dimension depends on the clinical procedures being performed. In modern systems, FPDs have replaced image intensifiers.

5.4.3.3   Fluoroscopic Flat Panel Detectors

Fluoroscopic FPDs principally use indirect conversion as described above and operate at frame rates up to 30 frames per second.

The input surface of FPDs is not circular but square or rectangular. Typical values range from 20×20 cm up to 40×40 cm. Detector dimensions are quoted as the edge or diagonal length.

FPDs provide various advantages over II:

- the imaging chain is less bulky, allowing more flexible movement during procedures;
- DQE (Detective Quantum Efficiency) is higher, and it is possible to obtain the same image quality with reduced dose, or improved image quality at the same dose level;
- Wider dynamic range, and consequently higher flexibility in the optimisation process;

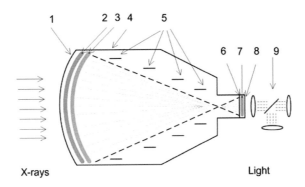

FIGURE 5.24   Block diagram of main component of an Image Intensifier (II); 1 – II entrance window; 2-input phosphor; 3-photocathode; 4-mu metal shielding; 5-focusing electrodes; 6-thin Al plate; 7-output phosphor; 8-output window; 9-optical system linking to TV camera or similar.

- Absence of the image distortion typically seen in II (pincushion S-wave distortion, vignetting, veiling glare);
- 3D imaging capabilities (using cone beam computed tomography – CBCT);
- Digital magnification to reduce dose (e.g. in conventional image intensifiers, decreasing the imaging field of view by a factor of two produces a zoomed image, but increases the dose rate by a factor of four).

5.4.3.4 Fluoroscopic AEC

Fluoroscopic equipment always operates under automatic exposure control (sometimes called automatic brightness control). The AEC is designed to maintain the dose rate to the image receptor, irrespective of patient thickness.

A basic fluoroscopic AEC system varies the kV and tube current in a manner programmed by the manufacturer. There may be a number of AEC programmes to permit some operator control, for example, to improve contrast or reduce dose.

In more complex systems, other parameters of the exposure may be included in the AEC programme, for example, controlling pulse length and introducing filtration.

5.4.3.5 Radiation Safety Issues in Fluoroscopic Examinations

Fluoroscopic procedures deliver the second highest patient doses in diagnostic radiology (after computed tomography). Usually this is associated with the long duration of these procedures. High patient doses may also lead to high staff doses, particularly to the eyes and extremities of operators. In rare cases, deterministic radiation effects to the skin of patients have been observed (early transient erythema, permanent epilation, dermal necrosis).

Doses and associated risks can be reduced by:

- adequate equipment selection and a good knowledge of operational modes of the equipment;
- regular quality control of the equipment;
- minimising exposure time, for example, by using the last image hold function;
- minimising the distance between the patient and image detector;
- maximising the distance of the X-ray tube from the patient;
- using image magnification sparingly;
- good collimation;
- varying the angle of exposure to ensure the same area of skin is not constantly in the X-ray beam.

5.4.4 DUAL ENERGY IMAGING AND BONE ABSORPTIOMETRY

A radiographic image is essentially a map of the attenuation coefficients of the tissues being imaged. Attenuation is energy dependent and so an image of the same structure taken with X-ray beams of different energies will produce images with quite different appearances. Subtracting one image from the other (often weighted linear or logarithmic subtraction) will serve to improve the visualisation of structures in the image.

Such a technique could be utilised in the chest, for example, to visualise lung tissue with the overlying rib structure suppressed. Dual energy image acquisition can be performed sequentially using a low energy beam (60 kV) followed by a high energy beam (120 kV) but this has the potential disadvantage of motion artefacts. Special 'stacked' FPDs are available to acquire dual beam images simultaneously. The two detectors have a copper filter placed between them. The first plate acquires an image using all the X-ray energies in the beam whilst the second plate forms an image from only the higher X-ray energies in the beam.

Dual energy imaging can also be used in CT scanning to improve tissue differentiation in the image. Usually this is achieved by scanning with 80 kV and 140 kV but dual-layered detectors can also be used. Another possibility is to acquire a single scan with a pulsed X-ray beam switching between the low and high energies.

Dual energy imaging is used in bone densitometry to quantify bone density (or bone mass), often for the assessment of osteoporosis. In this application, the technique is referred to as Dual Energy X-ray Absorptiometry (DEXA or DXA). Bone mineral density (BMD) is estimated in units of g/cm^2.

The DEXA method is normally used to estimate BMD in the lumbar spine and femoral neck. The method uses two scanning X-ray beams (e.g. 70 kV and 140 kV), produced either by a special X-ray generator, or by moving a metal filter in front of the X-ray tube (e.g. a cerium filter, which has k-edge at 38 kV). The method is very accurate (error approximately 3%) and the patient dose is low (1–3 µSv).

5.5 IMAGE QUALITY IN CR AND FPD SYSTEMS

5.5.1 Uniformity

The uniformity of response of individual pixels in a direct or indirect FPD is generally quite poor. Detectors may often have faulty detector lines or individual malfunctioning pixels, and the uniformity of response from the correctly functioning pixels may be poor.

Image processing can be used to improve uniformity. Initially a dead pixel map is applied in which the data from any malfunctioning pixel is replaced with the average of its neighbours (using manufacturer-specific algorithms). Once the dead pixel map has been created, the FPD is exposed to a uniform X-ray beam and scaling factors are applied to the remaining pixel data to correct for variations in sensitivity. Consequently, within the operational range of receptor doses, multiplicative structural noise associated with detector non-uniformities is reduced.

CR systems are unable to apply sensitivity correction on a pixel-by-pixel basis in the same way. This is partly because corrections would need to be made individually for a large number of CR cassettes, rather than just one detector. More importantly, it is difficult to register the exact position of an individual pixel to the same exact position on a CR plate for successive readings of the same plate. For these reasons, the uniformity of response for CR imaging is generally poorer for CR than DR and the contribution of multiplicative structural noise is greater.

5.5.2 Resolution

The resolution of digital detectors is fundamentally limited by the distance between the centres of adjacent pixels, known as the pixel pitch (p). This is subtly different from the pixel size which can theoretically be smaller than the pixel pitch if the pixels are surrounded by insensitive ('dead') areas used to accommodate the control electronics. At the time of writing, pixel pitches of FPD and CR systems are in the range from 50 µm to 200 µm, depending on their application. Typically, mammography detectors will have smaller pixel pitches than radiographic detectors which may in turn have smaller pixel pitches than fluoroscopy detectors. The maximum possible resolvable spatial frequency (Nyquist frequency) is equal to 1/2p. For example, a 50 µm pixel pitch detector will not be capable of achieving resolutions greater than 10 line pairs per mm.

Whilst the pixel pitch places a fundamental limit on the achievable spatial resolution, there are other factors that further limit the resolution properties of detectors. In CR systems, spread of the laser light when reading the CR plate means that the stimulated light emissions will not emanate from a clearly defined sharp-edged pixel and consequently resolution is degraded. Furthermore, the resolution properties of CR systems are not truly isotropic. Resolution is marginally poorer in the laser scan direction than in the subscan direction. This is a result of the finite time taken for the fluoroscopic light to be emitted from the detector after the laser light of the reader is incident on the plate. As the laser sweeps across the plate in the scan direction, the PMT tube is sampled to extract the data from pixel A. Subsequently the laser continues to sweep across the plate and the PMT is resampled to extract data from adjacent pixel B. When the PMT is sampled for pixel B, it is still receiving some residual light from pixel A as well as the light from pixel B.

For FPD systems with indirect detectors, lateral spread of the fluoroscopic light before it is detected in the photodiode array degrades the resolution properties. For FPD systems with direct detectors, there is very little spread of the electron hole pairs created before the charge reaches the capacitors. Consequently, the resolution properties of direct systems are generally superior to other digital imaging systems.

5.5.3 SENSITIVITY AND NOISE

The key factor determining detector sensitivity is the thickness of the detector element that captures the X-ray photons. Some proportion of the photons incident on a detector will pass straight through it, meaning that sensitivity is less than 100%. Designing detectors with thicker X-ray capturing elements decreases the proportion of X-ray photons that pass straight through the detector and therefore improves the sensitivity. However, increasing the thickness of the detector will come at the cost of degrading the resolution properties, as a thicker detector will allow for more spread of fluoroscopic light in an indirect FPD system or more spread of the laser light in a CR system.

The main type of noise to consider in CR and FPD systems is the primary quantum noise. This is the Poisson noise associated with statistical fluctuations in the number of X-ray photons interacting with each pixel. This noise increases in proportion to the square root of the photon intensity incident on the detector and therefore increases proportionally with the square root of the incident air kerma reaching the detector. The signal from the detector can be expected to increase linearly with increasing detector dose although image processing may mean that a different response is sometimes seen in practice. Therefore, if primary quantum noise is the only noise source, then the signal to noise ratio can be expected to increase in proportion to the square root of the incident air kerma.

In practice, at very low detector doses in FPD systems, electronic noise may be the dominant noise source, or at least a significant contributor to image noise. This may be particularly important when considering FPD detectors for fluoroscopy that are required to operate at very low input dose rates.

A detector generally provides a linear response for only a limited range of dose values, known as the dynamic range (Figure 5.25). A key advantage of digital detectors (both CR and flat panel) is their wide dynamic range – that is, the ability to produce images from widely varying X-ray fluences. This is markedly different from the earlier analogue film-screen systems which operated over a restricted dynamic range.

X-ray fluences outside the limited dynamic range of film screen systems would result in unusable images either under- or over-exposed. This problem is virtually non-existent with digital detectors but there is the disadvantage that unnecessarily high/sub-optimal X-ray fluences, resulting in higher than necessary patient doses, may not immediately be apparent.

5.5.4 ARTEFACTS

All imaging systems may suffer from false image representations – artefacts. CR is particularly susceptible to artefacts (well documented in the literature). Many of these result from mechanical damage to the plates from the readout process causing scratching or cracking. Dust on the plates or reader is also a very common cause of artefacts, requiring regular cleaning to reduce the problem.

FPD systems also suffer from artefacts. For example, a large FPD detector is often constructed from several small FPD detectors. Misalignment between the 'stitched' small detectors, or a defect in one of the elements, are potential sources of artefacts.

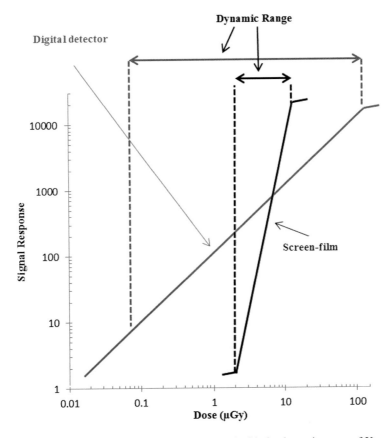

FIGURE 5.25   Dynamic range of an FPD detector, compared with the dynamic range of X-ray film.

5.6   COMPUTED TOMOGRAPHY (CT) SCANNING

Computed tomography is an X-ray imaging method, invented in mid-20th century, which allows the visualisation of cross-sectional images (also known as 'slices') of anatomical structures. This revolutionary new method resolved the main problem of radiography – the overlapping 'shadows' of anatomical structures above or below the investigated organ (e.g. ribs over the image of the heart). CT is based on collecting radiographic 'projections' at specific angles around the object, and using the data from these projections to calculate the elements (pixels) of a new image (the CT scan). The calculation process is based on the mathematical method of image reconstruction from projections. According to this method (developed initially by J. Radon in 1917), the projections around an object can be used to calculate the inner parts of this object (i.e. to reconstruct an image of the object from its projections). The method was first applied in medicine by Alan Cormac, while the first CT scanner was invented by Godfrey Hounsfield (with both receiving the Nobel Prize for Physiology or Medicine in 1979). CT was the first imaging method to use a digital image, that is, an image composed of pixels, each containing digital information about the X-ray attenuation of a specific part of the 'slice'. The calculation of pixel values in the 'slice' is very complex and requires a computer, hence the name of the method computed tomography (the Greek word 'tomos' means 'slice' or 'cut'). The CT 'backprojection reconstruction' method has subsequently been applied to various other tomographic methods in imaging and forms the basis of SPECT, PET etc.

5.6.1 INTRODUCTION TO CT

The human body is a three-dimensional distribution of various tissue types, each with its own X-ray attenuation properties. When acquiring a conventional planar radiograph, the three-dimensional structure of the body is reduced to a flat, two-dimensional representation. There is no depth information in a planar radiograph, making it difficult to determine where a structure lies within the patient. Perhaps more importantly, tissue contrast is reduced in a planar radiograph due to the superposition of other tissues above and below the region of interest (Figures 5.26 and 5.27).

CT images maintain the three-dimensional tissue distribution within the patient and apply a specific imaging method, 'windowing', which allows selective visualisation of structures with specific X-ray attenuation. As a result, structures can be seen with increased contrast when compared with planar radiographs. This makes CT a useful tool when lesions with small differences in their attenuation properties need to be visualised.

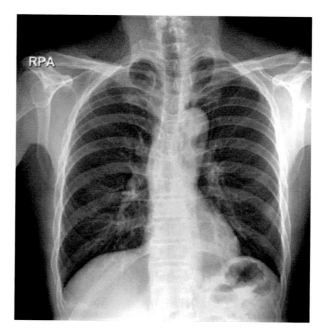

FIGURE 5.26   Typical conventional planar radiograph of the chest.

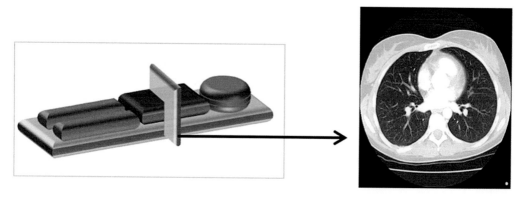

FIGURE 5.27   Scanning plane around the patient (left) and CT chest scan (right).

5.6.2   CONSTRUCTION OF A CT SCANNER

A typical diagnostic CT scanner consists of several key components: a rotating gantry with a circular aperture of around 70 to 90 cm diameter; a moveable couch to enable the patient to be positioned at the correct height and translated through the gantry; an X-ray tube and bank of X-ray detectors positioned opposite to one another on the rotating part of the gantry (Figure 5.28).

In modern CT equipment, slip-ring technology enables the rotating part of the gantry (X-ray tube and detectors) to turn continuously in one direction at high speed whilst supplying electrical power to the X-ray generator and detector bank. At the time of writing, gantry rotation times can be as low as 0.3 seconds for a full 360° rotation around the patient. However, CT scanners normally use longer rotation times (of the order of several seconds) to enable more projections to be obtained (measurements of X-ray attenuation from different angles) resulting in a CT scan image with better resolution. The measurements (signals) generated by the detector bank are also transferred to the computing reconstruction system through the slip-rings, although some CT scanners use an optical system to transfer this data.

X-ray tubes used for CT scanning require a high power generator (usually more than 100 kW). They must be capable of high power output for an extended exposure time. For example, a clinical chest/abdomen/pelvis scan of a large patient may require 500 mA for 15 seconds, equal to 7500 mAs. The X-ray tube heat capacity must also be high to enable a large number of scans to be carried out per day without the need for delays to allow for tube cooling.

Filtration is crucial in the CT scanner X-ray beam. Flat filters of metallic material (aluminium or copper) are used to remove low-energy X-rays. Other, non-flat filters (thinner in the central part), made of aluminium, Teflon or other dense materials with low atomic number, are used for beam shaping (these are known as bow-tie filters). Bow-tie filters compensate for the fact that human bodies in cross section are approximately elliptical in shape (thicker at the middle and thinner at the ends).

Pre-patient collimation defines the incident X-ray beam width in the z-direction (head to foot). In the x-y plane, the X-ray beam is fan-shaped, with a fan angle of approximately 60 degrees. The X-ray beam can be collimated to a range of user-defined widths along the length of the patient (the z-direction) – specifying slice width. Available collimation widths depend on the particular scanner model, but may be in the range of 1 mm to 160 mm (when the detector bank includes multiple rows of detectors), corresponding to a cone angle not larger than 2–3 degrees (Figure 5.29). The angular extent of the beam fan angle is also controlled to optimise the beam dimension at the isocentre: for example, 250 mm for head scanning and 500 mm for body scanning.

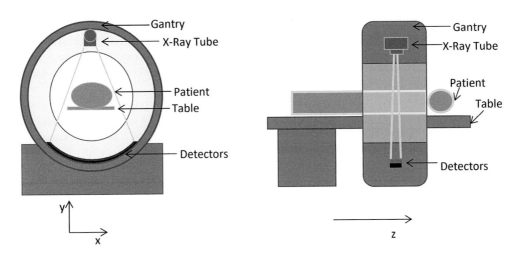

FIGURE 5.28   CT scanner key components.

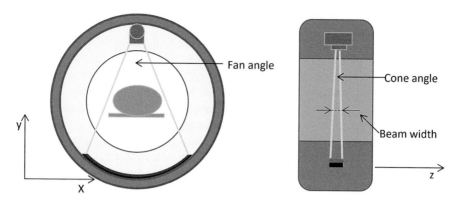

FIGURE 5.29   X-ray beam collimation.

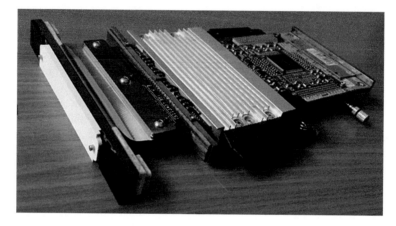

FIGURE 5.30   CT detector with associated electronics.

Detector design in CT is very demanding to ensure high DQE (detective Quantum Efficiency), accurate detection of the signal, wide dynamic range, high spatial resolution and other specialised technical requirements.

Most CT detection systems have the following common elements: a scintillation material (to convert X-rays to light), a photodiode (to convert light to electrical signals), data acquisition electronics and an anti-scatter grid (to reduce the influence of scattered photons on the detectors). Ceramic scintillators are best suited to provide the high performance required of CT detectors with gadolinium oxysulphide (Gd_2O_2S) often used for the purpose (Figure 5.30). Its high density, combined with the high atomic number of gadolinium, makes it an efficient X-ray absorber at the energies used for diagnostic CT scanning.

Most current CT scanners have a detector bank which includes a number of parallel rows of detector elements. The number of detectors in each row varies according to the manufacturer but is often around 1000. The number of detector rows is usually between 16 and 320. One rotation of the CT scanner obtains a number of slices equal to the number of detector rows. Naturally the larger the number of rows, the wider the cone angle of the X-ray beam.

5.6.3   DATA COLLECTION AND IMAGE RECONSTRUCTION

5.6.3.1   Basic Concept

An image of a slice of the body can be considered as a matrix of three-dimensional parallelepipeds (voxels, or volume elements). Conventionally, the slice lies in the x-y plane, and the z axis is the

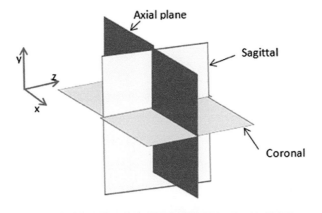

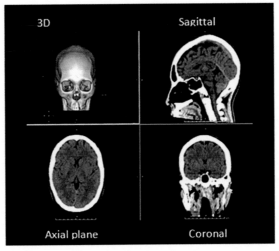

FIGURE 5.31 Axial, sagittal and coronal planes.

longitudinal axis of the patient. The voxel is square in the x-y plane, and presented within the image as a pixel (picture element), while the size of the voxel on the z axis is the slice thickness. The pixel is effectively a memory cell within the image matrix which quantifies the overall X-ray attenuation of the voxel. From axial images, coronal and sagittal planes can also be reconstructed. Volumetric (3D) reconstructions are available on newer CT scanners (Figure 5.31).

Ignoring scatter and considering only absorbed photons, the attenuation of an X-ray beam incident on a uniform object (Figure 5.32) can be defined as the ratio I/I_0, where I is the intensity of the X-rays that pass through the object and I_0 is the original intensity of the X-rays (before passing through the object).

According to Beer's Law:

$$I = I_0 e^{-\mu d}$$

where μ is the linear attenuation coefficient and d is the thickness of the object.

Linear attenuation coefficient is obtained using the equation: $\ln(I/I_0) = -\mu d$

For non-uniform objects (Figure 5.33), the X-ray beam intensity can be evaluated, by multiple application of Beer's Law, as

$$I = I_0 e^{-\mu 1 d1} e^{-\mu 2 d2} e^{-\mu 3 d3} \cdots e^{-\mu n dn}$$

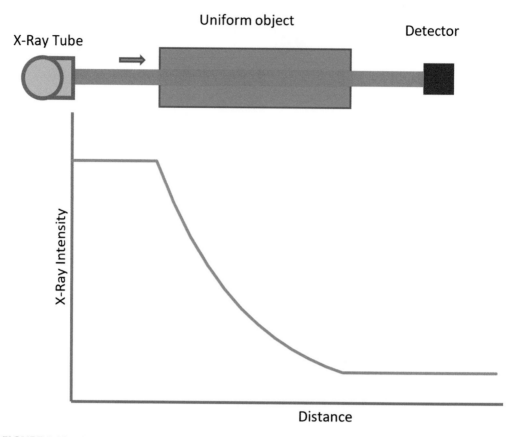

FIGURE 5.32   Attenuation in uniform object.

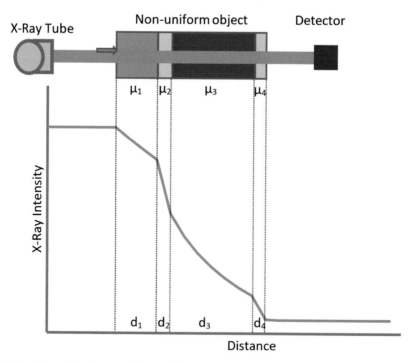

FIGURE 5.33   Attenuation in non-uniform object.

Thus, the intensity of the exit beam, passing through a composite object, is defined as

$$ I = I_0 \exp\left[-\int_0^d \mu\left(s;E\right)ds \right] $$

(s = position; E = effective energy of the beam).

The quantity $\ln\left(\dfrac{I}{I_0}\right)$ represents a line integral of linear attenuation coefficients. Data collection and image reconstruction in CT are based on this relationship.

5.6.3.2   Data Collection

CT backprojection can be illustrated by considering the acquisition of data from a 360° rotation of the X-ray tube and a single row of detector elements around a simple non-uniform object and the subsequent reconstruction of this data into an image. This mode of operation, where the patient is stationary whilst the X-ray tube and detectors rotate about them, is called axial scanning.

The first acquisition is usually made with the X-ray tube at 0° and the detector at 180° (directly opposite each other). For each element of the detector, the ratio of the measured intensity (I) to the intensity of a calibration scan (I_0) is recorded. The line integrals of linear attenuation coefficients $\left(\ln\left(\dfrac{I}{I_0}\right)\right)$ for each element of the detector form an attenuation profile, known as a projection.

To attribute the correct linear attenuation coefficient to each element of the object, measurements of the object's attenuation profile must be made at a number of X-ray tube/detector rotation angles around the object. Clinical scanners may acquire around 1000 projections per rotation.

5.6.3.3   Backprojection

The process of image reconstruction is based on the backprojection algorithm. Expressed non-mathematically, this algorithm is relatively straightforward. Each measured attenuation profile is placed at the edge of the image space at the angle at which it was acquired. Attenuation profile values from every angle are then projected across the image, into every pixel that they cross, and summed for all projection angles. A simple numerical example is described in Figure 5.34.

Figure 5.35 shows an example of backprojection of an object simulating the human head (left) – with backprojection from an increasing number of angles. A rough image of the object starts to appear after using six projections/views around the object. On the lower row, one can see further refinement of the image/scan of the object, based on increasing the number of projections/views (12 to 90). With only 12 projections, one sees significant artefacts but as the number of projections increases, image quality improves and artefacts reduce.

5.6.3.4   Filtered Backprojection

Simple back-projection cannot be used for clinical application as it introduces blurring at every object boundary resulting in a loss of image contrast. Applying a mathematical filter (e.g. convolution filter) to each projection before reconstruction can remove this blurring. The filtering applies a negative offset at every object edge in the projections. The resulting image has reduced blurring and improved contrast – Figure 5.36.

Mathematical filtering applies a convolution function (kernel) to the original projection. Different kernels can be selected, depending on the clinical application (soft tissue imaging, bone imaging, viewing small details, lung studies, cardiac investigations, etc). Images of the same examination reconstructed with different convolution kernels can significantly affect the visibility of low contrast and high contrast details.

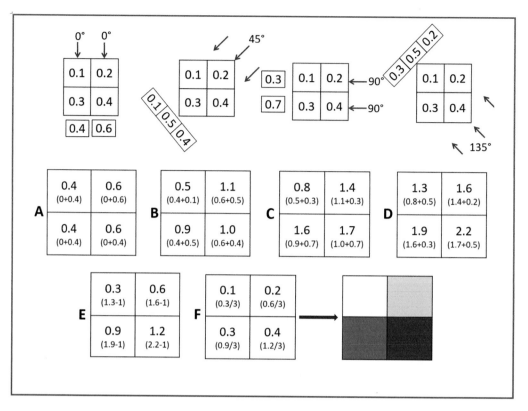

FIGURE 5.34 A simple numerical example of backprojection reconstruction – scanning from 4 angles (45 degree step). Explanation: Top row – an object made up of four voxels (each with attenuation coefficient μ from 0.1 to 0.4) is irradiated with the X-ray tube rotated to four angles (from 0 to 135 degrees); A, B, C, D: the absorption values are stored/built-up in the image matrix; E, F: matrix simplification and visualisation in grey levels (denser grey level corresponds to higher absorption). The final numbers in matrix F represent the object absorption units. When more projections are acquired (and used in the reconstruction), the resultant pixel numbers are closer to the μ values of the scanned object voxels.

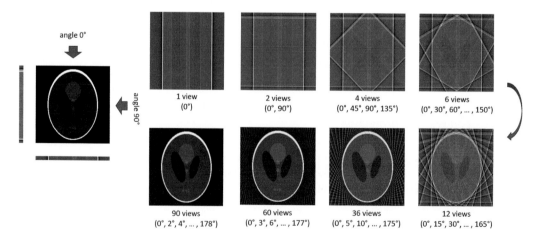

FIGURE 5.35 Forming a rough image of an object (on left) from its projections – upper row from 1 to 6 projections; lower row image quality improves with more projections (from 12 to 90). Increasing the number of projections around the object improves image quality. (Image courtesy of F. Brun, University of Trieste, Italy.)

CT Slice through thin wire – top: visualisation of projections, bottom: CT image and pixel values

CT Slice through human head – top: projections visualisation, bottom: CT image

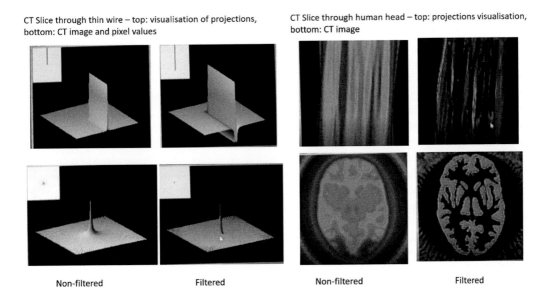

Non-filtered　　　　　　　Filtered　　　　　　　Non-filtered　　　　　　　Filtered

FIGURE 5.36 Comparison between backprojection and filtered backprojection: scanning of a wire and human head.

5.6.3.5　Helical Scanning Reconstruction

CT scanners have been capable of helical (or spiral) scanning since the late 1990s. This mode of operation involves continuously rotating the X-ray tube and detectors, whilst the patient is moved through the gantry aperture. The name is derived from the pattern of irradiation and data collection that arises from this type of scanning (Figure 5.37). The ability to continuously move the patient through the gantry aperture without pausing produces faster scan times.

To describe helical scanning requires an introduction to the concept of helical pitch. This is a measure of how fast the patient is moved through the gantry relative to the active detector width. For a single slice helical CT scanner, pitch is calculated according to the equation

$$\text{Pitch} = \frac{D}{T}$$

where D is the distance the couch is moved through the aperture in one gantry rotation (mm), and T is the slice thickness. Because slice thickness (T) and X-ray beam width (W) are equivalent in single slice helical CT, pitch is often defined as D/W.

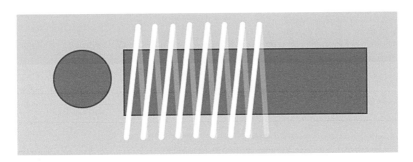

FIGURE 5.37　Helical scanning.

A pitch of 1 results in a contiguous spiral of data with no overlap or gap; a pitch more than one results in a spiral with a gap (e.g. with a pitch of 2 the gap has the same width as the active detector width). Pitches of less than 1 result in an overlapping spiral. (see Figure 5.38).

The backprojection method described earlier assumes that all projections used to reconstruct an image at a particular location along the patient were collected at that exact location. This is the case for axial scanning (the patient is stationary during the scan/data acquisition). However, when acquiring helical data, the patient is continuously moving through the gantry aperture so no two projections are acquired at the same patient position. Thus, when reconstructing an image at a particular position along the patient, there is only one projection at exactly that location: the projections at other X-ray tube to detector angles are located away from the required image plane (Figure 5.39). Projection data at all the required X-ray tube to detector angles have to be estimated by interpolating between pairs of measured projections at angles spanning the required image plane. This makes image reconstruction possible but results in a broadening of the slice profile due to the interpolation process. The higher the pitch, the further the interpolation distance, which leads to reduced accuracy of the projections, but also reduced patient dose. Note that this slice broadening does not occur in a multi-slice scanner (see Section 5.6.3.6).

The distance over which interpolation needs to take place can be reduced by making use of the concept of complementary projection data. It can be assumed that a projection measured at a 0° tube to detector angle is equivalent to that made at a 180° tube to detector angle. An additional complementary set of projections can be generated using this assumption (Figure 5.40). The resulting dataset interleaves with the original projection data, offset by half a rotation. Image reconstruction can make use of the original and complementary data, reducing the required interpolation distance.

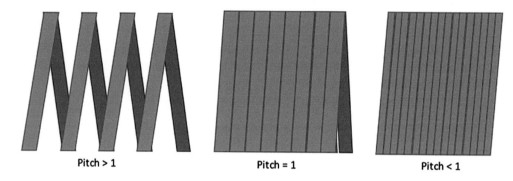

Pitch > 1   Pitch = 1   Pitch < 1

FIGURE 5.38   The effect of pitch selection on spiral.

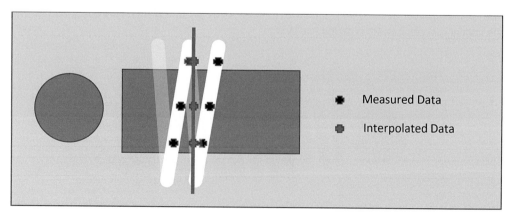

Measured Data

Interpolated Data

FIGURE 5.39   Interpolation in helical scanning.

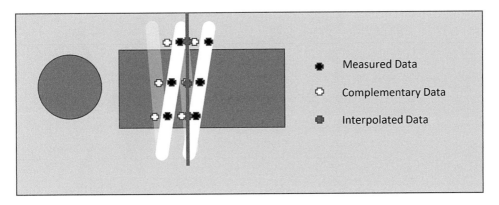

FIGURE 5.40 Interpolation using complementary projection data.

To avoid incomplete data collection and guarantee correct interpolation, helical scanners must irradiate beyond the boundaries of the volume to be imaged. Typically, an extra rotation is required at the beginning and end of the imaged volume. This additional exposure is known as **over-scanning**.

5.6.3.6 Multi-Slice Helical Reconstruction

Modern CT scanners acquire multiple interleaved helices of projection data during acquisition using an array with several parallel rows of detectors along the z axis (Figure 5.41).

The reconstruction process has access to more data than for a single-slice system. Rather than using all of the available data either side of the image reconstruction plane, multi-slice systems apply a filter of a certain width to the data: projection data within the filter is used for the reconstruction whilst data outside the filter is not. The filter width is set to the desired reconstructed slice thickness. In contrast to single-slice helical reconstruction, interpolation distance and slice thickness are not dependent on pitch.

The X-ray beam is collimated so that it extends several millimetres beyond the active detector width to avoid irradiating edge detector rows with the beam penumbra. If the penumbra were used, then the edge detector rows would receive less radiation dose than those in the centre, affecting image quality. This larger collimation is known as **overbeaming**. To reduce this unnecessary dose to the patient, manufacturers may provide dynamic or adaptive beam collimation.

The ratio of active detector width to collimated X-ray beam width is called the **geometric efficiency**. The additional beam width is fixed regardless of the collimation being used resulting in high geometric efficiency for large collimations and low geometric efficiency for very narrow collimations. Scanning with a low geometric efficiency collimation increases patient dose compared with high geometric efficiency collimation.

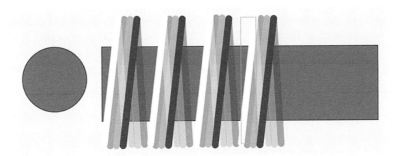

FIGURE 5.41 Multiple interleaved helices in MSCT (multi-slice CT).

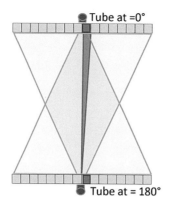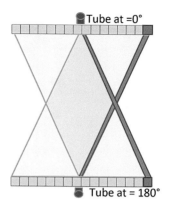

FIGURE 5.42 Wide beams require corrections for cone beam artefacts.

5.6.3.7 Issues with Wide Beams

The reconstruction methods described earlier are only valid when the X-ray to detector path is perpendicular to the scan plane. As the number of detector rows increases and larger cone angles are required to cover the detectors, this no longer holds true. Objects which are near the cone beam axis can be reconstructed without artefacts (the X-ray projections will lie within the same plane). Details positioned peripherally in the cone beam are susceptible to cone beam artefacts because the X-ray projections illuminating them lie in different planes (Figure 5.42).

If the angle is not accounted for in the reconstruction process, artefacts will occur. In practice, scanners capable of 16 slices and above apply corrections to the projection data before it is reconstructed to account for cone beam angle.

5.6.3.8 Iterative Reconstruction

Iterative reconstruction is now replacing filtered backprojection as the method of choice for CT scan data reconstruction. Iterative reconstruction begins by making an initial estimate of the final image. Forward projections are calculated for this estimate and compared with the measured projection data. The difference between the measured and estimated data is used to adjust the pixel values in the estimated image to produce a new estimate. This process is repeated until the difference between the measured and forward projected data falls below a pre-determined threshold. Contemporary systems often use filtered backprojection for their initial estimate in iterative reconstruction.

Iterative reconstruction methods can reduce the level of image noise compared with filtered backprojection. This also allows a reduction in patient radiation dose whilst maintaining a particular image noise level or a lower noise level if the radiation dose remains the same (Figure 5.43).

5.6.3.9 CT Image Presentation: The CT Number Scale

CT images are presented as a digital matrix of pixels, each representing one voxel from the object, and having a number (CT number) corresponding to the overall attenuation of the voxel. Each CT pixel number is expressed in Hounsfield Units (HU, named after G. Hounsfield). CT numbers are normalised to the attenuation of water and calculated according to the following equation, where μ_t is the X-ray attenuation coefficient of the tissue in question and μ_w is the X-ray attenuation coefficient of water.

$$CT \text{ number} = \frac{\mu_t - \mu_w}{\mu_w} \times 1000$$

There are two fixed points on the CT number scale: water has a value of 0 HU and air has a value of -1000 HU due to the negligible X-ray attenuation coefficient of air relative to that of water.

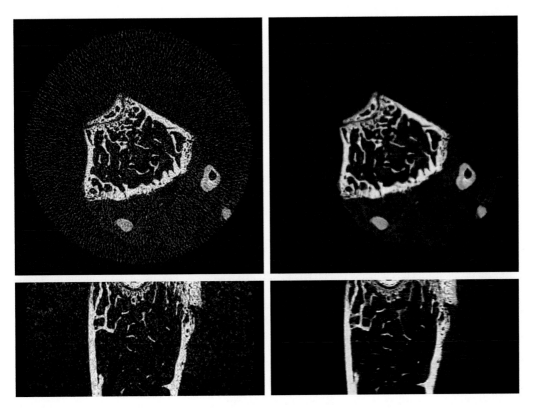

FIGURE 5.43 Comparison of low-dose CT scans of mouse bone: Filtered backprojection FBP (left) versus Iterative reconstruction IR (right). For low-dose images, IR algorithms outperform conventional FBP.

(Image courtesy of F. Brun, University of Trieste, Italy.)

Tissues with a greater attenuation coefficient than water will have a positive CT number whilst tissues with a lower attenuation coefficient than water will have a negative CT number.

Bone (relatively high linear attenuation coefficient/high density) has high, positive CT numbers. The most highly attenuating tissues in the human body, such as dense cortical bone, have CT numbers above 1000 HU. **Lungs** (low linear attenuation coefficient/low density) have low CT numbers (−800 to −500). **Soft tissues**, similar to water, have CT numbers ranging from −100 HU (fat) to +100 HU (liver).

5.6.3.10 CT Image Presentation: CT Image Size and Storage

CT images are generally reconstructed as a digital matrix of 512 × 512 (or 1024 x 1024) pixels. If the reconstruction field of view is 250 mm, this corresponds to a pixel size of 0.49 mm (250/512). Using a larger or smaller reconstruction field of view will scale the pixel size accordingly. Pixel values are stored as 12-bit integers: $2^{12} = 4096$ possible HU values representing a range between −1000 and +3096 HU, covering low-density clinical tissues such as lung air spaces up to high-density cortical bone.

Contemporary digital images used in diagnostic radiology (including CT) have a matrix depth of 16 bits, with 12 bits used to record image contrast and the other 4 bits used for supporting information (e.g. text or graphics displayed over the image). The 12 bits provide $2^{12} = 4096$ grey (or colour) levels which is more than the human eye can visualise.

The selection of 4096 grey levels is based on the set difference between the CT numbers of air and water, 1000. As cortical bone has an attenuation up to three times greater than that of water, this gives a CT number scale from −1000 (air) through 0 (water) to + 3096 (very high absorption bone), totalling 4096 levels.

5.6.3.11 CT Image Visualisation

Display monitors typically show up to several hundreds grey levels, however about 200 grey levels is the maximum that can be distinguished by the human eye. As there are up to 4096 CT number values, this would imply that each of the 200 visible grey level would represent about 20 HU which would impact on contrast resolution. The solution is 'windowing' in which only a specific range of CT numbers, the window width, is displayed around a central window level. CT numbers outside the window are displayed as black (below the window) or white (above the window).

The controls available to and adjusted by the operator are:

- window level, to select the CT number displayed as the mid-grey level on the monitor
- window width, to determine the range of CT number to display (e.g. −200 HU to +200 HU)

If, for example, a window level equal to 0 and window width of 400 is selected:

- all pixels with CT number > 200 (0 + 400/2) are **white**
- all pixels with CT number < −200 (0 − 400/2) are **black**
- all pixels within this window (−200 HU to +200 HU) are visualised using the available grey scale.

Differing window selections dramatically change the appearance of the same image. In Figure 5.44, a chest image is visualised using different WW (Window width) and WL (Window level) settings.

5.6.4 Scan Settings

Operational parameters of a CT scanner are selected based on the anatomical region to be investigated, the clinical question and the required image quality.

5.6.4.1 X-Ray Beam Collimation

The X-ray beam width can be varied by the user, typically in the range from 1 to 40 mm, although systems are available with collimations up to 160 mm at the time of writing. Using a wide collimation reduces the time required to cover a particular scan length and also maximises the scan length that can be covered before X-ray tube heat capacity limits are reached. However, as the voxels are very large, resolution in the z-direction (patient longitudinal axis) is decreased.

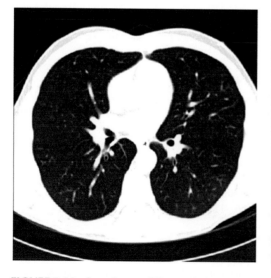 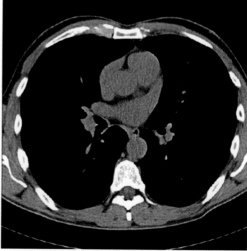

FIGURE 5.44   Same image, different windowing – left for visualising the lungs; right for visualising the muscles.

5.6.4.2 kVp

This is the peak kilovoltage used to generate the X-ray beam. Modern diagnostic CT systems have kVp values in the range 70 to 140 kVp and most have three or four set choices available to the user such as 80, 100, 120 and 140 kVp.

5.6.4.3 Rotation Time

This is simply the time required for a complete rotation of the gantry. The user has a choice of speeds typically in the range 0.3 to 2.0 seconds for a single rotation. Faster rotation speeds shorten the time required to scan a particular length along the patient and restrict the maximum radiation dose that can be delivered per rotation. They may also reduce the number of projections per scan leading to lower spatial resolution.

5.6.4.4 mAs per Rotation

The user is able to set the X-ray tube current used, which when multiplied by the rotation time gives the mAs per rotation. This is directly related to the number of X-rays produced hence a larger mAs will reduce image noise but increase patient dose and vice versa. Some scanners do not have a separate mA setting and instead require the user to set the mAs per rotation directly.

5.6.4.5 Pitch

Earlier in this chapter, the definition of pitch for a single-slice helical CT was given. For a multislice scanner, the denominator is replaced with the total thickness of all of the simultaneously acquired slices according to the equation;

$$\text{Pitch} = \frac{D}{n \cdot T}$$

where n represents the number of slices and T is the slice thickness T.

The term '**beam pitch**' is used when referring to pitch in multislice scanners whilst, in single-slice scanners, the term used is '**detector pitch**'.

5.6.4.6 mAs Adjusted for Pitch

The basic quantity describing the intensity of X-rays produced in scanners is mAs per rotation. This influences noise level in the image and patient dose. In helical scanning, the term used is *effective* mAs which is determined by dividing the actual mAs per rotation by the pitch. In multislice scanning, *effective* mAs is adjusted, usually automatically, to maintain a set noise level when the pitch is changed.

5.6.4.7 Scan Field of View (FoV)

Scan FOV defines the area being scanned. The FOV is circular with diameter measured at the CT system isocentre (Figure 5.45). After the scan (i.e. after the acquisition of data), differing FOVs can be selected for reconstruction, for example, to achieve a higher resolution over a smaller area.

5.6.4.8 Automatic Exposure Control

Control of patient exposure through the use of AEC systems has long been a feature in planar radiography (see Section 5.2.4). In the early days of CT scanning, the focus was principally directed towards achieving good-quality images and dose control was minimal, for example, by modifying tube current according to the body region being imaged; head, thorax, abdomen, etc. More recently, as patient dose in CT scanning became a cause for concern, increasingly sophisticated systems for dose control have been developed.

CT images are reconstructed from X-ray attenuation data as the gantry rotates around the body. Body section dimensions change but are basically elliptical. In the head, the major axis of the ellipse is in the anterior-posterior (AP) direction, whilst in the thorax and abdomen, the major axis would normally be in the lateral direction. The basic principle of CT AEC systems involves modulating

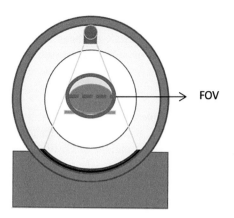

FIGURE 5.45 Field of view.

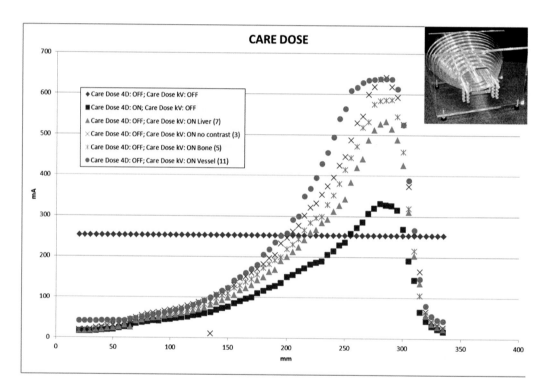

FIGURE 5.46 mA modulation with different AEC modes (the mode names are specific to the manufacturer).

the X-ray tube current as the gantry rotates – a projection through the major axis of a body section requiring a higher tube current than through the minor axis. With the gantry performing complete rotations in the order of a second or less, this is a requirement of significant complexity.

A range of approaches have been developed by manufacturers to achieve tube current modulation. The effect of one such approach is shown in Figure 5.46. The figure shows tube current from scans taken at different positions along an elliptical Perspex (polymethylmethacrylate, PMMA) test object for various operational modes of the AEC. Of particular relevance is the specific case where no AEC has been used. The tube current remains constant and independent of the test object dimensions which represents the way in which older CT scanners operated. The implications of this for patient dose are self-evident.

5.6.5   CT Artefacts

In medical imaging, artefacts are misrepresentations of human body structures. CT artefacts are common and can occur for various reasons. In general, artefacts lead to mispositioning of projections or miscalculation of the pixel values producing distortions of the final image. Knowledge of these artefacts is important because they can simulate pathology, obscure pathology or degrade image quality to non-diagnostic levels.

5.6.5.1   Patient-Based Artefacts

Voluntary or involuntary patient movement during image acquisition produces movement artefacts. This can be a particular problem in paediatric or non-cooperative patients and usually requires immobilisation or sedation. Beating of the heart, respiration and gastrointestinal peristalsis are the most common involuntary movements. Adequate acquisition parameters and special reconstruction techniques can help in minimising the effects on image quality.

A metal object in the scan field (dental filling, prosthetic device, surgical clip, jewellery) can cause streaking artefacts (Figure 5.47). A variety of interpolation techniques are available to reduce artefacts caused by metal objects.

5.6.5.2   Physics-Based Artefacts

Beam hardening and scatter can lead to more photons being detected than expected; producing streaking (dark bands) and cupping artefacts (reduced voxel values near the centre of the image). Beam hardening occurs with polychromatic X-ray sources. As the X-ray beam passes through the body, low energy photons are more easily attenuated than high energy photons. Due to this 'selective attenuation', the effective energy of the beam gradually increases. In other words, polychromatic beam transmission does not follow the simple exponential decay of a monochromatic X-ray source as assumed in reconstruction kernels.

Partial volume effects occur when tissues with significantly different attenuation are within the same CT voxel. The CT number is representative of the average attenuation of the materials within a voxel. This artefact is generally negligible in modern CT.

5.6.5.3   Hardware-Based Artefacts

Ring artefacts are equipment-related phenomena (Figure 5.48). Miscalibration or failure of one or more detector elements in a CT scanner may cause circular streaks. The distance of the streak from

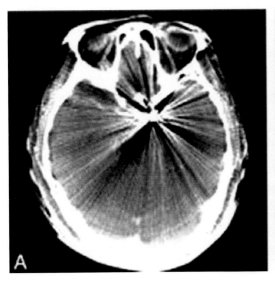
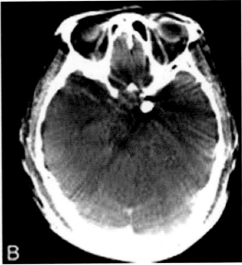

FIGURE 5.47   Metal artefact.

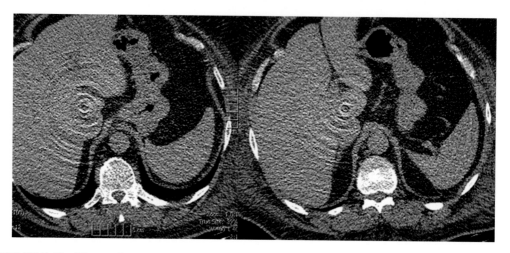

FIGURE 5.48   Ring artefact.

the isocentre depends on the position of the faulty detector along the detector arc. The remedy is often a simple recalibration of the scanner. Occasionally detector elements need replacement.

5.6.6   DOSIMETRY IN CT

Dose management is a key aspect in CT because this imaging modality is the largest source of diagnostic medical radiation exposure. Specific dose indicators, related to scanner output, have been defined to facilitate monitoring.

5.6.6.1   CTDI

CTDI is the acronym for Computed Tomography Dose Index. It is defined as the integral of a single scan radiation dose profile along the z axis, normalised to the thickness of the imaged section ('slice thickness').

$$\text{CTDI} = \frac{1}{T} \int_{-\infty}^{+\infty} D(z)\, dz$$

A 100-mm-long pencil ionisation chamber is specifically designed to make practical measurements, Figure 5.49.

Various indexes derive from the theoretical definition, each representing a specific measurement condition:

- $\text{CTDI}_{100,\,air}$ is obtained with a 100-mm-long pencil chamber in air (positioned in the centre of the gantry, aligned to z axis);
- $\text{CTDI}_{100,\,c}$ and $\text{CTDI}_{100,\,p}$ are measured, respectively, in the central and in the peripheral holes of standard CT dose phantoms (16-cm-diameter and 32-cm-diameter PMMA cylinders – Figure 5.50)

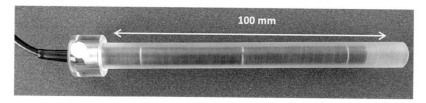

FIGURE 5.49   Pencil ionisation chamber.

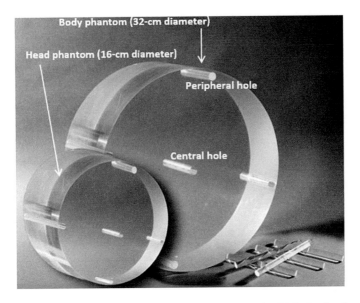

FIGURE 5.50 Standard CT dose phantoms (for head and body scans) with holes for the pencil ionisation chamber.

The indexes are reported in mGy and tend to be used for comparing different scanners or scanner modes.

5.6.6.2 CTDI$_w$

CTDI$_w$ is defined as

$$CTDI_w = 1/3\ CTDI_{100,\ c} + 2/3\ CTDI_{100,\ p}$$

where CTDI$_{100,\ p}$ is the mean value of the measurements in the four peripheral holes of the standard dose phantom.

The index is measured for a single slice but represents the mean dose in a slice for a large number of contiguous axial scans.

5.6.6.3 CTDI$_{vol}$

With helical CT, a new parameter, named CTDI$_{vol}$, was introduced, to consider the effect of pitch selection. The definition is

$$CTDI_{vol} = CTDI_w / pitch$$

This index is effective in describing the dose increase for overlapping slices (pitch < 1) and the dose reduction when there is a gap between slices (pitch > 1).

5.6.6.4 DLP

To quantify the total radiation dose to which a patient is exposed, a more effective index is the Dose Length Product (DLP), defined as

$$DLP = CTDI_{vol} \cdot Scan\ length$$

The units of measurement are mGy·cm.

Radiation Level	Effective Dose Range		Investigation
	Adult	Pediatric	
	0 mSv	0 mSv	MRI; ultrasound
	< 0.1 mSv	< 0.03 mSv	Chest and extremities radiography
	0.1 – 1 mSv	0.03 – 0.3 mSv	Pelvis radiography; mammography
	1 – 10 mSv	0.3 – 3 mSv	Head CT
	10 – 30 mSv	3 – 10 mSv	Abdomen CT with and without contrast agent
	30 – 100 mSv	10 – 30 mSv	Complex interventional procedure (as TIPS)

FIGURE 5.51   Typical ranges of Effective dose for common examinations.

5.6.6.5   Patient Dose

The indexes described above are not to be confused with measures of patient dose. They do not account for patient size and specific exposure parameters. Although the effective dose can be estimated approximately from the DLP value, accurate patient dose estimation is a complex process.

5.6.6.6   CT versus Other Imaging Modalities

CT scanning parameters are very flexible. Scan and reconstruction settings can be modified significantly to obtain the desired image quality. Although the limiting spatial resolution does not exceed 1.2–2.5 lp/mm (compared to 5–10 lp/mm for FPDs), CT is extremely powerful in the detection of low contrast details (i.e. contrast resolution) and guarantees higher image quality than planar radiography although this comes with the disadvantage of higher patient doses.

Figure 5.51 shows typical ranges of effective dose for different imaging modalities.

REFERENCES

1 United Nations Scientific Committee on the Effects of Atomic Radiation UNSCEAR, 2008 *Report Volume I: General Assembly*, Scientific Annexes, 2008
2 NHS Imaging and Radiodiagnostic activity in England, 2013 [access Sep 2017], https://www.england.nhs.uk/statistics/wp-content/uploads/sites/2/2013/04/KH12-release-2012-13.pdf
3 National Council on Radiation Protection and Measurements. Ionizing radiation exposure of the population of the United States. NCRP Report No. 160. Bethesda, Md: National Council on Radiation Protection and Measurements, 2009 (see https://pubs.rsna.org/doi/pdf/10.1148/radiol.12112678 [access Nov 2019]

FURTHER READING

Bushberg J T, Seibert J A, Leidholdt Jr. E M, Boone J M, *The Essential Physics of Medical Imaging*, International Edition, 2012, Lippincott Williams and Wilkins, Philadelphia
Dowsett D J, Kenny P A, Johnston R E, *The Physics of Diagnostic Imaging*, 2nd Edition, 2006, CRC Press
Dendy P P, Heaton B, *Physics for Diagnostic Radiology*, 3rd Edition, 2012, CRC Press
Tabakov S, Milano F, Stoeva M, Sprawls P, Tipnis S, Underwood T (Ed), 2021, *Encyclopaedia of Medical Physics*, 2nd Edition, CRC Press (available also as a free online resource at www.emitel2.eu)
Kalender W, *Computed Tomography: Fundamentals, System Technology, Image Quality, Applications*, 3rd Edition, 2011, Publicis
Sprawls Resources, free online resource at www.sprawls.org
EMERALD Resources, free online resource, at www.emerald2.eu
Encyclopaedia of Medical Physics, free online resource, at www.emitel2.eu

6 Nuclear Medicine Imaging

Elena De Ponti
ASST Monza, Monza, Italy

Luciano Bertocchi
Abdus Salam International Centre for Theoretical Physics, Trieste, Italy

CONTENTS

6.1 NUCLEAR MEDICINE FUNCTIONAL IMAGING

Diagnostic nuclear medicine imaging is often referred as molecular imaging to distinguish its ability to reveal metabolic and functional processes in organs and other structures from other radiological imaging procedures that provide mainly anatomical information (e.g. computed tomography, magnetic resonance imaging, ultrasound imaging) (see Figure 6.1).

DOI: 10.1201/9780429155758-6

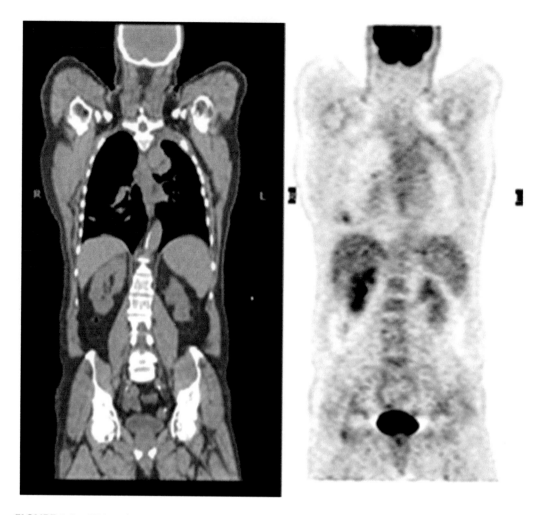

FIGURE 6.1   CT imaging (on the left) versus PET imaging (on the right) of the same patient.

The invention of the cyclotron by Ernest Orlando Lawrence in 1928 can be considered one of the starting points in the development of this form of diagnostic imaging. Several years earlier, in 1923, another crucial step had been achieved by Georg Charles de Hevesy who first used a naturally radioactive isotope of lead to study the uptake of labelled lead ions from dilute solutions through the roots, stem, leaves and fruit of plants. However, it was the invention of the gamma camera in the 1950s by Hal Anger that set nuclear medicine imaging on the path towards its modern role in medical diagnostics.

The basic principle of nuclear medicine imaging is the administration to patients of radioactive tracers or radiopharmaceuticals that distribute in the body according to specific metabolic processes. Administration can be by intravenous injection, inhalation, oral ingestion or direct injection into an organ. Tracer uptake times may take from a few minutes to a few hours before optimal distribution in the organ is achieved. The patient can then be scanned and gamma ray photon emissions from the tracer detected by a scintillator crystal detector coupled with a photomultiplier tube (PMT) or a solid-state detector. Energy and positional information from the emitted photons is collected and

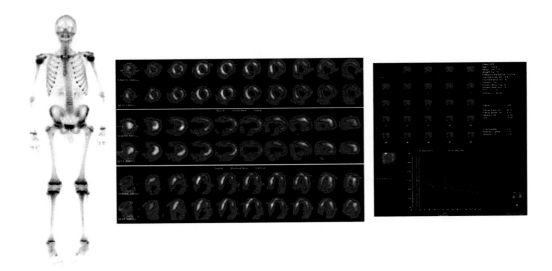

FIGURE 6.2   Example of static planar bone scintigraphy (left); axial, coronal and sagittal projections of cardiac SPECT (centre); dynamic renal scintigraphy (right).

used to create an image or images characterising the radiopharmaceutical (tracer) distribution inside the patient. Images can be acquired and reconstructed as static planar or tomographic images or can be collected over time in dynamic sequences (Figure 6.2).

There are two 'tomographic' acquisition modalities (i.e. modalities that acquire images slice-by-slice or from a whole volume, rather than as a projection through the body). Single Photon Emission Computed Tomography (SPECT) is performed using a rotating gamma camera designed to collect single gamma ray photons emitted by the tracer distribution. Positron Emission Tomography (PET) uses positrons (anti electrons, or positive beta particles, β^+) which quickly undergo annihilation reactions with electrons. Two gamma ray photons of identical energy (511 keV) are emitted 180° apart when annihilation occurs, and these are detected by a circular ring of detectors surrounding the patient in a PET scanner (Figure 6.3).

Sections 6.2, 6.3, 6.4 discuss radioisotope physics and the preparation of radiopharmaceuticals, necessary for nuclear medicine imaging. Sections 6.5 and 6.6 discuss equipment and image formation in nuclear imaging – the gamma camera, SPECT, PET and related forms of hybrid tomographic imaging.

6.2   NUCLEAR DECAY PROCESSES

Radioactive tracers that can be used for nuclear medicine imaging are selected based on the metabolic behaviour and radioactive decay properties of the isotopes used to label them.

Nuclear medicine imaging is essentially based on detection of gamma photons of an energy that can exit the patient's body and interact with detectors in the imaging system. The human body is, at least partially, transparent to electromagnetic radiation in two regions of the spectrum: the megahertz frequency interval (radio waves) and photon energies above about 15 keV (X-rays and γ rays). Electromagnetic radiation in both ranges can be used to produce images: MRI in the first case, X-ray radiology and nuclear medicine in the second case.

The decay half-life of a radioactive isotope is a significant characteristic that must be considered in the development of new radiotracers. Very short half-lives are not desirable because large

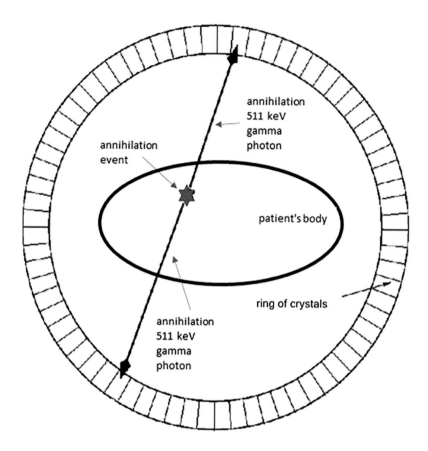

FIGURE 6.3 Detection of two coincident 511 keV photons from annihilation of a positron and electron in PET imaging.

quantities of radioactive material need to be injected into the patient to provide sufficient detected gamma ray photons (counts) to minimise statistical errors that result in image noise. Long half-lives are excluded for radiation protection reasons as significant activity remains for a long time after the scan, resulting in higher doses to patients and potentially to others who come into contact with them.

These characteristics are related to the nuclear decay process by which an unstable nucleus reaches a new and more stable energetic condition. An unstable nucleus can decay by emitting various kinds of particles: α rays (nuclei of $^4He_2$); positrons and electrons and γ rays.

However, charged particles (α particles, positrons and electrons) are absorbed in tissue within a very short distance and cannot escape from the patient's body to be detected. Consequently, they are not used in diagnostic nuclear medicine. Alpha particle and electron (β^-) emitters are useful in molecular radiotherapy, another important branch of nuclear medicine (see Chapter 11). Therefore, the only decay products that can exit from the body and be detected to form an image are γ ray photons.

'Weak decay' of a 'parent' isotope can result in electron emission (β^- decay) or positron emission (β^+ decay). These processes result in a transformation of the nucleus into an isotope of a different element, known as a 'daughter' nucleus. Electron capture (EC) is another form of decay also resulting in nuclear transformation. The daughter nucleus formed by β decay or EC may be in an excited state, and these excited energy levels can decay emitting γ rays that can be collected by gamma cameras for planar and SPECT imaging.

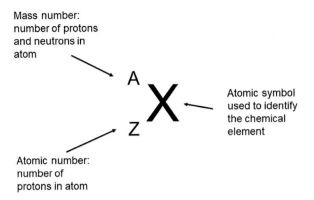

FIGURE 6.4 Notation used for a nucleus.

There is another process which produces useful γ rays. In positron (β⁺) decay, a positron is emitted, with some kinetic energy. The positron travels through matter and, after multiple interactions with electrons, slows down. At the end of its path, the positron annihilates with an electron in the medium through which it has travelled, emitting two γ rays. If the annihilation takes place when the positron has no residual kinetic energy, we say that annihilation occurs at rest and energy-momentum conservation tells us that the two γ rays are emitted with the same energy (511 keV) and opposite momenta: the same absolute value (511 keV/c) but in opposite directions. When the positron is emitted in the patient's body after the administration of a positron-emitting tracer, a fraction of the γ ray pairs emitted in positron-electron annihilation can escape from the body surface and be collected, forming an image of the unstable positron emitting isotope distribution. This is the basis of PET imaging. As the purpose of detection is to identify the position of positron annihilation, the two 511 keV photons must be detected at the same time and we speak of coincidence events recorded using a ring of crystals surrounding the patient's body (Figure 6.3).

In the following pages, we shall briefly describe the properties of the types of radioactive decay used in nuclear medicine.

In nuclear physics, it is customary to use notation in which the symbol of the element is accompanied by the atomic number (Z, the number of protons in the nucleus, which defines the chemical species) and the mass number (A, the total number of nucleons [protons and neutrons] in the nucleus) (Figure 6.4). N is also used to indicate the number of neutrons.

6.2.1 β⁻ Decay

β⁻ decay takes place in a nucleus which is neutron-rich, that is, the number of neutrons (N) is larger than the number of neutrons in the stable isotope with the same mass number (A).

The decay scheme is as follows.

$$_Z^A X \rightarrow\ _{Z+1}^A Y + e^- + \bar{v}$$

A parent nucleus X transforms into a daughter nucleus, which has one additional proton, and an electron and an antineutrino are emitted.

Nuclear decay reactions are subject to classical conservation laws for electric charge, momentum, angular momentum, and energy (including rest mass energies). An additional conservation law found in nuclear decay is conservation of the number of nucleons. The energy balance of any decay process can be summarised by the Q-value, defined as the mass-energy difference between the initial and final nuclear state. This is an important quantity to identify which nuclear transformations are permitted by energy conservation. The allowed nuclear transformations are those with a positive Q value, while those with negative values are forbidden.

The final nucleus can be either in its ground state or in an excited state (or, in some cases, in a number of excited states) which will undergo further decay, emitting γ rays.

In β⁻ decay, the kinetic energy spectrum of the electrons is continuous, ranging from zero to a maximum value, equal to the Q value.

6.2.2 β⁺ (Positron) Decay

β⁺ decay takes place in a nucleus which is proton rich, that is, the number of protons (Z) is larger than the number of protons in the stable isotope with the same mass number (A) (often instead of 'proton rich', the equivalent term 'neutron poor' is used.)

The decay scheme is as follows.

$$_Z^A X \rightarrow {}_{Z-1}^A Y + e^+ + \nu$$

There is an important difference between β⁻ decay and β⁺ decays. In the latter case, although the positron energy spectrum is continuous, the maximum positron energy is not Q but instead is given by

$$Q - 2mc^2 = Q - 1.022 \, MeV \tag{6.1}$$

where mc^2 is the electron and positron mass energy, 0.511 MeV. This difference arises because in the case of β⁻ decay the daughter nucleus with Z+1 protons is located in an atom with only Z electrons, an electron deficit offset in mass-energy terms by the additional electron generated in the reaction. In β⁺ decay, on the other hand, the daughter nucleus with Z−1 protons is located in an atom with Z electrons, and to this electron excess the positron generated in the reaction is added, thus giving a mass-energy of $2mc^2$ overall.

Therefore, β⁺ decay can take place only when the mass-energy of the parent nucleus exceeds that of the daughter nucleus by at least 1.022 MeV. We say that this type of decay has a *kinematical threshold*.

6.2.3 Electron Capture (EC)

Another reaction mechanism involving the same parent and daughter nuclear states as positron emission is as follows.

$$e^- + {}_Z^A X \rightarrow {}_{Z-1}^A Y + \nu$$

Here an orbital electron in the parent atom is 'captured' by the nucleus. This process is called electron capture (EC).

The captured electron usually originates from the K or L electron shell. It follows that the energy threshold for EC is equal to the electron binding energy in the atomic shell. In a hydrogen-like atom (i.e. a nucleus with a single orbital electron in the most external orbit), this can be approximated by the following Equation (6.2).

$$B = 13.6 \frac{Z^2}{n^2} \, eV \tag{6.2}$$

where Z is atomic number and n the principal atomic quantum number.

Taking $Z = 60$ and $n = 1$, one gets B ~ 50 keV, which is much smaller than the threshold for positron emission, 1,022 keV. Therefore, EC can occur even when positron emission is forbidden by energy conservation. When both β^+ decay and EC are allowed energetically, the ratio of the probabilities of EC and positron emission is proportional to Z^3. The daughter nucleus following EC is often in an excited state, leading to subsequent emission of γ rays.

Finally, we note that in EC no charged particles are emitted, only a neutrino. However, after EC has occurred, the electronic level from which the electron has been captured remains empty. This level will be filled by an electron from a higher orbit, emitting either characteristic X-rays or an Auger electron.

6.2.4 γ DECAY

γ decay takes place between two different energy levels of the same isotope. In this decay, there is no change in the Z, N or A of the nucleus, that is, no change of element or isotope. The γ ray energy is equal to the Q-value, namely the energy difference of the two levels.

6.2.5 INTERNAL CONVERSION (IC)

There is another mechanism which competes with γ decay, known as internal conversion (IC). In IC, instead of being emitted in the form of a γ ray, energy is transferred to an orbital electron. This electron is expelled from the atom with kinetic energy equal to the Q-value minus the electron binding energy in the atomic shell. The important feature of IC is that the emitted electrons are 'monochromatic', that is, they have only one energy value. This is in contrast to β^- decay, where the emitted electrons have a continuous energy spectrum. When the half-life for decay of an excited state formed following β^- decay is short enough, both a continuous spectrum of electrons and monochromatic electrons are observed effectively at the same time.

The relative positions of the monochromatic emission lines and the continuous spectrum will depend upon the Q value of the weak decay (i.e. the β^- emission), the Q value of the electromagnetic transition (i.e. the IC) and the binding energy of the electron level from which the IC electron has been ejected. The ratio of the probabilities of IC and γ emission is proportional to Z^3.

6.2.6 EXAMPLES OF DECAYS

6.2.6.1 Example of 'Pure' Positron Emission

To maximise the probability of positron emission, we choose proton-rich isotopes with Q-values larger than 1,022 keV, so that positron emission is kinematically allowed, but with a quite low value of Z, so that positron emission will be predominant over EC. Examples are the decays of light positron emitting isotopes used in PET applications, such as $^{11}C_6$ (99.8%); $^{13}N_7$ (100%), $^{15}O_8$ (99.9%), $^{18}F_9$ (97%) (Figure 6.5), where the percentage of positron emission is indicated in parentheses.

When Z increases, this percentage becomes smaller, as for the isotopes $^{68}Ge_{31}$ (90%) and $^{82}Rb_{37}$ (96%).

6.2.6.2 Example of Beta-Gamma and EC-Gamma Decays

$^{131}I_{53}$ is an example of β^--γ decay (i.e. a β^- decay followed by gamma ray emission) which is very important in nuclear medicine (Figure 6.6). Several iodine isotopes play key roles in nuclear medicine because of their selective metabolic uptake by the thyroid. The decay of $^{131}I_{53}$ is used in thyroid uptake imaging to visualise the thyroid gland when disease of the thyroid is suspected or for staging purposes. The same isotope is used in higher activities for radiometabolic treatment of thyroid cancer. This iodine isotope, which is neutron rich, decays with a half-life of about 8.02 days to the $^{131}Xe_{54}$ isotope through several β^- decay pathways.

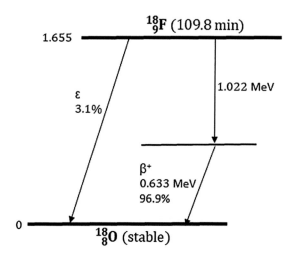

FIGURE 6.5 Decay scheme of ^{18}F.

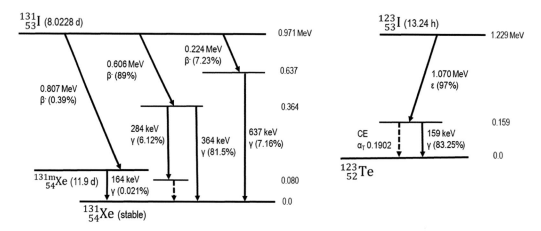

FIGURE 6.6 Decay scheme of ^{131}I (left) and ^{123}I (right).

From the $^{131}\text{I}_{53}$ ground state, several excited states of $^{131}\text{Xe}_{54}$ can be reached by β^- decay, but not the $^{131}\text{Xe}_{54}$ ground state. In Figure 6.6, the β^- transitions which are, at least in principle, possible are indicated.

From these excited states, one can finally reach, through several γ decay pathways, the xenon ground state. This leads to a very rich γ ray spectrum, ranging in energy from 723 keV down to 80 keV. Among these γ lines, the most important is the one at 364 keV, emitted in 80% of cases and usually quoted as the $^{131}\text{I}_{53}$ γ ray energy. It is a γ ray of medium-high energy, important in nuclear medicine imaging.

Another iodine isotope used in nuclear medicine application is $^{123}\text{I}_{53}$ which, after EC, emits a γ ray of 159 keV with a half-life of 13 hours, without emission of any charged particle. It is used in brain scintigraphy for the study of the most common neurodegenerative disorders.

A second example of a β^--γ decay is the $^{99}\text{Mo}_{42} - {}^{99}\text{Tc}_{43}$ decay chain (Figure 6.7). This shows some similarities with the previous example, but also some very important differences. $^{99}\text{Mo}_{42}$ is also neutron rich. The half-life of $^{99}\text{Mo}_{42}$ is 2.75 days. It decays through several β^- transitions into a number of excited states of $^{99}\text{Tc}_{43}$, and also here direct transition to the ground state of technetium is forbidden by the weak decay selections rules. From these excited levels, a number of γ rays of

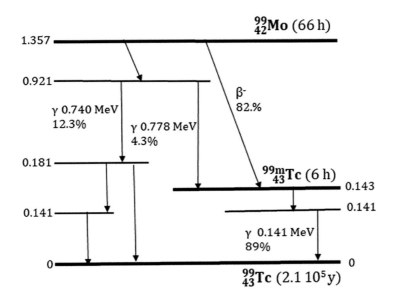

FIGURE 6.7 Decay scheme of $^{99}Mo_{42}$ to $^{99}Tc_{43}$ via the isomeric state $^{99m}Tc_{43}$.

different energies are emitted. In the case of the $^{99}Tc_{43}$ daughter isotope, the half-lives of almost all the excited levels are very short as compared with the weak (β^-) decay half-life, except for a level at 142 keV, which has a half-life of about 6 hours. It is an 'isomeric' or 'metastable' state, that is, a state with a long half-life, denoted $^{99m}Tc_{43}$.

The transition from the 143 keV state to the 141 keV state is dominant over direct transition to the ground state. The 141 keV state decays, with a half-life of 0.2 ns to the ground level, emitting a 141 keV γ ray, which to all intents and purposes is emitted after the half-life of the 143 to 141 keV transition, that is, 6 hours.

The 141 keV γ rays from ^{99m}Tc decay have extremely important applications in nuclear medicine imaging: more than 90% of examinations are performed using this isotope, labelling different tracers for cardiac, neurological and oncological applications.

6.3 PRODUCTION OF UNSTABLE ISOTOPES

The radioisotopes commonly used in medicine have short half-lives and so do not occur in nature. These artificial radioisotopes are produced by nuclear reactions from stable elements (X) that are bombarded with particles (p_1) resulting in the production of unstable elements that reach a stable condition (Y) via radioactive decay (emitting p_2)

$$_{Z_1}^{A_1}X\left(p_1,p_2\right)_{Z_2}^{A_2}Y$$

One of the most important processes is absorption of a neutron by an atomic nucleus, in which the mass number of the element concerned increases by one for each neutron absorbed. In this case, the atomic mass increases, but the element is unchanged. In some cases, the product nucleus is unstable and decays, typically emitting protons, electrons (beta particles) or alpha particles. Neutron irradiation is performed in a nuclear reactor.

Another method used to produce radioisotopes is proton bombardment. The protons are accelerated to high energies either in a cyclotron or a linear accelerator. The resulting unstable isotopes can be either neutron-rich, in which case they decay by emitting electrons in β^- decay, or proton-rich, in which case they decay by emitting positrons (β^+ decay), or by absorbing electrons (electron capture, EC).

In nuclear physics, the probability that a nuclear reaction will occur is represented by the reaction cross section, denoted σ and measured in barns (1 b = 10^{-28} m²). Nuclear reaction cross sections are often plotted as a function of the incident particle's energy as shown in the following paragraph.

6.3.1   PRODUCTION OF NEUTRON-RICH ISOTOPES

6.3.1.1   Neutron Radiative Capture Mechanism

The standard mechanism for production of neutron-rich isotopes is neutron radiative capture $X(n, \gamma)Y$

$$n + {}^{A}_{Z}X \rightarrow {}^{A+1}_{Z}X + \gamma$$

This is an exothermic reaction; the activation energy needed to initiate the reaction is less than the energy it releases, so that its cross section at low energy follows the so-called 1/v law, that is, the cross section is inversely proportional to the velocity of the neutron. Therefore, very low energy neutrons are employed, typically thermal neutrons (energy of 0.025 eV) from a fission reactor. Two examples of neutron radiative capture are described below.

The first example involves a stable samarium isotope

$$^{152}_{62}Sm\left(n, \gamma\right) {}^{153}_{62}Sm$$

The dot in Figure 6.8 corresponds to thermal energy, where the cross section for thermal neutrons is quite high, σ = 200 b. The half-life for the decay is 46.7 hours; therefore, even for a flux as high as Φ = 10^{14} neutrons/(cm².s), the production-decay equilibrium is a secular equilibrium, which means that the quantity of a radioactive isotope remains constant because its production rate due to decay of the parent isotope is equal to its decay rate.

The second, very interesting and useful, example is the reaction

$$^{98}_{42}Mo\left(n, \gamma\right) {}^{99}_{42}Mo$$

This reaction is used to produce $^{99}Mo_{42}$, that is, the parent of $^{99m}Tc_{43}$. In this case (Figure 6.9), the cross section for thermal neutrons is quite low, only σ ~ 0.13 b, much smaller than in the first example; again, the production-decay equilibrium is a secular equilibrium.

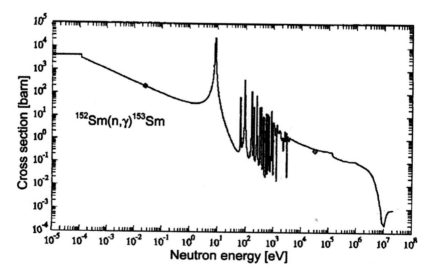

FIGURE 6.8   Neutron cross section for the reaction $^{152}Sm(n,\gamma)^{153}Sm$.

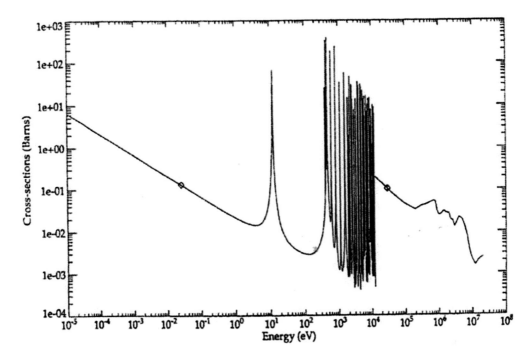

FIGURE 6.9   Neutron cross section for the reaction $^{98}Mo(n,\gamma)^{99}Mo$.

The mechanism of producing neutron-rich isotopes through radiative neutron capture has two disadvantages.

Firstly, the reaction is 'non carrier free'. The meaning of this sentence is that the produced isotope is a different isotope of the same chemical species as the target (same value of Z, different value of N). Therefore, the produced isotope cannot be separated chemically from the target isotope.

Secondly, the reaction is not very efficient. Take, for example, the case of production of ^{99}Mo. The lifetime of ^{99}Mo is 3.44×10^5 sec and the cross section for thermal neutrons is $\sigma \sim 0.13$ b. Even with a very large neutron flux value of $\Phi = 10^{14}$ neutrons/(cm^2.s), production and decay will be in secular equilibrium, and the fraction of the ^{98}Mo target nuclei transmuted into ^{99}Mo nuclei is very low.

$$\text{Efficiency} = 4.5 \times 10^{-6} \qquad\qquad (6.3)$$

6.3.1.2   Use of $^{236}U_{92}$ Fission Fragments

There is however a much more efficient mechanism to produce neutron-rich isotopes: to make direct use of the fission fragments produced by the uranium fission process in a fission reactor. When the $^{236}U_{92}$ nucleus is split into two heavy fragments during the fission process, the fragments are neutron rich. This can be simply understood, since the N/Z ratio in $^{236}U_{92}$ is 144/92 = 1.565. In an oversimplified situation where $^{236}U_{92}$ splits into two equal fragments, each with Z = 46, both would have the same N/Z ratio, 72/46 = 1.565, while the stable isotope with Z = 46, $^{106}Pd_{46}$, has a ratio N/Z = 60/46 ~ 1.3.

Even taking into account that in the fission process a number of free neutrons are liberated (on average about 2.5 neutrons per fission, because the fragments are so neutron rich), and that the most probable fission process is not symmetrical, it is still the case that the remaining heavy fragments are very neutron rich. They are therefore unstable, usually via β^- decay. This means that they are of little direct utility in biomedical imaging, but quite useful in molecular radiotherapy.

For the fissile element $^{235}U_{92}$, some examples of the isotopes produced in fission reactions are as follows.

$^{235}U_{92}(n,f)^{90}Sr_{38}$ (28.6 years)   Fission yield = 5.8%

$^{235}U_{92}(n,f)^{99}Mo_{42}$ (66.0 hours)   Fission yield = 6.2%

$^{235}U_{92}(n,f)^{131}I_{53}$ (8.0 days)   Fission yield = 2.9%

Here (n,f) indicates a fission reaction initiated by a neutron, figures in parenthesis indicate the half-life of the unstable isotope produced in the fission and fission yield represents the fraction of the fission product produced per single fission. It is interesting to see that the fission yield for the $^{99}Mo_{42}$ isotope is quite large, 6.2%, which is much more efficient than neutron capture.

6.3.2   PRODUCTION OF PROTON-RICH ISOTOPES

There are in principle two ways of producing proton-rich isotopes: either to increase the value of Z, that is, increasing the number of protons, or to decrease the number of neutrons, N. In the first case, the chemical element is changed; in the second case, the chemical element remains the same.

To increase the value of Z, beams of light, charged nuclei are directed onto a target of stable nuclei; the particles in the beam are usually protons, (Z = 1), deuterons $^2H_1$ (Z = 1) or $^3He_2$ and $^4He_2$ nuclei (Z = 2). The reaction will only be possible for energies higher than a kinematical threshold, which depends upon the masses in the initial and final states. Both the incident particle and the target nucleus are positively charged; therefore, a positive Coulomb potential tends to repel the incident particle.

For a nuclear reaction to take place, the incident particle must be 'trapped' inside the nucleus, so that classically its energy should be higher than the value of the electrostatic potential at the boundary of the nucleus, the value of which is of the order of at least a few MeV. Quantum mechanically, the incident particle could tunnel through the Coulomb barrier; but it can be shown that the net result is an exponential decrease of the cross section at very low energies. In other words, at the kinematical threshold, the cross section is very small, then increases with energy and usually reaches a maximum value; after the maximum, the cross section decreases again.

Charged particle beams produced by accelerators are used, with energies in the range of 1 to 100 MeV. The preferred reactions employed to produce light positron emitting isotopes are listed in Table 6.1. Proton beams in various energy intervals are used, usually between 5 and 16 MeV.

For heavier nuclei, the mechanism of decreasing the number of neutrons is normally used. Typical reactions using incident proton beams can be written as follows.

$$p + {}^A_Z X \rightarrow {}^{A+1-n}_{Z+1} X + n\,neutrons$$

where n neutrons are extracted. Since the incident proton remains bound in the final nucleus, the chemical element changes.

TABLE 6.1

Reactions and Energy Interval of Charged Particle Employed to Produce Light Positron Emitting Isotopes

β^+ Emitting Isotope	Reaction	Energy Interval
^{11}C	$^{14}N(p, \alpha)\ ^{11}C$	5 – 16 MeV
^{13}N	$^{16}O(p, \alpha)\ ^{13}N$	8 – 16 MeV
^{15}O	$^{14}N(p, n)\ ^{15}O$	5 – 14 MeV
^{18}F	$^{18}O(p, n)\ ^{18}F$	5 – 16 MeV

An alternative reaction is as follows.

$$p + {}^A_Z X \rightarrow {}^{A-n}_Z X + p + n \, neutrons$$

In this case, the incident proton does not become bound, so the chemical element does not change; again the number of neutrons decreases by n.

Here the key factor is that it takes an energy of the order of 8 to 10 MeV to extract each neutron from the potential well; therefore, by choosing the beam energy interval appropriately, production of one specific isotope can be maximised. This type of reaction is used, for example, to produce ^{123}I used for neurological SPECT and in strontium-rubidium and germanium-gallium generators for PET applications.

6.4   RADIOPHARMACEUTICALS

From the very beginning, it was evident that diagnostic nuclear medicine would depend on the availability of radioactive tracers with specific biochemical properties that are prepared in a pharmaceutical form suitable for administration *in vivo*.

Radioactive tracers suitable for labelling and use in nuclear medicine must simulate substances normally used by metabolic pathways in the human body. They will be chosen based on the physiological process or target organ to be studied. The term 'tracer' is derived from this typical behaviour. For example, to perform a bone scan, a molecule is needed that, when introduced into a patient's body, will be concentrated in the skeleton at sites of higher cellular activity that could be related to local bone metastasis in cancer patients. With cardiac tomography, the target organ is the left ventricular wall of the heart, where tracer molecules will be selectively taken up in relation to blood flow or perfusion. This can distinguish between healthy tissue and tissue that has been damaged by reduced blood flow. Another important example is fluorodeoxyglucose (FDG), which behaves like glucose in terms of uptake by cells. FDG is the most commonly used tracer in PET tumour imaging, based on the principle that glucose metabolism is increased in cancer cells, which are more dependent on anaerobic glycolysis.

Another factor in selecting tracers is the ability to label them with radioactive isotopes. The isotope is included in the molecule as an additional atom or replacing an atom within the original molecular structure, as for example in the FDG molecule used for PET (Figure 6.10).

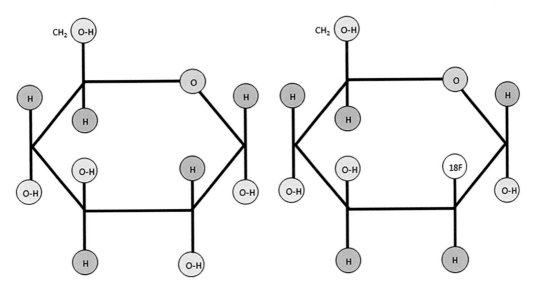

FIGURE 6.10   Glucose molecule (left) and 18F labelled fluorodeoxyglucose (FDG) used in PET scans (right).

TABLE 6.2

Physical Characteristics of Isotopes Used in Conventional Nuclear Medicine Imaging and in PET Applications

Radionuclide	$T_{1/2}$ (d)	Decay Mode	Main Gamma or X Emissions $E_{(g,x)}$ (keV)	Main Electron or Positron Energy Emissions (keV)
I131	8.00	β^-	284, 365, 637	606
Ga67	3.26	EC	93, 185, 300	93
In111	2.80	EC	23, 171, 245	219
I123	0.55	EC	27, 159, 529	158 (<1%)
Tl201	3.04	EC	71, 135, 167	153
Tc99m	0.25	γ	141, 21, 18	138
C11	0.01	β^+	511	960
Ga68	0.05	β^+	1077, 1883	1899, 822
Rb82	0.001	EC, β^+	776	3350
F18	0.08	β^+	511	634
O15	0.0014	β^+	511	

Radioactive isotopes used for labelling tracers (Table 6.2) can be gamma emitters for conventional nuclear medicine applications or positron emitters for PET. For gamma emitters, the gamma ray photon energy should be high enough to exit from the patient's body and then interact with the detector. For PET tracers, the positron energy is less important as imaging depends on coincidence detection of the two annihilation gamma ray photons, which always have the same energy of 511 keV.

Half-lives of isotopes used in PET applications range from a few minutes (e.g. ^{15}O, 2 minutes) to a few hours (e.g. 18F, 1.83 hours). For gamma emitters, they range from a few hours (e.g. 99mTc, 6.01 hours) to a few days (e.g. 67Ga, 3.26 days; 111In, 2.81 days, 131I, 8 days). PET tracers with very short half-lives can be used only at sites that are very close to where the isotopes are produced. Isotopes with long half-lives are not used because of radiation protection concerns for both the patient and their contacts.

6.5 GAMMA CAMERA PRINCIPLES AND CONSTRUCTION

The post-uptake, two-dimensional distribution of a radiopharmaceutical is detected using a gamma camera. In basic systems (Figure 6.11), gamma ray photons emitted from the patient pass through a collimator and interact in a scintillator crystal which is coupled to an array of PMTs or semiconductor detectors usually through a light guide. The scintillator acts as a transducer, converting the energetic gamma rays into visible light that can be translated into an electric signal which is then decoded and displayed on a monitor.

6.5.1 The Collimator

To understand the purpose of the collimator, consider a point source and the image that would be created if it was placed in front of an extended detector (Figure 6.12).

Gamma rays emitted from the source would interact with the whole detector and the image would consist only of an intensity distribution reflecting the distance between the source and points on the

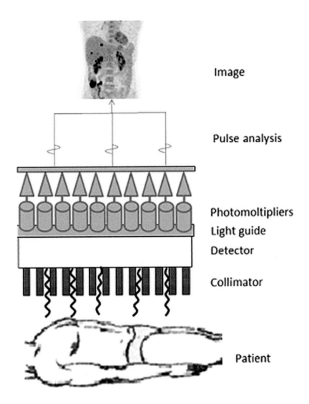

FIGURE 6.11 Gamma camera schematic.

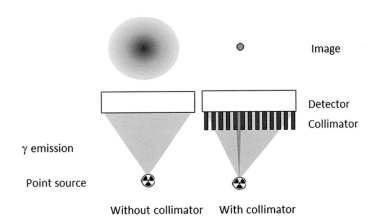

FIGURE 6.12 Imaging of a point source without the collimator (left) and with the collimator (right).

detector face according to the inverse square law. The insertion of the collimator has the effect of absorbing all the gamma ray photons not striking the detector in an orthogonal direction and so the source is more accurately imaged.

Similarly, consider a radioactive distribution within the body of a patient (Figure 6.13). Some gamma ray photons will be emitted in an orthogonal direction with respect to the collimator face (γ1 and γ2 in Figure 6.13), some will be incident at an angle (γ3) whilst others will scatter within the patient (γ4) before the secondary gamma ray photon is incident in an orthogonal direction (γ5).

The principal function of the collimator is therefore to reduce image blurring arising from gamma ray photons interacting with the scintillator that have not been emitted in an orthogonal direction towards the detector from the radiopharmaceutical distribution in the patient. However, collimators are not perfect. In Figure 6.13, the gamma ray photon detection at position A provides genuine data, the scattered gamma ray photon detected at position B is not conveying genuine information, the gamma ray photon interacting at position C (from γ6) is incident in a non-orthogonal direction and is not carrying genuine information but has penetrated the lead septum of the collimator and has been detected. Gamma ray photon γ2 would be useful for image formation but is absorbed in the collimator and does not reach the detector.

Collimator design requires compromises to be made (see Figure 6.14). Image resolution can be increased by reducing the diameter of the holes in the collimator but this will decrease sensitivity. Conversely, larger holes in the collimator increase sensitivity but decrease resolution. Higher radioisotope energies require an increase the thickness of the collimator, including the septa.

Typical materials used in collimator construction are lead ($Z = 82$) and tungsten ($Z = 74$). Collimators are classified in terms of image characteristics (high-resolution, high sensitivity or general purpose) and also according to gamma ray photon energy range (low, medium, high and ultra-high energy collimators).

The collimators used in most studies have parallel holes and septa to provide a direct correspondence between the radiopharmaceutical distribution and the reconstructed image. Holes and septa can also be divergent or convergent to magnify or compress image size, respectively.

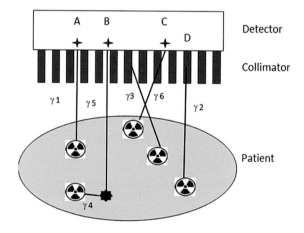

FIGURE 6.13 Event detection: influence of the collimator.

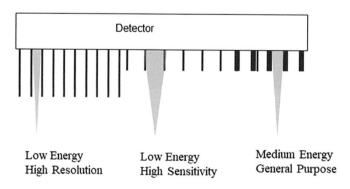

FIGURE 6.14 Characteristics of collimators according to energy range and spatial resolution.

6.5.2 THE SCINTILLATOR CRYSTAL

A scintillator crystal is an energy converter. Gamma ray photons interact with the scintillator crystal and an electron-hole pair is created. Small amounts of impurities (activators) are added to the crystal structure and create energy levels to which electrons are excited before returning to the valence band. The emission spectrum is in the near-UV or visible range to match the wavelengths where photomultipliers are most effective. De-excitation occurs quickly with typical half-lives of activator excited states in the region of 10^{-7} seconds (Figure 6.15).

Scintillators may be organic or inorganic compounds with added impurities to create the activation energy levels. Inorganic scintillators may be doped with thallium to produce the activation levels.

The principal characteristics of some common scintillator materials are shown below in Table 6.3.

Sodium iodide doped with thallium (NaI(Tl)) was one of the first scintillators developed and is still used in radionuclide detection. The attenuation coefficient at 140 keV is 2.44 cm^{-1}, sufficient to achieve good detection efficiency with a relatively thin crystal. The crystal thicknesses in a gamma camera is typically approximately 1 cm. NaI(Tl) also has a good light yield (38 photons/keV), a fundamental parameter associated with both energy and spatial resolution performance. Moreover, NaI(Tl) crystals can be grown to large sizes (>70 cm in diameter), allowing a detector to be shaped from a single crystal of sufficient size for body imaging applications. The scintillation light rise time is fast and the decay time is relatively short (<250 ns) enabling relatively high count rates (of the order 10^5 per second) without significant losses.

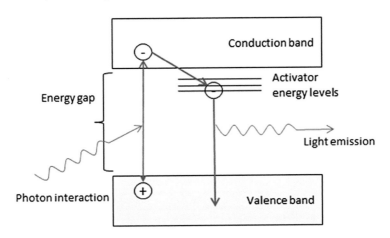

FIGURE 6.15 Light emission from crystal after interaction between gamma ray photon and electron in the valence band.

TABLE 6.3
Physical Characteristics of Crystals Used in Nuclear Medicine and PET Applications

	NaI(Tl)	$Bi_4Ge_3O_{12}$ (BGO)	Lu_2SiO_5 (LSO:Ce)
Light yield (photons/keV)	38	8.2	25
Emission peak (nm)	410	480	420
Decay time (ns)	230	300	40
Density (g/cm³)	3.7	7.1	7.4
1/μ (cm) – 140 keV	0.41	0.086	0.11
1/μ (cm) – 511 keV	3.1	1.1	1.2

6.5.3   SIGNAL ANALYSIS

Gamma ray photons stopped in the scintillator crystal emit light isotropically around the point of interaction. Usually this light is detected in more than one photomultiplier tube (Figure 6.16).

Output signals from all the detectors in the camera (up to 91) are summed to produce a 'Z pulse' representing the total energy deposited by the gamma ray photon. Energy discrimination circuitry determines whether the Z pulse is from a primary gamma ray photon. This is important in order to eliminate gamma ray photons which may have been scattered in the patient prior to being detected (interaction B in Figure 6.13) and are thus not representative of the true tracer distribution in the patient.

Interaction positions are identified by an analogue logic system using a weighted average of the light contribution from each detector (Figure 6.17).

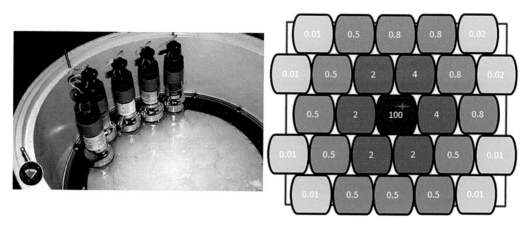

FIGURE 6.16   PMT assembly coupled through a guide light to the crystal (left); light intensity collected from PMTs decreases with increasing distance from the point at which the gamma ray photon interacts (right).

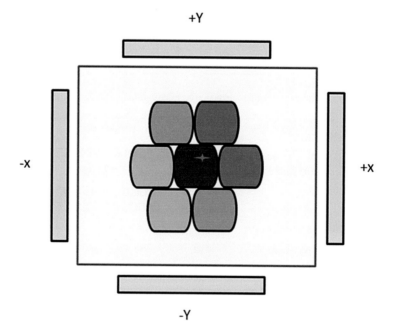

FIGURE 6.17   Signal combination to define x and y coordinates of event position.

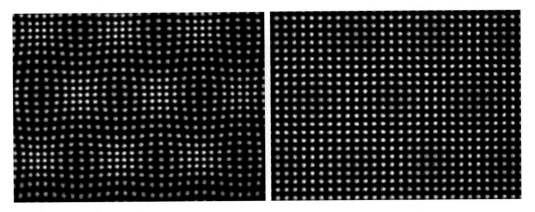

FIGURE 6.18 Point collimator image before (left) and after (right) linearity calibration.

Weights are attributed to each photomultiplier by a matrix of impedances that progressively attenuate the contribution of each PMT according to its position in the ±x and ±y directions relative to the centre of the scintillator crystal. The four components are then combined to give the position signals X and Y. In modern gamma cameras, all these operations are performed digitally using analogue-to-digital converters.

Linearity calibration is performed after installation and repeated if subsequent quality assurance measurements indicate increases in non-linearity. This calibration ensures accurate correspondence between the true gamma ray photon interaction point and its representation in the image. Calibration is performed using a line source moved sequentially across the image in the x and y directions or alternatively using a point source (Figure 6.18).

The resultant digital image is displayed with brightness levels associated with the numerical value of each matrix element.

Photomultiplier tubes used in gamma camera systems are selected according to a number of important characteristics, including the following.

- Size of the entry window
- Spectral sensitivity
- Gain
- Dynamic range and linearity
- Speed
- Sensitivity to magnetic fields

Another essential requirement is that the PM tubes are matched ('tuned') in terms of response and performance. Tuning requires adjustment of the voltages and gains in each individual tube to ensure near identical responses. Also, the camera head must be light tight to ensure no stray light enters the system from outside.

6.5.4 HYBRID PHOTODETECTORS AND SILICON-BASED PHOTODETECTORS

Solid state or semiconductor detectors are increasingly being used to replace photomultiplier tubes. Their advantages include small size, which improve spatial resolution performance but also allows greater versatility of configuration. They are also not susceptible to magnetic fields which is important for systems integrated with MRI scanners. The main hindrance to their more widespread use is cost.

A number of different detector types are available, usually referred to by their acronyms; SiPM (Silicon PhotoMultipliers), SSPM (Solid State PhotoMultipliers), and MAPD (Micropixel Avalanche PhotoDiodes) or alternatively GAPDs (Geiger Avalanche PhotoDetectors). Such detectors are usually organised in 2D-arrays composed of several layers with different doping concentrations in a common silicon substrate. Each structure is known as a microcell and acts as an independent detector. The typical dimension of a single microcell is 50×50 μm with a thickness of around 5 μm.

In these detectors, a small photodiode is biased several volts above breakdown voltage. When a gamma ray photon produces an electron-hole pair in the photodiode, the junction becomes conductive and a current flows through the diode until the bias voltage reaches the breakdown voltage, at which point the current stops.

6.5.5 ACQUISITION PARAMETERS

The two most important physical parameters defining the performance of nuclear medicine systems are sensitivity and spatial resolution, which are inter-related. In general, better spatial resolution reduces sensitivity and *vice versa*.

The simplest acquisitions made with a gamma camera are static planar acquisitions of the count distribution from the tracer. These do not provide depth or temporal information. Parameters characterising a planar image mainly relate to the acquisition matrix. The detection of smaller details increases the number of pixels required to display the acquired image. Generally, matrix sizes adequate for most clinical applications are 64×64, 128×128 and 256×256 pixels.

Another relevant parameter is acquisition time, which will be influenced by the matrix size required in the resultant image but also by the sensitivity of the system. During the study, patient movement must be minimised and this is also a consideration in terms of acquisition times. Technological improvements allow most modern systems to acquire images with larger matrix sizes (up to 256×256) within reasonable acquisition times.

Alternatives to static planar images include the acquisition of dynamic sequences, where multiple image frames which can be played sequentially in an image loop enable the demonstration of temporal trends in tracer distribution. Finally, tomographic acquisition allows two dimensional slices and three dimensional images to be acquired, thus providing depth information.

6.6 TOMOGRAPHIC ACQUISITION

Tomographic acquisition has been available for many years in nuclear medicine using the technique of Single Photon Emission Computed Tomography (SPECT). In this technique, the gamma camera rotates around the patient acquiring images at discrete angles typically $3° - 6°$ apart (Figure 6.19).

A more recent development is PET. This technique detects the gamma photons produced following annihilation of positrons. A positron travels a very short distance (of the order of a millimetre) in the tissue before annihilating with an electron. Two gamma ray photons, each of energy 511 keV, are emitted in almost opposite directions. The annihilation gamma ray photons are detected through coincidence counting in a ring of detectors around the patient (Figure 6.20).

6.6.1 SINGLE PHOTON EMISSION COMPUTED TOMOGRAPHY (SPECT)

SPECT is a technique that allows reconstruction of a three-dimensional image of the patient's body, as opposed to the planar projection images acquired in conventional gamma camera imaging. The general principle of this approach is to collect projections at multiple different angles along a circular path around the object and use these to reconstruct a three-dimensional image of the distribution of radioactive material within it. The characteristics of the object and the degree of detail that needs to be represented in the image determine the number of projection angles needed.

The Nyquist theorem tells us that if we require a spatial resolution (pixel size) d in the image, then the linear sampling frequency v must satisfy the following relationship.

$$v < \frac{1}{2d}$$

(6.4)

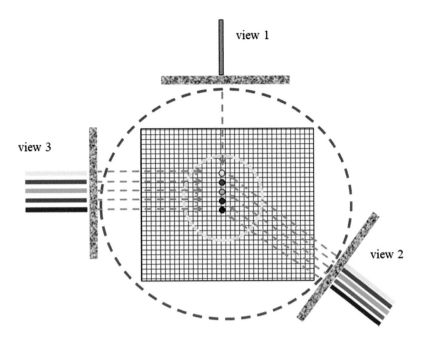

FIGURE 6.19 Example of three views of five point sources with the gamma camera head at different angles in a SPECT tomographic acquisition.

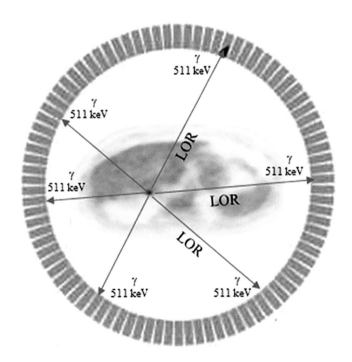

FIGURE 6.20 Example of PET emission and detection showing typical lines of response (LORs).

This sets a limit on the maximum spatial frequency that can be present in the acquired signal if aliasing is to be avoided. Filters are used to eliminate higher frequency signal components.

Once the linear sampling frequency needed to provide the desired level of detail has been defined, it is possible to calculate the corresponding angular sampling frequency and hence the required number of projection angles, N (Figure 6.21).

$$N = \frac{\pi D}{2d} \tag{6.5}$$

where D is the required image field of view.

As can be deduced from the formula, the number of views that must be sampled increases according to the field of view size and is inversely proportional to the size of the smallest objects that we want to distinguish in the image, which determines the required pixel size.

SPECT data is acquired in the form of a 'sinogram'. This is a two-dimensional representation of data collected from a slice of the patient's body as the gamma camera head is moved through different angles. It is therefore a two-dimensional image in which the horizontal axis represents the position of the count on the detector and the vertical axis indicates the angular position of the detector. Each line of the sinogram corresponds to the profile of counts acquired at each position across the detector face (each values of r) for at a given angle Θ (Figure 6.22).

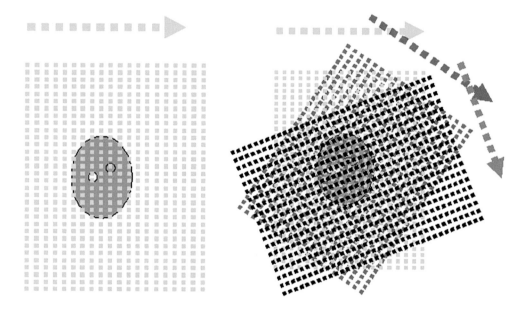

FIGURE 6.21 Linear sampling (left) and angular sampling (right).

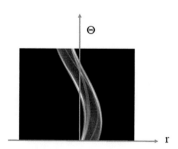

FIGURE 6.22 Sinogram: x-axis represents the position on the gamma camera head and the y-axis represents the angular position of the gamma camera head around the patient.

6.6.2 FILTERED BACK PROJECTION (FBP)

The process by which the sinogram is used to generate an image of the three-dimensional distribution of radioactivity in the object is called 'reconstruction'. Reconstruction algorithms are divided into two categories: analytical and iterative.

The most common analytical method is FBP. This method is very robust, although it is very sensitive to noise and there can be significant artefacts. Back projection allows us to obtain images of the object under examination starting from the acquired projections. For each angular projection, that is, each value of Θ in the sinogram, data from the corresponding profile is 'back projected' through the imaged object, that is, each pixel along a line running through the object, orthogonal to the face of the detector, is assigned the same signal value, depending only on the position of that line along the profile (the value of r in Figure 6.22). This is repeated for each value of Θ, that is, for each projection angle, and the overall signal intensity in each pixel is the sum of the intensities obtained from each of the separate back projections (see Figure 6.23).

This simple approach to back projection produces a poor image, because all pixels along each of the lines of back projection are assigned the same value, including those that are outside the imaged object. Therefore, typical artefacts called 'star artefacts' are produced in the reconstruction (Figure 6.24).

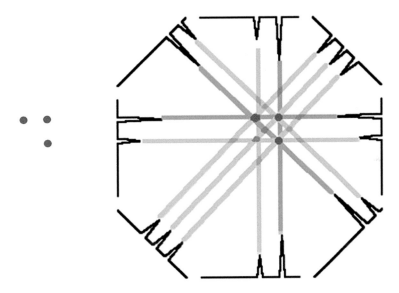

FIGURE 6.23 Three point sources comprising an imaged object (left); back projection of acquired profiles from eight angular projections (right).

FIGURE 6.24 Star artefacts in FBP using 2, 4 and 8 views.

By increasing the number of projections used (which is subject to a practical limit), star artefacts can be reduced in significance, but they still leave a blurred background that impairs image quality.

To overcome these problems, it is necessary to filter the images. This is done by carrying out a convolution operation between the image and a chosen filter function. This allows us to obtain an image as close as possible to the real spatial distribution of the radiopharmaceutical. The convolution process in the spatial domain is equivalent to a multiplication in the frequency domain, so for simplicity it is performed in the frequency domain following Fourier transformation. The function f(r) representing signal as a function of position is decomposed into its spatial frequency components, that is expressed as the weighted sum of trigonometric functions (sine and cosine) of different frequencies. Once the data has been transformed into the spatial frequency domain, it is filtered in such a way as to reduce the negative effects of the background. At this point, the filtered data is Fourier transformed back into the spatial domain, and the desired image can be obtained.

6.6.3 Iterative Reconstruction

The FBP remains the most widely used reconstruction technique, representing a good compromise between the quality of the data provided, the power required by the calculation system and the processing time. However, introduction of very powerful computers at ever-decreasing cost has seen the development of iterative reconstruction techniques to replace analytical ones. The main advantage of iterative methods compared to FBP is that they allow modelling of physical phenomena such as attenuation, gamma ray photon diffusion or system response, which vary according to distance. They can also take into account information known *a priori*, for example, patient geometry and attenuation. Moreover, more advanced statistical models than simple back projection of lines can be used.

The basic concept of IR, based on maximum likelihood methods, is simple: the distribution of activity in the reconstructed slice is assumed to be the distribution that has the highest probability of producing the observed projection data. Since no analytical solution is available, the reconstruction must be performed iteratively.

The reconstruction process starts with an estimate of the image, for example, an image reconstructed using FBP, or even a uniform distribution of activity. Numerical projection data are calculated from this estimate and compared with the measured projection data. The result of this comparison is used to adjust the image estimate, reducing the difference between these two data sets. This process represents an iteration. The process is then repeated multiple times until the difference falls below a threshold value and this final version of the image is accepted (Figure 6.25).

There are various IR algorithms, differing from each other depending on how the estimated and measured projections are compared and the type of correction applied to the current estimate at each iteration. The most common iterative techniques implemented on systems available on the market are the following.

Maximum Likelihood Expectation Maximisation (MLEM): an algorithm that searches for the statistically most probable image that reproduces the observed projections. It is very time-consuming because it updates the data after all the projections have been processed.

Ordered Subset Expectation Maximisation (OSEM): an accelerated version of MLEM which divides the projections into subsets, which have projections equally distributed in the space around the object, on which it works simultaneously.

Iterative methods have a great advantage over FBP as they can include some important physics phenomena that influence the formation of the image in the reconstruction process. For example, they take into account the Poisson statistical nature of radioactive decay, the system spatial resolution, the patient attenuation, the scatter correction, the penetration of the collimator septa and the energy resolution.

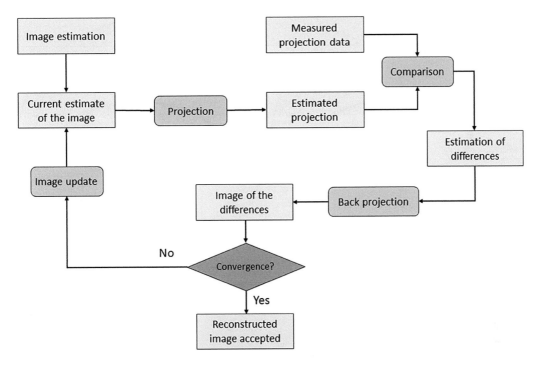

FIGURE 6.25 Flow chart of iterative image reconstruction scheme.

6.6.4 Positron Emission Tomography (PET)

In PET imaging, the position of positrons emitted by a tracer distributed in the patient's body is identified by recording the two 511 keV gamma ray photons produced when the positron annihilates with an electron. These are recorded as a coincidence if they reach the detector crystals within a narrow time window that depends on the properties of the crystals, in terms of conversion of the gamma ray photon energy into light. For example, bismuth germanate (BGO) has high effective atomic number as well as high density and hence high performance in detecting gamma ray photons, but its light output is fairly poor as it has a longer decay time constant and therefore enhanced dead time and a limited ability to cope with high count rates.

Lutetium oxyorthosilicate (LSO), on the other hand, provides considerably higher light output and hence permits development of high-resolution systems. Moreover, its decay time constant is shorter, so the dead time is reduced and a shorter coincidence time window is possible. This drastically curtails the likelihood of random events being incorrectly accepted as coincidences at higher activity levels.

To expand on this point, the high numbers of annihilation gamma ray photons emitted in PET can result in erroneous recording of events that are not true coincidences, as shown in Figure 6.26. Random coincidences and multiple coincidences involve gamma ray photons emitted in two different annihilation events, while in scatter coincidence one of the gamma ray photons is detected after being scattered in the patient's body and consequently losing information as to its original direction. All these situations result in incorrect information about the location of the positron annihilation and hence contribute to noise in the image.

PET image noise is characterised by the noise equivalent count rate (NECR) parameter, which takes into account these components and is given by the following mathematical formula.

$$\text{NECR} = \frac{T^2}{T + S + R} \tag{6.6}$$

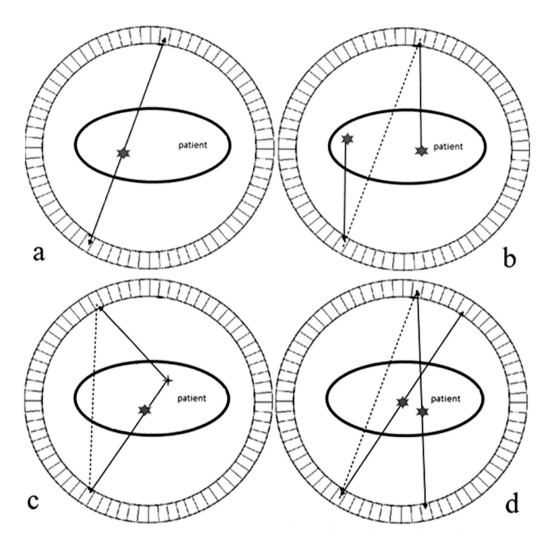

FIGURE 6.26 Coincidence events recorded in PET. a) true coincidence, b) random coincidence, c) scatter coincidence, d) multiple coincidence.

where T, R and S are the true, random and scatter coincidence count rates, respectively. NECR is often used to compare the performance of different PET scanners. Various corrections have been proposed by authors in the literature and are implemented by vendors, to estimate the contributions of S and R and to calculate T as a function of the total count rate detected by the scanner.

6.6.5 INTEGRATED SPECT/CT AND PET/CT

One of the main corrections that need to be applied to both SPECT and PET tomographic data is related to the attenuation of gamma ray photons emitted in the patient's body as they pass through tissues. This effect is proportional to the thickness of tissue crossed but also depends on the tissue type. Gamma ray photon attenuation, if not correctly considered, results in underestimated activity in central parts of the object in the reconstructed image (Figure 6.27).

In the past, various techniques were available for attenuation correction, the most common being to use attenuation data collected from a known external sealed source rotating around the patient. This allowed a correction to be calculated from transmission images, taking into account

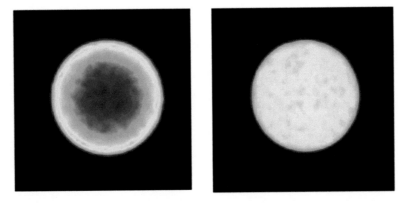

FIGURE 6.27 Cylindrical uniform phantom without (left) and with (right) attenuation correction.

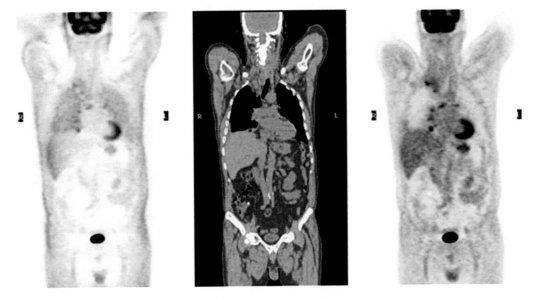

FIGURE 6.28 PET/CT acquisition: PET image without attenuation correction (left), CT image (centre) and PET image with attenuation correction (right).

the different thickness and materials crossed by gamma ray photons emitted by the sealed source. Typical isotopes used for this purpose were Cs137 for SPECT and Ge68 for PET. As an alternative method for symmetrical, round-shaped body sections, like the head, geometrical corrections were applied to reconstructed images using specific software.

Nowadays the most common way to correct for attenuation is with CT images acquired by integrated SPECT/CT and PET/CT systems. This is a very accurate correction as the real attenuation factor of each pixel is calculated from the corresponding CT numbers. In PET/CT systems, the attenuation coefficients obtained from the CT image, which apply at CT energies, are converted into the corresponding values for the annihilation gamma ray photon energies (511 keV) and then applied to correct each element in the image (Figure 6.28).

Hybrid SPECT/CT and PET/CT systems are replacing SPECT and PET scanners for another important reason. The poor spatial resolution and low contrast of SPECT and PET images sometimes make them difficult to interpret, particularly given the lack of anatomical references. Fusion with structural CT images alleviates this problem.

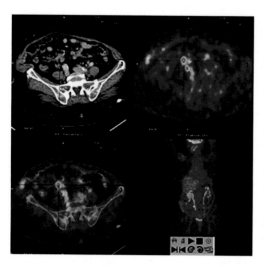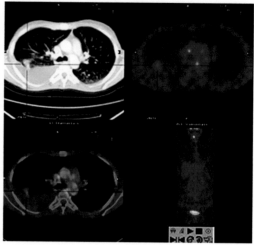

FIGURE 6.29 Examples of anatomical localisation of active lesions in PET imaging.

The uptake of radioactive tracer in a specific organ or structure can be an important part of the diagnosis, as pathological uptake can differ from normal physiological uptake. This can improve confidence of diagnosis and disease staging and aid the selection of appropriate treatment for the patient. A hybrid system allows anatomical assessment (CT) and functional assessment (PET/SPECT) in a single examination session, without moving the patient from the couch (and therefore in the same spatial reference system). Once reconstructed, the CT and nuclear medicine images are spatially co-registered, thus allowing the reader to see the precise localisation of the nuclear medicine data with respect to anatomy (Figure 6.29).

6.6.6 TIME OF FLIGHT CORRECTION

PET scanners based on crystals with a very short decay time and narrow coincidence window (for example, LSO or LySO), or based on digital detection that improves coincidence timing resolution, have become important for PET as they are able to measure the difference in the arrival times of the two annihilation gamma ray photons with high precision. This information can be used to localise the emission point within a small region of the object. This is called Time of Flight correction (TOF) (see Figure 6.30).

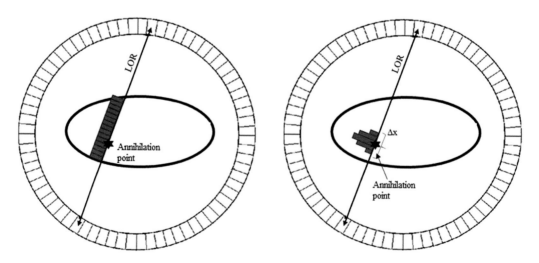

FIGURE 6.30 Signal distribution along LOR without (left) and with (right) Time of Flight correction.

The uncertainty in this localisation is characteristic of each system and is determined by its coincidence timing resolution, Δt, which is measured as the full width at half maximum of the histogram of TOF measurements from a point source (known as the timing spectrum). The corresponding uncertainty in spatial localisation (Δx) along the line of response (LOR) is given as:

$$\Delta x = c \frac{\Delta t}{2} \tag{6.7}$$

where c is the speed of light. TOF correction improves the performance of the PET scanner, particular sensitivity and spatial resolution. In clinical use, it can result in better contrast and lower noise, particularly when imaging heavier patients.

BIBLIOGRAPHY

Herzog, H.; "In vivo functional imaging with SPECT and PET"; *Radiochim Acta* 89, 203–214 (2001)

International Atomic Energy Agency, *Nuclear Medicine Physics*, IAEA, Vienna (2009)

International Atomic Energy Agency, *Cyclotron Produced Radionuclides: Guidance on Facility Design and Production of [18F]Fluorodeoxyglucose (FDG)*, IAEA, Vienna (2012)

International Atomic Energy Agency, *Quality Assurance for PET and PET/CT Systems*, IAEA, Vienna (2015)

Phelps, Michael E.; *PET Molecular Imaging and Its Biological Applications*, Springer editor (2004)

Qaim, S. M.; "Nuclear data for medical applications: an overview"; *Radiochim Acta* 89, 189–196 (2001)

Sorenson, James A., Michael E. Phelps; *Physics in Nuclear Medicine* WeB. Saunders Co. editor (1987)

Townsend, D. W.; "Physical principles and technology of clinical PET imaging"; *Ann Acad Med Singap* 33(2), 133–145 (2004)

7 Magnetic Resonance Imaging

Stephen Keevil
Guy's and St Thomas' NHS Foundation Trust, London, UK

Renata Longo
Università degli studi di Trieste, Trieste, Italy

CONTENTS

DOI: 10.1201/9780429155758-7

7.1   HISTORICAL INTRODUCTION

The discovery of nuclear magnetic resonance (NMR) in the 1930s and 1940s was an important step in the history of physics, confirming predictions of the quantum theory of the nucleus. It also secured Nobel Prizes for Isidor Rabi, Felix Bloch and Edward Purcell, the key scientists involved. Maybe Bloch had some inkling that NMR might one day have medical applications, having inserted his finger into his apparatus and obtained a strong signal (probably ill-advised even in those less health and safety conscious days!). But no one can have guessed that within a few decades NMR would have transformed medical diagnosis and made significant contributions to the health and quality of life of many millions of people.

Over the decades following its discovery, NMR became a standard tool in physics and chemistry, particularly for investigating the molecular structure of chemical compounds. It was applied to an ever-growing range of materials, including biological tissues and fluids. In 1971 the American physician and inventor Raymond Damadian discovered that NMR signals obtained from normal tissue and from cancerous tumours differed in certain respects, suggesting that a diagnostic test based on NMR might be possible and potentially of clinically value. But none of these early studies involved *imaging*: they lacked an effective means of mapping NMR signals in space so that a high-resolution image could be obtained. Damadian developed an approach to this, but the method used today was reported in 1973 by Paul Lauterbur, an American chemist, and further developed by the British physicist Peter Mansfield (later Sir Peter). Lauterbur and Mansfield were awarded the Nobel Prize in physiology or medicine in 2003 'for their discoveries concerning magnetic resonance imaging'. As so often in the history of science, the roots of the idea that Lauterbur and Mansfield exploited so successfully lay to some extent in earlier work, particularly by Herman Carr, Robert Gabillard and Vsevold Kudravcev. The fact that Damadian was not also awarded the Nobel Prize aroused strong passions at the time and remains controversial.

Since these early days, magnetic resonance imaging (MRI)[1] has gone on to revolutionise the practice of medicine, allowing high spatial resolution imaging of both structure and function throughout the body. In common with other imaging modalities, it is no longer purely a diagnostic technique, also having a growing role in planning and guiding treatment and in the evaluation of patients' response to therapy as well as in improving our understanding of basic physiology and neuroscience. Today, there are well over 30,000 MRI scanners in hospitals worldwide, and in the UK National Health Service alone several million patients are imaged annually. It is a modality of unparalleled flexibility, able to produce images containing a wealth of different physical, chemical and physiological information. Much of this flexibility can be exploited by programming existing scanners to collect and reconstruct data in new ways, rather than requiring expensive hardware upgrades (although these have a place too). Fifty years after Damadian's initial discovery, new MRI techniques and applications continue to emerge.

This chapter provides an introduction to the physics of NMR and MRI, focusing on the key concept of the 'pulse sequence', which gives MRI its power and flexibility. A range of specialised applications are included to illustrate how this flexibility has been exploited in specific medical disciplines and areas of research. There is also discussion of key aspects of MRI scanner hardware. The chapter includes a description of the biological effects and hazards of MRI, including current regulations and guidelines designed to mitigate the resulting risks and so ensure patient and staff safety in the challenging MRI environment. Although the physics of MRI cannot be appreciated properly without a certain amount of mathematics, the key ideas are presented conceptually and understanding all of the equations is not essential.

7.2 NUCLEAR MAGNETIC RESONANCE

7.2.1 THE NUCLEAR MAGNETIC MOMENT AND NMR-ACTIVE NUCLEI

A good place to start to understand NMR is to consider the meaning of the term 'nuclear magnetic resonance' itself.

> *Nuclear* – it is a phenomenon involving atomic nuclei.[2] As we shall see, not all nuclei can be studied using NMR, and almost all medical MRI involves only the nucleus of the hydrogen atom.
> *Magnetic* – this phenomenon involves the magnetic properties of nuclei, and specifically it occurs when nuclei are placed in a magnetic field.
> *Resonance* – NMR occurs when energy is applied to the nuclei at a specific *resonance frequency*. The nuclei absorb and then re-emit this energy, and analysis of the resulting signal can reveal various physical and chemical properties of the material under study.

NMR is fundamentally a quantum mechanical phenomenon. Fortunately that does not mean that it is necessary to have an advanced understanding of quantum mechanics in order to learn about and work with MRI: if that was the case, this would be a very different chapter, written by very different authors! For most purposes in medical MRI, it is sufficient to adopt a 'semi-classical' approach to the subject. By this, we mean that we will first need to establish what quantum mechanics tells us about the properties of atomic nuclei, because objects on such a small scale cannot be understood in any other way, but from then on we will proceed largely on the basis of classical (i.e. non-quantum) physics. This approach works well in most circumstances, but in some situations we will have to refer back to the quantum mechanical model to make sense of what is happening, and in others we will have to accept that our semi-classical approach is an approximation that doesn't bear too much scrutiny!

One of the key tenets of quantum mechanics is that when we consider very small objects, such as atoms and nuclei, many physical variables that appear to be continuous on the larger scale of everyday life are actually found to be *quantised*, i.e. they can only take certain discrete values. The intervals between these allowed values are extremely small, which is why quantisation usually only becomes apparent on the atomic scale. Variables that are quantised in this way include energy, electric charge and, crucially for our purpose, angular momentum. Specifically, the magnitude of the angular momentum of an atomic nucleus is given by the following equation:

$$|\mathbf{J}| = \hbar\sqrt{I(I+1)} \qquad (7.1)$$

where $\hbar = h/2\pi$, h is Planck's constant (6.63×10^{-34} J s), and I is a property of the nucleus known as the 'spin quantum number', or 'spin' for short. Quantisation of angular momentum arises because, under the rules of quantum mechanics, I can only take integer or half integer values (1/2, 1, 3/2, etc.), which in turn of course restricts the angular momentum to certain values too.

The value of I for a particular nucleus depends on the number of protons and neutrons that it contains. Protons and neutrons themselves both have spin $I = 1/2$. When they combine in a nucleus, protons pair off with each other in such a way that the spin of each pair of protons cancels out. If there is an odd number of protons, the unpaired proton contributes $I = 1/2$ to the overall spin of the nucleus. Neutrons pair off in a similar way. It follows that nuclei containing an even number of protons and an even number of neutrons have an overall spin $I = 0$. These nuclei do not undergo NMR and so are of no interest in the remainder of this chapter. But a nucleus that contains an odd number of protons, an odd number of neutrons, or both, will have nonzero spin. In simple cases with one unpaired proton or neutron, the nucleus has $I = 1/2$. However, more complex aspects of nuclear structure (specifically the existence of 'orbital angular momentum' as well as spin) mean that some nuclei have higher spin quantum numbers, although quantum mechanics dictates that the value is always an integer or half integer. The hydrogen nucleus, and most of the other nuclei that are of interest in biomedical applications of NMR, have $I = 1/2$, which is fortunate since the physics of nuclei with higher spin is considerably more complicated.

Table 7.1 lists some nuclei of potential biomedical interest, because of their role in tissue structure and/or metabolism, together with their spin. The 'relative NMR sensitivity' column indicates the amount of signal that would be obtained from samples containing equal numbers of each nucleus, relative to that of the hydrogen (^{1}H) nucleus. NMR is a relatively insensitive technique and the signals obtained are small, so it is advantageous to work with high sensitivity nuclei. The 'natural abundance' column indicates what percentage of nuclei of the element in question found in nature are of the specific isotope listed. For example, in the case of carbon only 1.11% of naturally occurring carbon nuclei are of the ^{13}C isotope, which has an unpaired neutron and hence $I = 1/2$. The remaining 98.89% are ^{12}C nuclei, which contain even numbers of protons and neutrons, have $I = 0$ and hence are insensitive to NMR. It follows that in a sample of pure carbon we would only be able to detect around 1% of the nuclei using NMR, further reducing the sensitivity by two orders of magnitude in addition to the low NMR sensitivity of the ^{13}C nucleus itself. The final column indicates the relative abundance of the element in the body; for example, 12% of atoms in the body are carbon atoms. It is clear that the ^{1}H nucleus combines the highest NMR sensitivity with natural abundance close to 100% and very high abundance in the body: this is why the ^{1}H nucleus is so dominant in medical

TABLE 7.1

Properties of Some Nuclei with Nonzero Spin of Potential Biomedical Interest

Nucleus	Spin	Relative NMR Sensitivity	Natural Abundance of Isotope (%)	Abundance of Element *in vivo* (atomic %)
^{1}H	1/2	1.00	99.98	62
^{13}C	1/2	0.0159	1.11	12
^{14}N	1	0.00101	99.63	1.1
^{15}N	1/2	0.00104	0.37	1.1
^{17}O	5/2	0.02291	0.04	24
^{19}F	1/2	0.83	100.00	0.0012
^{23}Na	3/2	0.0925	100.00	0.037
^{31}P	1/2	0.0663	100.00	0.22

MRI. These ^{1}H nuclei are of course predominantly found in water molecules, so when we obtain an MR image of the body it is essentially (with many caveats, as we shall see below) a map of the distribution of water. The concentration of water in the body is 10,000 times that of other ^{1}H-containing compounds (with the exception of lipids) and of compounds containing other NMR-visible nuclei. However, there are specific applications in which other NMR-visible nuclei are exploited, such as spectroscopic studies of energy metabolism using ^{31}P (albeit with poor spatial resolution on account of the low sensitivity and *in vivo* abundance) and molecular imaging using fluorine-containing compounds introduced into the body artificially.

In our semi-classical model of NMR, we picture a nucleus with nonzero spin as though it was a solid ball spinning on an axis, with the rate at which it is spinning determined by the quantum number *I*. It will be immediately apparent that this is at best a crude representation of a quantum-mechanical object that is at once a wave and a particle, but, as discussed above, it is adequate for our purposes. An atomic nucleus possesses an electric charge (recall that each proton carries a single quantum of electric charge and that neutrons are uncharged). We therefore picture a spinning ball of electric charge. This is where we can begin to apply classical physics. A moving electric charge of course generates a magnetic field, and our spinning nucleus is no exception. We define the *magnetic moment* of the nucleus as follows:

$$\mu = \gamma \mathbf{J} \tag{7.2}$$

where γ is a quantity known as the *gyromagnetic ratio*, which has a specific value for each nuclear species. For the ^{1}H nucleus, $\gamma = 42.57$ MHz T^{-1} or 2.68×10^8 rad s^{-1} T^{-1}.

NMR-sensitive nuclei are thus *magnetic dipoles*: in effect minute bar magnets, each with a north pole and a south pole and surrounded by a magnetic field. We are now ready to consider what happens when these nuclear magnetic moments are placed in a strong magnetic field.

7.2.2   THE EFFECT OF THE STATIC MAGNETIC FIELD

The most obvious feature of an MRI scanner is a large magnet. During imaging, the patient lying in the scanner, and hence the nuclei within the patient's body, is exposed to the magnetic field generated by this magnet, known as B_0. A magnetic field is of course a vector quantity, and so has both a magnitude (or strength) and a direction. In most (but not all) medical MRI scanners, the *direction* of the magnetic field inside the scanner is along the patient's foot-to-head direction. The *magnitude* of the field is expressed in tesla^3 (T), and for a clinical system the field strength over the imaging volume inside the scanner is usually 1.5 T or 3 T (unlike the other magnetic fields used in MRI it does not vary over time, and hence is known as the *static* magnetic field). There are an increasing number of whole-body research systems operating at 7 T, and a few at 9.4 T and higher. These are very strong magnetic fields by everyday standards: the average strength of the earth's magnetic field is 50 µT, and a strong fridge magnet may have a field of 5 mT close to its surface.

When a magnetic field is applied to a sample containing nuclear dipoles (in the case of medical imaging, the 'sample' is of course the patient!), the dipoles align themselves in specific orientations with respect to the applied field. The orientations depend on another quantum number, known as the *magnetic quantum number*, m_I. In general, m_I can take a range of values in integer steps from $-I$ to I. For our simple case where $I = 1/2$, therefore, we have $m_I = \pm 1/2$, and the nucleus can adopt two possible orientations with respect to the field. According to the rules of quantum mechanics, these orientations are such that the *z*-component of the nuclear magnetic moment (i.e. the component lying along the direction of the static magnetic field) is given by the following expression:

$$\mu_z = \gamma \hbar m_I = \pm \frac{1}{2} \gamma \hbar \tag{7.3}$$

The two possible values here correspond to a nucleus lying with a component of its magnetic moment parallel or antiparallel to the field direction (sometimes referred to as 'spin up' and 'spin down', respectively). Knowing the magnitude of the magnetic moment, μ, and the z-component μ_z, it is easy to calculate that the magnetic moment, in either orientation, lies at an angle of 54.7° to the magnetic field. This is left as an exercise for the reader.

According to classical electromagnetism, a magnetic moment lying at an angle to a magnetic field experiences a torque (or turning force). This causes the magnetic moment to rotate, or *precess* around the direction of the magnetic field (which, recall, is the z-axis). The frequency of this precession, known as the *Larmor frequency*, ω_0, depends only on the gyromagnetic ratio of the nucleus involved and the strength of the magnetic field[4]:

$$\omega_0 = \gamma B_0 \tag{7.4}$$

This will turn out to be the most important equation in NMR and MRI. The situation is illustrated in Figure 7.1, which shows nuclei in the two possible orientations with respect to the magnetic field. The broad white arrows represent the spin of the nucleus, generating the magnetic moment, and the black circular arrows represent precession around the z-axis. We define x- and y-axes to be perpendicular to the z-axis, but their orientation within the xy-plane (which is often known as the *transverse plane*) is of no particular significance. During precession, the x- and y-components of the magnetic moment are constantly changing, but the z-component remains constant, as given by Equation 7.3.

So far, we have not considered whether equal numbers of nuclei adopt each of the two possible orientations in the magnetic field. It turns out that they do not, and that if they did this would be a very short chapter!

The energy possessed by a nuclear magnetic moment with $I = 1/2$ in a magnetic field is given by the following equation:

$$E = \pm \frac{1}{2} \hbar \gamma B_0 \tag{7.5}$$

where again the two values arise from the two possible orientations of the magnetic moment with respect to the field.

It is therefore clear that there is a difference between the energy of nuclei in the two orientations, given by the following expression:

$$\Delta E = \hbar \gamma B_0 = \hbar \omega_0 \tag{7.6}$$

where ω_0 is the Larmor frequency of the magnetic moments, which we have encountered before.

The two nuclear spin orientations correspond to two different quantum mechanical energy levels. This makes sense: it takes more energy for a spin to be oriented against the field direction than with it. It also follows that we could excite a nucleus from the lower to the higher energy level by adding the correct amount of energy, ΔE, to the system. This observation will be important in the next section.

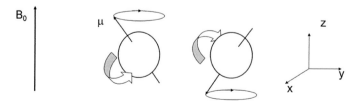

FIGURE 7.1 ^{1}H nuclei with spin $I = 1/2$ precessing in a static magnetic field B_0 oriented along the z-axis. The two orientations shown correspond to $m_I = \pm 1/2$.

Because nature tends to prefer the lowest energy configuration, we would expect to find more nuclei in the lower energy level than the higher one. It can be shown that the difference between the populations of the two energy levels is given by the following expression:

$$\frac{N\!\uparrow - N\!\downarrow}{N} \approx \frac{\hbar \gamma B_0}{2kT} \tag{7.7}$$

where $N\!\uparrow$ and $N\!\downarrow$ are the numbers of nuclei in the lower (spin up) and higher (spin down) energy levels, respectively, $N = N\!\uparrow + N\!\downarrow$ is the total number of nuclei in the sample, T is the absolute temperature and k is Boltzmann's constant (1.38×10^{-23} J K^{-1}). If we put numbers into this equation, we will find that the difference in population amounts to only a few nuclei per million. But it is this small excess of nuclei oriented along the direction of the magnetic field that gives rise to the NMR signal.

We can now move on from individual nuclei to consider the sample as a whole, which contains very many (typically $\gg 10^{18}$) nuclei. What do the phenomena that we have discussed above imply about the magnetic properties of this macroscopic object? We have seen that the sample contains a large number of nuclear magnetic moments with a component parallel to the z-axis (the direction of the applied magnetic field) and a slightly smaller number with a component antiparallel to this axis. If we add up all of the individual moments, it follows that the sample as a whole will have a small net magnetisation along the z-axis. This is termed the *bulk magnetisation*, **M**. What about the directions perpendicular to the field? One way to answer this is to recall that each nuclear magnetic moment is precessing around the field direction at the Larmor frequency, ω_0. At any given moment in time, each nuclear magnetic moment will be at a slightly different point in its precession, making a slightly different angle (or *phase*) in the xy-plane. Thus when we add up the x- and y-components of magnetic moment over the very large number of nuclei involved, they cancel to zero. So we have a situation in which the nuclear magnetisation within the sample (or patient) lies only along the z-axis, and there is no net x or y magnetisation.[5]

Figure 7.2 illustrates diagrammatically the distribution of nuclei between the two energy levels in the case where $I = 1/2$ (with the difference in population greatly exaggerated of course: it would be necessary to add another one million nuclei in each level to give an accurate impression!), the origin of the net magnetisation along the z-axis and the lack of any net magnetisation in the xy-plane.

7.2.3 THE EFFECT OF THE RADIOFREQUENCY FIELD

So far, we have considered the *nuclear* and *magnetic* aspects of NMR. Now we turn to the *resonance* aspects. This concerns the use of additional magnetic fields, which vary in time rather than remaining static like B_0, to manipulate the direction and size of the bulk magnetisation vector, **M**, that we have generated as described in Section 7.2.2. As we will see, this allows us to collect a signal from the spin system. This signal may contain a variety of physical and chemical information about the sample, and in the case of MRI it is used to form an image. We will approach this using classical electromagnetism, and then consider briefly what is happening in quantum-mechanical terms.

FIGURE 7.2 Distribution of precessing nuclei between energy levels with an energy gap ΔE. The difference in population has been greatly exaggerated, but gives rise to net magnetisation along the z-axis, while x- and y-components average to zero.

First of all, let us consider the effect of an additional magnetic field, B_1, on the bulk magnetisation, $\mathbf{M}$, that we have generated using B_0. We will assume that B_1 is orthogonal to B_0, i.e. it lies in the transverse plane. It is much weaker than B_0 (typically of the order of microtesla rather than tesla) and is applied for a short period of time (typically hundreds of microseconds, whereas B_0, the *static* magnetic field, is present continuously).

We saw earlier that nuclear magnetic moments experience a torque in the presence of B_0 which causes them to precess. In just the same way, there will now also be precession of the bulk magnetisation around the direction of B_1, causing the magnetisation vector to tip (or *nutate*) away from the z-axis and towards the transverse plane. But there is a complication. The bulk magnetisation is made up of numerous individual nuclear magnetic moments, all precessing at the Larmor frequency. In order for nutation to happen, it is necessary for the B_1 field to keep 'in step' with these precessing nuclei, i.e. the B_1 field must itself be rotating in the transverse plane at the Larmor frequency. For the magnetic field strengths used in NMR and MRI, the Larmor frequency lies in the same frequency range as the radiofrequency (RF) part of the electromagnetic spectrum (10s–100s MHz). Therefore, B_1 is often termed the *radiofrequency magnetic field*. We can generate such a field by using two RF coils aligned along the x- and y-axes, generating sinusoidal magnetic fields that are 90° out of phase with each other and so add up to produce a circularly polarised magnetic field. This is the *resonance* part of NMR: nutation only occurs if B_1 is applied at the correct resonance frequency, which is to say the nuclear Larmor frequency. This fact allows us to choose which nuclear species we wish to interrogate in the sample: B_0 generates bulk magnetisation from all nuclei that have nonzero spin (^{1}H, ^{13}C, ^{31}P etc.), but choice of the frequency of B_1 allows us to select which of these we wish to study (in the case of MRI, almost always ^{1}H).

Figure 7.3(a) shows bulk magnetisation, $\mathbf{M}$, aligned along the B_0 field. Once B_1 is applied, $\mathbf{M}$ tips away from the z-axis because of precession around B_1, but of course it will also precess around B_0 and so the overall motion is quite complicated, with the magnetisation vector following an expanding helical trajectory (Figure 7.3(b)). And recall that the B_1 field itself is rotating in the transverse plane as well! This complexity is reflected in the mathematics needed to describe the situation. Fortunately, there is a conceptually straightforward mathematical 'trick' that allows us to simplify things greatly. As a thought experiment, let us imagine that we could run on a circular track around the sample at the same rate as the bulk magnetisation is precessing, i.e. at the Larmor frequency. This is rather an unlikely scenario given that the precession rate is tens or hundreds of megahertz, but it is after all a thought experiment! From this perspective, the B_1 field would appear stationary, and the precession of $\mathbf{M}$ around B_0 would disappear as well. We could in effect neglect the effect of B_0 altogether, and all that would be left would be the nutation of $\mathbf{M}$ as a result of B_1. For those who are more comfortable with mathematical rigour than with a somewhat implausible thought experiment, what we have just done is to effect a change from the *laboratory frame of reference*, i.e. the 'real world' in which the NMR experiment is performed, to a *rotating frame of reference* moving at the Larmor frequency, ω_0, relative to the laboratory frame. This new frame of reference

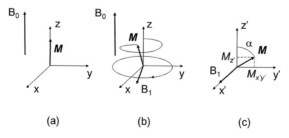

FIGURE 7.3 (a) Magnetisation vector in a static magnetic field B_0, (b) precession of magnetisation around both B_0 and B_1, and (c) motion simplified to nutation around B_1 in the rotating frame.

has coordinate axes x', y', z', where z' is actually exactly the same as z but x' and y' are rotating at ω_0 relative to the laboratory frame axes x and y.

In the rotating frame of reference, the motion of **M** is simplified to nutation from the z'-axis towards the $x'y'$ plane (Figure 7.3(c)). This nutation continues for as long as the B_1 field is present, and so by judicious choice of the strength of the B_1 field and of its duration, t_{RF}, it is possible to nutate the magnetisation vector through any arbitrary 'flip angle' α, given by the following expression:

$$\alpha = \gamma B_1 t_{RF} \tag{7.8}$$

The net effect of such an RF pulse is to reduce the longitudinal magnetisation (M_z) while generating transverse magnetisation for the first time, according to the following equations:

$$M_{z'} = |\mathbf{M}|\cos\alpha = |\mathbf{M}|\cos\left(\gamma B_1 t_{RF}\right) \tag{7.9}$$

$$M_{y'} = |\mathbf{M}|\sin\alpha = |\mathbf{M}|\sin\left(\gamma B_1 t_{RF}\right)$$

$$M_{x'} = 0$$

Here we assume that B_1 lies along the x'-axis in the rotating frame, so that the generated transverse magnetisation lies along the y'-axis.

A simple, and common, example would be to apply an RF pulse of the appropriate amplitude and duration to tip **M** completely into the transverse plane, leaving no magnetisation along the z'-axis. This is known, prosaically enough, as a '90° pulse'. A pulse that tips magnetisation through 180° into the $-z'$-axis is known as a '180° inversion pulse'. Later in the chapter, we will also encounter the '180° refocusing pulse', which is applied after a 90° pulse to 'flip' magnetisation over in the transverse plane.

The astute reader will be wondering what is going on here in terms of the individual nuclear spins that make up **M**. In quantum mechanical terms, an electromagnetic field of frequency ω_0 carries energy $E = \hbar\omega_0$. This is exactly the same as the energy gap between the two energy levels given in Equation 7.6. Nuclear spins in the lower energy level are therefore able to absorb energy from the RF field, exciting them to the higher energy level. Just as in the classical approach, this works only if the RF frequency matches the nuclear Larmor frequency. The effect is to reduce the difference between the populations of the two energy levels, and hence the size of the longitudinal magnetisation. Also, and less obviously, the excited nuclei are made to precess in phase with each other, so that net magnetisation is generated in the transverse plane. So the quantum-mechanical result accords with the classical model, and all is right with the world.

7.2.4 Signal Detection

The key effect of the RF pulse is to generate transverse magnetisation, whereas previously only longitudinal magnetisation existed as a result of the B_0 field. In the rotating frame this magnetisation is stationary and orthogonal to the direction along which B_1 was applied. But of course, we are observing the situation in the laboratory frame, rotating at frequency ω_0 relative to the rotating frame. In the laboratory frame Equations 7.9 become as follows:

$$M_z = |\mathbf{M}|\cos\left(\gamma B_1 t_{RF}\right) \tag{7.10}$$

$$M_y = |\mathbf{M}|\sin\alpha\cos\omega_0 t = |\mathbf{M}|\sin\left(\gamma B_1 t_{RF}\right)\cos\omega_0 t$$

$$M_x = |\mathbf{M}|\sin\alpha\sin\omega_0 t = |\mathbf{M}|\sin\left(\gamma B_1 t_{RF}\right)\sin\omega_0 t$$

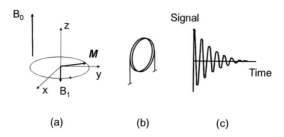

FIGURE 7.4 (a) Magnetisation vector rotating in the transverse plane, (b) radiofrequency receive coil, and (c) detected free induction decay signal.

So in the laboratory frame, we have a time-varying magnetisation. Faraday's law of induction tells us that a changing magnetisation will generate an electromotive force (EMF, or voltage) in a nearby electrical conductor. If this conductor is connected to a circuit, an electric current will flow, which we can detect and analyse. We can imagine placing a loop of copper wire next to the sample on which we are conducting our NMR experiment, resulting in generation of what is known as a *free induction decay* signal (FID). This is our first NMR signal. This simplistic picture is not that far removed from reality: the detectors (or 'receive coils') used in MRI scanners are in their simplest form literally copper loops.

Figure 7.4 illustrates rotating magnetisation following a 90° pulse, a receive coil and a typical FID signal. The signal oscillates at the Larmor frequency, matching the rate of change of the magnetisation that is generating it. It also decays over time, which we will explore in the next section. In the context of chemical NMR, analysis of signals like this yields information about molecular structure. In the context of MRI, the signals contain the information that we will use to form our image, but we are some way from that point as yet!

7.2.5 NUCLEAR RELAXATION: T_1 AND T_2

Figure 7.4 shows that the FID signal does not persist indefinitely, but decays exponentially; it has usually disappeared within a few hundred milliseconds of the excitation pulse. This is an example of *nuclear relaxation*. Once again, we will start with a description of this phenomenon on a macroscopic scale, and then pause to consider what is happening at the level of individual nuclei.

When discussing relaxation, we need to consider the behaviour of longitudinal magnetisation and transverse magnetisation separately. Let us imagine that we have applied a 90° pulse. Immediately after the pulse, there is no longitudinal magnetisation, but if were able to observe M_z over time, t, we would find that it gradually returned to its original size according to the following equation:

$$M_z = |\mathbf{M}| \left(1 - e^{-\frac{t}{T_1}} \right) \tag{7.11}$$

Here, T_1 is a parameter known as the *longitudinal relaxation time* or (for historical reasons) the *spin-lattice relaxation time*, or simply as the T_1 relaxation time.

Turning to the transverse magnetisation (M_{xy}), over time following the 90° pulse we would observe an exponential decay according to the following equation:

$$M_{xy} = |\mathbf{M}| e^{-\frac{t}{T_2}} \tag{7.12}$$

This is almost (but not quite, as we will see below) what we are seeing in the decay of the FID signal in Figure 7.4. Here T_2 is another parameter known as the *transverse relaxation time*, the *spin–spin relaxation time*, or simply the T_2 relaxation time.

T_1 and T_2 are determined by the environment in which magnetisation is located, and particularly by water content. As a result, relaxation times vary between different body tissues and also between healthy and diseased tissue. This is the main source of contrast (difference in signal intensity) in the majority of conventional MR images (although we can produce images that reflect a whole host of other tissue properties and physiological parameters as well). Table 7.2 shows relaxation time values in different brain tissues measured on a 1.5 T clinical MRI system, and Figure 7.5 shows relaxation curves for the three tissue types based on mean values taken from this table. The variation seen largely reflects differences in water content, with CSF, composed primarily of water, having by far the longest relaxation times and hence the slowest T_1 recovery and T_2 decay. As well as varying between tissues, T_1 is found to vary depending on the static field strength of the magnet used, while

TABLE 7.2
¹H NMR Relaxation Times of Brain Tissues Measured at 1.5 T

Tissue	T1 (ms) (mean ± SD)	T2 (ms) (mean ± SD)
Grey matter	1078 ± 53	100 ± 12
White matter	684 ± 21	79 ± 12
Cerebrospinal fluid (CSF)	3959 ± 306	914 ± 422

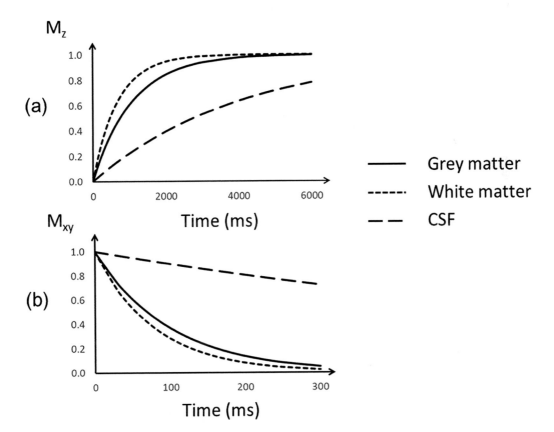

FIGURE 7.5 (a) Relative intensity of longitudinal magnetisation during T_1 recovery and (b) relative intensity of transverse magnetisation during T_2 decay in grey matter, white matter and cerebrospinal fluid in the brain following a 90° excitation pulse.

T_2 is largely independent of field strength. These observations reflect the biophysical mechanisms underlying relaxation, which are beyond the scope of this chapter.

There is a common misconception that during relaxation the magnetisation vector returns to the z-axis by retracing the trajectory that it followed during nutation. It is clear from the fact that T_2 is typically much shorter than T_1, and from the very different timescales on the x-axes in Figure 7.5(a) and (b), that this cannot be the case. For most tissues, transverse magnetisation decays away completely within a few hundred milliseconds of the initial excitation pulse, while longitudinal magnetisation takes quite a lot longer to recover. It follows that we must consider recovery of longitudinal magnetisation and decay of transverse magnetisation as different processes with different timescales.

In Section 7.1, it was noted that, back in 1971, Damadian observed differences in NMR signals obtained from healthy tissues and from cancerous tumours. These were in fact differences in relaxation times. Early in the history of MRI there was a view that the future of the technique lay in mapping T_1 and T_2 and using this quantitative information for diagnostic purposes, rather than simply producing images that would be interpreted by eye. Things did not turn out that way, but there are clinical applications in which relaxation time mapping has been shown to be of value.

In the quantum mechanical approach to NMR, nutation of the magnetisation vector results from absorption of energy from the RF field, exciting nuclei from the lower energy level to the higher one. Relaxation occurs when an excited nucleus releases the 'packet' (or 'quantum') of energy that it has absorbed, returning as a result to the lower energy level. If the quantum of energy is lost from the spin system, there is a net increase in the population of the higher energy level relative to that of the lower, resulting in an increase in longitudinal magnetisation. These events therefore contribute to T_1 relaxation. It is assumed that the emitted quantum of energy is absorbed by the surrounding environment, known for historical reasons as the *lattice*, hence spin-lattice relaxation. Alternatively, the emitted quantum may be absorbed by another nucleus in the lower energy level, so that two nuclei swap places and there is no net increase in M_z. However, the newly excited nucleus will not be precessing in phase with the other nuclei that were excited by the initial RF pulse, and so the net transverse magnetisation vector will be reduced in size. These events in which energy is exchanged between spins therefore contribute to T_2 (spin–spin) relaxation.

With this account of nuclear relaxation, our description of the behaviour of magnetisation in a simple NMR experiment is complete. However, there is a complication to consider, which appears to be inconvenient but in fact leads us to fertile new territory.

7.2.6 T_2* AND THE SPIN ECHO

In the initial discussion of T_2, it was noted that Equation 7.12 almost, but not quite, describes the decay of the FID signal following RF excitation. In fact, the exponential decay of the FID, and of the underlying transverse magnetisation, is faster than T_2 alone would suggest, and is described by the following equation:

$$M_{xy} = |\mathbf{M}| e^{-\frac{t}{T_2*}} \tag{7.13}$$

where T_2* (read 'T-2-star') is a modified transverse relaxation time, shorter than the actual tissue T_2. The cause of the difference lies in microscopic inhomogeneities in the static magnetic field, B_0. MRI equipment manufacturers go to great lengths to make the field as uniform as possible over the imaging volume. The field strength of a well set-up 1.5 T scanner will usually be within 1 part per million (0.0001%) of 1.5 T over a 40 cm diameter spherical volume (DSV) at the centre of the bore. However, and as soon as the patient's body is introduced into the scanner the field is distorted on a microscopic scale because of the differing magnetic properties (technically 'magnetic susceptibilities') of different body tissues and structures. As a result, elements of the transverse magnetisation in different places experience slightly different magnetic field strengths and hence precess at slightly different frequencies. The resulting dephasing causes the transverse magnetisation to decay more

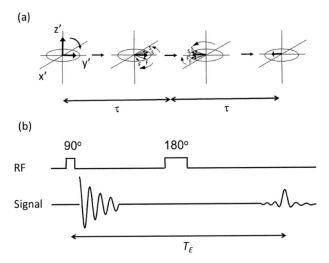

FIGURE 7.6 (a) Accelerated decay of transverse magnetisation due to magnetic field inhomogeneities, reversed by a refocusing pulse; (b) pulse sequence showing RF pulses, FID and spin-echo signal.

quickly than T_2 alone would predict. In MRI, loss of signal due to field inhomogeneities can be particularly marked close to boundaries between different tissues, for example, soft tissue and bone, and close to air-filled sinuses, and can result in significant image degradation and loss of diagnostic information.

Figure 7.6(a) illustrates the situation, and also the solution. The 90° excitation pulse generates transverse magnetisation, which is stationary in the rotating frame. But elements of this magnetisation experiencing different local B_0 field strengths due to inhomogeneities precess at different rates, some slower than average (labelled 'S') and some faster (labelled 'F'). Groups of spins precessing at different frequencies are known as *isochromats*. At a time τ after the 90° pulse, we apply another RF pulse, this time nutating the magnetisation through 180°. The 'fan' of isochromats that has developed as a result of inhomogeneity is 'flipped' to the other side of the transverse plane by this pulse, so that now the 'slow' isochromats are in front of the net magnetisation vector and the 'fast' ones are behind. The 'fan' therefore begins to close, and after another interval τ the magnetisation is back in phase. Note that the rephased (or refocused) transverse magnetisation is smaller than it was immediately after the 90° pulse: the 180° pulse reverses dephasing, but not true T_2 decay which has taken place during the time interval 2τ!

Figure 7.6(b) shows the sequence of RF pulses that have been applied, a 90° pulse followed after an interval τ by a 180° 'refocusing' pulse (alluded to previously in Section 7.2.3). It also shows the FID signal generated immediately after the excitation pulse, decaying with relaxation time T_2^*, and another signal generated at time 2τ when the magnetisation rephases. This latter signal is known as an *echo* and is smaller than the FID because of T_2 decay. The interval between the 90° pulse and the peak of the echo signal is known as the *echo time* (T_E); clearly, $T_E = 2\tau$. This is our first example of a *pulse sequence*: a series of RF pulses applied to generate a signal with particular properties. The pulse sequence concept is central to NMR and MRI, and we shall encounter many more examples.

Referring back to Figure 7.5(b), it is clear that the intensity of the NMR signals collected from different tissues at a given echo time depends on the T_2 values of those tissues, and that the contrast in signal intensity between tissues will depend on T_E as well. It follows that in medical MRI we can control the extent to which T_2 influences contrast (known as 'T_2-weighting') by varying the interval τ between the two RF pulses, and hence T_E. It is normal in MRI to collect a delayed echo signal, rather than an FID, using our receive coil, and as we will see below as well as allowing us to control image weighting the time interval this creates also allows us to add the extra pulse sequence elements required for imaging.

7.3 FROM SIGNALS TO IMAGES: ENCODING SPATIAL INFORMATION IN NMR

7.3.1 CONCEPTS IN MRI SPATIAL LOCALISATION

We have reached an important stage, and are now ready to consider how NMR, an important technique in analytical chemistry, can be transformed into MRI, a uniquely powerful medical imaging modality.

In the NMR experiments that we have considered so far, a signal (FID or echo) is collected from the entire sample under study, with no means of differentiating signals from different parts of the sample. For the purposes of imaging, we are going to have to find a means of 'tagging' (or *encoding*) parts of the signal according to where they originate from in space, and then using these tags to 'deconstruct' the acquired signal and map it to spatial locations. Since our starting point is a signal originating from a three-dimensional volume (the portion of the patient's body within the imaging volume at the centre of the magnet and the sensitive volume of the transmit and receive RF coils), we could, if we wished, encode signals in three dimensions and generate three-dimensional image data. This is indeed sometimes done, as we will see later, but it is more usual to first restrict the region from which signal is acquired to slices ('slice selection') and then encode data in two dimensions within these slices. Slice-by-slice, or 'tomographic', imaging is the norm in many medical imaging modalities. An important difference with MRI is that the means by which slice selection is achieved (discussed below) allows us to select slices in any orientation within the body: we are not restricted to transaxial slices, or even to a choice of transaxial, coronal and sagittal slices, but have a completely free choice to select oblique slices that align with anatomical structures of interest. For example, in neuroimaging it is usual to align nominally transaxial slices with the anterior commissure–posterior commissure (AC-PC) line in the brain, regardless of the tilt of the patient's head inside the scanner, and we can collect images aligned with the long and short axes of the heart, which are not conveniently oriented along standard anatomical planes.

In spite of this flexibility, to make things easier we will assume in what follows that we are going to select transaxial slices, parallel to the z-axis of the MRI scanner (defined by the direction of B_0), and that two-dimensional spatial encoding in the xy-plane is going to be needed to map signal to the correct location within these slices. It is important to be aware though that this is a convenient simplification and significantly underplays the capability of MRI in this respect!

A two-dimensional MR image is essentially a rectangular (not necessarily square, although we will assume that initially) array of voxels, the number of voxels in each direction determining the *matrix size*. The size of the image in each direction is known as the *field of view* (FOV), and the nominal spatial resolution on a given axis (the *pixel size*) is simply the FOV size divided by the matrix size, which is determined by aspects of the image acquisition process as we shall see later. In MRI, the actual in-plane spatial resolution is usually very close to the nominal value (typically 1–2

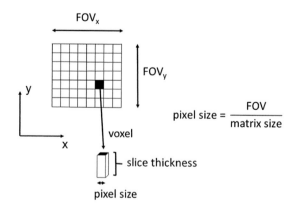

FIGURE 7.7 Definitions of some geometric parameters in two-dimensional MRI.

mm). However, voxels are of course three-dimensional objects, and spatial resolution in the through-plane direction is usually not as good as in-plane, typically 5 mm nominally and less well-defined because of slice profile issues that are discussed later.

7.3.2 MAGNETIC FIELD GRADIENTS IN SPATIAL LOCALISATION

As discussed in Section 7.1, the breakthrough that allowed NMR to be developed into a useful imaging technique was due to Lauterbur and Mansfield, albeit foreshadowed in the work of earlier researchers. This key concept is the use of *magnetic field gradients*.

A magnetic field gradient (also known as a switched gradient) is an additional magnetic field that is applied for a short period of time (known as a 'gradient pulse', typically lasting 1–2 ms) during the imaging process. Like all magnetic fields, it is a vector. Its *direction* is along the z-axis, aligned with B_0; its *magnitude*, or strength, varies linearly with position along some chosen direction within the scanner bore.

The terminology used in relation to gradients can be confusing: when we talk about an 'x-gradient', we mean a gradient field that varies linearly in *strength* along the x-axis: the *direction* of the field remains the z-axis. The term *gradient strength* can also be confusing, because it refers to how steeply the field strength varies with position, expressed in mTm^{-1}. Typical values are of the order of 10 mTm^{-1}, so the field strength variation due to the gradient is very small relative to B_0 (usually 1.5 or 3 T). Figure 7.8(a) shows an x-gradient field diagrammatically. We note that the field is antiparallel to B_0 for $x < 0$, passes through zero at $x = 0$ (the centre of the magnet bore, known as the *isocentre*) and is parallel to B_0 for $x > 0$.

During a gradient pulse, the total field experienced by magnetisation at a given position along the x-axis is the sum of B_0 and the additional gradient field (see Figure 7.8(b) where the spatial variation in field strength is greatly exaggerated of course). We can write the following equation:

$$B(x) = B_0 + G_x x \tag{7.14}$$

where $B(x)$ is the total magnetic field strength at position x and G_x is the strength of the x-gradient.

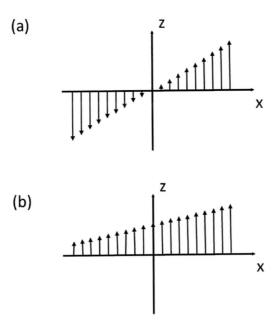

FIGURE 7.8   (a) Magnetic field gradient with the field oriented along the z-axis but varying in strength along the x-axis (an 'x-gradient'); (b) overall spatial variation in field strength along the x-axis due to B_0 and the gradient field.

The gradient fields are generated by specially designed magnetic field coils within the housing of the MR scanner. There are sets of coils to generate linearly varying fields along each of the x-, y- and z-axes. Gradients are generated when needed simply by passing an electric current through the relevant gradient coils: if we wish to produce gradient fields in oblique directions (in order to image oblique slices), this is done by passing currents through two or three coils simultaneously, with the strength and direction of the resultant field being the vector sum of the x-, y- and z-components.

We know from Equation 7.4 that the Larmor frequency depends on magnetic field strength. It follows that in the presence of a gradient (in this case along the x-axis, as an example) it also depends on position according to the following equation:

$$\omega(x) = \gamma B(x) = \gamma \left(B_0 + G_x x\right) \tag{7.15}$$

The Larmor frequency plays many important roles in NMR. It is the frequency of the RF pulse that we need to apply to nutate longitudinal magnetisation into the transverse plane, the frequency at which transverse magnetisation precesses around the z-axis, and the frequency at which the FID or echo signal oscillates. Perhaps we can see in this ability to make the Larmor frequency vary with position using magnetic field gradients a hint as to how we might encode spatial information into the NMR signal!

Figure 7.9 illustrates the effect of a gradient on the precession of transverse magnetisation, pictured from the perspective of an imaginary observer looking down onto the transverse plane. The arrows represent transverse magnetisation at different points along the x-axis and at different points in time during a gradient pulse. Before the gradient field is switched on, magnetisation precesses at the same frequency (determined by B_0), and therefore has the same phase (i.e. the same angle in the transverse plane) regardless of position along x (to keep things simple, for now we are going to neglect the dephasing effect of static field inhomogeneities discussed in Section 7.2.6). We then switch on an x-gradient of strength G_x, so that the Larmor frequency varies with position according to Equation 7.15. We can clearly see the resulting difference in precession frequencies in the figure, and that as a result the elements of magnetisation get out of phase with each other over time.

We can therefore view the effects of a gradient pulse on transverse magnetisation in two complementary ways.

- During the gradient pulse, the precession *frequency* of magnetisation varies linearly with position along the gradient direction according to Equation 7.15. This frequency variation is *constant* (magnetisation at a given location precesses at the same frequency throughout the gradient pulse) but *transient*: at the end of the gradient pulse, magnetisation goes back to precessing at the Larmor frequency determined by B_0 alone, regardless of position along the x-axis.

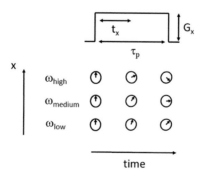

FIGURE 7.9   Variation in frequency and phase of transverse magnetisation in the presence of a magnetic field gradient.

- Because of this frequency difference, a difference in phase develops during the gradient pulse, so that at any point in time (for example, time t_x in Figure 7.9) there is a variation in *phase* along the x-axis, $\varphi(x)$, given by the following equation:

$$\varphi(x) = \omega(x)t_x = \gamma B(x)t_x = \gamma(B_0 + G_x x)t_x \qquad (7.16)$$

- The phase variation increases with time during the gradient pulse, but is *persistent*: the phase distribution that has built up at the end of the gradient pulse (given by substituting $t_x = \tau_p$ in Equation 7.16) is not suddenly reversed when the pulse is switched off, but remains until the transverse magnetisation decays away, or another gradient is applied which alters the phase distribution.

The spatially dependent phase variation imposed on magnetisation by a gradient depends on the product of the gradient strength and its duration, $G_x \tau_p$. This quantity is often rather strangely referred to as the gradient 'area', which makes a bit more sense once you see gradients depicted in pulse sequence diagrams.

Having explored the effect that magnetic field gradients have on the behaviour of transverse magnetisation, we are now in a position to apply this to the problem of spatial localisation. We will show how the spin-echo sequence that we discussed previously can be modified to produce an imaging sequence. We shall do this in three stages: first selecting a transaxial slice (as explained previously this is for illustrative purposes; in MRI we can image slices in any orientation) and then encoding two-dimensional spatial information into the signal obtained from that slice.

7.3.3   SLICE SELECTION

In Section 7.2.3, we discussed how we can nutate longitudinal magnetisation into the transverse plane by applying a time-varying magnetic field at the Larmor frequency, $\omega_0 = \gamma B_0$. We are now going to extend this principle to selectively excite transverse magnetisation within a transaxial slice of the patient's body.

Figure 7.10 shows diagrammatically a patient lying within the bore of an MRI scanner. The B_0 field, which lies along the bore axis in a conventional MRI system, generates longitudinal

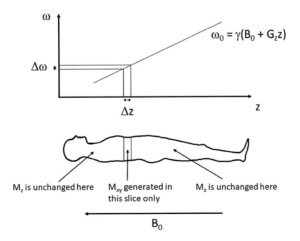

FIGURE 7.10   Slice selective excitation using a magnetic field gradient and RF pulse.

magnetisation throughout the patient's body. We now apply a gradient of strength G_z along the z-axis so that the Larmor frequency varies with position according to the following equation:

$$\omega(z) = \gamma B(z) = \gamma \left(B_0 + G_z z \right) \tag{7.17}$$

If, while the gradient is switched on, we also apply a 90° RF pulse, not at a single frequency but containing a range of frequencies $\Delta\omega$, we will generate transverse magnetisation within a slice of the patient's body in which the frequency content of the RF pulse matches the spatially varying Larmor frequency. It is easy to see that the thickness of this slice is given by the following equation:

$$\Delta z = \frac{\Delta\omega}{\gamma G_z} \tag{7.18}$$

So we can control the slice thickness by varying the frequency content of the RF pulse or the gradient strength.

What *shape* of RF pulse do we need to apply to achieve this? Well, assuming we want the selected slice to have a sharp 'top hat' shaped profile, the pulse shape in the time domain needs to be the Fourier transform of a 'top hat' function. This is the sinc function, given by the following equation and shown in Figure 7.11(a) together with an ideal slice profile, Figure 7.11(b).

$$\text{sinc}(x) = \frac{\sin(x)}{x} \tag{7.19}$$

Unfortunately, a sinc function is infinitely long, and so inevitably we have to truncate it for practical purposes (Figure 7.11(c)). This results in degradation of the profile of the excited slice (Figure 7.11(d)), and so reduction in spatial resolution in the through-slice direction and generation of artefacts. To improve this, the sinc function is often multiplied by a Gaussian function, but the result is still not perfect. More often these days, numerically optimised pulse shapes are used.

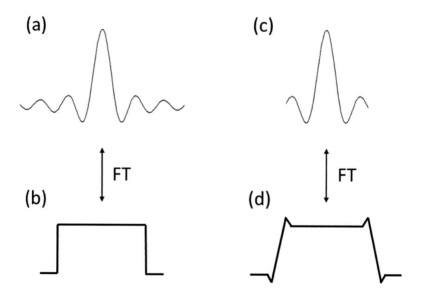

FIGURE 7.11 RF pulse shapes and corresponding slice profiles: (a) sinc pulse and (b) rectangular slice profile; (c) truncate sinc pulse and (d) degraded slice profile.

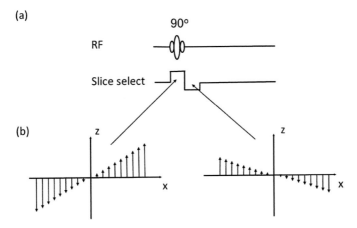

FIGURE 7.12   (a) Frequency-selective RF pulse, with associated slice selection and rephasing gradient, as they appear in a pulse sequence diagram; (b) amplitudes and directions of the two gradient fields.

As soon as magnetisation is generated in the transverse plane, elements of that magnetisation at different positions along the z-axis (i.e. at different positions within the thickness of the slice) begin to precess at different frequencies and so get out of phase with each other (recall the effect of gradients on phase, Equation 7.16). This leads to an undesirable loss of signal, but fortunately we can correct for it to a large extent. Remember that phase imparted by a gradient persists *until we apply another gradient*. Let us assume that nutation occurs all at once in the middle of the gradient pulse, so that accumulated phase is given by $\varphi(z) = \frac{1}{2}\gamma G_z z \tau_{ss}$ where τ_{ss} is the slice selection gradient pulse duration and the factor of ½ arises because of the assumption we have made that nutation occurs at $t = \frac{1}{2}\tau_{ss}$. To reverse this dephasing we can apply another gradient in the opposite direction to the slice selection gradient and with half the 'area' (recall this odd concept from Section 7.3.2). Of course, the assumption that we have made about the time at which nutation occurs is not quite correct (although it is a reasonable approximation, given the shape of the sinc pulse), so there is always some residual dephasing in the through-slice direction.

Figure 7.12(a) shows the slice selection gradient and RF pulse and the rephasing gradient, as they might appear in a pulse sequence diagram. Figure 7.12(b) illustrates the relationship between the directions and 'areas' of the slice selection and rephasing gradients.

We now have a situation in which transverse magnetisation, and therefore signal generation, is restricted to a single slice in the patient's body. There will be no signal from elsewhere, where magnetisation remains along the longitudinal axis. Our remaining challenge is to encode the signal collected from this slice in such a way that we can map it in two dimensions (x and y) and hence form an image.

7.3.4   FREQUENCY ENCODING

Having used the excitation process at the beginning of the pulse sequence to achieve slice selection, we next turn our attention to the opposite end of the sequence, when the spin echo is acquired. Recall that a spin echo is formed when magnetisation that has been dephased because of inhomogeneities in the static magnetic field is 'flipped' in the transverse plane and brought back into focus using a 180° refocusing pulse. The precession frequency of this refocused magnetisation is of course the Larmor frequency, ω_0. Let us imagine that we apply a gradient along the x-axis while the echo signal is being acquired. The precession frequency of transverse magnetisation will now depend on position along the x-axis, according to Equation 7.15. Therefore, the acquired echo signal will also contain a range of frequencies, with the strength of the signal at each frequency corresponding to the amount of signal coming from different locations along *x*. So we have used difference in frequency to encoded, or 'tag' elements of the signal according to their location in space. This technique is known as 'frequency encoding', and the gradient that we use as a 'frequency encoding gradient'.

There is a complication, however. Remember that differences in *frequency* due to magnetic field gradients result in accumulation of differences in *phase*: the frequency differences that we are using intentionally for spatial encoding will result in unwanted rapid dephasing of magnetisation at different locations along the *x*-axis, destroying the echo signal that we want to collect. Fortunately, once again there is a way to compensate for this. Before frequency encoding, we apply a gradient in the opposite direction to the frequency encoding gradient and of half the 'area'. The dephasing caused by this gradient is reversed during the first half of the frequency encoding gradient pulse, so that an echo forms at this point in time.

What we have done here is to use gradients to intentionally dephase magnetisation and then rephase it again, generating what is known as a 'gradient echo'. In many books on MRI, spin-echo and gradient-echo pulse sequences are presented separately, and the point is missed that what we conventionally call a spin-echo imaging sequence in fact involves collection of a simultaneous spin *and* gradient echo. The timing of the gradient echo is determined by the amplitudes and durations of the gradients on the frequency encoding axis (it occurs at the point in time at which the net gradient 'area' experienced by magnetisation is zero); the timing of the spin echo is determined by the time interval between the 90° and 180° pulses: it is the task of the pulse sequence designer to ensure that the two echo conditions coincide in time.

Figure 7.13(a) shows the frequency encoding gradient and preceding dephasing gradient as they appear in a pulse sequence diagram. Figure 7.13(b) shows how the phase of magnetisation evolves during these gradient pulses and the relationship to the echo signal. Once again for convenience, we neglect dephasing due to field inhomogeneities, which will be refocused anyway at the centre of the spin echo. The echo signal is collected using a receive coil, digitised to typically 256 or 512 samples, and stored in computer memory for analysis. The number of samples collected determines the matrix size, and hence the spatial resolution of the image, along the *x*-axis.

If we could somehow 'decompose' this echo signal to determine the amount of signal at each frequency, that would reveal the amount of signal at each position along the *x*-axis. Fortunately, a mathematical technique exists to do just this: the Fourier transform! We have achieved part of our aim, but we do not as yet have an image: Fourier transformation of the echo signal would give us a 'squashed' projection of a slice of the imaged object along the *x*-axis, with no spatial discrimination at all along the *y*-axis. Another step is needed.

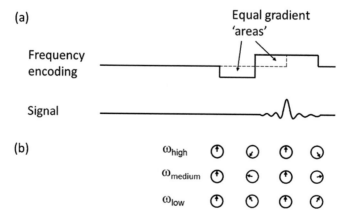

FIGURE 7.13 (a) Dephasing and frequency encoding gradients as they appear in a pulse sequence diagram; (b) phase evolution of transverse magnetisation and generation of a gradient-echo signal in the presence of these gradients.

7.3.5   PHASE ENCODING

Students of MRI sometimes ask at this point why we cannot simply apply two frequency encoding gradients along orthogonal axes at the same time to achieve two-dimensional spatial encoding. Because magnetic fields are vectors, this would simply result in a one-dimensional projection along an oblique axis at an angle determined by the strengths of the two gradients, not two-dimensional information at all. Instead, a mathematical 'trick' is needed, and to understand how this works we turn our attention back to Figure 7.13(b), and consider how the phase of magnetisation develops in the presence of the frequency encoding gradient.

Each of the phase distributions shown in the figure results from precession of transverse magnetisation in a constant gradient for a different period of time. The timepoints shown are illustrative of the typically 256 or 512 digital samples collected from the frequency-encoded echo: each sample contains a different phase distribution as the spatially dependent phase of the magnetisation evolves in the presence of the gradient according to Equation 7.20 (which is simply Equation 7.16 seen from the rotating frame of reference), where t_x is a timepoint during the acquisition of the echo:

$$\varphi(x,t_x) = \gamma G_x x t_x \tag{7.20}$$

We know from the discussion in Section 7.3.2 that we can impose a chosen phase distribution onto magnetisation using a gradient of the appropriate 'area' (i.e. product of strength and duration), and that this phase distribution will persist once the gradient is switched off. What if, in between excitation of transverse magnetisation and collection of the echo, we were to apply a gradient on the y-axis of strength G_y and duration t_y? Clearly, this would impose a variation in phase on the transverse magnetisation given by the following equation:

$$\varphi(y,t_y) = \gamma G_y y t_y \tag{7.21}$$

But, with the exception of the fact that we are now working on the y-axis rather than the x-axis, this looks exactly like the phase distribution that is present in a sample of the frequency-encoded echo (Equation 7.20). If we went on to collect an echo after 'preparing' transverse magnetisation in this way, this phase information would be 'locked up' inside the signal: inaccessible (since the Fourier transform can only extract frequency content, not phase), but nevertheless there.

What if we repeated this entire experiment many times (typically 256 or 512), applying the whole pulse sequence each time with a different y-gradient amplitude? A different variation in phase along the y-axis would be 'locked up' in each resulting echo. If we then lined up the signals next to each other, the evolution of phase across the set of signals would be exactly the same as the evolution of phase between samples of the frequency-encoded echo, i.e. it would appear as if we had a single signal containing a range of frequency components as a function of position along the y-axis.

Figure 7.14 perhaps illustrates the situation more clearly. In this example, we perform an NMR experiment three times. During each repetition, we apply a gradient along the y-axis while magnetisation is in the transverse plane; this gradient has a different amplitude each time. We collect a spin-echo signal each time. Locked up inside each signal is a different phase distribution along the y-axis. We cannot extract this phase information directly, but if we line up the signals next to each other and look across them (horizontally in the diagram) there appear to be high-, medium- and low-frequency components. The appearance is not just graphical: mathematically as well, there appears to be frequency information that we can extract using the Fourier transform. The three repetitions shown are of course purely illustrative: in practice, we repeat the process typically 256 or 512 times, and the number of repetitions determines the matrix size, and hence the spatial resolution of the image, along the y-axis.

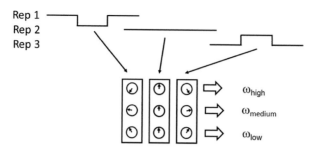

FIGURE 7.14 Phase distributions imparted to transverse magnetisation as a function of position along the *y*-axis during pulse sequence repetitions with phase encoding gradients of different amplitudes. The effect of many such repetitions is to simulate frequency variation along the *y*-axis.

This process, known as 'phase encoding', has a reputation for being one of the more difficult concepts to grasp in MRI physics. But actually, once it is realised that all we are doing is 'tricking' the Fourier transform algorithm by replicating by other means the phase evolution that occurs during a frequency encoding gradient, it is not really so much of a challenge!

7.3.6 BRINGING IT TOGETHER: THE PULSE SEQUENCE CONCEPT

Figure 7.15 shows our first complete pulse sequence diagram, for a two-dimensional spin-echo imaging sequence. It is worth taking some time to work through this diagram, reiterate what each component is doing and understand how they come together to achieve imaging.

The sequence begins with a 90° pulse applied together with a slice selection gradient to achieve slice-selective nutation. This is followed by a rephasing gradient to correct (as far as possible) for through-slice dephasing. On the next line, we have the frequency encoding gradient, which uses precession frequency to encode or 'tag' transverse magnetisation with information about position on the *x*-axis. But we notice something odd: the dephasing gradient that we previous showed immediately before the frequency encoding gradient has moved earlier in time, and what's more it has changed direction! Time is often at a premium in MRI pulse sequences, and it is not a problem to speed things up by applying this gradient at an earlier point during the sequence, at the same time as other gradients: the phase that it imparts to magnetisation will persist anyway until the frequency encoding gradient is applied. The change in sign is because we have moved it to before the 180° refocusing pulse: the phase that the gradient imparts to spins is reversed when this pulse flips

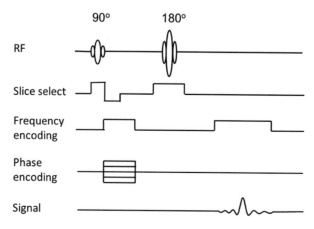

FIGURE 7.15 Two-dimensional spin-echo pulse sequence.

magnetisation across the transverse plane. Turning to the 180° pulse itself, we note that this also has been made frequency selective, so that only magnetisation within the selected slice is refocused. This time no separate gradient is needed to correct for through-slice dephasing: the 180° pulse itself corrects for that in the same way that it corrects for the effects of field inhomogeneity. The interval between the two RF pulses has been set so that the spin echo occurs at the same point in time as the gradient echo. Finally, we have the phase encoding gradient: the curious 'ladder' appearance is pulse sequence shorthand, telling us that the whole sequence is repeated multiple times, and each time this gradient has a different amplitude. On each repetition, magnetisation is encoded or 'tagged' with a different amount of phase depending on its location along the y-axis. It is only when all of the typically 256 or 512 signals have been collected that we have the appearance of frequency variation along the y-axis.

Having to repeat the entire pulse sequence so many times can be very time-consuming, and historically MRI was therefore a relatively slow imaging modality. But we will see in Section 7.5 that techniques have been developed to address this.

7.4 *K*-SPACE

7.4.1 THE *K*-SPACE APPROACH TO SPATIAL LOCALISATION

Figure 7.16(a) shows a 'stack' of echo signals, each collected following a separate repetition of the pulse sequence. The central echo corresponds to the middle frame in Figure 7.14, i.e. the phase encoding gradient amplitude is zero, so there is no variation in phase as a function of position along the y-axis. We therefore see a symmetrical frequency-encoded echo, peaking at the centre at which point magnetisation throughout the selected slice is in phase. Echoes either side of this are increasingly dephased as we work out from the centre, reflecting use of phase encoding gradients of increasing amplitude (positive towards the top of the diagram, negative towards the bottom). If we were to join up the points along the dotted line, the result would resemble another symmetrical echo, reflecting the way in which the phase encoding process simulates frequency encoding.

Each echo is digitally sampled, so the result of our complete set of pulse sequence repetitions is a two-dimensional array of data. A real example is shown in Figure 7.16(b). We can at last reconstruct our first image from this data by applying a two-dimensional Fourier transform, which extracts frequency content (and hence spatial information) from the actual frequency-encoded echoes (x-axis of the image) and from the echoes that we have simulated using phase encoding (y-axis of the image) (Figure 7.16(c)).

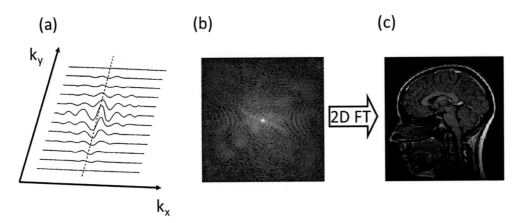

FIGURE 7.16 Array of data collected in k-space shown (a) diagrammatically and (b) as actual MRI raw data; (c) image reconstructed from this data using two-dimensional Fourier transformation.

It is important to realise that each data point in the raw data array contains information from the entire selected slice. The phase information within each data point reflects the 'area' (amplitude and duration) of the gradients that magnetisation has experienced on each axis at the point in time when the data point was collected. We can use this fact to label the axes of the data array as in Figure 7.16(a), defining k_x and k_y as follows:

$$k_x = \gamma G_x t_x \tag{7.22}$$

$$k_y = \gamma G_y t_y$$

We have now arrived at the idea of 'k-space', another concept that has a difficult reputation but we can see is simply a representation of the raw data collected in MRI, before it is Fourier transformed to produce the image. Each frequency-encoded echo collected during the imaging process represents a different 'line' in k-space, and each line is collected following a phase encoding gradient with a different amplitude.

7.4.2 NAVIGATING k-SPACE USING GRADIENTS

In order to be able to reconstruct an MR image, we need to fill k-space with data points and perform a two-dimensional Fourier transform on that data. One way of viewing the role of gradients in spatial localisation is that they allow us to move around in k-space, filling it with data points as we go. Figure 7.17 illustrates this, where for the sake of simplicity we have omitted the refocusing pulse (so actually we have jumped the gun a little: such 'gradient echo' sequences are discussed in Section 7.5.3). Part (a) shows the pulse sequence, one of numerous repetitions with different phase encoding gradient amplitudes. Following excitation, the phase encoding gradient (labelled A) takes us to a specific location on the k_y-axis, while the dephasing gradient in the frequency encoding direction (B) takes us to the extreme negative end of the k_x-axis. The frequency encoding gradient (C) is then switched on and moves us gradually along the selected line in k-space, with digital samples of the echo providing data points with different k_x values along this line. The whole sequence is then

(a)   **(b)**

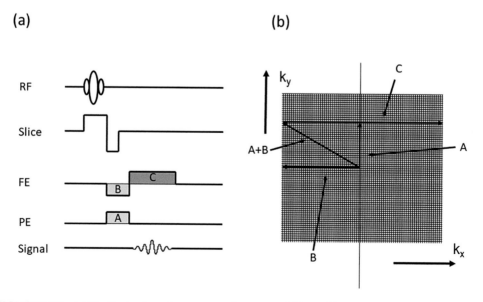

FIGURE 7.17   (a) Gradient-echo pulse sequence diagram and (b) resulting k-space trajectory.

repeated many times with different phase encoding gradient amplitudes, each taking us to a different line in k-space and followed by a frequency encoding gradient to allow samples to be collected along that line.

We have said nothing as yet about the *order* in which lines of k-space are collected during imaging. In most cases, it doesn't really matter, and it is convenient from the perspective of pulse sequence programming to start at one extreme of the k_y-axis and work through to the other end by incrementing the amplitude of the phase encoding gradient on each repetition. This is known as *linear phase encoding*. However, for some applications it is beneficial to acquire the central line of k-space first and work out towards the edges, alternating between positive and negative k_y values. This approach is known as *centric phase encoding*. It is also possible, by switching the gradients appropriately, to generate 'exotic' trajectories that traverse k-space for example radially or in a spiral. These topics are beyond the scope of this book, but once the basic principle of using gradients to navigate in k-space is grasped they will be easier to understand if and when the time comes!

7.4.3   *k*-SPACE AND IMAGE PROPERTIES

We have established that every data point in k-space contains information about the whole image, but the *nature* of the information varies depending on where data is located in k-space. Looking at Figure 7.16(b), it is clear that distribution of signal across k-space is not uniform, but rather is sharply peaked at the centre. This should not be surprising, as this is the point at which magnetisation throughout the imaged slice is in phase. It follows that signal from this part of k-space will have the greatest influence on the signal intensity and contrast properties of the image. On the other hand, data from the edges of k-space contain the greatest amount of spatial information (the highest *spatial frequency*), because it is here that variation in phase with position is steepest (as we can see from Equations 7.20 and 7.21).

We already know that the number of data points collected along each axis in k-space (k_x and k_y) determines the matrix size of the image along the corresponding spatial axis (x and y, respectively). To translate matrix size into pixel size, and hence spatial resolution, we also need to know the size of the image field of view (FOV). These image properties also relate to aspects of the k-space data acquisition process.

- Going further out in k-space (acquiring more of the echo, or more lines in the phase encoding direction) brings in higher spatial frequency data and increases spatial resolution;
- Increasing the density of data points in k-space (increasing the rate at which the echo is digitally sampled, or adding phase encoding steps so as to decrease the interval between lines in the phase encoding direction) increases the FOV.

Either of these things is easy to achieve in the frequency encoding direction, simply by increasing the duration of the echo sampling window or the sampling frequency (which is routinely set high anyway to prevent artefacts). In the phase encoding direction, acquiring more lines of k-space means performing more repetitions of the pulse sequence and hence prolonging the image acquisition period. As often in MRI, there is a trade-off between aspects of image quality and imaging duration, which impacts on both productivity and patient experience: we will explore some possible solutions to this in Section 7.5.5.

Many students of MRI initially find k-space rather a daunting concept, feeling that it is a mathematical abstraction dreamed up by physicists. Hopefully you can see now that it is nothing of the sort, it has a concrete reality as the array of data collected during MR signal acquisition and stored in a computer memory prior to reconstruction of the image. k-space is a really important tool, which helps us to design new pulse sequences, understand and compare the properties of existing ones, understand how modifying acquisition parameters will affect the properties of the resulting image, and understand and eliminate certain types of image artefact.

7.5 PULSE SEQUENCES AND CONTRAST MANIPULATION

7.5.1 CONTRAST MANIPULATION IN THE SPIN-ECHO PULSE SEQUENCE

The 'pulse sequence' concept was introduced right at the start of this chapter as the source of the power and flexibility that MRI has as an imaging modality. In Section 7.2.6 we encountered our first pulse sequence, a simple combination of 90° and 180° RF pulses to generate a spin-echo signal. Addition of imaging gradients in Section 7.3 allowed us to develop our first full imaging sequence, summarised in Section 7.3.6 and depicted in Figure 7.15. The properties of the RF pulses and gradients in this sequence determine the geometrical properties of the resulting image: the slice thickness, field of view, spatial resolution, etc. But pulse sequence properties also determine the information content of the image: in simple terms the contrast (difference in signal intensity) between different structures depicted in the image, which can be made to depend on a wide range of tissue properties and physiological parameters simply by modifying the pattern and timing of RF and gradient pulses within the pulse sequence. In this section, we will introduce this concept of contrast manipulation in the context of the simple spin-echo pulse sequence that we have developed so far. In later sections of the chapter, we will see how wider possibilities can be opened up by adding or removing additional pulse sequence elements.

Repetition of the pulse sequence shown in Figure 7.15 multiple times with different phase encoding gradient amplitudes, followed by two-dimensional Fourier transformation of the resulting k-space data set, results in a two-dimensional image made up of an array of voxels. But what determines the intensity of the signal in each of these voxels? As we learned in Section 7.2.1, the simple answer is that signal intensity reflects water distribution in the body, or more accurately the distribution of hydrogen nuclei. This is known as a *proton density-weighted* image (remember that the hydrogen nucleus consists of a single proton). However, water distribution varies relatively little between soft tissues, and so a proton density-weighted image is not the most useful from a medical imaging perspective. We noted in Section 7.2.5 that, right back at the beginning of the history of MRI, Raymond Damadian found differences in relaxation times between tissues that might be of diagnostic use. Can we somehow incorporate information about these relaxation time differences into image contrast?

In Section 7.2.6 we learned that we can do exactly this: by varying the echo time (T_E) in a spin-echo sequence we can control the extent to which differences in T_2 are reflected in the NMR signal and hence in the resulting image (the degree of 'T_2-weighting'). With reference to Figure 7.5(b), we can see that if we wanted to maximise the contrast between grey matter and white matter in the brain, collecting a spin echo at a T_E of around 90 ms would be a good choice. As a general rule, optimal T_2 contrast occurs if the T_E is set equal to the mean of the T_2 values of the tissues of interest. At very long T_E values there is very little signal from grey or white matter, and we would collect an image dominated by CSF.

The apparently inconvenient fact that we must repeat the pulse sequence multiple times to cover k-space and hence reconstruct an image actually provides a convenient way of weighting the image to reflect differences in T_1. The interval between successive 90° pulses during the series of repetitions is known as the *repetition time* (T_R). During T_R, longitudinal magnetisation undergoes T_1-recovery according to Equation 7.11. We might consider recovery to be essentially complete at $T_R = 5T_1$, at which point $M_z = 0.993|\mathbf{M}|$. If we use a shorter T_R than this, then the amount of longitudinal magnetisation that has recovered for a specific tissue will be given by Equation 7.11, resulting in recovery of different amount of longitudinal magnetisation in different tissues as we can see in Figure 7.5(a). The maximum difference in recovery between grey matter and white matter in the figure occurs at a T_R value of about 850 ms, the mean of the T_1 values of the two tissues. The importance of this to contrast manipulation is that the amount of longitudinal magnetisation that has recovered in a specific tissue at T_R determines the amount of transverse magnetisation that will be generated by the 90° pulse that is applied at that time. So we can manipulate T_1-weighting in an image by varying T_R in much the same way that we can manipulate T_2-weighting by varying T_E.

Figure 7.18 illustrates the behaviour of magnetisation and the development of T_1- and T_2-weighting over two consecutive repetitions of a spin-echo pulse sequence. During the first repetition, we illustrate longitudinal magnetisation, reduced to zero by the initial 90° pulse and then recovering through T_1 relaxation. During the second repetition, we illustrate transverse magnetisation. The 90° pulse in this repetition of the sequence converts the longitudinal magnetisation that was present at T_R into transverse magnetisation, which undergoes T_2 decay. An echo signal is collected at T_E. (For simplicity, we have neglected the impact of field inhomogeneities on the decay of the transverse magnetisation in this diagram.) It is clear that the signal intensity, I, from each tissue type acquired in the echo will in general depend on both T_1 and T_2 relaxation. We can summarise this behaviour in Equation 7.23, which follows from Equations 7.11 and 7.12. Here, $\rho(x,y)$ is proton density as a function of position within the imaged slice:

$$I \propto \rho\left(x,y\right)\left(1-e^{-\frac{T_R}{T_1}}\right)e^{-\frac{T_E}{T_2}} \tag{7.23}$$

In practice, we usually collect images that are either predominantly T_1- or T_2-weighted. The clinical information provided by these two types of image is complementary: it is often stated that T_1-weighted images display anatomy primarily and T_2-weighted images pathology, which is an oversimplification but a reasonable starting point. Table 7.3 summarises the choice of sequence timing parameters required for different types of image weighting. Note the important point that to increase T_1-weighting we must *shorten* T_R, whereas to increase T_2-weighing we must *lengthen* T_E:

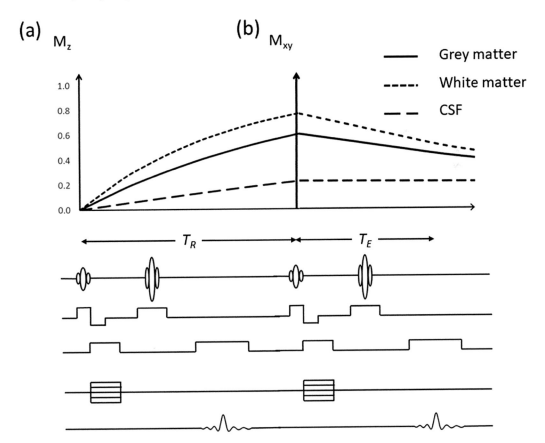

FIGURE 7.18 (a) T_1 recovery of z-magnetisation following a 90° excitation pulse; (b) the recovered magnetisation is nutated into the transverse plane by the next excitation pulse and undergoes T_2 decay.

TABLE 7.3

Timing Parameters Required for Different Spin-Echo Image Weighting

Weighting	T_R	T_E
Proton density	Long	Short
T_1	Short ($\approx$ mean of tissue T_1 values)	Short
T_2	Long	Long ($\approx$ mean of tissue T_2 values)

this reflects the fact that longitudinal relaxation is a *recovery* process while transverse relaxation is a *decay* process. For the same reason, tissues with a long T_1 appear *dark* on T_1-weighted images, whereas tissues with a long T_2 appear *bright* on T_2-weighted images. We avoid a combination of short T_R and long T_E, which would result in competitive T_1- and T_2-weighting that cancel out, resulting in poor contrast.

Although we talk of 'T_1-weighted' and 'T_2-weighted' images, Equation 7.23 shows that in reality any image inevitably contains proton density information, and also some T_1-weighting (unless T_R is unfeasibly long) and some T_2-weighting (unless T_E is unfeasibly short). Because T_R is the interval between repetitions of the pulse sequence, it heavily influences the time taken to acquire the MR image data (which will be $256T_R$ if we are collecting 256 lines of k-space). Therefore, there is a strong incentive to minimise T_R from the perspectives of patient experience and throughput, and a degree of T_1-weighting is pragmatically tolerated in a nominally T_2-weighted image. It would certainly be impossible to achieve complete recovery of CSF longitudinal magnetisation given $T_1 \approx 4$ s!

Figure 7.19 shows a series of spin-echo images of the same slice of a volunteer's brain collected with different sequence timing parameters, illustrating the points we have discussed in this section. Note in particular that CSF appears dark in the T_1-weighted image (but also in the proton density-weighted image due to residual T_1-weighting) and increasingly bright as T_2-weighting is increased. It is also worth noting that fat around the scalp is the brightest tissue in the T_1-weighted image, reflecting the short T_1 of fat (≈ 250 ms at 1.5 T). The bone of the skull is not seen at all: this is because solid materials have such short T_2 values that they cannot be seen other than with very specialised pulse sequences developed for the purpose. MRI is therefore not particularly useful for imaging bone; this is one of its few major limitations as an imaging modality.

7.5.2 ENHANCING T_1 CONTRAST: THE INVERSION RECOVERY PULSE SEQUENCE

Simply by varying timing parameters within the spin-echo pulse sequence, we can produce images that vary greatly in appearance and clinical application. But we can do much more by adding additional pulse sequence components. As an example, we will consider the *inversion recovery* sequence. As shown in Figure 7.20, this consists of a spin-echo sequence with an additional slice selective 180° pulse before the 90° excitation pulse. The purpose of this pulse is to nutate longitudinal magnetisation into the $-z$-direction, and hence it is also known as an 'inversion pulse'. The inverted longitudinal magnetisation undergoes T_1 recovery: this does *not* involve retracing the pathway followed during nutation (recall that discussion in Section 7.2.5!); instead, the negative longitudinal magnetisation vector gets shorter over time, passes through zero and then grows along the positive z-axis until it has fully recovered.

Evolution of longitudinal magnetisation over time is given by Equation 7.24. Importantly, no transverse magnetisation is generated during this process:

$$M_z = |\mathbf{M}| \left(1 - 2e^{-\frac{t}{T_1}} \right)$$

$$(7.24)$$

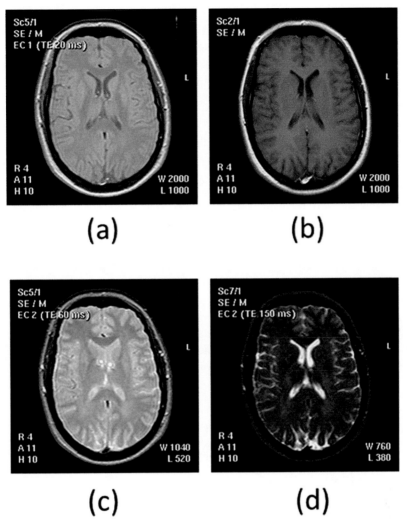

FIGURE 7.19 Contrast in a spin-echo pulse sequence. (a) Proton density weighting (T_R = 2000 ms, T_E = 20 ms); (b) T_1-weighting (T_R = 500 ms, T_E = 30 ms); (c) T_2-weighting (T_R = 2000 ms, T_E = 60 ms); (d) heavy T_2-weighting (T_R = 2000 ms, T_E = 150 ms).

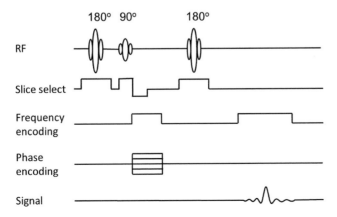

FIGURE 7.20 Two-dimensional inversion recovery pulse sequence.

At a chosen point in time, known as the inversion time (T_I), a 90° pulse is applied, tipping longitudinal magnetisation into the transverse plane. Differences in T_1 between tissues are reflected in different degrees of recovery at this time point and hence in different amounts of transverse magnetisation following the 90° pulse. Because M_z can take both positive and negative values, there is potential for much greater T_1 contrast than in a conventional spin-echo sequence where only the positive z-axis is used.

Figure 7.21(a) illustrates the recovery of longitudinal magnetisation during T_I, and 7.21(b) shows the transverse magnetisation generated by a 90° pulse applied at time point B (we shall return to time points A and C below). Note that because the longitudinal magnetisation vectors in CSF and in grey matter are still along the negative z-axis at this point, the 90° pulse tips these magnetisation vectors to the opposite side of the transverse plane as compared to magnetisation in fat. But where is the white matter magnetisation? We can see from Figure 7.21(a) that at time point B this vector is just passing through the zero point on its recovery from −z to +z, so there is no longitudinal magnetisation at this point, and hence no transverse magnetisation after the 90° pulse. This fact leads to a very useful additional application of inversion recovery sequences.

We can reconstruct inversion recovery images in two different ways, depending on whether or not we pay attention to the 180° phase difference between magnetisation that has and has not passed through the zero point on the z-axis at T_I. Figure 7.22(a) shows a *real* or *signed* inversion recovery image, collected with a T_I corresponding to time point B in Figure 7.21, where this phase difference is taken into account. Areas in which there is zero signal (for example outside the head) appear midgrey, tissue with long T_1 (such as CSF) appear darker and those with short T_1 (such as fat) appear brighter. There is therefore a much wider dynamic range of signal intensities than in a conventional T_1-weighted spin-echo sequence, resulting in greater soft-tissue contrast.

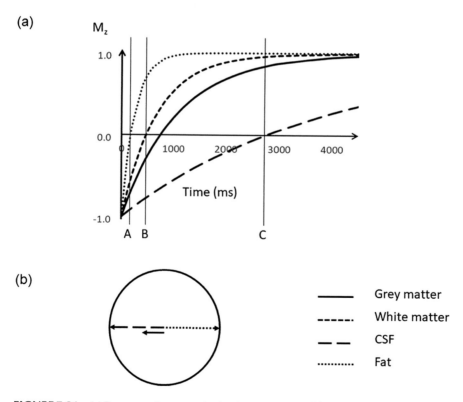

FIGURE 7.21   (a) Recovery of z-magnetisation in grey matter, white matter, CSF and fat following an inversion pulse; (b) distribution of magnetisation in the transverse plane following a 90° pulse applied at time point B.

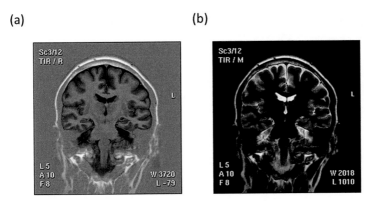

FIGURE 7.22 (a) Real (or signed) inversion recovery image; (b) magnitude inversion recovery image.

Figure 7.22(b) shows a *magnitude* inversion recovery image, collected with the same T_I, in which phase difference is neglected and signal intensity reflects only the magnitude of the magnetisation vectors. In this image, areas outside of the head where there is no signal appear black, as does white matter (we know that longitudinal magnetisation in this tissue is passing through the zero point when the 90° pulse is applied). We also note that signal intensity in CSF and in fat are the same, despite their very different T_1 values, and can see from Figure 7.21 that this arises because the *magnitude* of the longitudinal magnetisation in these two tissues at T_I is the same. It may initially be difficult to imagine when an image with such odd contrast properties may be of use, but the secret lies in the ability to *null* signal from specific tissues by appropriate choice of T_I. In Figure 7.22(b), signal from white matter has been nulled. It can easily be shown from Equation 7.24 that more generally signal from tissue with a specific T_1 value can be nulled by setting:

$$T_I = \ln 2T_1 \approx 0.693T_1 \qquad (7.25)$$

There are many clinical situations in which bright signal from fat can obscure useful detail in a T_1-weighted image. By setting $T_I \approx 170$ ms, we can null this bright signal so that contrast between other tissues is more visible. This is known as a STIR (Short Tau Inversion Recovery) sequence. Equally, bright CSF in the brain in a T_2-weighted image can make pathological changes in signal intensity difficult to see. In this case, we can use an inversion recovery sequence with $T_I \approx 2500$ ms to null the CSF signal, known as a FLAIR (Fluid Attenuated Inversion Recovery) sequence. The T_I values needed in these two situations correspond to time points A and C, respectively, in Figure 7.21. We can see that in each case the longitudinal magnetisation vector in the tissue that we want to null is passing through the zero point when the 90° pulse is applied.

So an inversion recovery sequence can be used either to produce an image with enhanced T_1 contrast, or to 'edit out' signals in a way that improves the diagnostic utility of images. All of this is achieved simply by adding an additional RF pulse! It is a good example of the flexibility that can be achieved through pulse sequence design, and a simple example of a whole class of pulse sequences in which magnetisation is manipulated, or *prepared*, prior to the excitation pulse in order to change the contrast properties of the resulting image.

7.5.3 Speeding Things Up: the Gradient-Echo Pulse Sequence

In the preceding section we looked at the effect of adding an additional element to the pulse sequence; here we will instead consider removing one!

We saw in Section 7.3.4 that in what we conventionally call a spin-echo sequence we are in fact acquiring a simultaneous spin echo (generated by the 180° refocusing pulse) and gradient

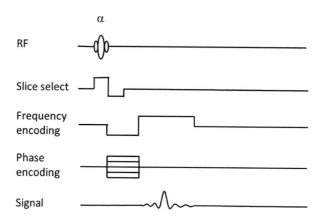

FIGURE 7.23 Two-dimensional gradient-echo pulse sequence.

echo (generated by the frequency encoding gradient and the dephasing gradient that precedes it). Figure 7.23 shows a sequence in which the refocusing pulse is omitted, so that the echo we collect is generated only by switching of gradients on the frequency encoding axis.

This is known as a 'gradient-echo' sequence, and there are two other differences as compared to the spin-echo sequence diagram in Figure 7.15. Firstly, in the absence of the refocusing pulse, the dephasing gradient that precedes the frequency encoding gradient is now in the opposite orientation to the frequency encoding gradient itself. Secondly, the excitation pulse is labelled 'α' rather than '90°'. This reflects the fact that in a gradient-echo sequence the flip angle of the excitation pulse is usually less than 90°. This of course means that as well as generating transverse magnetisation it leaves some residual longitudinal magnetisation (see Equations 7.9). In a spin-echo sequence, we would have to worry about the effect of the 180° refocusing pulse on this longitudinal magnetisation. That is not an issue in a gradient-echo sequence, and as we will see below the reduced flip angle has some advantages.

In the absence of a refocusing pulse, the echo collected is of course T_2^* rather than T_2-weighted, according to Equation 7.13. This means that there will be less signal than in a spin-echo sequence collected with the same echo time, and also makes the sequence particularly sensitive to image distortion due to magnetic field inhomogeneities, for example at the interface between different tissues or where there are air-filled sinuses or cavities in the body.

We saw in Section 7.5.1 that in order to avoid T_1-weighting in a spin-echo sequence it is necessary to employ a long repetition time so that recovery of longitudinal magnetisation is complete. This results in lengthy image acquisition times (to collect 256 k-space lines with $T_R = 3$ s takes almost 13 minutes). However, if we consider Equations 7.9 with a flip angle $\alpha = 10°$ we see that immediately after excitation $M_z = 0.98|\mathbf{M}|$, i.e. the longitudinal magnetisation has only been reduced by 2% from its initial value. Clearly we do not need to wait anything like as long for recovery before we can apply our next excitation pulse, so T_R can be very much shorter, resulting in much quicker image acquisition. There is a downside of course: $M_{xy} = 0.17\ |\mathbf{M}|$ in the scenario we are discussing, so we have sacrificed most of our signal. As so often in MRI, there is a trade-off between image quality and acquisition time. It can be shown that the signal in a gradient-echo sequence follows Equation 7.26. By varying the flip angle and repetition time, we can strike the right compromise in each clinical setting:

$$I \propto \rho(x,y)\frac{\left(1-e^{-\frac{T_R}{T_1}}\right)}{1-\cos\alpha\, e^{-\frac{T_R}{T_1}}}\sin\alpha\, e^{-\frac{T_E}{T_2^*}} \tag{7.26}$$

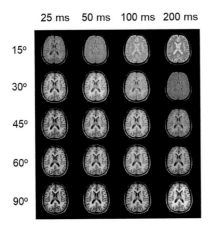

FIGURE 7.24   Contrast in a gradient-echo pulse sequence, effect of variation of T_R and flip angle.

Figure 7.24 shows an array of gradient-echo images, collected with different combinations of α and T_R. With large α and short T_R we observe T_1 weighting as expected (note the dark CSF). However, as α is reduced weighting switches to proton density and then to $T_2{}^*$ (bright CSF), even with T_R as short as 100 ms. This is clearly a huge improvement on 3000 ms to minimise T_1-weighting, as would be required in a spin-echo sequence, and reduces the imaging time to 26 s, again acquiring 256 lines in k-space.

The spin-echo, inversion recovery and gradient-echo pulse sequences are simple but widely used examples of how pulse sequences can be modified to change the information content of the resulting images. We will encounter others in this chapter, including sequences that allow us to encode tissue motion, blood perfusion and many other parameters into the signal. The fact that all this can be done simply by reprogramming the scanner, with no need for expensive additional hardware, is one of the major strengths of MRI as a clinical imaging modality and research platform.

7.5.4   Multislice and Three-Dimensional Imaging

So far, we have concentrated on imaging a single slice of the patient's body, selected using slice selective excitation and mapped in two dimensions using frequency encoding and phase encoding. This is of course rarely going to be adequate for clinical imaging, where we generally need to cover the whole of an organ or body area of interest. It might be thought that to image N slices would take N times as long as imaging a single slice, which particularly for T_2-weighted spin-echo sequences acquired with long T_R would take an unacceptable length of time. However, there is a 'trick' that we can employ to dramatically reduce acquisition time.

Remember that the excitation pulse at the start of the pulse sequence generates transverse magnetisation only within a slice defined by the gradient strength and the frequency content of the pulse. We have to wait for the appropriate T_R interval, determined by the amount of T_1-weighting that is desired before we can excite magnetisation within the selected slice again. However, magnetisation outside this slice remains along the z-axis: there is nothing to stop us from applying another excitation pulse immediately after the first, with a different frequency content so that magnetisation in a different slice is excited. We can continue to do this as many times as possible during the T_R period, so that we can collect data from multiple slices in parallel rather than having to wait until all the data from one is collected before moving on to the next. This is illustrated in Figure 7.25 for a gradient-echo sequence. In this case, we are able to image three slices in parallel, and in general the number is T_R / T_{slice}, where T_{slice} is the time taken to acquire one line of data from one slice (and is slightly longer than T_E as it includes the whole of the echo acquisition time whereas T_E is measured to the centre of the echo).

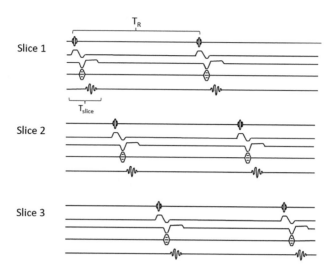

FIGURE 7.25   Interleaved imaging of multiple slices within a single T_R interval.

When acquiring data from multiple slices using this 'interleaved' approach, we must bear in mind the *actual* profile of excited slices, which as discussed in Section 7.3.3 deviates somewhat from the ideal rectangular slice profile. For this reason, we need to leave a small gap (typically 10–20% of the slice thickness) between slices that are excited within the same T_R interval: otherwise material close to the edges of the slices will be excited at an interval much shorter than T_R, resulting in artefacts. If we wish to eliminate this gap and acquire *contiguous* slices, for example in clinical situations where there is a risk of missing small lesions that may fall between slices, we can repeat the entire set of interleaved acquisitions acquiring odd-numbered slices first and then even-numbered slices. This of course doubles the length of the acquisition, and there is a decision to be made as to whether the clinical need for contiguous acquisition justifies this.

In the previous section, we looked at how T_R can be drastically reduced without incurring heavy T_1-weighting by using a gradient-echo sequence. Clearly a shorter T_R means that fewer interleaved acquisitions can be accommodated, and so fewer slices imaged in parallel. That might mean that it is necessary to run the whole set of repetitions twice to image enough slices to cover the area of clinical interest. A modest increase in T_R might be sufficient to avoid this, without significantly changing image weighting. It is another trade-off situation, requiring skill and experience on the part of the MR system operator!

As noted in Section 7.3.1, rather than perform two-dimensional tomographic imaging it is possible to collect image data from an entire volume simultaneously and encode it in three dimensions. Figure 7.26 illustrates a gradient-echo pulse sequence modified for this purpose. Here the slice select axis has become 'slab select', with a gradient and selective excitation pulse used to generate transverse magnetisation within a thick slab of the patient's body. Signal from this slab is spatially encoded using frequency encoding on one axis and phase encoding on the other two: hence the diagram shows the distinctive phase encoding gradient 'ladder' on the slab select axis as well as the usual phase encoding axis. We can picture the data acquisition process in this case in a three-dimensional k-space with axes k_x, k_y, k_z, whereby analogy with Equations 7.22:

$$k_z = \gamma G_z t_z \tag{7.27}$$

In order to acquire data from the whole of this k-space volume, we must acquire a frequency encoded echo (a line of data along the k_x axis) for every *combination* of k_y and k_z. This means that for *each* value of the phase encoding gradient on the z-axis we must acquire a separate echo with

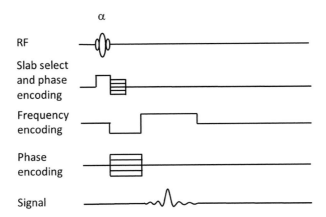

FIGURE 7.26 Three-dimensional gradient-echo pulse sequence.

each value of the phase encoding gradient on the y-axis. Such a lengthy acquisition is really only feasibly using a gradient-echo sequence, or some of the tricks we will encounter in the next section.

Three-dimensional imaging allows us to overcome a key drawback of multislice imaging, where the thickness of a selected slice is frequently considerably greater than the spatial resolution within the slice. 'Isotropic' imaging, i.e. with equal spatial resolution in all three dimensions, is readily achievable, which is an advantage in many clinical applications that require imaging of small and complex anatomical structures.

7.5.5 ADVANCED PULSE SEQUENCES AND FAST IMAGING

In an introductory book like this, it is not possible to do justice to the huge array of pulse sequences that have been developed by applying the principles discussed in previous sections in ever more innovative ways. Some advanced applications are discussed later in the chapter. In this section, we will briefly discuss approaches to speeding up data acquisition in an attempt to overcome the trade-off between acquisition time and image quality.

As we have seen, in a spin-echo sequence it is necessary to use a long repetition time if T_1-weighting is to be avoided. Given the need to repeat the pulse sequence typically 256 times to achieve sufficient k-space coverage, it appears that lengthy data acquisition is inevitable. The time can be put to good use by interleaving imaging of multiple slices, but still we may be left with considerable 'dead time' and a patient experience challenge. However, what if instead of acquiring a single line of k-space following each 90° pulse we were to acquire several? Clearly, there is the potential to reduce acquisition time considerably. Figure 7.27 shows a way of doing this, known as a *fast spin-echo* (FSE) or *turbo spin-echo* (TSE) sequence. Instead of a single refocusing pulse producing a single spin echo, here we have a series of such pulses, refocusing transverse magnetisation repeatedly to generate a train of echoes, each of course with its own frequency encoding gradient. The trick is that before each echo (other than the first) there is an additional phase encoding gradient, which of course means that each echo is collected from a different line in k-space. In this example we collect three echoes following each excitation pulse, reducing the overall image acquisition time by a factor of three. In a real sequence, the number of echoes collected (known as the *echo train length* (ETL) or *turbo factor* (TF)) is typically 4–32, but may be much higher up to and including 'single short' imaging where data from the whole of k-space is acquired following a single excitation pulse.

In practice, the ETL is limited by the repetition time, which determines how many echoes can be accommodated before the next excitation pulse, and there is a trade-off with the number of slices imaged since we also have to accommodate interleaved excitation of multiple slices during the T_R

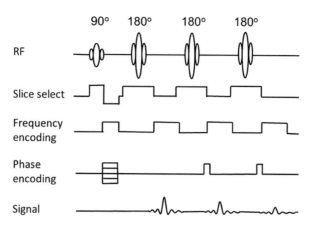

FIGURE 7.27 Two-dimensional fast spin-echo pulse sequence.

interval. Each echo of course is acquired at a different echo time, and this also sets a limit to the achievable ETL since echoes acquired at very long T_E in a long echo train will contain very little signal.

We need to consider a further complication arising from this last point. Since the echo time determines the T_2-weighting of the acquired image, how much weighting is there when different lines in k-space have been acquired at different echo times? The answer lies in Section 7.4.3, where we saw that data collected from the centre of k-space, where magnetisation is in phase throughout the imaged object, dominate the contrast properties of the image. The degree of T_2-weighting in a fast spin-echo image is therefore dominated by echoes close to the centre of the k_y-axis, and careful pulse sequence design ensures that all of these echoes are collected at the same T_E value (known technically as the *effective echo time* or *pseudo echo time*, but often in practice simply as the echo time). Conversely, echoes acquired at long T_E are typically positioned towards the edges of k-space, and at high ETL this can become an issue because of the loss of high spatial resolution data and consequent image blurring. It follows from all of this that the appearance of the image is determined by the k-space trajectory in a way that we have not encountered before: otherwise, identical sequences with different phase encoding patterns can produce very different images!

The development of fast spin-echo sequences revolutionised T_2-weighted imaging: with an echo train length of 8, an image with a resolution of 256 pixels on the phase encoding axis and a T_R of 2s can be acquired in 1 minute as opposed to 8.5 minutes with a conventional spin-echo sequence. FSE has become the default to the extent that an inexperienced operator selecting a T_2-weighted sequence on a scanner may be unaware that, behind the scenes, it is in fact an FSE sequence.

The gradient-echo sequence has already been presented (in Section 7.5.3) as a means of speeding up imaging, since we are able to reduce T_R dramatically without incurring heavy T_1-weighting. If T_R becomes so short that it is comparable to the T_2 values of tissues, we need to consider the effect of subsequent excitation pulses on residual transverse magnetisation as well as on recovering longitudinal magnetisation. Things can become very complicated at this point, and the simplest approach is to add a strong gradient pulse after acquisition of each echo signal, before the next excitation pulse. This so-called 'spoiler gradient' completely dephases the residual transverse magnetisation, so that once again we only have to consider longitudinal magnetisation. The resulting 'spoiled gradient-echo' sequences follow standard rules on image weighting as discussed in Section 7.5.3.

But what if we omit this spoiler gradient and face the consequences? In this case, each excitation pulse will tip some residual transverse magnetisation onto the z-axis, as well as tipping some longitudinal magnetisation into the transverse plane. After several repetitions, the conserved transverse magnetisation is brought back into the transverse plane where it generates a spin-echo (i.e. T_2-weighted) signal alongside the usual gradient-echo (T_2*-weighted) signal generated from nutated longitudinal magnetisation. A variety of these so-called steady state or refocused gradient-echo sequences exist, in which the various echo signals are manipulated in different ways to produce diverse contrast properties. Within this category lie for example 'balanced gradient-echo' sequences, widely used for cardiac imaging, where contrast turns out to depend on the ratio T_2/T_1! It is also possible to apply various types of preparation pulses prior to data acquisition in order to introduce desired contrast into the image, for example, an inversion pulse may be applied initially so that each repetition of the gradient-echo sequence samples longitudinal magnetisation at a different point during its recovery, with overall image contrast determined by the T_1 of the echoes closest to the centre of k-space. To make this complex situation even more confusing, for commercial reasons different MR equipment manufacturers use their own proprietary names for sequences within this broad family, so that completely different terminology may be used for essentially the same sequence on different scanners.

At the extreme of fast gradient-echo sequences, we find *echo planar imaging* (EPI), in which rapid switching of the frequency encoding gradient is employed to produce a train of gradient echoes after a single excitation pulse, just as a series of 180° pulses is used to produce an echo train in a fast spin-echo sequence. Again, preparation pulses and appropriate k-space navigation can be used to give a range of desirable contrast properties. In the extreme case of 'single shot EPI', the whole of k-space can be covered and a single-slice image acquired in 100 ms or less! EPI was actually invented by Peter Mansfield as long ago as 1977, although it was many years before gradient technology made it realisable in a consistent way on clinical imaging systems.

All of the pulse sequence-based approaches that we have discussed so far in this section aim to reduce image acquisition time either by decreasing T_R (while somehow managing the impact on T_1-weighting) or by collecting data from multiple lines of k-space within a single T_R interval. There is another approach, which is to reduce the number of k-space lines collected.

In Section 7.4.3, we explored the relationship between data acquisition in k-space and the resulting image properties. From that discussion, we know that we can reduce the number of k-space lines in two ways. Firstly, we can increase the spacing between lines, resulting in a reduced image field of view (FOV) along the phase encoding direction. We can maintain the original FOV size along the frequency encoding axis of course, resulting in a rectangular FOV which we can align so that the phase encoding axis lies along the shortest dimension of the body part being imaged. There is no such thing as a free lunch in MRI, and reducing the amount of data acquired in this way leads to a loss of image signal-to-noise ratio (SNR). Secondly, we can reduce how far we go out in k-space along the phase encoding axis, a technique variously known as 'reduced k-space' or 'reduced scan percentage'. This time we sacrifice high spatial frequency data, and hence there is a loss of spatial resolution along the phase encoding direction. A third approach is to exploit the symmetry that exists in k-space. Remember that every point in k-space contains information about the entire image, with the spatial frequency content at each point depending on its k_x and k_y values. But k-space displays *Hermitian symmetry*, that is the say the information content of the point at coordinates $(-k_x, -k_y)$ is the same as that at coordinates (k_x, k_y). Hence in principle we can leave out one entire half of the phase encoding axis without losing any information. We do however sacrifice SNR, and in practice collect data from more than half of k-space for this reason as well as to ensure that phase information is captured properly in the image data. This is known as the *half-scan* or *partial Fourier* technique. Figure 7.28 shows these various approaches to speeding up imaging by reducing k-space coverage diagrammatically. The techniques can be used separately or in combination, and alongside pulse sequence-based approaches as well.

Yet another approach to scan time reduction has emerged over the past 20 years, known as *parallel imaging*. In these techniques, data are acquired using an array of receive coils rather than a single coil. Spatial information about the imaged object can be obtained mathematically by exploiting the fact that each coil in the array has a different 'view' of the object. Versions exist that work in either the image domain or in k-space. In either case, they reduce the need for phase encoding by a factor limited by the number of coils in the array. More recently this approach has been generalised to a wider class of techniques that exploit redundancy in the acquired data (partial Fourier acquisition is

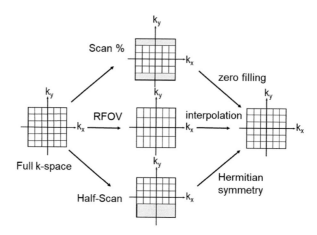

FIGURE 7.28 Approaches to accelerating imaging by reducing k-space coverage.

a simple example of this) known as *compressed sensing*. This approach has recently become available on clinical imaging systems and promises to revolutionise data acquisition in MRI.

7.5.6 ARTIFICIAL CONTRAST AGENTS IN MRI

The ability to manipulate image contrast and information content using pulse sequences is a unique feature of MRI. Nevertheless, there are situations where it is advantageous to introduce exogenous agents into the body which affect contrast in a way that reflects some pathological or physiological process. These agents generally work by reducing the T_1 and/or T_2 relaxation times of tissue in which they reside, resulting in local changes in signal intensity. Their applications fall into the following categories.

- To change (usually increase) signal from selected tissues or areas of pathology, usually on the basis of vascularity.
- To enhance signal from flowing blood in contrast-enhanced MR angiography.
- To eliminate signal from organs that might otherwise obscure structures of interest or cause artefacts, such as the bowel.
- For dynamic studies of perfusion or function, or for tissue characterisation.
- In emerging areas such as molecular imaging.

Some of these applications are discussed later in this chapter.

Despite a great deal of research and development in this area over many years, the contrast agents in clinical use are almost exclusively intravascularly injected agents based on the gadolinium ion (gadolinium-based contrast agents, or GBCAs). Gadolinium is a paramagnetic ion, which affects signal primarily by shortening T_1. Because it is a toxic heavy metal, it is used in a chelated form, in which each gadolinium ion is embedded in a molecular 'cage' to reduce toxicity. A variety of agents are available from different manufacturers, each using different chelate molecules.

Figure 7.29 illustrates an example of the use of GBCA in cancer staging. In a low-grade glioma, T_1-weighted images collected before and after intravascular administration of GBCA are very similar (although some enhancement can be seen due to the presence of GBCA in blood vessels). However, in a high-grade glioma, where the blood-brain barrier is compromised, GBCA crosses into the tumour and causes signal enhancement due to T_1 shortening. Being able to differentiate tumour grade in this way makes a great difference to the patient's prognosis and management.

GBCAs were for many years regarded as extremely safe contrast agents, with a very favourable safety profile compared to agents used in x-ray imaging. However, in 1997 a new condition

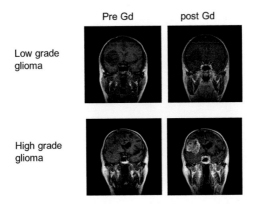

Pre Gd   post Gd

Low grade
glioma

High grade
glioma

FIGURE 7.29   Effect of gadolinium-based contrast agent injection on image contrast in a low-grade glioma and a high-grade glioma.

was identified, now known as nephrogenic systemic fibrosis (NSF), in which fibrous plaques form initially on the skin, sometimes spreading to solid organs resulting in disability or death. This condition was found to occur predominantly in patients with severe renal failure who had received large or multiple doses of GBCA. It appears that renal failure delays excretion of the GBCA from the body, and hence increases the likelihood that the metal-chelate complex will break down, exposing the patient to toxic gadolinium ions. It was found that some of the chelates used in GBCAs are intrinsically less chemically stable than others, and that NSF is associated mainly with the less stable agents. Introduction of guidance limiting the use of GBCAs in renal failure patients led to effective elimination of new cases of NSF. However, in 2014 it emerged that there are permanent changes in signal intensity in some parts of the brain following repeated GBCA administration, believed to result from retention of gadolinium in these tissues. This again occurs primarily with less stable agents, but regardless of the patient's renal function. Although there are no known clinical consequences, retention of toxic heavy metal ions in the brain is clearly best avoided, and restrictions on the use of GBCAs have been introduced in many jurisdictions around the world. It should be noted however that this issue only seems to arise following repeated administration of GBCAs over a number of years, and only with less stable agents: one-off administration of GBCA as part of an MRI examination appears to be safe.

7.6   ARTEFACTS: PROBLEMS, SOLUTIONS AND NEW IDEAS

Artefacts are features appearing in an image that are not present in the investigated object. In MRI, artefacts test our understanding of the underpinning physics; if this is understood the artefacts can be avoided, reduced or compensated for. Moreover, it is interesting that a number of artefacts, or the solutions developed to minimise them, provided the seeds of advanced MRI techniques. In the following sections, major MRI artefacts are discussed from this perspective.

7.6.1   MOTION ARTEFACTS IN MRI

The acquisition time of an MRI dataset is typically a few minutes, due to the repeated echoes acquired to fill the k-space matrix. As discussed in Sections 7.3.6 and 7.4.1, each data point in the raw data array contains information from the entire selected slice (or selected volume when three-dimensional imaging is applied, see Section 7.5.4). Moreover, interleaved multislice imaging is usually used to cover the whole of an organ or body area of interest. As a result, the effect of movement during acquisition affects all the slices in the investigated volume, not just a limited number of them. Accordingly, it is of utmost importance to optimise patient positioning; an uncomfortable position increases the probability of movement during the exam.

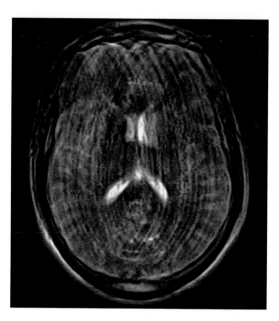

FIGURE 7.30   Example of a motion artefact; the phase encoding gradient is in the left–right direction.

Patient movement during MRI disrupts the relationship between phase and position discussed in Section 7.3.5. This can result in faint displaced images, known as ghosts. The severity of these motion artefacts is related to the extent of the displacement compared to the voxel size and to its timing, whether the movement is continuous during the whole acquisition or more limited in duration. Figure 7.30 shows an example of a severe motion artefact. In medical imaging, some types of motion cannot be avoided (such as breathing, the cardiac cycle, and blood flow). Nevertheless, there are some techniques to minimise the resulting artefacts.

Cardiac motion artefacts are avoided by filling k-space with signals acquired in the same temporal window of the cardiac cycle by triggering data acquisition using an MRI-compatible ECG system or a peripheral pulse unit (PPU). The use of electrodes that are safe in the MR scanner is vital: generic ECG electrodes are very dangerous, and can result in burns to the patient's skin. Moreover, they affect image quality by inducing metallic artefacts, as discussed in the following sections. Peripheral gating is safe and fast to apply, it detects the arterial pulse which is broader and delayed compared to the ECG wave. Both ECG and PPU are used, the choice is based on the specific clinical application. Using ECG, precise synchronisation of echo acquisition with the cardiac cycle is possible. This is exploited in cardiac imaging, a very important application of MRI discussed in Section 7.7.1. When the ECG signal is used to trigger the excitation pulse of the sequence, the cardiac rate of the patient defines the repetition time and hence the degree of T_1-weighting.

Breathing artefacts in radiology are frequently controlled using breath-holding techniques, and this approach is applied in MRI when the acquisition time is short (less than half a minute). Unfortunately, many MRI sequences require a longer acquisition time than this. The respiratory cycle can be monitored using a 'breathing belt' strapped around the thorax: this generates an electric signal dependent on the pressure of the thorax on the belt, allowing breathing to be monitored. Respiratory gating can then be applied, similar to cardiac triggering, in order to perform acquisition of k-space lines during the end-expiration phase when chest motion is minimal.

Another successful approach is respiratory compensation or ROPE (Respiratory-Ordered Phase Encoding): k-space lines are acquired during the whole respiratory cycle and similar phase encoding gradient amplitudes are used with the thorax at similar positions. This avoids abrupt variation in the position of chest structures between closely-spaced k-space lines, minimising artefacts. ROPE is more effective than respiratory gating because data are acquired throughout the respiratory cycle, while in respiratory gating only the quasi-stationary phase of the cycle is used for acquisition.

Respiratory movements can also be monitored using 'navigator echo imaging' without any external device: the basic idea is that the position of the diaphragm, in the cranio-caudal direction, can be automatically detected by software due to the high contrast between liver and air in the lung. In navigator echo imaging, this interface is detected using one-dimensional MR data acquisition. This information is used in ROPE or for respiratory gating.

In MRI, flowing blood and cerebrospinal fluid (CSF) represent large numbers of moving spins. Moreover, sometimes they undergo pulsatile flow, which is more complex than flow at a constant velocity. When spins are moving in the time interval between the RF excitation pulse and echo acquisition, they accumulate a phase shift different from the phase expected for a stationary spin. This means that in the phase encoding direction we observe ghosts of the vessel; if the motion is pulsatile the intensities of the ghosts have a periodic modulation. The phase encoding gradient waveform can be modified to cancel the additional phase shift associated with flow, removing the artefacts. The drawback is that the duration of a flow-compensated phase-encoding gradient is longer than that of a conventional phase-encoding gradient, limiting the minimum echo time that can be achieved. Sometimes the flow artefacts overlie the region of clinical interest in an image. In such a case the simplest solution is to swap frequency and phase encoding directions and repeat the acquisition: the flow artefacts are still present, appearing as a stripe across the vessel, but the direction will be rotated by 90° and no longer obscure the desired clinical information.

Blood signal in MR images is frequently hyperintense or hypointense relative to the surrounding tissue, due to what are known as 'in-flow effects'. Hypointense signals are observed in spin-echo images because, while stationary spins are affected by both the excitation and refocusing pulses and create the echo signal, for moving blood the echo signal is created only by the fraction of spins that is in the same slice during both pulses. If the echo time is long and blood flow is fast, no echo is created within the blood vessel (Figure 7.31(a)). Hyperintense signals are observed in gradient-echo sequences. In such sequences, stationery spins experience a series of excitation pulses at short T_R, and so signal from them is suppressed (they are said to be 'partially saturated'). However, there is a lack of partial saturation effect in flowing blood within vessels crossing the imaged slice or stack of slices, where spins are refreshed (partially or completely) between subsequent excitation pulses (Figure 7.31(b)). These effects are used in MR angiography, discussed in Section 7.7.2. If the hyperintense signal decreases the diagnostic quality of the image, it can be cancelled by applying a 'saturation slab' outside the volume to be imaged to saturate the blood signal before its arrival in the imaged slices (Figure 7.31(b)). When a saturation slab is selected by the operator the pulse sequence is modified so that a selective 90° pulse, known as a saturation pulse, is applied in a user-defined region immediately before the RF excitation pulse. This is followed by strong spoiler gradients to completely dephase the transverse magnetisation generated by the saturation pulse, so that negligible signal is obtained from spins if they flow from the saturated region into the imaged slices during data acquisition. Flow artefacts are not the only application of saturation slabs: they may also be used if a moving organ in the FOV prevents correct visualisation of the region of diagnostic interest.

7.6.2 Water and Fat Frequency Shift

The resonance frequency of nuclei is defined by the gyromagnetic ratio and the intensity of the static magnetic field that they experience. MRI makes use of hydrogen nuclei located mainly in water and fat molecules. However, the resonance frequencies of water and fat protons are slightly different, a phenomenon known as *chemical shift*. The cloud of moving electrons surrounding each molecule creates a magnetic field of its own, and as a result the local magnetic field experienced by the hydrogen nucleus differs slightly from B_0. This difference depends on the structure of the molecule. The frequency shift between different molecules is proportional to magnetic field strength, and at 1.5 T the water–fat shift is about 220 Hz. Chemical shift is also the basis of NMR spectroscopy, discussed

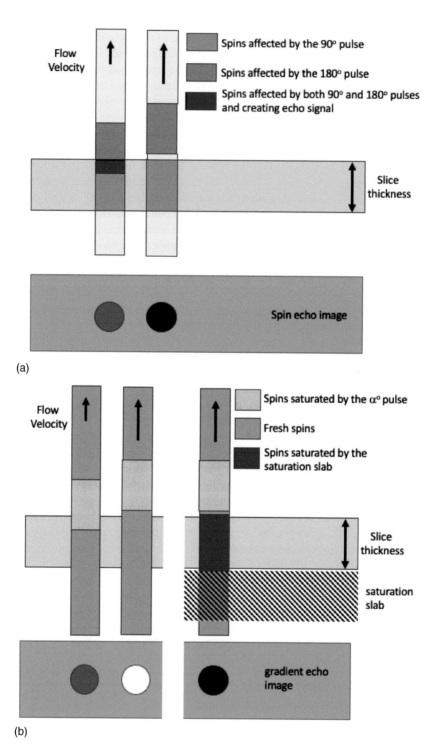

FIGURE 7.31 In-flow effects in spin-echo (a) and gradient-echo (b) images. The vertical boxes represent blood vessels, the lengths of the arrows are proportional to blood velocity and the pink rectangles represent the image slice. The grey rectangles at the bottom represent the MR images and blood signal intensities obtained in the various conditions shown.

in Section 7.7.5. The water–fat shift can be exploited in order to eliminate fat signal in images: a selective 90° pulse is applied at the fat proton frequency only, immediately before the imaging pulse sequence, and the resulting transverse magnetisation is completely dephased using a spoiler gradient. Therefore there is no fat magnetisation for the subsequent excitation pulse to nutate, and hence no fat signal. This is known as Chemical Shift Selective Saturation (CHESS), and is an alternative to STIR for fat suppression (see Section 7.5.2).

The water–fat shift is also a source of an artefact visible in the frequency encoding direction at the interface between water and fat in an image (e.g. kidney and surrounding adipose tissue). Fat and water signals are displaced from each other along the frequency encoding direction, resulting in a hyperintense edge where the signals from water and fat overlap and a dark edge on the opposite side of the organ where there is a signal void. The displacement may be expressed in pixels, and is calculated as the ratio between the water–fat shift and the pixel bandwidth (the range of frequencies contained in each pixel, which depends on the gradient strength used). It is worthy of note that the pixel bandwidth influences the signal to noise ratio (SNR) of the image. Using a smaller bandwidth improves the SNR, but also results in a larger water–fat shift artefact (Figure 7.32). The balance between these effects must be optimised, taking into account the anatomical region being studied (for example, water–fat shift is crucial in abdominal imaging but negligible in brain studies). If there is doubt as to whether an image feature is a water–fat shift artefact, the simple way to solve the impasse is to re-acquire the image switching phase and frequency encoding directions, as shown in Figure 7.32. This artefact is known as the chemical shift artefact of the first kind.

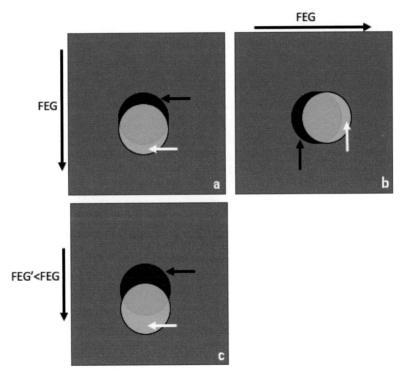

FIGURE 7.32 Water–fat displacement in the direction of the frequency encoding gradient (FEG). The light grey disk represents fat and dark grey in the water background. The black arrow indicates the area of signal void due to displacement between water and fat signals (see text). The white arrow indicates the hyperintense signal due to superposition of fat and water signals. 7.32 (a) and (b) show the dependence of the displacement direction on the frequency encoding direction, (a) and (c) show larger displacement with a weaker gradient (and hence smaller pixel bandwidth).

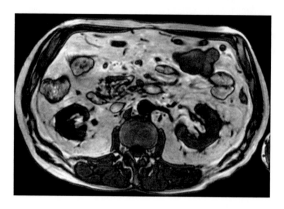

FIGURE 7.33 Chemical shift artefact of the second kind. The lack of signal at the interfaces between adipose tissue and organs is due to water and fat spins having opposite phase at the selected echo time.

When both water and fat are present in the same voxel, the signal acquired depends on the relative phase of water and fat signals at the echo time: if the relative phase is zero the signal is maximum, if the relative phase is 180° the signal is the difference between water and fat signal intensities; this reduction in signal is known as a chemical shift artefact of the second kind. The echo time needed to obtain in-phase or opposed-phase signals depends on B_o. At 1.5 T, in phase signals occur at T_E = 4.55n ms and opposed phase at $TE = (2.3 + 4.55n)$ ms, where n is an integer. In Figure 7.33, the dark edges are due to an opposed-phase condition in voxels containing a mixture of water and fat. Comparing in-phase and opposed-phase imaging allows quantification of the fat component in tissue (e.g. in liver steatosis), and is known as the Dixon method.

7.6.3 WRAP-AROUND ARTEFACTS

As discussed in Section 7.3, the spatial encoding process in two-dimensional MRI uses techniques known as frequency and phase encoding. In both cases, position is encoded into the acquired MRI signal in the form of phase. Phase is a periodic variable between 0° and 360°. When the phase value reaches 360° it starts again from 0°, and in general phase $\phi > 360°$ maps to a value $\phi - 360°$. This has consequences in MRI: if the field of view (FOV) does not contain the whole extent of the body in the phase encoding direction, for example, tissues outside of the FOV, where the phase due to the imaging gradients would be > 360°, will be reconstructed in the position corresponding to a phase $\phi - 360°$, and superimposed onto signal from tissue actually in that location (see Figure 7.34). This is known as wrap-around or fold-over artefact. It can be avoided by accurate selection of the FOV in terms of dimension and positioning. It is interesting that in parallel imaging the acquisition of 'folded' images using a reduced FOV is done intentionally to reduce acquisition time (fewer k-space lines). Data acquired using multiple coils are then integrated and the unfolded image is reconstructed (see Section 7.8.3).

7.6.4 SUSCEPTIBILITY ARTEFACTS

Magnetic field inhomogeneity affects the image in terms of geometrical distortion and loss of signal. Assuming good homogeneity of the B_0 field, the cause of magnetic inhomogeneity is the patient's body itself, due to the differing *magnetic susceptibilities* of tissues, bones, air and implants (for example dental implants, metal plates and screws). Magnetic susceptibility is a measure of how much a material becomes magnetised in an external magnetic field, and hence how much it distorts that field.

MRI spatial encoding is based on use of linear magnetic field gradients, resulting in linear variation in the Larmor frequency. The difference in magnetic susceptibility of materials perturbs this

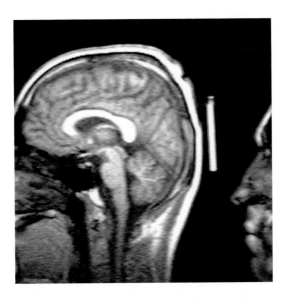

FIGURE 7.34   Wrap-around artefact. In this example the dimension of the field of view is correct but does not contain the whole head, consequently the nose appears in the posterior of the image.

expected linear variation resulting in distortion in the reconstructed images and a decreased T_2^* at the interface between materials. The spatial distortion affects the slice profile too. Spin-echo sequences (see Section 7.2.6), designed to compensate for static magnetic field inhomogeneities, are less affected than gradient-echo sequences. Moreover, images acquired with short T_E are better than those acquired with long T_E: the longer the echo time, the longer the time available for dephasing due to magnetic field inhomogeneity.

Magnetic susceptibility may not only be a source of artefacts, but also an indicator of specific pathological or physiological conditions. In *Susceptibility-Weighted Imaging* (SWI) the highest visibility is obtained for structures with significantly different susceptibilities compared to surrounding tissues, such as deoxygenated blood in veins, hemosiderin and methaemoglobin in haemorrhage or iron-rich brain nuclei. See Section 7.7.2 for discussion of the use of SWI in MR angiography. The different magnetic susceptibilities of oxy and deoxy-haemoglobin is the basic mechanism behind functional MRI (fMRI) of the brain, allowing investigation of brain activity (see Section 7.7.3).

7.6.5   TRUNCATION ARTEFACTS

As discussed in Section 7.4, MR image data is acquired in k-space, which is the Fourier transform of image space. In order to fully recover spatial frequency information, data should be collected from a large matrix in k-space, requiring a very long acquisition time. Practical considerations limit the maximum spatial frequency that it is possible to collect. This is the origin of the truncation or 'Gibbs artefact' that appears at high-contrast boundaries, appearing as a series of dark and light lines superimposed on the image. Sometimes the number of phase encoding steps is reduced to shorten the acquisition time (see Section 7.5.5); in such a case truncation artefacts become visible in the phase encoding direction. A possible solution is to filter data in k-space prior to reconstruction: the artefact is reduced, but spatial resolution is reduced too.

7.6.6   EQUIPMENT-RELATED ARTEFACTS

Up to now we have discussed artefacts that are due to the human body (motion, susceptibility), MR physics (chemical shift) or inappropriate selection of acquisition parameters (wrap around, Gibbs artefacts). There is another family of artefacts that arise when some component of the scanner is not working optimally. If these artefacts are very evident, clinical use of the scanner may have to be stopped until it is repaired. A good programme of quality assurance (QA) may prevent this through early detection of small deterioration in image quality which can be addressed without interruption of clinical activity.

7.6.6.1 Gradient Nonlinearities

Accurate reproduction of image geometry depends on linear variation of the strength of the gradient fields with position. There are limits to what is technically achievable, particularly towards the edges of a large imaging volume. Usually, for a large FOV the gradients are not linear at the periphery of the image, resulting in distortion which can be corrected automatically by the image reconstruction software. An appropriate QA programme can detect very small linear distortions, which can then be addressed during scanner maintenance.

7.6.6.2 Zipper and 'Corduroy' Artefacts

Data errors in k-space, or detection of unwanted RF signals due to external interference, result in characteristic artefacts related to the Fourier transform process. A single abnormal bright pixel in k-space is transformed into sinusoidal noise in image space: this is the source of the corduroy (or herring-bone) artefact, often caused by static electrical discharge. One possible cause the humidity in the scanner room being too low (see Figure 7.35).

If an extraneous RF signal is detected by the RF coil during signal acquisition, it is stored in k-space along with genuine image data. After Fourier transformation, it appears as an intense thin bright line through the image in the phase-encoding direction. The source may be radiofrequency noise from other instrumentation in the scanner room or an external signal not stopped by the RF shielding of the room due to a defect or a partially closed door (see Section 7.8).

7.7 ADVANCED TECHNIQUES

A large family of techniques is available to address specific clinical challenges exploiting the physics of MR imaging. Many novel solutions, innovative contrast mechanisms and applications have been developed and new ones are constantly under development. In this section, a brief overview is presented, hopefully giving some of the flavours of these intriguing imaging techniques. A detailed discussion of the following topics is out of the scope of this chapter, but there are a lot of very well-written books about the topics covered in each of the following sections.

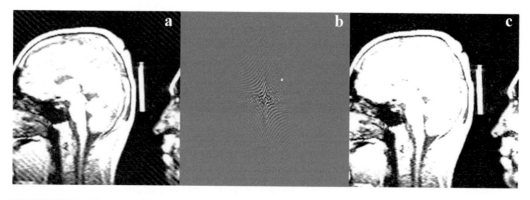

FIGURE 7.35 Zipper artefact; (a) shows a zipper artefact due to an abnormal bright pixel in k-space visible in (b); (c) is the image reconstructed with this bright pixel removed.

7.7.1 Cardiac Imaging

Cardiac imaging is a 'holy grail' of medical imaging. The heart is obviously a moving organ, and unlike breathing its motion is not under voluntary control. Both cardiac morphology and cardiac motion (mechanical function) are of high clinical interest. The main problem in cardiac imaging is motion artefacts, addressed using ECG monitoring for synchronisation of signal acquisition with the desired cardiac phase (see Section 7.6.1). The orientation of the heart in the chest is not parallel to any of the gradient axes. Therefore, cardiac imaging planes are usually double oblique, orientated to the heart chambers and valves. A lot of practice is required for the radiographer to correctly plan slice orientation on the basis of preliminary (scout) images. Pulse sequence design is used to manipulate blood signal so as to optimise heart muscle visibility, for example by applying saturation slabs as discussed at the end of Section 7.6.1.

A high-quality morphological image is not sufficient in cardiology, functional information is also necessary. As an example, cardiologists want to know the ejection fraction, which is the percentage of blood ejected by the heart in each heartbeat. This is obtained by measuring the blood volume in the left ventricle in the end-diastolic and end-systolic phases. These measurements are made by segmenting the heart muscle in all images slices, which is a time-consuming procedure requiring expert supervision. In a functional cardiac study, images are collected in different stages of the cardiac cycle, to be viewed in a cine loop. This provides information about contractile uniformity, which may be impaired after a heart attack, resulting in injury to part of the heart muscle. In order to acquire images of the heart at different stages of its cycle, data must be acquired in different temporal windows using ECG monitoring. Unfortunately, many patients requiring cardiac imaging have arrhythmia, which means that their heart rate may vary during the scan resulting in variable repetition time and impaired reproducible of heart position in each temporal window. A number of techniques have been developed to address this problem (e.g. prospective and retrospective cine imaging, definition of an arrhythmia rejection period). If low spatial resolution is acceptable, single-shot techniques allow an almost real-time approach.

Another clinical requirement is study of myocardial perfusion and viability based on administration of gadolinium contrast agent. In perfusion studies, a series of multislice images are acquired before and after bolus administration of a contrast agent. The arrival of the contrast agent bolus in the tissue increases the signal: qualitative analysis of the image series gives information about muscle areas that have reduced perfusion. However, in quantitative analysis perfusion indices (the change in signal intensity over time, the time between injection and signal peak, etc.) are obtained for accurate evaluation of the myocardial muscle condition. Software to evaluate perfusion indices is available on all commercial MRI scanners; however, segmentation of myocardial muscle under expert supervision is required.

A cardiac MRI exam usually includes study of blood flow in the aorta. In the next section, the extraction of information from flowing spins in blood vessels is discussed.

7.7.2 Flow Measurement and Magnetic Resonance Angiography

Blood flow is a cause of artefacts in MRI as discussed in Section 7.6.1. However, the effects of flowing spins on signal phase or intensity are also the basis of MR angiography. Stationary and moving spins acquire different phases in the presence of the imaging gradients. In conventional imaging sequences, flow compensation gradient schemes can be used to maximise echo signal and avoid flow artefacts. However, the phase shift due to the spin movement is exploited in phase-contrast imaging, to quantitatively study flow velocity (magnitude and direction).

Assuming that spins are moving along the gradient direction, the phase shift of moving spins is proportional to gradient intensity and duration and increases linearly with flow speed. To evaluate the flow velocity two images are acquired, one with flow-encoding gradients and the other with flow-compensation gradients. Quantitative comparison of the two phase maps allows extraction of pixels containing moving spins and measurement of the flow-induced phase-shift. The MR signal is a complex number, with a phase as well as a magnitude, and the Fourier transform operates in complex space; therefore the phase map can easily be obtained. If only one vessel is investigated and is orthogonal to the slice plane, two images are sufficient. However, a complete phase-contrast study of a volume requires application of flow-encoding gradients in three orthogonal directions. Therefore, four images are necessary including the reference flow-compensated image. Knowledge of the sequence parameters allows conversion of the phase shift to linear velocity, including the direction of flow indicated by a positive or negative sign. Because phase is a cyclic variable, phase shift must be between $-180°$ and $180°$. Consequently, gradient properties have to be selected in order to avoid errors in phase shift evaluation, as can happen in the case of fast motion and strong gradients. The user defines a 'maximum expected velocity'. If the true velocity exceeds this value, the corresponding phase shift exceeds the range $-180°$ to $180°$, and the velocity will be misrepresented as a lower value within the permitted range. This is similar to the wrap-around artefact discussed in Section 7.6.3. Quantitative flow measurements of this kind are mostly used in cardiac studies.

Angiography is very important in diagnostic imaging, and digital subtraction angiography using x-ray imaging is a powerful technique, especially during a heart attack since it can be used to guide coronary angioplasty. Nevertheless, it is an invasive procedure; CT angiography is an alternative method, but still using ionising radiation and iodinated contrast agent. MRI offers a family of angiographic techniques based on flowing spin physics or use of a contrast agent, with no ionising radiation hazards.

The flow-related enhancement observed in gradient-echo sequences when a vessel is crossing the imaged slice was discussed in Section 7.6.1. The MR angiography technique exploiting this hyper-intense signal is known as time-of-flight MR angiography (TOF-MRA). To saturate the stationary tissues a short T_R and large flip angles are used, resulting in bright vessels against a very low signal background. Therefore, three-dimensional angiographic images are obtained using the Maximum Intensity Projection (MIP) algorithm, as in CT angiography. Three-dimensional imaging results in high spatial resolution; however, the volume thickness must be limited in order to guarantee inflow of fresh blood between excitation pulse: the speed of the flowing spins needs to be higher than the ratio between volume thickness and repetition time. An extended volume of interest may be divided into multiple thin slabs to prevent loss of signal. An important limitation of TOF-MRA is the signal void that occurs in case of turbulence, for example, downstream of a stenosis, resulting in an artefactual overestimation of the stenosis dimension.

Flow-induced phase shift is exploited in phase-contrast angiography (PC-MRA), using phase-sensitive pulse sequences. The number of phase-sensitive directions to use is decided by the user, but as blood vessels follow tortuous paths it is generally necessary to encode along three independent directions. In an image obtained using phase sensitisation in a single direction, positive flow is bright while negative is dark, and the background is grey. The angiographic image is the magnitude sum of the individual phase images obtained with sensitisation along each direction. High-quality brain angiographic images can be obtained using PC-MRA.

Another mechanism that allows the visualisation of the venous system in the brain is the difference in magnetic susceptibility between brain tissues and deoxygenated blood in veins, due to the paramagnetic properties of deoxyhaemoglobin. Susceptibility-Weighted Imaging (SWI) combines conventional imaging and phase mapping, since differences in magnetic susceptibilities result in phase differences between magnetisation in neighbouring tissues. The clinical applications of SWI, including angiography, benefit from more widespread availability of MRI scanners with 3 T magnetic field strength or higher. It is interesting that in a high field scanner magnetic susceptibility artefacts (see Section 7.6.4) are also more evident, so magnetic susceptibility has two expressions at high field: a source of artefacts and a useful additional contrast mechanism.

The TOF-MRA and PC-MRA techniques are very elegant from a physicist's perspective, and the absence of contrast agent allows their wide application. However, contrast-enhanced MR angiography (CE-MRA) also has an important role. In CE-MRA, vessels are visualised after a rapid injection of a bolus of gadolinium-based contrast agent and angiographic projections are obtained by MIP analysis, since the shortened blood T_1 causes vessels to be very bright. Unlike TOF-MRA, which is also based on bright vessel signal, CE-MRA sequences can be applied without limitations in terms of slice orientation, field of view or stack thickness. Furthermore, images with high spatial resolution and SNR are obtained. To achieve the optimal vessel visibility, images must be acquired as the bolus of contrast agent transits the region of interest. Thus the acquisition protocol is optimised in terms of the delay between contrast agent injection and image acquisition (this is a good example of a situation in which centric phase encoding is useful (see Section 7.4.2), so that the centre of k-space (which determines image contrast) can be acquired first, at the peak of the bolus). In principle, using repeated scans it is possible to follow the bolus in the patient in order to study arterial and venous flow, directionality and regions of delayed filling. A number of acceleration techniques are available on a modern scanner, allowing good SNR and high spatial resolution.

7.7.3 BOLD CONTRAST AND fMRI

Differences in magnetic susceptibility perturb magnetic field homogeneity and decrease $T_2{}^*$ values; in Section 7.6.4 this mechanism has been discussed as a source of artefacts at the boundary between different tissues. On the other hand, in the previous section, susceptibility-weighted imaging (SWI) was described as a useful additional MRI contrast mechanism and an MR angiography technique. Susceptibility is also a powerful mechanism for non-invasive *in vivo* studies of human brain activity. A brief introduction to physiological changes during brain activity is necessary. When neurons are activated during a specific task (either sensory or motor) their oxygen consumption increases, and a haemodynamic response is activated within a few seconds resulting in a local increase in blood flow and volume. During activation, the local concentration of oxyhaemoglobin is therefore increased, and the $T_2{}^*$ increases compared to the basal condition. Therefore, from an MRI perspective the main difference between task and rest conditions is a slight increase in $T_2{}^*$ during the task period in the activated brain region. Functional MRI (fMRI) based on BOLD (blood oxygen level dependent) contrast aims to map brain activity by studying the local variation in MR signal between different task and rest periods. The idea is quite simple and elegant, and was experimentally validated in pre-clinical studies in the 1990s. The susceptibility change in the activated region is very small, therefore the experimental protocol, including both data acquisition and statistical processing, must be carefully designed.

Acquisition sequences for fMRI are characterised by very high sensitivity to magnetic susceptibility (exactly the opposite of the principle applied in the optimisation of morphological sequences!), and gradient-echo EPI sequences are best. Since the susceptibility effects are proportional to magnetic field strength, fMRI studies are feasible at 1.5 T or higher. Moreover, to increase BOLD contrast visibility, the magnetic field homogeneity must be optimised within the field of view before image acquisition; this is done by iterative adjustment of the shimming coil currents (see Section 7.8.1). BOLD studies are based on comparison between images acquired at rest and during task conditions. However, changes in signal intensity due to the BOLD phenomenon are comparable to the image noise and simple comparison of the two images does not produce significant results. An fMRI experiment consists of rapid imaging carried out continuously while the subject performs various tasks, known as the experimental paradigm, commonly arranged in a block design with periods of activity alternated with periods of rest. A block length is typically about 30 s and perhaps three or four complete cycles are acquired, resulting in a few minutes of scan time and more than 100 brain volumes obtained, each volume containing 20 slices or more: the amount of acquired data is impressive and data processing is another fundamental requirement in fMRI.

The basic idea of fMRI data processing is simple: to study the correlation of signal variation over time with the experimental paradigm in each pixel separately, assuming that signal increases during tasks in the activated regions. If the correlation is statistically significant the pixel is assumed to be in an activated brain region. An important step in this procedure is translation of the experimental paradigm into an associated haemodynamic response function, which peaks about 4 s after the beginning of the task. Moreover, it is necessary to check the alignment of all the collected brain volumes and correct for misalignments occurring during the long acquisition time; a movement of a few pixels between successive images significantly affects the results. Statistical analysis is optimised to the experimental paradigm: a block design is a simple and robust approach but in cognitive studies (e.g. generating words, doing mental arithmetic, image recognition) more complicated protocols are required. The pixels that are found to be statistically significant are represented by coloured clusters overlaid on morphological images or three-dimensional rendered views (Figure 7.36).

A very common clinical application of fMRI is pre-surgical evaluation for brain tumour resection. The problem is to determine how the surgeon should work in order to minimise damage to functional regions of the brain close to the tumour. In general, fMRI is fascinating for the unexpected insights it provides into brain activity in a completely non-invasive manner. To obtain rigorous results an interdisciplinary team is recommended. Neuroscientists, radiologists, neurologists, physicists, statisticians, psychologists and biomedical engineers may all be collaborators within the fMRI team, depending on the institution and the aim of the fMRI programme. Usually, physicists are responsible for optimisation of the image acquisition protocol and are involved in the data analysis. However, collaborative discussion of the protocol is mandatory. As an example, even in a simple task-and-rest block design there can be problems: the definition of the rest condition is not obvious, and correct instruction of the subject is very important. If during the rest period the subject concentrates on the scanner noise and counts its cycles we have an unwanted extra source of brain activation in our experiment!

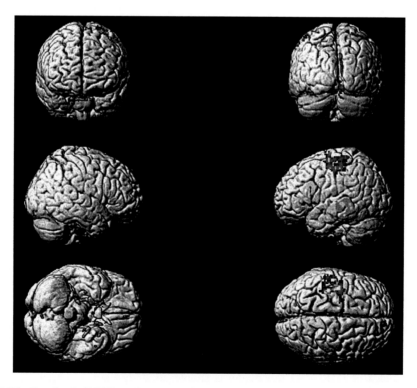

FIGURE 7.36 Result of a BOLD study presented as coloured clusters on a three-dimensional rendering of the brain. The task is finger tapping.

7.7.4 Diffusion Imaging

Another group of powerful techniques aims to study water diffusion in human tissues. In a glass of water, the random molecular motions are described by a scalar quantity, the diffusion coefficient of the water (D), which is a function of temperature. In the human body, temperature is constant, but diffusion is restricted by physical barriers (e.g. cell membranes). In this situation diffusion is represented by another coefficient, the apparent diffusion coefficient (ADC), that reflects both the fluid diffusion coefficient and environment characteristics that reduce effective diffusion. By applying an appropriate gradient scheme, it is possible to weight MR signal intensity according to the ADC: in diffusion-weighted imaging (DWI) bright voxels indicate reduced molecular mobility. An important clinical application is early detection of ischemic stroke, because well before anatomical damage there is marked reduction of water diffusion compared to normal brain tissue.

From the same data set, including images acquired with and without diffusion weighting gradients, both ADC maps and DWI images can be obtained. Similar information is encoded in both images, but they display opposite grey scales because a high value in the ADC map indicates high diffusion and hence low signal in DWI. The study of diffusion using NMR is older than MRI and allows us to investigate processes in tissues at a dimension scale much smaller than the voxel, for example, cell swelling in the case of stroke. Even more intriguing is the study of anisotropic diffusion in tissues, because it allows the reconstruction of axonal bundles in brain and spinal cord, giving important information about the connections between different regions of the brain. From a physics point of view, the problem is measurement of the diffusion tensor that describes the anisotropic diffusion in white matter due to tightly packed and coherently aligned axons. In diffusion tensor imaging (DTI) a diffusion tensor is calculated for each voxel in the investigated volume. The tensor is a 3×3 symmetric matrix, and therefore six independent DWI images are required, plus an image without diffusion sensitivity. The DWI images are independent of each other if the orientations of the diffusion gradients are different. In order to evaluate the diffusion tensor robustly, a large number of DWI images are acquired, typically 30–60 or even more in research studies. After calculation of the diffusion tensor in each voxel, a fractional anisotropy (FA) map can be created. FA is defined as equal to zero in the case of isotropic diffusion and one in the case of monodirectional diffusion; intermediate values demonstrate degrees of anisotropic diffusion. FA maps show the degree of anisotropy (a bright pixel indicates high FA); moreover, information about the preferential diffusion direction can be added using colour coding: in Figure 7.37, red indicates left-right, green anterior-posterior and blue cranio-caudal diffusion direction.

For the next step in DTI, we assume that diffusion along nerve sheaths defines nerve tracts, so that the preferred diffusion direction corresponds to tract orientation. We then perform three-dimensional reconstruction of the fibres by following the preferred diffusion direction voxel-by-voxel (Figure 7.38). A number of models and numerical techniques have been developed to obtained better fibre reconstructions, including taking account of regions where fibres cross or bifurcate. Amazing reconstructions of axon fibres have published: this is another powerful and non-invasive technique and an invaluable tool for the neuroscience community to study brain connectivity. Since a number of physiological and pathological changes, including cancer, affect extracellular water diffusion, the field of application of DWI and DTI is growing. An important clinical application is pre-surgical evaluation of brain tumours: with fibre tracking it is possible to see if the tumour has displaced or infiltrated nerves.

7.7.5 Spectroscopy: In Vivo Biochemistry

In Section 7.6.2 the water–fat shift was discussed as a source of artefacts. However, chemical shift is exploited in MR spectroscopy (MRS), which had an important role in physics and chemistry well before the development of MR imaging. Because of chemical shift, hydrogen nuclei in different

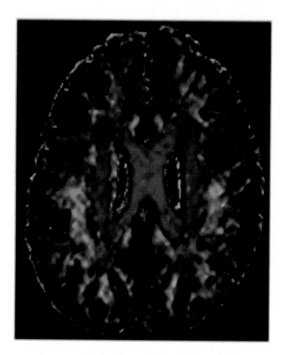

FIGURE 7.37 Fractional Anisotropy (FA) colour map obtained from a 32-direction DTI data set. The red structure in the centre is the corpus callosum.

chemical groups within a molecule resonate at slightly different frequencies. The frequency spectrum of the MR signal obtained from a sample therefore allows the study of its chemical composition. As well as hydrogen (or proton, ^1H) spectroscopy, phosphorus (^{31}P) spectroscopy is feasible in MR scanners due to relatively high sensitivity, isotopic abundance and concentration in the body. However, dedicated RF hardware is required to work at the ^{31}P resonance frequency. In proton MRS, because of the very high concentration of water in tissue compared to other hydrogen-containing compounds (several order of magnitude), it is necessary to modify the pulse sequence to include a water suppression step during acquisition in order to detect signals from metabolites.

In proton spectroscopy, the main metabolites studied are choline (Cho), creatine (Cr), N-acetylaspartate (NAA) and lactate (Lac), moreover, using short echo time and high B_0 it is possible to detect myo-inositol (mI), glutamine, glutamate (Glx) and glucose. In phosphorus spectroscopy, energy metabolism can be investigated by detecting phosphocreatine (PCr), adenosine triphosphate (ATP) and inorganic phosphate (Pi). Moreover, in high-quality spectra more signals are detected: nicotinamide adenine dinucleotide phosphate (NADP), phosphoethanolamine (PE), phosphocholine (PC), glycerophosphoethanolamine (GPE) and glycerophosphocholine (GPC). A limitation of *in vivo* MRS is the poor signal-to-noise ratio due to the low concentration of the observed metabolites. Better spectra are obtained at higher magnetic field strength, since the SNR increases. Moreover, chemical shifts and the frequency separation of metabolic peaks are larger too. A requirement of *in vivo* clinical MRS is the ability to collect a spectrum from a region within the body selectively, without contamination from the surrounding region: a spectrum containing undifferentiated signal from several organs and tissues would not be of interest. Techniques have been developed to acquire spectra from volumes selected on MR images, using slice selection gradients and selective RF pulses similar to those used in imaging. The voxels used in MRS are of the order of 1 cm^3 or more, due to the poor SNR (Figure 7.39). Two approaches are available: single voxel and multivoxel. The latter is also known as chemical shift imaging (CSI) or MR spectroscopic imaging (MRSI) and applies phase-encoding techniques to segment a selected volume into voxels.

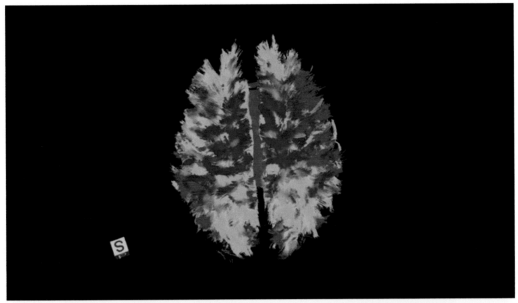

a)

b)

FIGURE 7.38   Fibre tracking showing (a) all detectable fibres and (b) the cortico-spinal tract. The usual colour code for the preferential diffusion direction is applied (red: left–right, blue: cranio-caudal, green: antero-posterior).

(Courtesy of Dr Luca Weiss, San Camillo Hospital, Venice.)

Spectral analysis is based on the evaluation of peak areas, and peak ratios are used for comparison between healthy and pathological areas of the same organ (e.g. left and right brain hemispheres, cancer and surrounding healthy tissues). The peak area depends on the repetition time (T_R) and echo times (T_E) used in the acquisition sequences, and the peak ratio is also affected by T_R and T_E since each metabolite has different relaxation times. Calibration of peak areas in order to calculate absolute concentration is a difficult task, requiring a calibrated reference object and correction for T_R and T_E on the basis of the relaxation times of the metabolites.

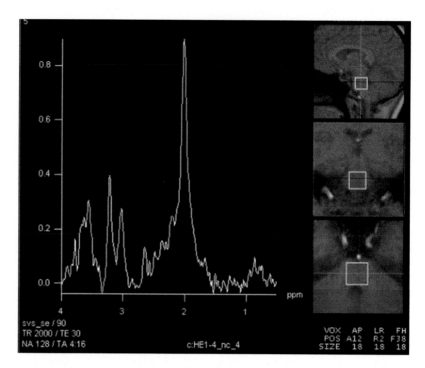

FIGURE 7.39 A proton spectrum obtained from a cubic volume of side 1.8 cm. On the right the position of the selected volume is visible on brain images. The spectrum is the sum of 128 acquisitions to improve SNR. Peaks of choline (3.2 ppm), creatine (3.0 ppm), and N-acetylaspartate (2.0 ppm) are visible. (Frequency in MR spectra is expressed in parts per million (ppm) relative to an agreed standard.)

7.8   MRI INSTRUMENTATION

An MR scanner consists of three main components: a magnet generating a strong magnetic field to create tissue magnetisation, a gradient coil system for signal localisation and the RF transmitter and receiver coils, tuned to the Larmor frequency, to create the RF pulses and collect the echo signal. Moreover, a patient support and a physiological monitoring system are required to move the patient into the scanner and for monitoring cardiac and respiratory movement, when required. To avoid artefacts due to external RF signals (see Section 7.6.6.2) the room containing the scanner is enclosed by a Faraday cage, or RF cage, a copper or aluminium enclosure that prevents entry of extraneous RF radiation. Obviously, the door must be closed carefully during imaging so that the integrity of the cage is maintained: actually, the scanner is so noisy during data acquisition that it is impossible not to notice if the door is open! Access to the scanner room is strictly controlled because the magnetic field is very strong in the area close to the magnet and accidents can happen if persons approach the magnet without adequate preparation (see Section 7.9). A powerful computer system is used to control gradient and RF systems during data acquisition, and to perform image reconstruction, display and archiving. A technical room is nearby, and the connections of the gradient and RF coils to their associate electronics in this room must be carefully designed to avoid spurious RF signal contamination.

7.8.1   THE MAGNET AND THE QUESTION OF FIELD STRENGTH

The B_0 field strength used in MRI scanners is a factor of more than 10^4 greater than the earth's magnetic field (~0.05 mT). Moreover, the field must be uniform over a large volume, comparable to the FOV of the images. In everyday life, we are familiar with permanent magnets and electromagnets. MRI scanners using permanent magnets are possible with field strengths of up to about 0.3 T, while

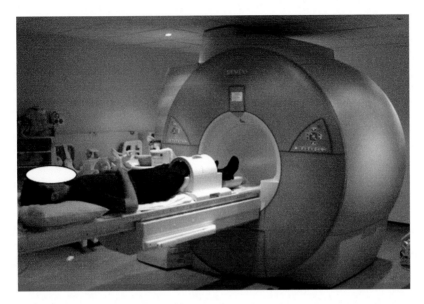

FIGURE 7.40   An MRI facility at work. A subject ready for a knee examination is visible, with a knee coil in position and additional coils on the shelf behind. A vent pipe can be seen running between the superconducting magnet and the ceiling, necessary for safety in case of a quench.

(Courtesy of Dr Geoff Charles-Edwards, Guy's and St Thomas' NHS Foundation Trust.)

with conventional electromagnet technology up to 0.6 T is feasible. Today the field strength used clinically is usually 1.5 T or 3 T, with higher field scanners increasingly widely used in some clinical applications and in research. State-of-the-art superconductor technology is used to achieve the large electric current needed in the magnet coils to obtain intense magnetic fields (>1 T) over a large volume of interest.

Once an electric current is running in a loop of superconducting wire, it will continue to circulate indefinitely without resistive losses. In a superconducting magnet, the electric current is induced by a portable DC power supply that is removed when the expected magnetic field intensity is obtained: the supply will not be needed again during the lifetime of the scanner. Superconductivity means zero electrical resistance (which is fundamentally different from negligible resistance!). You may be not familiar with superconductivity because of the low temperatures involved. In the niobium-titanium (NbTi) alloy used in most MRI scanner magnets, the superconducting transition happens at 7.7 K. Different alloys have different transition temperatures, but they are always close to absolute zero (0 K). In summary, superconducting magnets are permanent magnets, provided the coils remain below their critical temperature of a few kelvin. Superconductivity is another phenomenon described by the quantum physics that underlies NMR.

The magnet structure consists of a cryogenic storage dewar containing the superconducting magnet coils, immersed in liquid helium. A vacuum chamber surrounding the helium chamber reduces heat transfer. The helium continuously evaporates and is recycled through a refrigeration system; you can usually hear the cyclic noise of the compressor in the magnet room. A sudden loss of superconductivity results in a magnet 'quench'. If the temperature of the superconducting wires increases over its critical value, the magnet coils enter the normal resistive state and energy is dissipated as heat, due to the Joule effect: the field collapses and large amounts of helium boil off as gas. The amount of helium in a scanner is about 1500 litres and its liquid-to-gas expansion ratio is 1 to 754: quench pipes are necessary to transport the gas out the scanner room, to prevent asphyxia and damage due to pressure build-up. In the normal life of an MR scanner a quench is performed during decommissioning, but occasionally a quench happens due to defects in the magnet, and a wall-mounted button is provided to quench the magnet deliberately in an emergency. During a quench all the helium is lost to the atmosphere, and to restart the magnet it must be refilled with liquid helium, which is very expensive. The quench button is for use in emergencies only! Some manufactures are developing a new generation of magnets requiring a very limited amount of helium, about 7–20 litres. In this new technology, magnet coils lie in the vacuum and are cooled by cooling tubes thermally connected to them; liquid helium circulates in the tubes themselves. In case of a quench all of the helium is fully contained within the system, therefore no vent pipes are necessary for safety.

The magnetic field extends for some distance beyond the magnet bore, and for safety reasons this fringe field is contained so as to limit the area around the magnet where the field strength exceeds 0.5 mT. The first MR units were surrounded by passive shields made up of iron plates. Modern magnets use active shielding: additional shield coils outside the main magnet coils with current flowing in the opposite direction, resulting in a partial cancellation of the fringe field.

The magnet is defined by the field strength at the centre of the bore, bore diameter (60–70 cm) and field homogeneity. The homogeneity required is of the order of 1 part per million (ppm) over a 40 cm diameter spherical volume at the centre of the bore. This homogeneity is obtained by a process known as 'shimming' during installation. Shimming can be done by adding small iron plates around the magnet bore (passive shimming) or by adjusting the currents in additional magnetic field coils known as 'shim coils' (active shimming). Shimming corrects for magnet defects and for field distortions caused by any ferromagnetic structures present around the scanner. The homogeneity achievable in practice is limited because human body susceptibility differences cause inhomogeneity of about 1–5 ppm. Patient-specific active shimming is required in some applications to compensate for inhomogeneity within the field of view due to this effect.

We close this section with some comments about optimal field strength. Increased B_o field strength brings with it exciting improvements in signal-to-noise ratio, reduction of the voxel size and improvement in techniques such as fMRI brain mapping and spectroscopy. However, high magnetic fields do not allow optimal imaging in high susceptibility regions because the severity of susceptibility artefacts is proportional to magnetic field strength. Recent studies suggest that low-field MRI (~0.5 T) combined with state-of-the-art hardware and software has advantages for some applications including imaging anatomy near air-tissue interfaces. The cost of the MRI system and its installation increases with field strength too. In summary, the choice of field strength must be carefully evaluated on the basis of the requirements of the specific clinical site: biggest is not always best!

7.8.2 THE GRADIENT SYSTEM

When MRI entered clinical practice, almost 40 years ago, it was the only imaging technique that allowed arbitrary orientation of the slice plane, completely unlike CT scanners that allowed axial slices only. The flexibility of the acquisition plane is due to the gradient system. This system is simple in principle: three orthogonal gradient fields (G_x, G_y and G_z) are generated by the currents flowing in three coil systems. Oblique and double oblique slices are created using combinations of G_x, G_y and G_z, as discussed in Section 7.3.2. The main requirement is that the gradients should be linear with position over as large an FOV as possible. Lack of linearity results in image distortion, especially at the periphery of the image.

Gradient pulses are usually trapezoidal in shape as a function of time, and gradient performance is defined by maximum gradient value (G_{max}), rise time (time to go from zero to G_{max}) and slew rate (ratio between G_{max} and rise time), see Figure 7.41.

Typical values are: G_{max} = 1–50 mT m^{-1}, rise time = 200–1000 μs and slew rate = 20–200 Tm^{-1}s^{-1}. A greater slew rate allows faster application of the phase encoding gradient and hence shorter echo time. High slew rates and short rise times are required to switch gradients quickly and allow ultrafast imaging sequences such as EPI. Moreover, strong gradients allow thinner slices and smaller voxels. The gradients need an efficient cooling system because the coils heat up rapidly due to the large currents driven by the gradient amplifiers located in the technical room. If the temperature

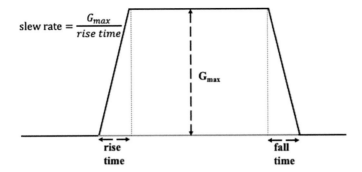

$$\text{slew rate} = \frac{G_{max}}{\text{rise time}}$$

G_{max}

rise
time

fall
time

FIGURE 7.41   A schematic presentation of key gradient parameters.

increases over a certain value, the acquisition is stopped to prevent malfunction, this may happen in examinations that make extensive use of EPI sequences (e.g. fMRI or DTI).

To obtain a trapezoidal gradient shape, the current intensity in the gradient coil has to be optimised to compensate for the effects of eddy currents induced in adjacent conducting materials (coils, electric wires, cryostat, etc.) due to rapid gradient switching. These eddy currents oppose the gradient fields and cause a deterioration in the gradient profile.

The gradient coils carry electric currents and therefore experience forces due to the surrounding static magnetic field. As the gradients are switched on and off these forces vary, causing vibrations in the gradient coils and generating a significant amount of acoustic noise. The sounds are tones, knocking or banging noises depending on the frequency and waveform of the applied gradient pulses. Acoustic isolation of the magnet room is very important for MRI workers, and hearing protection must be worn at all times by patients!

7.8.3   The RF System and Coil Design

The radiofrequency (RF) system includes both transmitter and receiver antennae, or coils. In principle, the same coil can be used as a transmitter and receiver, but usually the transmitter coil is a large coil within the magnet bore, known as the body coil, and a variety of dedicated coils are used to collect signals from different anatomical regions. The RF coils are tuned to the Larmor frequency, which of course is proportional to B_0.

The transmit coil and the RF pulse are designed to generate uniform rotating B_1 fields, at right angles to B_0, over the imaging volume. According to the imaging sequence, the RF pulses are shaped in terms of centre frequencies, bandwidths, amplitudes and phases: the centre frequency determines slice position, the bandwidth controls the slice thickness, the amplitude controls the flip angle and the phase defines along which axis the magnetisation is flipped in the rotating frame. The stability and reproducibility of the RF amplifiers and associate electronics is a key requirement for good image quality and minimisation of artefacts.

Even a simple plain coil, linearly polarised, with electric current oscillating at the Larmor frequency can be used for signal excitation. But a circularly polarised coil is better because it creates a rotating B_1 field, instead of an oscillating one, and requires less power to provide equivalent excitation. A circularly polarised RF field can be obtained by a combination of two linearly polarised coils with orthogonal axes: when sinusoidally oscillating currents flow through the two coils, with one shifted by 90° relative to the other, they create a rotating magnetic field in the plane of the coil axis. (This is a simplified picture: quadrature transmission coils are actually more complicated than this.)

The signal acquired in MRI is very small, and therefore considerable efforts have been spent in the development of receiver coils and associated electronics to preserve and improve signal-to-noise ratio. Receiver coils are electrical circuits tuned to the Larmor frequency to collect the signal and efficiently transfer it to the amplifier before analogue to digital conversion.

MR signal amplitude mainly depends on field strength and acquisition sequence. However, electrical noise is generated by the patient's body and the analogue part of the electronics chain. The simplest strategy to increase the detected signal is to position the receiver coil as close as possible to the investigated volume, which can be done using surface coils or small volume coils (e.g. head coils (Figure 7.42(a)) or knee coils (Figure 7.40)). The sensitivity of the surface coil decreases with distance from the surface; this is a limitation when deep organs are investigated, but beneficial when surface regions are being imaged and deeper moving organs can create artefacts as in spine studies at the level of the heart, for example.

Quadrature coils, like those used for RF transmission, can also be used in receive mode. They are more effective than single surface coil in terms of volume sensitivity, and independent acquisition of two orthogonal signals increases the SNR by a factor $\sqrt{2}$. So-called phased-array coils are state-of-the-art coil technology, but the idea is simple: an array of small coils, each with an independent amplifier and receiver. In this approach, the SNR is as high as achievable with a small surface coil but by combining together signals from multiple coils, a larger FOV is obtained. An example of a phased array coil for the torso is shown in Figure 7.42(b)). The phased array systems need to be designed carefully, because interaction and coupling between elements of the array can result in a reduction in SNR. Phased arrays are available as rigid volume coils, for example for the head and neck, or as large flexible arrays suitable for wrapping around the torso of an adult.

It is important to remember that surface coils must be orientated with their axes perpendicularly to B_0; if the coil axis is parallel to the static magnetic field, no signal is induced in the coil by the rotating transverse magnetisation. A phased array coil designed for spine studies consists of a unidimensional array, while a large coil for the torso may be a 2×6 matrix of elements. In a phased-array system signal from the imaged volume is collected by each array element, each with a different sensitivity profile due to their different orientations and distances from the object. The final image is obtained by combining signals from the same voxel collected by each coil element.

Phased array coils stimulated the development of parallel imaging, a family of very powerful techniques that permit a reduction of the acquisition time of a factor of two or three without affecting spatial resolution. The basic idea is that the sensitivity profile of each coil element within the array contains spatial information that can be used to reduce the amount of phase encoding needed, allowing undersampling of k-space and consequent speeding up of image acquisition. The disadvantage is a reduction in SNR compared with image acquisition with the same parameters but no acceleration factor. As discussed at the end of Section 7.5.5 a class of techniques known as *compressed sensing* successfully combines different approaches and promises to revolutionise data acquisition in terms of acquisition time and image quality.

7.9   MRI SAFETY

One of the main attractions of MRI as a medical imaging modality is absence of ionising radiation and the associated risks. The risk of stochastic effects of ionising radiation is assumed to scale linearly with dose, with no lower threshold. Some degree of risk is therefore unavoidable whenever imaging is performed. In MRI, potential direct biological effects are deterministic in nature, and can be avoided by keeping exposure below the relevant effect threshold. MRI has been in clinical use since the 1980s, with no evidence of *adverse health effects*. However, there are certainly *biological effects* of the static magnetic field, switched gradients and RF field, as discussed in this section. Moreover, unlike x-ray imaging there are *indirect* effects, notable strong attraction of ferromagnetic objects towards the magnet, which has been the cause of some dramatic accidents. Patient safety is assured through careful development of pulse sequences to ensure that exposure is below the thresholds for direct effects, and protocols for avoidance of indirect effects including screening of patients and staff before admission to the scanner room. When accidents have occurred in MRI, it has usually been because operators are not well trained and patients not carefully prepared.

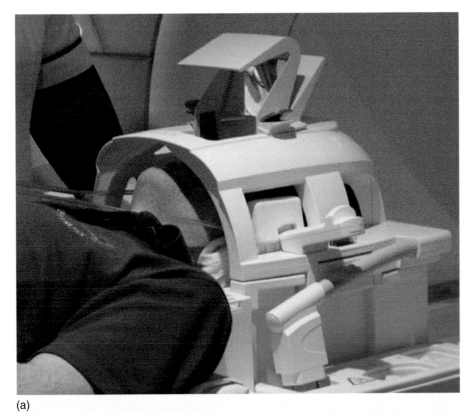

(a)

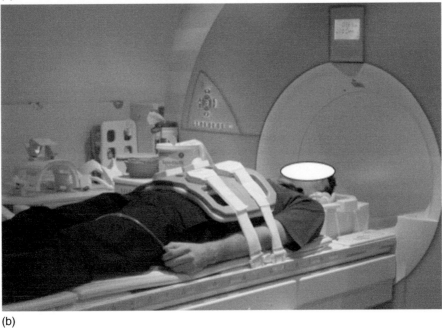

(b)

FIGURE 7.42   (a) A phased array head coil; note the mirror, which is helpful for claustrophobic patients and presentation of visual stimuli in fMRI. (b) A torso phased array placed on the subject ready to be moved into the scanner. The subject is holding an alarm ball: squeezing this alerts the operator that the patient needs attention.

International Electrotechnical Commission (IEC) standard 60601-2-33 defines three operating modes for MRI scanners, in terms of the risk of direct biological effects.

1) Normal operating mode – the usual mode of operation, with no risk of physiological stress.
2) First level controlled operating mode – in this mode physiological stress is possible, requiring supervision and a medical assessment of the risk versus benefit for the patient having the scan. The operator is required to explicitly approve moving into this mode, for example by pressing a key on the scanner control console.
3) Second level controlled operating mode – this mode may involve significant risk, and requires ethics committee approval (or local equivalent).

Limits for each of these modes are defined for the static magnetic field, gradient fields and RF field in order to limit the direct physiological effects described below.

7.9.1 Static Magnetic Field Hazards

When MRI is explained to patients, the very strong magnetic field seems the most extraordinary element of the scanner. A magnetic field that is tens of thousands of times as strong as the earth's magnetic field is impressive. Static magnetic fields used in clinical MRI are in the range 0.3-3 T, while advanced clinical research scanners may be in the range 7-11 T.

The static magnetic field exerts a torque on ferromagnetic objects, tending to align them with the direction of the magnetic field (the compass needle is a good example of this twisting force). It also exerts a translational force, attracting ferromagnetic objects toward the magnet. Both of these effects increase dramatically with field strength. The translational force is stronger where B_0 is increasing mostly steeply with distance; in the MRI room the force increases very rapidly close to the magnet, especially for actively shielded systems where the fringe field is contained close to the magnet. Reference is often made to the *spatial gradient* of the field, which must be carefully distinguished from the switched gradient fields used during imaging! In terms of direct biological effects, the static magnetic field can induce mild sensory effects (including vertigo and taste sensations) when a person moves close to a scanner with B_0 higher than 1.5 T. This a harmless transient effect and its perception is very subjective.

A major MR safety concern for many years has been that the magnetic field may interrupt the function of cardiac pacemakers. Unfortunately, fatal accidents have occurred due to unintentional scanning of persons with pacemakers. Recently, 'MR conditional' pacemakers have become available, and it is also possible to scan patients with older pacemakers if safety protocols are followed carefully. This is usually best left to specialist centres with considerable MR safety experience and expertise available. Pacemakers are not the only implanted devices that may be a source of serious hazards for the patients due to interaction with B_0. Serious injuries have occurred to patients with neural stimulators. Moreover, an intrauterine device can be displaced by the magnetic field; this presents no serious hazard for the patient and is not a definitive contraindication MRI, but a gynaecological examination may be necessary to confirm that the device remains correctly positioned. Intracranial aneurysm clips, orthopaedic implants and stents are nowadays nonferromagnetic, but it is necessary to carefully evaluate each patient to make sure. Implants in use before MR safety was a consideration in device design could be ferromagnetic. Some patients are at risk due to ferromagnetic fragments in the body, particularly in the eyes. These may include metal workers and people who have been injured during military service.

Because of the translational force exerted on ferromagnetic objects close to the scanner, they can become dangerous projectiles, and excluding such objects from the vicinity of the scanner is one of the most important aspects of MR safety. An internet search yields many images of projectile accidents: wheelchairs, patient beds, floor polishers, oxygen tanks and so on. There is a particular risk if healthcare staff approach the magnet with objects such as pens, scissors and scalpels in their pockets. Obviously, radiologists and radiographers know the risks and how to work safely, but if another specialist accompanying a patient enters the magnet room a tragic accident could be possible.

In Section 7.8.1, superconducting magnet technology is summarised, including reference to quenches. Because of this risk, the magnet room must incorporate 'quench pipes' for expelling the gas outside the building to prevent asphyxia or cold burns. Leakage of helium gas into the magnet room can result in a dangerous depletion of oxygen, and consequently oxygen level monitors are installed in scanner rooms.

7.9.2 SWITCHED GRADIENT FIELD HAZARDS

As discussed in this chapter, switched magnetic field gradients are used as part of the imaging process. They are switched on and off rapidly, producing time-varying fields in the kilohertz range. The potential biological effect of time-varying fields in this range is peripheral nerve stimulation (PNS). The gradients have a trapezoidal waveform (Figure 7.41) and electrical fields and currents are induced during the ramp-up and ramp-down periods, according to Faraday's law. If the electrical fields exceed the nerve depolarisation threshold PNS results, leading to tactile sensations and, at higher levels, muscle contraction and severe pain. At higher levels still, cardiac stimulation is possible, which must of course be avoided. However, PNS would become extremely painful well below that level. The most demanding sequence in terms of fast gradient switching is EPI. During imaging of oblique planes, when two or three gradient coils are used simultaneously, a very high slew rate can be reached. The scanner control software calculates the effective slew rate and warns the user if the first level controlled operating mode is reached, which may result in *mild* PNS.

The most characteristic effect of the gradients is acoustic noise (see Section 7.8.2). In order to avoid acoustic trauma, earplugs or ear defenders (or both) are used by patients. Since noise is proportional to the gradient switching rate, EPI is the technique with the highest noise level. Reduction of acoustic noise is an area of continuous technical development, in order to prevent it from becoming a limiting factor for further development of MRI sequences.

7.9.3 RADIOFREQUENCY FIELD HAZARDS

RF excitation results in deposition of energy in body tissues, leading to resistive heating. Cells are not damaged unless the temperature exceeds 42°C, and localised heating is dissipated by blood flow, thus a moderate heating effect is well tolerated. The IEC standard includes limits for temperature rise of 0.5°C in normal operating mode and 1°C in first-level controlled operating mode.

The rate at which RF energy is deposited in the body is known as the specific absorption rate (SAR), measured in watts per kilogram of body weight (W kg^{-1}). SAR increases with the square of the Larmor frequency, and hence of the static field strength. This is one of the major limitations in the development of very high field systems. SAR is very difficult to measure *in vivo* and numerical models have been developed for SAR calculation to verify the safety of the sequences during each exam. The scanner control system calculates SAR based on patient weight, which must therefore be entered correctly during patient registration. If SAR exceeds the limits the exam is stopped by the control system, and this may also happen if a series of high SAR sequences are applied successively without interruption.

The force exerted by the static magnetic field on non-ferromagnetic metallic implants is negligible. However, they absorb RF energy due to their electrical conductivity, which may result in localised heating and even burns (think about what can happen with a gold necklace!). Caution is required even with tattoos; some inks contain metallic particles that may heat up dramatically during the exam, resulting in skin burns. Obviously appropriate electrodes must be used for physiological monitoring. Special attention is required with unconscious patients arriving at the MRI unit with monitoring devices attached.

7.9.4 REGULATIONS AND RISK MANAGEMENT IN MRI

As discussed above, IEC standard 60601-2-33 sets limits for patient (and worker) exposure to the static magnetic field, switched gradients and RF field. These limits are built into the scanner itself and are intended mainly to avoid uncomfortable physiological effects. There are also regulations at international and national levels in some jurisdictions limiting exposure to electromagnetic fields

(EMF). An example is the European Union Physical Agents (EMF) Directive, which includes an exemption from occupational exposure limits for MRI subject to certain conditions being met.

However, in most situations the main hazards in MRI are not due to direct physiological effects of EMF exposure, but to indirect effects such as projectiles and RF heating of implants.

The IEC 62570 standard categorises implants and other devices that may be brought into the MR environment as either MR safe, MR conditional or MR unsafe. Unfortunately, most implanted medical devices fall in the MR conditional category, meaning that the operator has to perform a careful evaluation to ensure that condition set by the manufacturer for safe scanning can be met (these can include limits on static field strength, spatial field gradient and SAR, as well as, for example, the location of the device in the body).

A key element of MR safety is ensuring that there is adequate control of access to the MR scanner room, and that people entering the room have either been properly trained or have undergone sufficient screening to ensure safety. This involves physical measures such as locks and barriers, as well as administrative controls such as local rules and training and information programmes. In some jurisdictions, responsibility for ensuring MR safety is delegated to specific role holders, for example, the MR Responsible Person in the UK. The terminology and exact responsibilities vary in different countries, but one role that is increasingly recognised internationally is that of MR Safety Expert, a medical physicist specialising in MRI who provides scientific and technical advice on safety issues to the institution.

NOTES

1 Physicists can occasionally be pedantic, and the term 'magnetic resonance imaging' is sometimes challenged on the grounds that it is insufficiently precise: there are other magnetic resonance phenomena in physics that do not involve nuclei. The word 'nuclear' was dropped, in the context of medical MRI, very early on in the development of the technique. Some people claim that this was to reassure patients worried by the implied link with nuclear weapons and radiation (worried unnecessarily, since MRI has no connection with ionising radiation at all). Others claim that it was engineered by radiologists as the opening (and decisive) salvo in a turf war with nuclear medicine physicians. It may be that both accounts are true, since the possibility of patient anxiety appears to have been raised initially by radiologists…

2 This may appear to be stating the obvious, but I recall being told by a colleague some years ago that she had got quite some way into a lecture about NMR before realising that many in her audience of biological scientists had wrongly assumed that she was talking about the nuclei of *cells*!

3 Strictly speaking, tesla is the SI unit of *magnetic flux density* rather than *magnetic field strength*. However, the former term is used ubiquitously in MRI, to the extent that one of the authors once had a paper returned for amendment because 'magnetic flux density' would not be understood by readers!

4 The Larmor frequency is expressed here as an *angular frequency*, in radians per second (rad s^{-1}), indicating the angle that the magnetic moment moves through per second. It is also often expressed as a *frequency*, v_0, in megahertz (MHz), indicating the number of revolutions that the magnetic moment completes per second. Since there are 2π radians in a circle, $\omega_0 = 2\pi v_0$. The distinction doesn't matter very much in equations like this, but care is needed when calculating numerical values, for example using the form of the gyromagnetic ratio in the correct units (see Section 7.2.1), or a factor of 2π error can easily be made.

5 This is a plausible semi-classical explanation: readers with a grounding in quantum mechanics will appreciate that at a deeper level it is a manifestation of Heisenberg's uncertainty principle since orthogonal components of angular momentum are complementary, and so cannot be simultaneously determined, and the z component is defined by equation 7.3.

FURTHER READING

Runge, V. M., Nitz, W. R. Trelles, M. and Goerner, F. L. 2014. *The Physics of Clinical MR Taught through Images*. 3rd edition. New York: Thieme.

McRobbie, D. W., Moore, E. A., Graves, M. J. and Prince, M. R. 2017. *MRI from Picture to Proton*. 3rd edition. Cambridge: Cambridge University Press.

EMITEL e-encyclopaedia of medical physics and dictionary (n.d.). http://www.emitel2.eu/emitwwwsql/encyclopedia.aspx

EMAIOS: MRI step-by-step, interactive course on magnetic resonance imaging (n.d.) https://www.imaios.com/en/e-Courses/e-MRI

8 Ultrasound Imaging and Therapy

Raffaele Novario
Università degli Studi dell'Insubria, Varese, Italy

Sabina Strocchi
ASST dei Sette Laghi, Varese, Italy

CONTENTS

8.1 INTRODUCTION

The term 'ultrasound' refers to mechanic waves with frequency higher than the audibility threshold of the human being, which is about 20 kHz.

The idea of deriving information from the reflection of waves (mechanical or electromagnetic) can be traced back to Daniel Colladen in 1822 and to Lord Rayleigh, who in 1877 published his fundamental work on sound propagation. In Japan and the United States of America in the 1940s and

DOI: 10.1201/9780429155758-8

1950s, pioneering work was done attempting to obtain information through reflection of ultrasound waves within the human body. In 1949, working in Denver, Colorado, Howry obtained the first anatomical image of a volunteer, who was immersed in a cylindrical tank full of water, by means of a prototype ultrasound system with a moveable transducer (Novario et al. 2003).

Ultrasound imaging was the first technique for cross-sectional imaging of the living human body, and since its establishment as a medical imaging modality in the 1960s, there have been many technological innovations, with typically one or two major advances per decade (Bercoff 2011).

8.2 INTERACTION OF ULTRASOUND WITH TISSUES

8.2.1 PROPERTIES OF THE ULTRASOUND BEAM

The intensity of an ultrasound beam can be expressed in terms of power per unit area, measured in watts per square metre (Wm^{-2} or more usually $mWcm^{-2}$). On the other hand, the amplitude of an ultrasound wave at a point in a medium can be expressed as a pressure, a density, or a displacement. It can be demonstrated that the intensity is related to the pressure amplitude A_p, to the density ρ and to the velocity of ultrasound in the medium c according to the following equation:

$$I = \frac{A_p^2}{\rho \cdot c} \tag{8.1}$$

Therefore, the intensity of an ultrasound wave (in $mWcm^{-2}$) is proportional to the square of its pressure amplitude.

The interaction of ultrasound with tissues, and with materials in general, is related to the mechanical properties of the medium. The mechanical properties of materials also determine the type of mechanical waves that travel in them, i.e. longitudinal or transverse. Ultrasound in the megahertz range generally propagates in tissues (with the exception of bone) in the form of longitudinal waves. The wavelength in normal tissue is around 1.54 mm at 1 MHz and 0.154 mm at 10 MHz.

The velocity of ultrasound depends on several quantities, mainly the density and compressibility of the medium in which it is travelling. The velocity c in a medium is inversely proportional to the square root of the density ρ:

$$c \propto \frac{1}{\sqrt{\rho}} \tag{8.2}$$

The velocity is also inversely proportional to the square root of the compressibility K, which quantifies the fractional decrease in the volume of a material under pressure. K is the inverse of the bulk modulus β of the medium, if it does not have time-dependent response:

$$c \propto \frac{1}{\sqrt{K}} \propto \sqrt{\beta} \tag{8.3}$$

Moreover, the velocity is dependent on frequency (a phenomenon known as dispersion) and temperature, but in practice these are minor effects in the body. In summary, the above expressions can be written as follows (the Newton–Laplace equation):

$$c = \frac{1}{\sqrt{\rho \cdot K}} \quad \text{or} \quad c = \frac{\sqrt{\beta}}{\sqrt{\rho}} \tag{8.4}$$

In reality, density and compressibility are related: a change in density is normally coupled with a larger and opposite change in compressibility. As a result, it is often found that as density increases, the velocity of ultrasound increases. Table 8.1 lists density, ultrasound velocity and acoustic impedance values for several body tissues and other materials.

The dependence of ultrasound velocity on temperature varies in magnitude between different media. For example, the velocity in water is 1480 ms^{-1} at 20°C and 1520 ms^{-1} at 37°C. However, in tissues temperature is regulated and varies by only a few degrees Celsius. As a result, temperature effects on velocity in the body are less than 1% and can normally be neglected (however, they do have to be taken into account in quality assurance phantoms). Velocity is also similar between different soft tissues (excluding bone and lung), ranging between 1450 ms^{-1} in fat tissue and 1580 ms^{-1} in muscle, with variations of the order of a few percent. The variation of velocity with frequency (known as dispersion) in diagnostic ultrasound is very small, less than 1%, and can also normally be neglected. Taking all of this into account, we can assume that ultrasound propagates with almost the same velocity in all the tissues of interest, assumed to be 1540 ms^{-1}. This is a key assumption in sonography.

TABLE 8.1

Typical Density, Ultrasound Velocity and Acoustic Impedance of Selected Media

Material	Density (kg.m^{-3})	Velocity (ms^{-1})	Acoustic Impedance (MPa.s.m^{-1})
Air	1.20	330	0.0004
Water	1000	1480	1.48
Mercury	13600	1450	20
Soft tissue average	1060	1540	1.63
Liver	1060	1550	1.64
Muscle	1080	1580	1.70
Fat	952	1459	1.38
Brain	954	1560	1.55
Kidney	1038	1560	1.62
Spleen	1042	1570	1.64
Blood	1057	1575	1.62
Bone	1912	4080	7.8
Lung	400	650	0.26
Lead zirconate titanate (PZT)	7650	3791	29
Example acoustic lens	1142	1620	1.85
Aqueous humour	1000	1500	1.5
Vitreous humour	1000	1520	1.52
Lucite	1180	2630	3.16
Polystyrene	1060	2350	2.49
Castor oil	969	1477	1.43

Source: From Novario et al. 2003, with kind permission of Società Italiana di Fisica.

An important quantity that is fundamental to our description of the interaction of ultrasound with matter is the acoustic impedance, z, of a medium. This is the product of the density ρ and the speed of ultrasound in the same medium c:

$$z = \rho.c \tag{8.5}$$

Acoustic impedance appears in several equations that describe the interaction of ultrasound with matter. The interactions that must be considered in order to understand ultrasound imaging are reflection, refraction, scattering, absorption, interference and diffraction.

8.2.2 REFLECTION

Reflection of ultrasound follows the same laws that hold for light, which is that when a beam of ultrasound is reflected by an interface between two different materials, the angle of reflection is the same as the angle of incidence, and

$$I_r = I_0 \cdot \alpha_r \tag{8.6}$$

where I_0 is the intensity (energy per unit time and per unit area) of the incident beam, I_r is the intensity of the reflected beam, and α_r is the reflection coefficient:

$$\alpha_r = \left(\frac{z_2 - z_1}{z_2 + z_1} \right)^2 \tag{8.7}$$

where z_1 and z_2 are the acoustic impedances of the first and second media, respectively. This relation implies that there is also a transmitted beam, the intensity of which is the difference between those of the incident and reflected beams.

The intensities of the transmitted and reflected beams therefore depend only on the acoustic impedances of both media. Note that the greater the difference in impedance, the stronger the reflected intensity. This is why ultrasound is almost completely reflected at a tissue/air interface (see Table 8.1), and why backing and matching are so important in ultrasound probes (see Section 8.3)

8.2.3 REFRACTION

Refraction describes what happens to a beam transmitted throughan interface between two different materials. Again as in the case of light, refraction of ultrasound follows Snell's law:

$$\frac{\sin(\varphi_1)}{\sin(\varphi_2)} = \frac{c_1}{c_2} \tag{8.8}$$

where c_1 and c_2 are the ultrasound velocities in the two materials and ϕ_1 and ϕ_2 are the angles between the normal at the incidence point and the direction of the incident and transmitted beams, respectively.

As the speeds of ultrasound propagation in different tissues vary only slightly, the refraction phenomenon is usually of little significance. But under some circumstances, it can become significant and cause artefacts.

This is also the physical principle on which acoustic lenses, which are used both in ultrasound probes and acoustic microscopes, are based.

8.2.4 SCATTERING

Scattering, or diffusion, is also called 'non-specular reflection' and occurs when all the dimensions of the reflective surfaces (interfaces) are shorter than the ultrasound wavelength. Each of these interfaces behaves as if it was a new radiant isotropic source.

Scattering allows ultrasound imaging of tissue boundaries that are not necessarily perpendicular to the direction of the incident ultrasound. It also allows imaging of tissue parenchyma as well as organ boundaries. It is relatively independent of the direction of the incident ultrasound and, therefore, is more characteristic of the scatterers than of the ultrasound beam. As most surfaces in the body are rough, effectively consisting of arrays of small reflectors, this phenomenon is of paramount importance in providing useful information about tissue textures.

Diffusion depends strongly on frequency, f, with the dependence ranging from f^2 to f^6 depending on wavelength and the type and structure of tissues. This has significant importance in the case of high-frequency ultrasound, making tissue characterisation theoretically possible.

Diffusion also contributes to increasing the divergence of the ultrasound beam as it travels through the medium.

Because ultrasound encounters a number of scatterers at any point along its path, several echoes are generated simultaneously. These may arrive at the transducer in such a way that they reinforce (constructive interference) or partially or totally cancel each other (destructive interference). This results in a dot pattern on the ultrasound image that does not directly represent scatterers, but rather represents an interference pattern generated by the scatterer distribution within the imaged object. This phenomenon is called acoustic speckle. Note that scattering is a major contributor to the attenuation of signal too.

8.2.5 ABSORPTION

Absorption describes how an ultrasound beam loses energy as it travels through matter, due to the fact that the beam causes particles of matter to vibrate and absorb energy.

This is usually modelled as an exponential decrease of the beam's intensity I as a function of the thickness of the material passed through, x, as follows:

$$I(x) = I_0 \cdot e^{-\mu \cdot x} \tag{8.9}$$

where I_0 represents the initial intensity and μ is the empirical linear attenuation coefficient of the material (typically in cm^{-1}), which depends on the frequency of the beam and the material itself. Absorption is strongly dependent on the wavenumber. For instance, water has an absorption coefficient of 2.17×10^{-3} dB.MHz^{-2}cm^{-1} (Poelma 2017). The non-linear nature of absorption results in a change of the pulse shape, not only its amplitude.

On the other hand, the attenuation coefficient in human tissues increases almost linearly with frequency; therefore, the attenuation coefficient for a given tissue at 20 MHz is about four times greater than that at 5 MHz.

The half-value layer (HVL) is defined as the thickness of a material that halves the intensity of the ultrasound beam. It is easy to demonstrate that HVL and μ are related as follows:

$$HVL = \frac{0.693}{\mu} \tag{8.10}$$

The HVL is therefore almost inversely proportional to frequency.

As can be seen from Table 8.2, at 2 MHz we need about 2.1 cm of normal tissue to reduce the intensity of the ultrasound beam to half, whereas at 20 MHz only 0.22 cm is required. Therefore, even echoes coming from just a few centimetres deep are greatly attenuated when working with frequencies above 10 MHz.

TABLE 8.2

Half Value Layers (HVL) in cm at Different Frequencies and in Various Tissues

Material	Frequency					
	1 Hz	**2 Hz**	**5 Hz**	**10 Hz**	**15 Hz**	**20 Hz**
Air	0.25	0.06	0.21			
Water	1360	340	54	14	6.2	3.4
Blood	17	8.5	3	2	1.2	0.85
Bone	0.2	0.1	0.04			
Brain	3.5	2	1	0.35	0.24	0.18
Fat	5	2.5	1	0.5	0.33	0.25
Liver	3	1.5	0.5	0.3	0.19	0.15
Muscle	1.5	0.75	0.3	0.15	0.1	0.08
Tissue (average)	4.3	2.1	0.86	0.43	0.3	0.22

Source: From Novario et al. 2003, with kind permission of Società Italiana di Fisica.

In all clinical applications, the highest possible frequencies are used that are compatible with the maximum depth being examined because increasing the frequency leads to an overall improvement of image quality.

The energy absorbed by the medium is usually converted into heat. Of course, in diagnostic use of ultrasound, the induced temperature rise in the patient must be kept to a minimum, particularly in foetal scanning. This is why a thermal index (TI) is shown on ultrasound scanner displays. TI is intended to give a guide to the likely maximum temperature rise that might be produced after long exposure. Three forms of TI may be displayed, according to the application: TIS assumes that only soft tissue is insonated, TIB assumes that bone is present in the focal zone, TIC assumes that bone is very close to the front face of the probe.

TI is defined as the ratio between the acoustic power needed to deliver the maximum temperature rise found in an insonated object and the power needed to deliver a temperature rise of 1°C. It is calculated at the location at which the maximum temperature rise occurs (Church 2007). However, note that errors in calculating TI values, and the limitations of the simple models on which they are based, means that TI values can underestimate (or overestimate) the temperature elevation by a factor of up to two.

8.2.6 INTERFERENCE

Ultrasound waves that are superimposed on each other exhibit interference phenomena. If waves with the same frequency are in phase, they undergo constructive interference that generates an increase in amplitude (and therefore intensity). If waves with the same frequency are out of phase, they undergo destructive interference resulting in a decrease in amplitude. Every combination, from completely constructive to completely destructive interference, can occur, resulting in a complex wave summation. Interference is important in the design of an ultrasonic transducer because it affects the uniformity of the beam intensity throughout the ultrasonic field.

Focusing of the ultrasound beam in real-time imaging is also based on the principle of wave interference. Another application is Doppler ultrasound velocimetry, which makes use of a 'beat' phenomenon that occurs when waves of similar but slightly different frequencies are combined.

In the case of Doppler velocimetry, the superimposed waves are the transmitted wave and a wave reflected from a moving interface.

8.2.7 DIFFRACTION

Huygen's principle is valid for ultrasound as well as for light. This states that every point along a wavefront can itself be treated as a point source emitting a spherical wave. This means that the wave properties at any other point in an ultrasound field can be calculated by adding the contributions from spherical wavelets from all points on a particular wavefront. For example, the ultrasound field on the far side of a small (similar in size to the wavelength) aperture insonated by a plane wave can be calculated by summing the contributions at each point in the field from all those points across the aperture. There will be constructive interference at those points where the wavelet fronts are coincident. If this calculation is carried out, we find that the wavefronts on the far side of the aperture take a particular shape, forming a diffraction pattern. They are flat for a short distance from the aperture and then become more convex, leading to ultrasound beam divergence, and far from the aperture they closely approximate spherical surfaces centred on the aperture. The point at which the beam starts to diverge, and the degree of divergence, depends on the ratio of the aperture dimensions to the wavelength of the ultrasound.

These effects must be considered in order to establish the shape of the beam generated by a transducer. The beam can be divided into 'near field' and 'far field' regions, the characteristics of which are different.

8.2.8 CAVITATION

When an ultrasound wave interacts with a material and the local variation of pressure is high, cavitation can occur. This is a phenomenon in which the rapid change of pressure, combined with characteristics of the medium such as fluid inertia, viscosity and surface tension, causes the formation and growth of bubbles in the medium itself. It is a complex phenomenon that must usually be avoided in diagnostic ultrasound. Therefore, in the 1990s an index was developed to quantify the characteristics of ultrasound interaction with tissue that are known to influence the formation of bubbles. This is known as the Mechanical Index (MI), defined as follows (J. Ultrasound Med. 2000):

$$MI = \frac{p_{r3}}{\sqrt{f}} \tag{8.11}$$

where p_{r3} is the peak rarefactional pressure (in MPa) measured in water, reduced by an attenuation factor equal to that which would be produced by a medium having an attenuation coefficient of 0.3 dB cm^{-1} MHz^{-1} and f is the centre frequency of the ultrasound beam (in MHz). As this index is related to the formation or growth of voids, it is of course proportional to negative peak pressure. Its relationship with frequency is more complex, but considering the dynamics of bubble formation, it becomes apparent that the time period of the negative half-cycle of the pressure waveform is a critical determinant of the extent of growth. The shorter this interval, the less likely it is that there will be extensive bubble growth and, hence, cavitation. We can therefore conclude that cavitation is less likely to occur with higher transducer frequencies, so it makes sense that frequency appears in the denominator in this expression.

MI is related to the protection of the patient from excessive mechanical stress. For this reason, it is shown on the ultrasound system display and the MI of diagnostic scanners is limited. However, in some applications MI is directly related to the intended effect. This is the case for instance in harmonic imaging (see Section 8.4.5) and targeted drug delivery (see Section 8.5.5), where localised cavitation is required for the technique to work. Note that if bubbles are already present, the minimum acoustic negative pressure amplitude required to cause significant bubble growth and subsequent collapse is strongly influenced by the initial size of these bubbles, which provide cavitation nuclei. With smaller nuclei, higher negative pressures are required to overcome the effect of stronger surface tension. Alternatively, with larger cavitation nuclei, inertial and viscous effects begin to dominate and, once again, higher negative pressure amplitudes are needed. For any given initial bubble dimension, the threshold pressure for bubble disruption is frequency dependent. In some situations, bubbles are injected into the patient for diagnostic or therapeutic purposes, and this behaviour is then an important consideration.

8.3 GENERATION OF ULTRASOUND

Ultrasound is generated by the transformation of electrical energy into mechanical energy. This is usually done by means of transducers based on the principle of piezoelectricity. Some materials exhibit the propriety that, under applied mechanical stress, electric charges of opposite sign accumulate on opposite surfaces. Examples are lead zirconium titanate (PZT) and lead-free piezoelectrics, composites, domain-engineered single crystals, piezoelectric gels and capacitive micromachined ultrasonic transducers (CMUTs). The converse is also true, i.e. application of an electrical potential to the material results in mechanical stress. The piezoelectric properties of these materials can therefore be used during both transmission and reception of ultrasound.

Historically, the classical single-element transducer for medical imaging applications is based on a piezoelectric plate or disc, the thickness of which determines the resonance frequency of the device, in that it is a multiple of half the wavelength of the ultrasound that is to be generated. In a typical transducer the velocity of ultrasound is about 4000 ms^{-1}, and the transducer thickness ranges from 1 mm at 2 MHz to 0.2 mm at 20 MHz.

The lateral dimensions of the transducer vary accordingly to the application, from one centimetre (for continuous-wave Doppler probes) to a tenth of a millimetre and even smaller, for example in linear probes.

The transducer has an acoustic impedance much higher than of biological tissues. When the transducer is placed in contact with the body, this large difference leads to an acoustic mismatch and poor axial resolution due to refraction. Consequently, matching layers are added in front of the transducer. The thickness of a matching layer is generally around a quarter-wavelength at the resonance frequency, and its acoustic impedance is intermediate between those of the piezoelectric material and tissues. Since air has an acoustic impedance that is very different from both the tissue and the materials that the transducer is made of, even the thinnest layer of air between the transducer and the patient's skin would be enough to reflect the ultrasound beam almost entirely. For this reason, a material is used, generally a watery gel, to eliminate air between the probe and patient.

For imaging purposes, we are interested in generating short pulses of ultrasound rather than continuous waves. This is achieved by applying very short voltage pulses across the transducer. This causes the material to ring at the resonant frequency. It should be noted that the movement of the surface is a very small fraction of the thickness of the transducer. The time taken for the ringing to fade (the pulse length) is determined by how much energy is lost during each half cycle of oscillation. The energy is lost by attenuation in the transducer material itself and by transmission of energy into the media in contact with the front and rear faces of the transducer. We can reduce the pulse length (a process known as *damping*) by ensuring that as much energy as possible is lost from the rear surface of the transducer. On this face a thick backing layer is usually added, the acoustic impedance of which should be as close as possible to that of the active piezoelectric element, to maximise energy loss. The attenuation coefficient and thickness of the backing layer must be sufficient that no energy can be radiated back into the active layer, which would produce parasitic echoes. Also the shape of the backing material is important. It is usually wedge-shaped, in order to restrict multiple reflections of ultrasound within it, so increasing path length and attenuation (see Figure 8.1).

Damping can also be achieved using a procedure called *dynamic damping*, which is an electronic means to suppress ringing. Immediately following the excitation pulse, a voltage pulse of opposite polarity is applied to the transducer. This counteracts the expansion and contraction of the transducer stimulated by the first pulse, and ringing is inhibited (see Figure 8.2).

The same considerations regarding backing, damping and matching apply when the transducer works as an ultrasound receiver. High damping leads to short pulses and therefore to high spatial resolution, but unfortunately it also reduces energy transmission and reception and leads to low sensitivity.

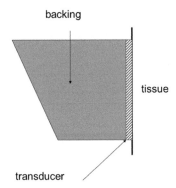

FIGURE 8.1   Structure of an ultrasound transducer with its backing layer. The backing layer is usually wedge-shaped in order to restrict multiple reflections of ultrasound, increasing path length and attenuation.

(From Novario et al. 2003, with kind permission of Società Italiana di Fisica.)

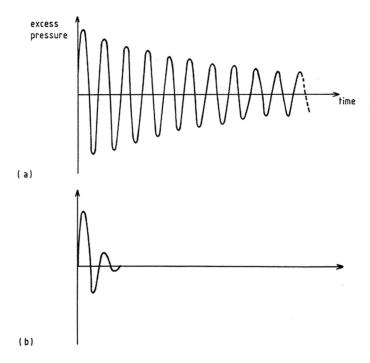

FIGURE 8.2   The waveform produced by an ultrasound transducer. (a) free ringing transducer, (b) damped transducer.

(From Novario et al. 2003, with kind permission of Società Italiana di Fisica.)

A quantity known as the mechanical coefficient (Q) characterises the frequency response of the transducer. It is a major consideration when selecting a transducer for a particular application. The Q value relates to two essential characteristics of the pulsed ultrasound beam: axial resolution (spatial resolution in the image in the direction of the ultrasound beam) and frequency bandwidth.

The Q value can be defined as follows:

$$Q = \frac{\text{Energy stored per cycle}}{\text{Energy lost per cycle}} \tag{8.12}$$

A high-Q transducer stores energy and therefore loses very little energy each cycle. After being stimulated by the voltage pulse, it vibrates for a long time, producing a long pulse. A low-Q transducer generates a short pulse after excitation because most of its energy is lost during the first few vibrations. It can be demonstrated using Fourier transformation that the previous formula is equivalent to the following:

$$Q = \frac{\text{Centre frequency}}{\text{Frequency Bandwidth}} \tag{8.13}$$

Diagnostic pulsed-wave ultrasound uses low-Q transducers, the Q value being typically 2–3. High-Q transducers (500 or greater) are good for continuous-wave ultrasound. Extremely high-Q transducers (more than 20000) are used for therapeutic applications.

The electromechanical coupling coefficient (k) is a quantity that describes the efficiency of the transformation of electrical energy into ultrasound energy (and vice versa). It is the product of two other quantities: the transmission coefficient and the reception coefficient. The transmission coefficient (h) is the fraction of electrical energy that is converted into acoustic energy. The reception coefficient (g) is the fraction of returning acoustic echo energy that is converted into electrical energy:

$$k = h \cdot g \tag{8.14}$$

The thickness of a transducer is related to the frequency emitted. A transducer produces its largest output when the wavelength is equal to twice or one half the thickness of the piezoelectric disc. This is because the forwards and backwards waves as the material pulses backwards and forwards reinforce due to being exactly in phase. The size of the transducer is related to the power that can be emitted. A diagnostic ultrasound probe is generally made up by a number of piezoelectric single transducers, to obtain many scan lines that together constitute an image.

To obtain the different image lines, each transducer is used to emit and receive ultrasound alternately. This requires that emission of ultrasound is pulsed rather than continuous. Each emission is made up of a number of complete cycles of pressurisation and depressurisation. Figure 8.3 shows a series of pulses each composed of three complete cycles and separated from each other by an interval. The whole timespan, including the pulses and the interval between them, is called the pulse repetition period (PRP). The inverse of the PRP is the pulse repetition frequency (PRF). PRF values typically range from a few hundred to a few thousand hertz and represent the number of pulses emitted by the transducer every second.

Duty factor is defined as the percentage of time that the transducer is emitting ultrasound. Typically, this is in the range 0.1–1%. This means that the transducer spends most of its time receiving.

An ultrasound scanner builds images from the echoes received during the time that the transducer is not emitting. It follows that if the PRF is fixed the maximum scanning depth is also fixed.

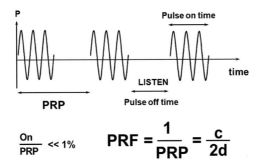

FIGURE 8.3   A series of pulses each composed of three complete cycles and separated from each other by an interval.

(From Novario et al. 2003, with kind permission of Società Italiana di Fisica.)

Fixing the pulse repetition period (PRP) means that the maximum scanning depth is determined by the maximum time available for the ultrasound pulse to travel from the transducer to that depth and back, which is exactly equal to the PRP.

The following relation applies:

$$PRF = \frac{c}{2 \cdot d} \tag{8.15}$$

where c is the average velocity of ultrasound propagation in tissue and d is the maximum scanning depth. As a consequence, to increase depth d, the PRF must be decreased. This causes a loss in the temporal resolution of the images (or actually in the frame rate as multiple images, or frames, are usually collected sequentially).

The dimensions of the emitting surface of the transducer are particularly important with respect to formation of the emitted beam. An ideal point transducer would emit isotropically in all directions, whereas an ideal parallel plane transducer would emit a flat wavefront. The actual situation is somewhere in between these extremes, and therefore the smaller the transmitting surface of a transducer the more isotropically it emits, with greater beam divergence; whereas the greater its surface area, the flatter the wavefront.

This situation is illustrated in Figure 8.4a: near to the transducer (known as *near field*) a flat wavefront is shown, characterised by constructive and destructive interference (Figure 8.4b), whereas beyond a certain distance D (known as *far field*) the beam diverges with an angle θ.

The boundary distance D between the near field and far-field regions and the divergence angle θ are described by the following relations:

$$D = \frac{a^2 \cdot f}{4 \cdot c} \tag{8.16}$$

$$\sin \theta = \frac{1.22 \cdot c}{a \cdot f} \tag{8.17}$$

where c is the speed of ultrasound, a is the size (perpendicular to thickness) of the transducer and f is the ultrasound frequency.

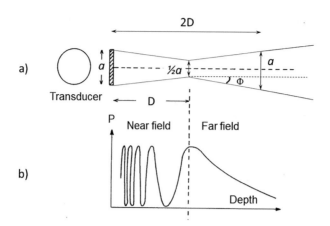

FIGURE 8.4 Near field and far field. (a) The approximate longitudinal section of the beam generated by a single flat round transducer of diameter a. (b) In the near field a flat pressure wavefront is shown, characterised by constructive and destructive interference, in the far field the beam diverges.

(From Novario et al. 2003, with kind permission of Società Italiana di Fisica.)

For higher frequencies, the boundary between near and far fields is deeper and the divergence angle is smaller. Smaller divergence angles guarantee better image quality. For example, a 1 MHz transducer one millimetre wide has a D value of 0.16 mm. This means that normally the probes used for imaging in two-dimensional real-time mode operate in the far field and therefore beam divergence becomes a limiting factor for the quality of the image.

It might be assumed that, in the near field, the beam width is equal to the transducer crystal diameter. However, a plane single-crystal transducer has a self-focusing effect, so the beam width actually decreases to a minimum value at the transition point between the near field and the far field and then begins to diverge. In fact, the beam width at the transition point is close to half the diameter of the crystal. The maximum acoustic pressure occurs at the transition point because the same power is distributed over a smaller area (see Figure 8.4). At a distance equal to 2D, the beam diameter diverges to a size approximately equal to the crystal diameter. The following approximate formulae describe the near field beam width (w_N) and the far-field beam width (w_F):

$$w_N = a - \frac{2 \cdot \lambda \cdot z}{a} \tag{8.18}$$

$$w_F = \frac{2 \cdot \lambda \cdot z}{a} \tag{8.19}$$

where a is the crystal diameter, λ is the wavelength of the ultrasound and z is the depth of interest.

Although the surface shown in Figure 8.4a contains most of the ultrasound intensity, there is often significant ultrasound energy outside this surface. The variation of intensity, particularly in the far field, is often illustrated by a polar intensity graph as shown in Figure 8.5. In this case, the length of the line from the origin indicates the intensity (I) measured at the angle φ from the beam axis. It can be seen that most of the ultrasound power is contained within a main lobe, the extent of which would correspond to the surface shown in Figure 8.4a, but there are smaller side lobes present outside this surface. These side lobes can give rise to spurious images on the ultrasound scanner screen and measures are often taken to reduce them.

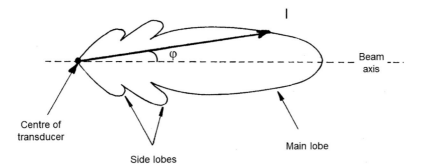

FIGURE 8.5   Polar graph of an ultrasound beam used to represent the geometry of the emitted power through an iso-power line.

(From Novario et al. 2003, with kind permission of Società Italiana di Fisica.)

8.4   DIAGNOSTIC MODALITIES

8.4.1   A-MODE

Historically, the first method used to obtain information about the interior of the human body through ultrasound was the so-called A-mode, which is now outdated. A-mode used a probe with a single crystal to scan the patient's body, acquiring echoes from a single direction at a time and hence gaining information about the depth of reflecting surfaces in that direction. The deeper the reflecting surface, the longer the time needed for the echo to travel back to the transducer. Echoes from deeper surfaces are attenuated more than those from shallower surfaces, and so it is necessary to apply depth-dependent amplification in order to correctly compare signal intensities. This is done using a technique known as Time Gain Compensation (TGC), which the user can adjust in order to correctly amplify the intensities of echoes returning with different delays (see Figure 8.6).

The spatial resolution that can be achieved in the direction of the ultrasound beam (axial resolution) in A-mode is directly related to the length of the pulse train used. Keeping in mind that two echoes have to be received separately to be interpreted as originating from two separated reflecting interfaces, and that an ultrasound pulse train must travel twice the depth of the interface to be received, the best axial resolution that can be obtained is half of the ultrasound pulse train length.

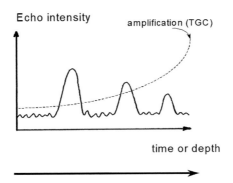

FIGURE 8.6   Echoes returning to the transducer, carrying information about three reflectors at three different depths. Those coming from deeper surfaces are attenuated more than those coming from shallower ones, so it is necessary to amplify them more to correctly compare their intensities.

(From Novario et al. 2003, with kind permission of Società Italiana di Fisica.)

8.4.2 B-Mode

To build a two-dimensional image, known as B-mode ultrasound, information from multiple directions must be acquired. This is achieved by using several narrow crystals arranged together in the same probe or transducer. Ideally, each crystal can work like an A-mode transducer, building the sequence of reflecting surfaces of a single column of a two-dimensional image. The problem with small single transducers is poor focusing of the ultrasound beam, due to the divergence angle. In reality, each scan line of the image is not acquired using a single crystal, instead several crystals are used together to produce an electronically focused beam (see Figure 8.7). Electronic focusing is employed both in transmission and in reception. In transmission, the focus is selected by the user and pulses are generated accordingly by a suitable group of crystals. If several foci are used in the same image, each of them needs accurately timed pulses generated by a group of crystals. So multiple focussing leads to a lower image frame rate. A complete image (frame) is made up of all the scan lines of the individual crystals. For each line, and for each transmit focus, a pulse must be emitted. Temporal resolution depends on frame rate, that is the number of complete images shown on the display per second. Putting together all this information, the minimum necessary pulse repetition frequency (PRF) is given by

$$PRF = NF \cdot L \cdot FR \tag{8.20}$$

where NF is the total number of foci, L is the number of lines of view and FR is the display frame rate.

Remembering that the maximum PRF used is related to the maximum depth that can be imaged, a compromise is needed. This is described by the following empirical relation:

$$P \cdot NF \cdot L \cdot FR \leq 77000 \tag{8.21}$$

where P is the maximum imaged depth (in centimetres) and the other quantities are as before.

How the single transducers are arranged within the probe and how they are driven to scan the volume of interest and to focus the beam define the shape and type of ultrasound probe. Examples

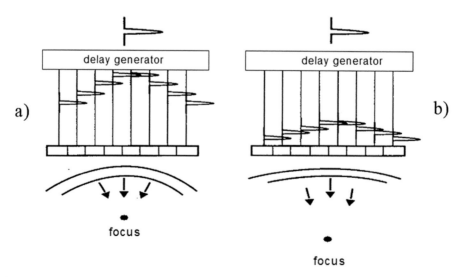

FIGURE 8.7 Electronic focusing of a beam, by means of timed driving of a single transducer in an array. Signal delays arranged to have (a) shallow focus or (b) deep focus.

(From Novario et al. 2003, with kind permission of Società Italiana di Fisica.)

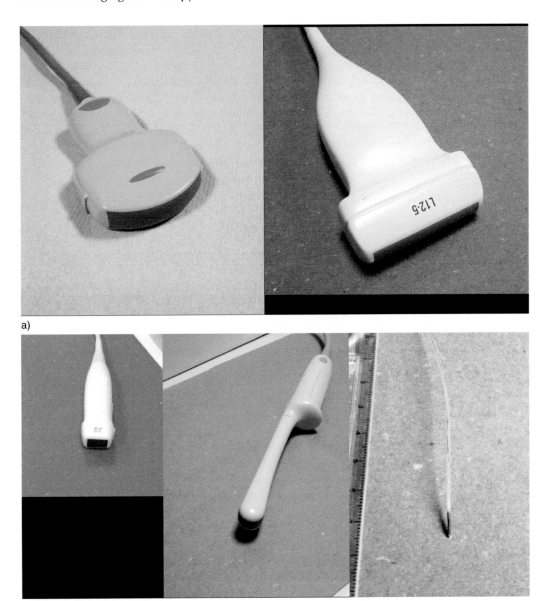

a)

b)

FIGURE 8.8 Examples of ultrasound US probes. (a) The most common probes, from the left a convex probe and a linear probe. (b) Specialised probes, from the left a sector array probe designed to scan the heart through the ribs, an endocavity probe and an endovascular probe (the scale is different for the three probes and the endovascular probe is greatly enlarged).

of probes are depicted in Figure 8.8. Convex and linear probes, which are the most common types, are shown in Figure 8.8a. Figure 8.8b depicts more specialised probes: a sector array probe designed to scan the heart through the ribs, an endocavity probe and an endovascular probe (the scale is different for the three probes and the endovascular probe in particular is greatly enlarged). Note that individual transducers are nowadays usually electronically driven, but there are also probes in which the transducers are mechanically driven.

With increasing computational speed, parallel computing and graphical processing units, the point at which the analogue signal is digitised is moving closer to the probe. In this situation, it can be advantageous for functions previously carried out in hardware to instead be performed by software algorithms, achieving higher speed and/or new goals. An example is increasing B-mode frame rate using 'Ultrafast Compound Imaging' (Bercoff 2011). If focusing is not used as outlined earlier in this section, and if the whole field is insonated simultaneously, with traditional image reconstruction a very poor image quality would result. But with advanced image reconstruction algorithms good image quality can be achieved with a very limited number of whole field insonations (typically 15) with plane waves at different angles, achieving a frame rate much higher than that obtained with physical focusing. The frame rate of ultrafast ultrasound imaging can be hundreds or thousands of frame per second, compared to tens of frames per second with traditional focusing.

8.4.3 Ultrasound Velocimetry

This section concentrates mainly on ultrasound techniques used to evaluate motion of tissues (usually blood) on the basis of Doppler shift of the ultrasound frequency. However, recently other techniques have been developed to evaluate motion, not based on the Doppler effect, and some of these are also described here.

Doppler is nowadays an important part of an ultrasound examination. As is well known, the Doppler effect takes place when a wave source and receiver are in motion relative to each other. As a general rule, the frequency received is higher than that transmitted if source and receiver are approaching each other, while it is lower when they are moving apart.

In ultrasound devices, if the ultrasound generated by the transmitter is reflected by moving objects, such as blood cells, it will be received with a modified frequency. The following equation holds:

$$f_D = \frac{2 \cdot v \cdot f \cdot \cos\varphi}{c} \tag{8.22}$$

where f_D is the frequency difference between the transmitted and the received beam, v is the blood velocity, f is the ultrasound frequency, c is the velocity of ultrasound in tissue and φ is the angle between the direction of the beam and blood flow. As f_D and φ can be measured, v can be calculated.

During typical operating conditions, the frequency variations f_D (from this point on referred to as the Doppler frequency) are within the audible frequency range, typically between 200 and 15000 Hz, and the Doppler signal is sometimes used to drive a speaker to make it directly audible.

Red blood cells are mainly responsible for the reflection of echoes within blood vessels; this is due to their number and size. Blood within vessels does not move at a constant velocity, because its velocity is modulated by the heart cycle and also because, at a given time, the red blood cells which travel close to the centre of a vessel move faster than those travelling close to the vessel walls, with a parabolic distribution of velocities. The smaller the vessel diameter, the steeper the parabola.

Continuous-wave Doppler was the first method used to exploit the Doppler effect in ultrasound. In this technology, two transducers are used: one to transmit and one to receive. Due to the Doppler effect, the received wave has a frequency that is slightly different to that of the transmitted wave, and if the two waves are added a typical *beat* is produced, which exactly represents the Doppler frequency. This operation is accomplished by a demodulator. As mentioned before, in reality a range of velocities are present, so a Doppler spectrum is actually received rather than a single Doppler frequency. The spectrum of the Doppler signal represents signal amplitude (quantity of red blood cells) as a function of frequency (velocity of red blood cells) present in the signal itself.

In continuous-wave Doppler there is obviously no spatial information regarding the volume in which the velocities are measured. Moreover, all of the flows within the sensitive volume, even those distinct and different from each other, contribute to the formation of the Doppler signal.

Finally, there is no information regarding the angle φ. Therefore, this technique remains qualitative. However, despite these drawbacks, continuous-wave Doppler has the advantage of not having limitations on the maximum measurable velocity, which is an issue with other Doppler ultrasound techniques.

In *pulsed-wave Doppler* the same transducer is used alternately to transmit and to receive ultrasound signals. This configuration offers the advantage of allowing temporal gating in order to define a sample volume. If only echoes received during a fixed interval of time are considered valid, and only their frequency content is analysed, a certain volume (sample volume), at a certain depth (depending on the time at which signal reception starts) and with certain dimensions (depending on the time at which signal reception stops) is sampled. This volume has a cross-section that corresponds to the size of the transmitting surface of the transducer, and a height depending on the temporal duration of ultrasound signal acquisition, which determines the range of depths from which valid echoes can be received.

Generally, pulsed Doppler analysis can be carried out simultaneously with acquisition of a real-time two-dimensional image (see Figure 8.9). The operator can choose the dimensions of the sample volume and its position on the two-dimensional image. In addition to this, simple graphical techniques are used in order to communicate to the scanner the angle φ between the direction of the beam and the flow. The operator rotates a cursor until it coincides with the direction of the axis of the vessel seen on a B-mode image (see Figure 8.9). This guarantees that the measured velocities are absolute values, not relative as in the case of continuous Doppler.

As the size of the sample volume is usually a few millimetres, the Doppler sampling time is typically around a microsecond. If we keep in mind that the Doppler frequencies involved can be ten kilohertz or more, we can estimate that the periods of the Doppler signals do not fall below a few hundred microseconds. This means that the temporal window of the transducer during reception is very short compared to the period of the Doppler signal to be measured. In other words,

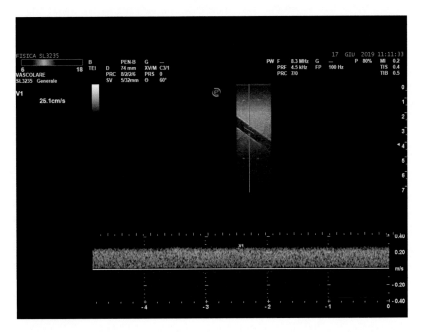

FIGURE 8.9 Display of an ultrasound scanner in pulsed Doppler mode. Both a B-mode image and the Doppler spectrum are displayed. The operator chooses the dimensions of the sample volume and its position on the two-dimensional image and selects the flow angle used by the scanner to calculate the absolute value of the flow speed.

instantaneous samples can be taken of this signal with a sample frequency (number of measurements of the signal taken every second) that is equal to the PRF. By collecting enough of these samples, the Doppler signal can be reconstructed and therefore it is possible to estimate its frequency content by means of Fourier analysis. Practically, this is achieved by sampling the signal typically 128 times, each sample separated from the following one by the pulse repetition period (PRP=1/PRF). So all of the samples are acquired within a few milliseconds, a time very short comparing to the dynamic blood flow phenomena under analysis. The Fast Fourier Transform (FFT) is used to convert the sampled time-domain data to frequencies. The frequencies represent the velocities present in the sampled volume, the amplitude of each frequency component represents the quantity of red blood cells moving at the corresponding velocity.

Normally the Doppler signal is displayed as a 'Doppler spectrum', with time on the x-axis and velocity on the y-axis (see Figure 8.9). For each point displayed, the x-axis value corresponds to an instant of time (each group of 128 samples corresponds to a point with a certain width and position) and the y-axis value corresponds to the blood flow velocity at that time. If there are several velocities present, there will be multiple data points placed vertically, and the vertical width of the Doppler spectrum will be an indicator of the distribution of velocities in the part of the vessel interrogated. Generally, small vessels have wide spectra and large vessels narrower spectra. The quantity of red blood cells at each data point is represented by the luminosity of the point: the greater the luminosity, the greater the quantity of cells.

It can be shown that when a continuous signal is sampled, it is necessary to collect at least two samples during every complete period in order for the frequency of the signal to be correctly represented. This is known as the Nyquist theorem. In other words, the maximum frequency measurable by sampling a continuous signal is half the sampling frequency. In the case of pulsed Doppler, it follows that, once a PRF is selected, there is a maximum measurable velocity, corresponding to a frequency equal to half the PRF. In Doppler velocimetry, failure to meet this criterion (due to an insufficient sampling frequency or insufficient PRF) results in 'aliasing': incorrect underestimation of blood velocities that exceed the maximum measurable velocity. To avoid aliasing, one can increase the PRF (with field depth limitations), change the baseline on the y-axis (if flow is predominantly in one direction), lower the transmitted ultrasound frequency (with a loss of image resolution) or change the angle ϕ between the beam and the flow (bearing in mind that angles higher than 60° lead to lower accuracy in velocity measurement).

Problems arise also for low velocities. Blood is not the only structure moving in the human body: there is also beating of the heart, respiration, vessel wall movement. This noise is called 'clutter' and can be eliminated by using a high pass filter called a wall filter, the properties of which can be changed by the operator.

Colour Doppler is a technique where, instead of a complete quantitative spectrum of the velocities present in a sample volume, the user obtains information only about the direction of the moving interfaces. This information is superimposed on B-mode image in a large volume by means of colour coding (for example, red for structures that are approaching the probe, blue for those that are moving away). Gross quantitative information about speed is added through the saturation of the represented colour (see Figure 8.10). This is achieved by superimposing 'Doppler' pixels containing velocity information onto part of a B-mode image. The velocity information is gathered rapidly using advanced techniques to achieve an acceptable frame rate. There are several approaches to this. *Spectral methods* analyse the frequencies of the signals whereas *correlation methods* analyse the signals as a function of time. Strictly speaking, the correlation method in the time domain does not use the Doppler equation but is based on different algorithms. It is not strictly correct to describe a device of this type as a colour Doppler ultrasound scanner, although such terminology is often used. More correctly, one would speak of 'colour velocimetry', keeping in mind that nowadays the calculation speed of computers has given rise to many alternative methods to grossly quantify the mean velocity of particles in a voxel, with accuracy that is limited but sufficient to attribute a colour and brilliance to it.

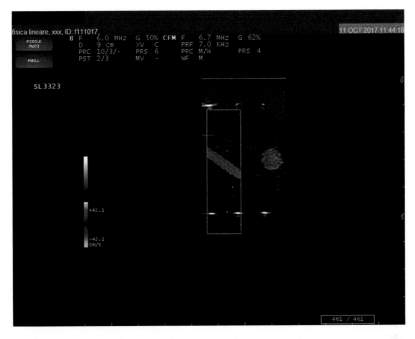

fisica lineare, xxx, ID: f111017,
B F   6.0 MHz   G 50% CFM F   6.7 MHz   G 62%
D   9 cm   XV C   PRF 7.0 KHz
PRC 10/3/-   PRS 6   PRC M/H   PRS 4
PST 2/3   MV -   WF   M

11 OCT 2017 11:44:10

SL 3323

+42.1

-42.1
cm/s

461 / 461

FIGURE 8.10 Display of an ultrasound scanner working in colour Doppler mode. Both the B-mode image and the colour Doppler sampled zone are displayed. The moving pixels are colour coded. Gross quantitative information about speed is added through colour saturation. The scale on the left indicates colour coding of blood speed.

Several spectral methods are available. Most of them are based on an approach known as 'quadrature detection', which can be applied in various ways including autocorrelation and autoregression methods.

In the classic autocorrelation technique, the phase difference between each time point in the signal and the next is calculated. Knowing that the time interval between these points is equal to the PRP, the Doppler frequency (and therefore the velocity) is obtained by calculating the ratio between the phase difference and the PRP to obtain the phase variation per unit time.

In 'zero crossing' methods, the number of times the Doppler signal crosses the axis from negative to positive is counted. This number is obviously mathematically correlated with frequency. These methods tend to underestimate velocity, have lower temporal resolution (velocity is calculated only when the zero signal axis is crossed) and are not very accurate when the signal-to-noise ratio is low.

In other methods, the time derivative of the Doppler signal is obtained, yielding a function that represents phase variation per unit time. This mathematically corresponds to the previously described autocorrelation technique, but is simpler and more economical to perform.

Autoregression is a 'maximum entropy' method, derived from information theory and based on knowledge of the response function of the system. If it is known how the system responds to different flows and velocities in every pixel, this information can be used to reconstruct the image in the case of unknown flows and velocities. The solution that the maximum entropy algorithm chooses is that which minimises the difference between the measured system response (a series of echoes received in a given time frame) and a calculated response (a series of echoes that would be obtained based on various combinations of flows and velocities).

These methods can be either direct or iterative, but in both cases the system response must be known for any flow and velocity. In direct methods, the solution is calculated using algebraic matrix operations. In iterative methods, the output of the nth calculation is used as input to the $(n+1)$th, and the iterative process is stopped, and therefore the velocity is identified, when the difference between the results of the two consecutive outputs is below a certain threshold. Iterative methods are more accurate but, as will be evident, need longer calculation times and consequently lead to a lower frame rate.

Other techniques are based on comparison of a larger number of echoes received from the same line of view. This comparison is carried out in the time domain, by comparing the signals received as a function of time. If there is no movement along a certain line of view within a two-dimensional image, the echoes returning to the transducer following a series of consecutive transmissions have the same temporal profile. However, if a reflector is in motion with respect to the transducer, the part of the echo signal that comes from that reflector is translated in time, and the velocity and the direction of the movement can be calculated.

The procedure can be described as follows. The wave train received following the first transmission is segmented into multiple components. Each component has its own unique and recognisable characteristics. The same procedure is followed for the second wave train to determine whether each of the components has changed position. If the position has not changed, then the reflector is stationary. If the position has changed then the velocity can be obtained by measuring the shift and dividing it by the elapsed time between the two trains (equal to the PRP). Obviously, even in this case, the assessment of velocity depends on the cosine of the angle between the direction of flow and the propagation of the beam.

Theoretically this method does not present aliasing problems (i.e. underestimation of velocities due to undersampling). However, a maximum measurable velocity does exist, because to measure a velocity the reflector must be on the same line of view during the transmission of two consecutive pulses. This condition allows higher velocities to be measured as compared to Doppler methods. Another advantage derives from the fact that, since it is not a spectral technique, shorter wave trains are possible, comparable with those used for two-dimensional imaging, with consequent improvement in spatial resolution. The main disadvantage is a decrease in sensitivity to slow flow.

In all the methods described above, accurate assessment of velocity in pixels along a certain line of view requires consecutive transmission of n pulses before moving to the next line of view. The value of n is known as the *packet size*, and is generally between 5 and 15. For every sample volume there are $n-1$ measurements of velocity, since information on velocity is obtained by comparing two consecutive echoes along the same line of view. The value that is displayed on the colour-coded image is the average of these $n-1$ values. The standard deviation of this mean velocity has a physical meaning: it is a measure of the dispersion of velocities in the sample volume. A large standard deviation (or more commonly variance) indicates a wider range of velocities (i.e. turbulent motion). Low variance indicates very homogeneous velocity. To depict the level of turbulence, a colour coding can be assigned to the variance, different from the coding used for velocity.

Because information about the angle φ is not available in colour Doppler imaging, artefactual variations in measured velocity can arise from one sample volume to another, for example over the course of a tortuous vessel where the direction of flow varies in relation to the ultrasound beam. The flow velocity within the vessel may appear to be higher in segments where φ approaches $0°$ or $180°$, lower in segments where φ approaches $90°$, and there may even appear to be zero velocity when φ is exactly $90°$.

The issues of aliasing and use of a wall filter to eliminate 'clutter' arise with colour Doppler as well as with pulsed Doppler. Colour aliasing can arise when the sampling frequency (PRF) is insufficient, resulting in an image with colours mixed together randomly, appearing similar to turbulent flow even though turbulence may not exist. Use of a wall filter with a high setting causes a loss in sensitivity to flow but eliminates noise from the coloured part of the image (coloured noise). Conversely, a low setting on the wall filter increases colour sensitivity but also increases noise.

Power Doppler or *Energy Doppler* was designed to overcome the main drawbacks of traditional colour Doppler: aliasing and sensitivity to the angle between the flowing blood and the ultrasound beam. This mode uses the 'power', i.e. the integral of the Doppler signal, rather than the mean frequency shift of the signal. Of course, only the magnitude of the local velocity is shown, not its direction. Therefore, it is a representation not of average velocities, but of the amount of red blood cells moving in the different sample volumes (regardless of their velocities) in which brighter colours represent higher flow and darker colours lower flow. This technique is potentially suitable for functional imaging, in that it can visualise changes in blood flow patterns (e.g. in the brain) as a result of external stimuli.

Various *Vector Flow Imaging* techniques have been introduced to obtain not only axial velocity components in the direction of the ultrasound beam, but also in-plane velocity components. There are various approaches, sometimes used in combination (Jensen 2006).

- Multi-angle Doppler analysis;
- Biaxial phase shift estimation from acoustic fields with transverse oscillations;
- Inter-frame blood speckle tracking;
- Directional cross-correlation analysis.

The majority of these methods take advantage of the fact that, with high-throughput computational methods and hardware, movement information can be acquired from different points of view by electronic steering of the beam generated by an array of ultrasound transducers. Moreover, as mentioned in Section 8.4.2, full-field insonation can achieve high-quality images by means of complex algorithms for combination, correlation and regularisation of the information gathered, both in the image domain and in the frequency domain. For example, the ultrasound pulse can be modulated in the lateral direction by means of beam steering. This modulation, on top of the original pulse shape, creates a pulse with a two-dimensional intensity pattern. The velocity component in the beam direction is obtained using conventional Doppler methods. The lateral component is obtained by cross-correlation of the signals from adjacent transducer elements. A lateral movement of the scatterer shifts the modulated pattern, and this can be represented on the image by arrows of different colours, directions and lengths.

8.4.4   3D ULTRASOUND

Ultrasound imaging is a tomographic technology, and it is straightforward to place images from separate slices side by side to obtain three-dimensional information. This has been done since the 1990s. However, 3D acquisition is also possible. At first, this was achieved mechanically, either by linear translation of the transducer, or by rotation or tilting of the array. Early systems acquired data slowly due to the limited frame rate (Jensen 2006). The volume of data was then post-processed before display and manipulation. Newer generations of scanners have introduced faster parallel beam formation to increase the acquisition rate. In parallel beam formation a fairly broad ultrasound beam is emitted, and the received signals are fed to several receive beam formers working in parallel. Nowadays, 3D scanners are based on two-dimensional (2D) arrays of transducers. The serious technological challenges of fabricating 2D arrays with at least $64 \times 64 = 4096$ small elements and handling the processing of gigabytes of data per second have been overcome.

These scanners are often used in cardiology, where the view of the heart is restricted by the rib cage. Acquiring data from a 3D volume in real-time makes it possible to visualise new image slices that could not be obtained with 2D systems, and the functionality of, for example, the heart valves can be studied in detail in real time. Gynaecology is another field in which 3D ultrasound scanners can be beneficial.

8.4.5   HARMONIC IMAGING

Harmonic imaging is based on the fact that the propagation of ultrasound waves in tissue is generally nonlinear, because the speed of sound depends on the acoustic pressure. High pressure results in a slight increase in tissue density, and low pressure decreases the density. Positive pressure parts of an ultrasound wave will thus propagate faster than negative pressure parts. The velocity of sound can be written as follows:

$$c^2 = \frac{\partial p}{\partial \rho} \tag{8.23}$$

where p is the acoustic pressure and ρ is the density. For a deeper discussion on how the properties of media lead to development of nonlinear waves, see Beyer (1974).

The differences in the speed of sound will gradually distort the wave and thereby give rise to higher harmonics in the propagating pulse. The received signal will therefore consist of both the emitted pulse and these higher harmonics. This increases the contrast in the image and can have a significant impact on image quality. Note that the harmonic signals, i.e. the signals whose frequencies are higher than the transmitted frequency, are normally significantly lower in intensity than the signal at the fundamental frequency. Therefore, scanner vendors have developed various method to 'blend' the received components at various frequencies in order to maximise the quality of the image and minimise the loss of sensitivity. Harmonic imaging usually uses a band of frequencies, rather than a single frequency.

In early forms of harmonic imaging, the received signal was filtered using a bandpass filter centred on the second harmonic. As modern transducers are normally broad-band, the emitted pulse has to be longer to reduce bandwidth and avoid overlap of the first and second harmonic spectra. As a consequence, the axial resolution may be reduced. Techniques such as pulse inversion overcome this problem. In this approach the reflected signal is measured twice, firstly using a normal pulse and secondly using an 'inverted' pulse, which has phase $m\pi$, where m is the harmonic number. For all odd harmonics the phase is inverted, and for all even harmonics the phase is the same as for the first pulse. Addition of the signals results in only even harmonics, essentially generating a second harmonic image. Subtraction enhances the odd harmonics. Smart signal combination can weight image characteristics of contrast and/or resolution as desired.

The original use of harmonic imaging was with exogenous contrast agents present in flowing blood. These contrast agents consist of gas-filled 'microbubbles' which produce a very high harmonic response in the reflected wave. Microbubbles consist of a chemical microsphere shell (albumin, galactose, etc.) containing gas, and increase the echogenicity of the tissues in which the contrast agent is contained. The typical dimensions of these microbubbles are of the order of a few microns. For this reason, they pass through the lung's filtering system and into the circulating blood, therefore acting as intravenous contrast agents. The composition of the shell can be tailored to promote uptake of the agent in a specific tissue, rather than remaining in the vasculature. Contrast agents can be employed to optimise two-dimensional images and in all Doppler methodologies.

It is noteworthy that the contrast agent does not behave in a linear manner. The elastic resonant behaviour of the microbubbles, caused by an incident ultrasound beam, generates a distorted echo that contains many higher frequencies than the incident beam, resulting in anomalous peaks within the reflected sinusoid. The harmonic behaviour of the contrast agent gives rise to harmonic components in the reflected spectrum that are much stronger than those from tissue. If only strong echoes are retained, signal from tissue that is not perfused by the contrast agent is suppressed and we obtain a higher quality display of flow with a higher signal-to-noise and signal-to-clutter ratio. This greatly increases the sensitivity of the technique.

8.4.6 Elastography

Ultrasound elastography is a common term for imaging techniques that aim to evaluate and visualise tissue elasticity and stiffness, so providing images and quantitative evaluation of what physicians may sense by palpation (Mulabecirovic 2016). The main methodologies used to achieve this are strain elastography and shear wave elastography.

Strain elastography uses autocorrelation of image lines collected under repeated episodes of applied stress, creating an image of local strain superimposed on the B-mode image in real time. The operator, by means of the ultrasound probe itself, applies stress to the patient's body using freehand compression, and in the ultrasound image it is possible to represent the stiffness of different structures in the form of the strain ratio (SR). The strain ratio is a relative measure of how much different tissues are deformed by the applied stress. Absolute quantification is not possible as the stress is not known in absolute terms:

$$SR = \frac{\left(\text{Mean strain in reference area}\right)}{\left(\text{Mean strain in lesion}\right)} \tag{8.24}$$

An SR higher than one represents a tissue stiffer than that used as the reference. A prerequisite is that the tissue under examination and the reference tissue have been subject to similar stress. The relationship between the local stress and the resulting strain is defined by Young's modulus (E) and quantifies tissue stiffness:

$$E = \frac{\Delta strain}{\Delta stress} \tag{8.25}$$

Shear-wave-based elastography, on the other hand, use physical excitation of tissue, often provided by an acoustic pulse of adequate energy, to create local shear waves travelling perpendicularly to the longitudinal ultrasound waves. Shear waves are transverse waves and their velocity in tissue is much lower than that of ultrasound. So it is possible to track them by continuous insonation in order to quantify their velocity in the ultrasound field. With this method, elasticity is quantified on the basis of the speed of these shear waves. Shear modulus can then be calculated, making two assumptions: that the tissue density (ρ) is a constant in the whole imaged field and that the soft tissue behaves like an incompressible material with a Poisson ratio of about 0.5 (the value for rubber). The Poisson ratio is defined as the ratio of transverse strain to axial strain; for small deformations it is simply the ratio of transverse expansion to axial compression.

With these assumptions, on the basis of biophysical analysis of elasticity in tissues, the simplified formula used in elastography is the following:

$$E \cong 3\rho \left(c_{sw}\right)^2 \tag{8.26}$$

where c_{sw} is the speed of shear waves in tissue.

In supersonic shear wave elastography (Bercoff et al. 2004) shear wave creation is combined with high frame rate sampling. A transient cone of shear waves is created by several acoustic pulses deposited at different depths with a small time lag. This creates an expanding 'acoustic cone' that can be tracked using plane ultrasound waves at high frame rate.

8.4.7 TISSUE CHARACTERISATION

Use of ultrasound in the human body continues to reveal new interactions and contrast mechanisms that are promising from the point of view of tissue characterisation and, potentially, diagnosis of disease. Detailed in vitro and in vivo testing is needed to evaluate which promising techniques are reproducible and useful in the clinic. Here we consider kidney stone characterisation as an example of such ever-growing ultrasound applications.

Kidney stone detection is traditionally done using x-ray computed tomography. Stones are visible on B-mode ultrasound as echogenic foci that may be accompanied by posterior acoustic shadow, but this method lacks sensitivity and specificity. However, recent studies have revealed that on Doppler ultrasound kidney stones demonstrate typical and useful 'twinkling' (Dai et al.2019). Twinkling is present for 43–96% of stones, and has been proposed as a potentially useful adjunct for stone detection, highlighting the presence of a stone not immediately evident on B-mode imaging. The prevailing explanation for twinkling is that small bubbles trapped in stone surface crevices oscillate and generate random backscatter when struck by incident Doppler pulses. These signals are interpreted and displayed as noise.

In addition to detection, careful use of information gathered by ultrasound can help with quantification of the dimensions of kidney stones, which is crucial to selection of the right therapy strategy. It is well known that on B-mode images the size of a stone is usually overestimated, particularly on compound imaging, because the averaging of multiple images generates a smoother overall image but also blurs the stone and shadow borders. Nevertheless, more accurate quantification of dimensions can be achieved with techniques that use the posterior acoustic shadow of the stone as an adjunct to other image information.

A striking aspect of the aforementioned techniques is that something that may be perceived as an artefact in reality carries additional information that can be of use, if correctly studied and managed. Twinkling is an example of this.

8.5 ULTRASOUND THERAPY

8.5.1 INTRODUCTION

Ultrasound consists of mechanical waves that carry energy. This energy can be used not only to gather information about the human body, but also to interact with it to treat diseases. This is an ever-growing field within medical ultrasound, and in the following sections some aspects are discussed.

8.5.2 LYTHOTRIPSY

Shock wave lithotripsy is the most common treatment for kidney and urethral stones. It is a non-invasive procedure introduced in the 1980s that replaced surgical removal for the majority of stones with dimensions less than about 2.5 cm. Shock waves from outside the body are targeted at a stone causing it to fragment. The stones are broken into tiny pieces that may be small enough to pass in urine. A shock wave is a short pulse of about 5 μs duration (Cleveland & McAteer 2007) with a peak positive pressure of about 50 MPa (for reference, in ultrasound imaging the peak pressure can be about 2 MPa). Most of the energy of the pulse lies between 100 kHz and 1 MHz. The peak positive pressure varies between 30 and 110 MPa and the negative pressure between −5 and −15 MPa. The fast transition in the waveform, with a pressure that rises to maximum value almost vertically (in less than 5 ns) is referred to as a 'shock'.

This steep edge often originates from the nonlinear acoustic properties of body tissues. As explained previously in the context of harmonic imaging, the speed of a mechanical wave is not constant but depends on the local compression of the medium. The phase speed of an acoustic wave is as follows:

$$C_{\text{phase}} = c_0 + \beta \frac{p_a}{\rho_0} c_0 \tag{8.27}$$

where C_{phase} is the phase velocity of the wave, c_0 is the reference velocity of sound in that medium, p_a is the local sound pressure, ρ_0 is the reference density of the medium and β is the nonlinearity coefficient, solely dependent on the medium itself. For water, β is about 3.5, for tissues it normally increases as tissue structure becomes more complex, but a reasonable average value for healthy tissue is 5. So an originally sinusoidal wave changes in shape while travelling through tissue and its higher pressure front becomes steeper and steeper until it is almost vertical. At this point, the wave has become a shock wave. Any acoustic wave can result in a shock wave if it can propagate for a long enough distance. For most sound waves in everyday life, however, the shock formation distance is so long that the wave is absorbed before it can form a shock. In lithotripsy the sound waves are intense enough that the wave typically does form a shock in the approximately 10 cm propagation path to the kidney.

Lithotriptors produce a powerful acoustic field that results in two mechanical forces: direct stress associated with the high amplitude shock wave and stresses and microjets associated with the growth and violent collapse of cavitation bubbles. Acoustic cavitation refers to the generation of cavities in a fluid (i.e. bubbles) when the tensile phase (negative pressure) of the acoustic wave is sufficiently strong to rip the fluid apart (see Section 8.2.8). In lithotripsy, the tensile phase of the shock wave is large enough ($\approx$10 MPa) to generate violent cavitation events.

Unfortunately, years of enthusiastic use of this technology have shown that those forces have disruptive effects not only on the targeted stones but also on living tissues, with effects currently under study. So a totally safe, yet effective, lithotripter has yet to be developed. Indeed, there is compelling evidence to suggest that a recent trend toward the development of lithotriptors that produce very high amplitude and tightly focused shock waves has led to increased adverse effects and higher re-treatment rates.

The issue of acoustic coupling is very important here, as modern lithotripters are 'dry-head' devices, in which the treatment head is brought into direct contact with the patient (Lingeman et al. 2009). Lithotripters are equipped with a cushion full of liquid that contains the treatment head. Coupling is achieved by applying a handful of gel to the cushion and the patient's skin, then the treatment head is pressed against the skin and the treatment started.

Three different technologies are used to generate shock waves. The electrohydraulic lithotriptor has a spark source, which directly generates a shock wave that is focused by an ellipsoidal reflector. With such a technology, the pressure pulse originates as a shock wave and remains a shock wave at all times during its propagation from the spark source to the reflector, and then as it focuses onto the target.

The electromagnetic lithotriptor uses an electrical coil in close proximity to a metal plate as an acoustic source. When the coil is excited by a short electrical pulse, the plate experiences a repulsive force, and this is used to generate an acoustic wave. If the metal plate is flat, the resulting acoustic wave is a plane wave that can be focused by an acoustic lens.

The piezoelectric lithotriptor uses piezoelectric crystals to form an ultrasonic wave. The crystals are placed on the inside of a spherical cap and the acoustic wave focuses at the centre of curvature of the sphere. As before, the acoustic waveform starts as an acoustic pulse, and a shock wave is created by distortion due to nonlinear propagation. For most clinical settings, a shock is produced before the wave reaches the focus.

8.5.3 PHYSIOTHERAPY

Therapeutic ultrasound is widely and frequently used in physical therapy to treat people with pain, musculoskeletal injuries and soft tissue lesions. However, the biophysical basis is not well understood, and clinical results not yet proven (Baker et al. 2001;Robertson & Baker 2001). Biophysical effects have been studied both in vitro and in vivo, and are usually divided into thermal and nonthermal effects (such as cavitation and cell lysis or increasing the permeability of membranes). Often results observed in vitro have not been studied in vivo, neglecting the completely different reaction of cells within an organism rather than in cultures (for instance, in terms of thermal regulation due to vasculature). Results of clinical studies are often inconsistent, probably due to different patterns of treatment and the use of different frequencies and power levels. Therefore, use of ultrasound in physical therapy is not considered further in this text.

8.5.4 High-Intensity Focused Ultrasound

High Intensity Focused Ultrasound (HIFU) is a therapy modality that is gaining more and more support amongst clinicians for the thermal treatment of benign and malignant lesions, for targeted drug delivery and for the disruption of thrombi (sonothrombolysis) (Jenne et al. 2012). These treatments are delivered under imaging control and often planned using ultrasound imaging itself or magnetic resonance imaging (MRI).

In the thermal ablation of tumours, high-intensity ultrasound waves are focused on the treatment area, with sudden temperature increase that can achieve 60°C or higher in seconds, thus disrupting the diseased cells. Moderate pressure amplitudes (about 10 MPa) and use of continuous-wave ultrasound (for a period of the order of seconds) lead to tissue heating. Frequently self-focusing spherical piezoceramic transducers are used, with fixed aperture and focal length. Phased array transducers composed of multiple elements (sometimes thousands) can be driven electronically in order to achieve different focal depths, beam steering and beam forming. In general, the ablated region is elliptical in shape and small, with a volume between 50 and 300 mm^3. By combining several such regions, larger target volumes can be treated. An adequate waiting time is needed between sonications to avoid tissue boiling and bubble formation, which can reflect and distort the ultrasound field, so a complete treatment is time-consuming, and can last hours. To overcome this problem, techniques have been developed to sonicate in patterns that can shorten the waiting time and treat a volume by electronically steering the beam along an optimal trajectory.

Image guidance is of importance in all stages of thermal ablation: planning, targeting, monitoring, controlling and assessment of treatment response. Image guidance by ultrasound (USgFUS) is the most commonly used method. During ablation ultrasound imaging can show a cloud of gas bubbles due to cavitation and tissue boiling within the focal region, but it cannot reliably be used to assess and control lesion formation. Today several groups are working to improve lesion detection and ultrasound-based temperature mapping, using several promising parameters such as speed of sound, attenuation or reflection coefficient.

MRI has some favourable attributes for HIFU therapy guidance, in that it can provide morphological images, reliable thermal mapping and assessment of the size and shape of the induced tissue lesion at the end of the therapy. The major disadvantages of this MRgFUS approach are the requirement for an MR compatible HIFU therapy device, together with the cost of the MRI itself.

Several MR parameters are temperature sensitive and are candidates for temperature mapping. To date, the most promising approaches seem to be mapping based on T_1 and T_2 relaxation times and the proton resonance frequency shift (PRFS) method. This last method relies on the fact that the proton resonance frequency has a temperature dependence which is largely independent of tissue type. Moreover, using MR-Acoustic Radiation Force Imaging (MR-ARFI), it is possible to measure tissue displacement with a resolution of micrometers or less. This technique makes it possible to detect the position of the ultrasound focus without significant tissue heating.

8.5.5 Drugs and Ultrasound in Therapy

Here we describe some recent developments in combined use of high intensity focused ultrasound and drugs, which are very promising for the treatment of cancer although to date they have only been studied in pre-clinical research.

One such approach is Targeted Drug Delivery (TDD), where focused ultrasound is used to increase cell membrane permeability, in order to achieve localised drug delivery or genetic material transfection only in cells present in the focal area. The underlying mechanism of sonoporation is not fully understood, but it is likely to be related to mechanical effects of ultrasound, such as microstreaming, microbubble oscillation (stable cavitation) and strong microbubble oscillation with subsequent bubble collapse and the formation of microfluid jets (inertial cavitation). After administration of gas-filled microbubbles (normally used as ultrasound contrast agents), the sonoporation effect is greatly enhanced. Alternatively, drug-loaded gas-filled microbubbles can be mechanically destroyed by the ultrasound focused in the targeted region, releasing the drug payload locally. An alternative approach uses thermo-sensitive liposomes, releasing their payload in regions where the local tissue temperature is slightly elevated (39–40°C) by HIFU as an external source of hyperthermia.

Another highly innovative therapeutic approach is to use pulsed focused ultrasound at moderate pressure levels to achieve transient reversible localised opening of the blood-brain barrier. The purpose of this barrier is to isolate the central nervous systems from potentially harmful toxic substances. Unfortunately, it also prevents the delivery of beneficial drugs, for example for chemotherapy of the brain. Focused ultrasound in combination with gas-filled microbubbles has been demonstrated to temporarily open the blood–brain barrier in mice for 4–70 hours without long-term effects.

8.6 CONCLUSION

In this chapter on ultrasound, covering both imaging and therapy, it has only been possible to provide a smattering of the physics and applications of ultrasound in medicine.

However, it is probably clear that the interaction between ultrasound and the human body is so physically complex that even today new aspects are being recognised, study of which can give rise to new methods to investigate the functioning of the body and to interact with it beneficially.

The physical principles of the interaction between ultrasound and body tissues are the basis of this constant flourishing of new techniques, but they are not enough to account for all of them. In fact, modern computing methods such as neural networks are playing an ever-growing role in ultrasound and can add new knowledge of the human body sometimes even before the underlying detailed phenomena are well understood. So, we can be sure that the coming years will bring us new interesting and amazing advances. Medical physicists must play an active part in that process and can contribute to assessment of which proposed technical advances are truly clinically beneficial.

FURTHER READING

Baker, K. G., Valma J. Robertson, Francis A. Duck 2001 A review of therapeutic ultrasound: Biophysical effects. *Physical Therapy* 81.7: 1351–1358.

Bercoff, J. 2011, Aug 23 *Ultrafast ultrasound imaging. Ultrasound Imaging-Medical Applications.* 3–24.

Bercoff, J., M. Tanter, M. Fink 2004 Supersonic shear imaging: A new technique for soft tissue elasticity mapping. *IEEE Transactions on Ultrasonics, Ferroelectrics, and Frequency Control* 51.4: 396–409.

Beyer, R. T. 1974 *Nonlinear acoustics.* Monterey: Department of the Navy.

Church, Charles C. 2007 A proposal to clarify the relationship between the thermal index and the corresponding risk to the patient. *Ultrasound in Medicine & Biology* 33.9: 1489–1494.

Cleveland, R. O., James A. McAteer 2007 The physics of shock wave lithotripsy. *Smith's Textbook on Endourology* 1529–558.

Dai, J. C., et al. 2019 Innovations in ultrasound technology in the management of kidney stones. *Urologic Clinics* 46.2: 273–285.

J. Ultrasound Med 2000- Section 7—Discussion of the mechanical index and other exposure parameters. *Journal of Ultrasound Medicine* 19.2: 143–168.

Jenne, J. W., Tobias Preusser, Matthias Günther 2012 High-intensity focused ultrasound: Principles, therapy guidance, simulations and applications. *Zeitschrift für Medizinische Physik* 22.4: 311–322.

Jensen, J. A. 2006 Medical ultrasound imaging. *Progress in Biophysics and Molecular Biology* 93.1–3: 153–165.

Lingeman, J. E., et al., 2009 Shock wave lithotripsy: Advances in technology and technique. *Nature Reviews Urology* 6.12: 660.

Mulabecirovic, A. et al. 2016 In vitro comparison of five different elastography systems for clinical applications, using strain and shear wave technology. *Ultrasound in Medicine & Biology* 42.11: 2572–2588.

Novario, R., A. Goddi, F. Tanzi, L. Conte, G. Nicolini 2003 Physics and technology of medical diagnostic ultrasound. *Rivista del Nuovo Cimento* 26.2: 1–64.

Poelma, C. 2017 Ultrasound imaging velocimetry: A review. *Experiments in Fluids* 58.1: 3.

Robertson, V. J., Kerry G. Baker 2001 A review of therapeutic ultrasound: Effectiveness studies. *Physical Therapy* 81.7: 1339–1350.

9 External Beam Radiotherapy

Tony Greener, Emma Jones, and Christopher Thomas

Guy's and St Thomas' NHS Foundation Trust, London, UK

CONTENTS

DOI: 10.1201/9780429155758-9

9.1 INTRODUCTION

9.1.1 WHAT IS RADIOTHERAPY?

Radiotherapy is the therapeutic use of ionising radiation to treat cancer. Following Roentgen's discovery of X-rays in 1895 it was soon recognised that ionising radiation could cause biological damage, such as reddened skin and ulcers, when they acted on normal tissues. Within just one year of their discovery, it was suggested that these X-rays might also be useful in treating cancer. There are many different forms of cancer, which can be triggered by a range of possible causes, but it is characterised by an abnormal growth of malignant cells. This uncontrolled growth of cells, which can form a tumour, increases in physical size over time and will invade into, or push against, adjacent structures. The local effect of this will therefore depend on the size and location of the tumour within the body. The initial site, called the primary cancer site, will continue to grow, and depending on its type and proximity to other structures may remain locally confined or may eventually spread to other sites within the body by depositing cancerous cells into the bloodstream or lymphatic system. This will result in metastatic spread of the disease where secondary cancers appear at remote locations within the body.

Cancer cells are sensitive to ionising radiation. To successfully treat cancer using radiotherapy, the ionising radiation must be directed towards the defined volume of malignant tumour cells, or target volume, accurately, and must be of sufficient magnitude to deliver a therapeutic dose of radiation to the target while minimising the dose to any surrounding, uninvolved healthy tissues.

In broad terms, the intent of radiotherapy is either radical or palliative. Radical (or curative) treatment aims to destroy all cancer cells within the target volume and is appropriate when there is a single primary site of disease. In cases where the disease has spread beyond the primary site, treatment will normally be palliative. Palliative radiotherapy is delivered to one or more sites within the body and is delivered with the intent to control symptoms associated with the spread of the disease, such as pain, rather than eradicate the disease itself. Radiotherapy is a local treatment with the desired anti-cancer effect only apparent in tissues directly irradiated and is often given in combination with other therapies, either after surgery and/or complementary to other whole-body (systemic) treatments such as chemotherapy and hormone therapy. Table 9.1 summarises the main methods of cancer treatment currently available.

TABLE 9.1
Types of Cancer Treatments

Method	Process	Comments
Surgery	Physical removal	Invasive but very effective if all of the tumour can be located and removed.
Chemotherapy	Anti-cancer drugs	Inhibits the reproduction of cancer cells. Drugs are often combined for synergistic[a] effect. Side effects to some healthy cells but these can recover.
Radiotherapy	Ionising radiation	Direct action on irradiated cancer cells which are sensitive to ionising radiation. Includes external beam, brachytherapy and molecular therapy.
Hormone therapy	Block or lower hormone production	Some cancers such as breast and prostate use hormones to grow and develop and are therefore hormone sensitive. Medicines used to block or lower these hormones slow down or stop the growth of cancer.
Immunotherapy	Boost natural defence mechanisms	'Wakes up' or helps the immune system to fight the cancer. Antibodies can be designed to send signals to the immune system to destroy.

[a] Where the combined effect of multiple drugs is greater than the sum of the individual effects.

This chapter covers the physics of external beam radiotherapy. External beam radiotherapy is the delivery of radiotherapy using ionising radiation that is externally generated and then directed towards the tumour. This is sometimes referred to as teletherapy. Chapters 10 and 11 cover the other forms of radiotherapy: brachytherapy and molecular radiotherapy, respectively. For the interested reader, further references and resources are provided in Section 9.7.

9.1.2   How Does Radiotherapy Work?

To destroy the cancer cells, the ionising radiation deposits energy into the target volume. The total amount of energy deposited is very small in absolute terms, typically <100 joules. The cancer cells are destroyed at a sub-cellular level due to the ionising radiation causing irreparable damage to the DNA, the most vulnerable part of a cell to the radiation. However, the ionising radiation can also cause damage to the DNA of healthy cells. Finding a balance between radiation dose to tumour cells and to normal healthy tissues is paramount in effective radiotherapy. The relationship between probability of curing cancer termed the tumour control probability (TCP), and the probability of damaging normal healthy tissues defined as the normal tissue complication probability (NTCP) is demonstrated using response curves as shown in Figure 9.1.

Examining the response curve for tumour control and normal tissue damage, it is clear that accurate determination and delivery of dose is crucial to the success of radiotherapy. Errors in dose delivery can lead to a failure of tumour control and/or lead to unacceptable normal tissue damage, which can cause serious short-term and long-term side effects. Quantification, measurement and calculation of the dose delivered to the patient in radiotherapy are described further in Section 9.3.

To prevent the destruction of healthy cells, radiotherapy treatment is usually delivered in multiple treatment sessions called fractions. This involves dividing the prescribed total radiation dose into

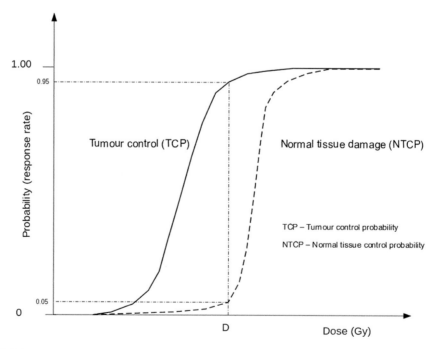

FIGURE 9.1   Dose–response curve. The vertical dotted line for a given dose D (Gy) indicates a tumour control probability of 95%. However, there is also a 5% probability that this dose will lead to complications due to normal tissue damage.

smaller doses of equal magnitude. This enables a higher therapeutic radiation dose to be delivered to the tumour than would be possible with just a single treatment by allowing normal healthy tissues to repair between treatments. The study of the biological effects of radiation on cells and how to maximise this effect on tumours without incurring serious damage to normal tissues lies within the realm of radiobiology (Chapter 4).

9.1.3 THE RADIOTHERAPY PROCESS – THE PATIENT PATHWAY

The patient is at the centre of the radiotherapy process. Safe and effective radiotherapy treatment relies on advanced technology and many complex and technical processes throughout the course of their treatment. One way to introduce these is to look at the chronology of the patient experience in terms of the key stages in this process; often referred to as the patient pathway. Table 9.2 outlines the key stages in this pathway which will follow on from the initial diagnosis and decision to treat the patient with radiotherapy.

The first stage is pre-treatment simulation; this requires the acquisition of 3D volumetric data (images) of the internal patient anatomy in order to localise the volume of tissue to be treated with ionising radiation (target) as well as the normal tissues to be avoided, referred to as organs at risk (OAR). These data are generally acquired on a CT scanner with, most importantly, the patient in the intended treatment position. Acquired CT scans are transferred electronically across the hospital computer network from the CT scanner and imported into the treatment planning system (TPS). The TPS consists of one or more computers running dedicated software to perform the treatment planning process. The imported CT image data set can be viewed, typically as a series of 2D transverse cross-sectional slices, within the TPS. Drawing tools supplied within the TPS software are used to delineate the target and the organs at risk. This 2D slice by slice contour information is combined along with the CT Hounsfield unit data to produce a geometrically accurate 3D model of the patient's internal anatomy from which the treatment plan is generated. These data represent the 'ideal' position of the patient for subsequent comparisons and is often called the reference data set. The treatment plan defines the precise arrangement of radiation beams required to deliver the calculated radiation dose distribution to the patient and is described further in Section 9.4. Once the treatment plan is produced and approved for treatment by the radiation oncologist it will go through one or more checking processes to ensure everything is as intended. The treatment plan information is then transferred electronically across a network, to a treatment machine in readiness for the patient's first treatment fraction. At treatment, the patient is set up on the treatment machine

TABLE 9.2
Key Stages in the Patient Pathway

Stage of Pathway	Description
Pre-treatment simulation	Patient is set up on a CT scanner in the intended treatment position using immobilisation equipment appropriate for the site being treated. Images are acquired and sent to the treatment planning system.
Planning	Tumour volume to be treated and the organs at risk to be avoided are defined. Optimal radiation beam arrangement is determined and treatment planned. Radiation dose distribution is evaluated and final treatment plan approved.
Verification	Images are acquired before treatment to confirm the patient is in the intended position. Patient position adjusted as required.
Treatment	Treatment delivered

in a position identical to their position on the CT scanner during the pre-treatment simulation stage. Prior to treatment, the patient is imaged in this treatment position using on-board imaging systems on the treatment machine. This verification image is crucial as it confirms that the patient is in the correct position and the radiation beams will be directed correctly towards the intended target. The acquired verification image is compared against the reference images derived from the original CT image data set used for planning. The patient position may need to be adjusted prior to treatment if there is a mismatch between the verification and reference images. This is achieved by moving the treatment couch on which the patient is lying, typically by a few millimetres in one or more orthogonal directions. This careful adjustment repositions the tumour target to the correct position relative to the beam, or beams, of radiation about to be delivered. Treatment verification is covered in further detail in Section 9.6.2. Once the patient position on the treatment machine has been verified treatment can commence by delivering the ionising radiation to the patient. The patient will normally return for subsequent treatments which will follow a similar pattern of verification, positional adjustment and then treatment delivery. Treatment will normally be delivered on a treatment machine called a linear accelerator, of which there are several types as outlined in Section 9.2. This patient pathway describes treatments using high energy megavoltage (MV) photons, the treatment of choice for tumours lying deep within the patient's body. For superficial lesions and tumours that are not as deeply positioned within the patient's body, lower energy kilovoltage (kV) X-rays or electron beams are more appropriate treatment modalities. The equipment and methods used to generate and treat these lower energy beams and treatment techniques are covered in Sections 9.2 and 9.5.

9.2 BEAM THERAPY EQUIPMENT

A wide range of equipment is used within radiotherapy from pre-treatment simulation through to treatment. Pre-treatment simulation is achieved using more familiar diagnostic imaging equipment adapted for use in radiotherapy and is described in Section 9.6.1. This section focuses on the equipment used to generate and deliver a clinically useful radiation beam for treatment. An important role of the physicist in radiotherapy is to commission, calibrate and perform routine quality control (QC) checks on this equipment. Commissioning involves all the initial measurements and checks that are made following installation of a new piece of equipment and establishes a baseline against which subsequent QC checks are compared. If a subsequent QC check on a specific parameter is outside a defined tolerance then a re-calibration will be required. Some of the ancillary equipment used to perform these tasks are introduced in Section 9.3.6. The design of radiotherapy treatment machines depends on the type of radiation (modality) such as photons or electrons, the energy required, and intended treatment site.

9.2.1 KILOVOLTAGE UNITS

The basic radiotherapy kV X-ray tube consists of a cathode and stationary anode assembly inside an evacuated tube (Figure 9.2). The cathode is a tungsten wire filament which, when heated by passing a current through it, emits electrons via thermionic emission. The filament is set into a focusing cup that focuses the electrons onto a small area of the target just a few centimetres away called the focal spot. The anode will typically consist of a tungsten target embedded within a block of copper. A high-voltage generator capable of producing the desired treatment kV, with minimal voltage variation, is connected across the cathode and anode of the X-ray tube. X-rays are produced by bremsstrahlung (braking) radiation as the electrons are decelerated in the vicinity of the target nuclei. Most of this energy is deposited as heat with only around 1% being converted into X-rays. The generated X-rays form a spectrum with the peak X-ray energy corresponding to the applied kV across the tube. The beryllium exit window has a low atomic number and physical density to maximise the transmitted radiation output available to treat the patient. A range of applicators or cones defining different field sizes are attached to the tube to alter the size of the useful radiation beam (Figures 9.2 and 9.3).

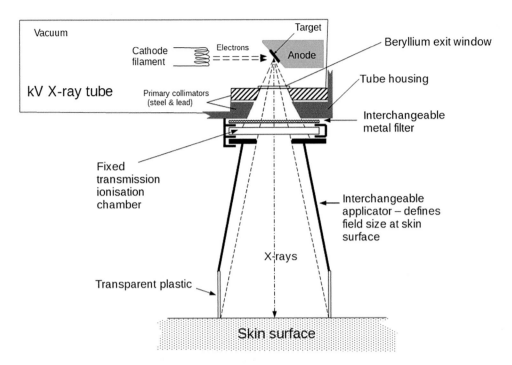

FIGURE 9.2 Schematic cross-section of a kV X-ray tube for superficial therapy.

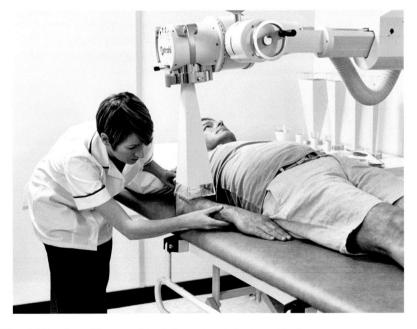

FIGURE 9.3 A kilovoltage X-ray machine being set up to treat a patient's arm.

(Courtesy of Xstrahl Ltd.)

9.2.2 Linear Accelerators

9.2.2.1 Conventional C-arm Type

The conventional C-arm type linear accelerator (linac) is the main treatment machine within most radiotherapy departments, delivering a wide range of treatments to most sites within the body using photon or electron beams. The linac consists of a moveable gantry supporting the treatment head that can be rotated 360° around the patient positioned on a moveable treatment couch. A key design characteristic of the linac is the isocentre, a point in space about which the linac motions are centred. For example, as the gantry rotates around this point the distance from the radiation source in the treatment head and the isocentre remains constant, normally 100 cm for most modern linacs. This fixed movement geometry around a single point makes patient set-up, treatment delivery and dose calculation simple and consistent. Radiation is generated in a linac using radiofrequency micro-waves to accelerate electrons produced from a hot cathode filament, called the electron gun. These electrons are accelerated along the linac waveguide where they are bent, using strong electromagnets, onto a target. MV X-rays are generated via bremsstrahlung (braking) radiation as the electrons are decelerated in the vicinity of the transmission target nuclei. Alternatively, the X-ray target can be moved out of the path of the electron beam and the electrons allowed to travel straight through the treatment head to deliver electrons instead of MV photons. The block diagram (Figure 9.4) shows the key design features of a linac set to deliver photons. The ion (ionisation) chamber fitted inside the treatment head is of a large pancake-shaped construction and measures the amount of transmitted radiation passing through it. Its main function is to terminate the beam when the prescribed dose has been delivered. In addition, it also has a secondary independent system that terminates the beam if the primary measurement channel fails. Modern linear accelerators are fitted with multileaf collimators (MLC). MLC typically consist of two opposed banks of individually controlled tungsten leaves, each thick enough to shield the radiation beam (Figure 9.36). A separate motor drive is attached to each leaf enabling independent movement and positioning. MLC facilitate the delivery of intensity-modulated radiotherapy (IMRT) where the incident 2D beam intensity is modified in order to produce the desired radiation distribution at depth. MLC typically project a leaf width of ≤1 cm at the isocentre. The smaller the projected leaf width the greater the ability of the MLC shape to conform to the required target (Figure 9.5). Some specialist linacs designed for very small volume treatments such as intra-cranial treatments have MLC widths ≤5 mm.

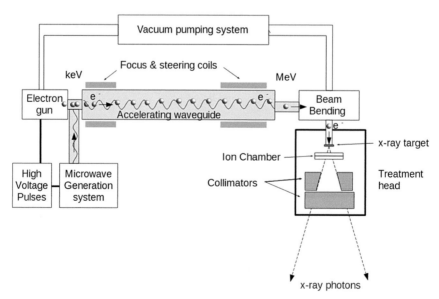

FIGURE 9.4 Block diagram showing the key components of a linear accelerator required to produce a beam suitable for clinical photon treatments.

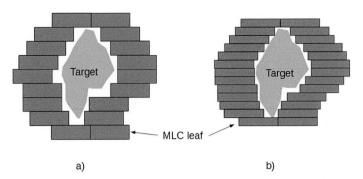

a)   b)

FIGURE 9.5 Target shape conformance for the smaller MLC leaf widths in b) is better than that achievable for the larger leaf widths in a).

Modern linacs come with additional on-board imaging equipment that enables verification imaging to be performed to ensure the patient is in the required position prior to delivering the treatment. Most linacs are fitted with two imaging options. The first, an MV imaging panel, is fitted in line with the radiation beam enabling the MV radiation exiting the patient to be imaged. The second system is a kV cone-beam CT system mounted orthogonally to the treatment beam direction consisting of a kV X-ray tube and imaging panel. Amorphous silicon is used for the active detector elements within the imaging panels, providing appropriate sensitivity and image resolution (Figure 9.6).

9.2.2.2 TomoTherapy® Helical Delivery

Helical TomoTherapy represents a completely different approach to the delivery of external beam radiation compared to the conventional C-arm type linac described above. In concept its origins are more like a CT scanner than a linear accelerator, delivering radiation in a helical manner using a narrow field length in the patient's superior-inferior direction. Helical radiotherapy describes a continuous treatment delivery where the radiation beam rotates round the patient as they are slowly moved through the narrow beam on the moving treatment couch. As the linac moves around the patient the primary radiation beam shape is continuously modified by 64 interleaved adjustable tungsten leaves which are in either a fully closed or open position, each projecting a transverse width of 6.25 mm at the machine isocentre. The pneumatically driven leaves are synchronised with both the rotational gantry position and translational couch movement to deliver the complex beams required for IMRT. The continual helical delivery enables the treatment of long treatment sites in just one single exposure, for example, the irradiation of the whole spine. The TomoTherapy unit also includes an on-board MV CT scanner. To enable this, the linear accelerator waveguide is detuned from the normal treatment energy of 6 MV to around 3.5 MV and this beam is collimated down to a field length of just 1 mm to create a fan beam suitable for CT on-board imaging of the transmitted MV beam. A bank of detectors reconstructs the acquired fan-beam projections to produce the MV CT image. CT images produced at MV energies do not exhibit as good tissue contrast compared with kV energies produced in diagnostic CT scanners, but are good enough to provide verification and image-guided radiotherapy (IGRT). MV CT has an advantage over kV diagnostic CT in that it is able to correctly image high atomic number materials such as artificial hips and dental fillings. At kV energies, high atomic materials attenuate the diagnostic X-rays to such an extent that there is insufficient signal reaching the image detectors. This leads to an inaccurate reconstruction of the CT image with incorrect high and low HU values being generated in the vicinity of the object. At MV energies the beam is attenuated much less by a high atomic number object so the detector signal is large enough for the image to be reconstructed correctly. The latest TomoTherapy Radixact platform is now available with a separate kV cone-beam imaging system mounted orthogonally to the treatment beam direction, similar to the C-arm type linacs. The difference on TomoTherapy is that this system is enclosed within the machine covers enabling a faster rotation and therefore acquisition speed as there is no danger of collision with the patient (Figure 9.7).

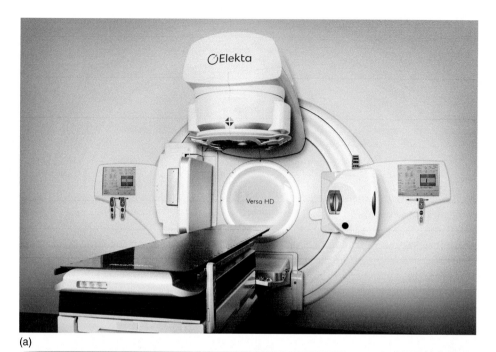

(a)

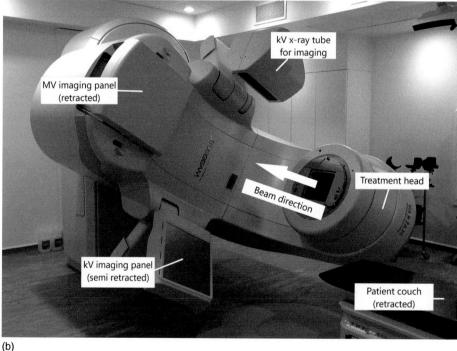

(b)

FIGURE 9.6 (a) Versa HD™ linac. The treatment head is at gantry 0° with the radiation beam directed vertically down towards the couch. The kV cone-beam imaging system is mounted orthogonally to this with the kV X-ray tube on the right-hand side and the imaging panel on the left. (Image supplied courtesy of Elekta Limited). (b) Photograph of a Varian TrueBeam® linac. The treatment gantry is rotated to an angle of 120 degrees with the radiation beam direction as indicated. The carbon fibre flat treatment couch on which the patient is positioned is retracted out of the beam in this picture.

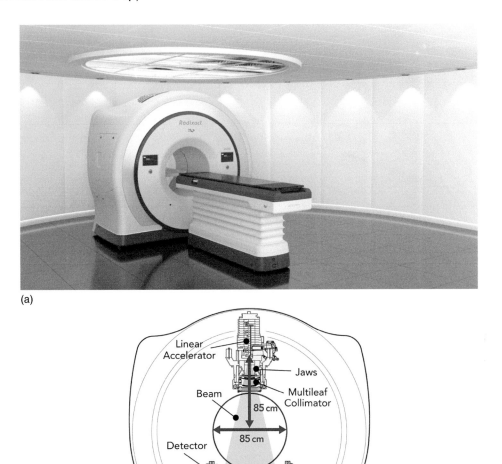

(a)

(b)

FIGURE 9.7 (a) TomoTherapy Radixact® Unit (Courtesy of Accuray). (b) Schematic of the beamline of the standard TomoTherapy Hi-Art Unit (Courtesy of Accuray). Radiation is generated in a short standing waveguide linac with integrated X-ray target as shown (~35 cm long). This is mounted in line with the beam and rotates around the patient during helical delivery.

9.2.2.3 The MR Linac

In the last few years, the first commercial magnetic resonance (MR) linacs have become available (Figure 9.8). An MR linac combines a standard linac with MR imaging capability, enabling images with much improved soft-tissue contrast compared to the cone-beam CT images acquired on conventional linacs. MR images can be acquired before and even during treatment delivery enabling treatment to be adjusted before and during treatment. The development challenges surrounding this equipment have been immense as the two systems are essentially unsuited to one another. The large magnetic field associated with the MR imaging system is undesirable in the vicinity of electrons generated in the linac (and those by photon interactions in the patient). Conversely, moving metal parts on the linac can lead to unpredictable distortions of the magnetic field uniformity which can

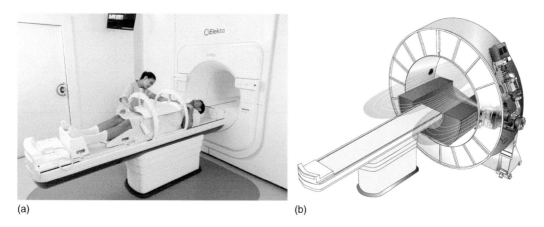

(a)   (b)

FIGURE 9.8   Elekta Unity MR linac.

(Images supplied courtesy of Elekta Limited)

affect MR image quality. The commissioning and QC of this equipment is a new challenge for the medical physicist. MR compatible equipment must be used and the effect of the strong static MR field on radiation beam measurements understood and quantified.

9.2.3   Radiosurgery Machines

9.2.3.1   Cyberknife®

Radiosurgery is the term used to describe radiotherapy delivered in a single, high-dose treatment, analogous to conventional surgery which is a single operation. The Cyberknife system (Accuray Inc, Sunnyvale, USA) is a radiosurgery device consisting of a 6 MV linear accelerator mounted onto a robotic arm. This arm is able to move around the patient with six degrees of freedom enabling the delivery of beams from almost any direction. During treatment, image guidance is provided using stereoscopic planar X-ray images automatically registered with the initial treatment planning scan to continuously correct the position of the robot arm (Figure 9.9).

9.2.3.2   Gamma Knife®

The Gamma Knife unit is different to all of the machines described previously as it uses radioactive sources to deliver the treatment as opposed to machine-generated radiation. It is designed specifically for intra-cranial radiosurgery and consists of a large concentric hemispherical helmet containing around 200 individual Cobalt-60 radiation sources. The small individual beams from each of these sources are focused onto a target with the aid of an integral collimation system enabling a range of small treatment spheres of varying diameter to be produced at the focus. A special frame is attached rigidly to the patient's head for the planning and treatment stages, normally completed on the same day, enabling very precise set-up and treatment (<0.5 mm). Multiple separate individual lesions can be treated by moving the patient on the couch so that each lesion is aligned in turn with the beam focus. The latest Gamma Knife Icon™ system now enables fractionated treatment with the aid of a thermoplastic head shell (see Section 9.4.2) in addition to the usual headframe for single-fraction radiosurgery. During treatment, the patient's position is actively monitored and the treatment interrupted if there is any deviation from allowed tolerances. The half-life of Co-60 is 5.25 years and therefore all of the radioactive sources need replacing every 5–10 years to ensure treatment times remain of an acceptable duration.

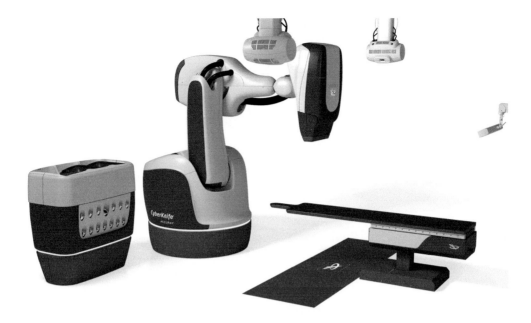

FIGURE 9.9 Cyberknife system showing the robotic arm, treatment head and couch.

(Courtesy of Accuray.)

9.2.3.3 Linac-Based Radiosurgery

Linacs are increasingly being used for radio surgical treatments for lesions down to 10 mm or even smaller. The small fields and high conformity can be achieved either using specifically designed collimators that define small circular fields of varying size, or more commonly now the use of high-resolution MLC leaves (≤5 mm). Treatment can be delivered as multiple non-coplanar static fields analogous to the Gamma Knife method above, or alternatively as non-coplanar arc treatments with either static or dynamically moving MLC. Immobilisation, pre-treatment imaging, as well as accurate and precise treatment positioning is especially crucial to ensure the patient is accurately positioned for these highly targeted, high dose, normally single-fraction treatments. Treatment couches now come with rotational as well as translational adjustment, which means that all 6 degrees of possible motion can be corrected to ensure the patient is in the correct location prior to treatment delivery.

9.2.4 PROTON THERAPY

Proton therapy is external beam radiotherapy delivered with protons, the positively charged particle found in the nucleus of an atom. The dose deposition characteristics of protons are quite different from photons, with the majority of the dose deposited within a narrow distance surrounding the Bragg peak (Figure 9.11). This makes proton therapy well suited for highly conformal treatments, with the possibility of reduced integral dose to the patient and greater avoidance of healthy tissues compared with photon beams. Protons, unlike photons, are completely stopped within the patient, with little or no dose deposited beyond the Bragg peak. This is of particular clinical importance for healthy tissues located directly adjacent to target structures. The generated Bragg peak is mono energetic and must first be adapted before clinical use, specifically the position (i.e. depth) and spread of the peak. The proton beam energy must be modified so that it does not deposit dose

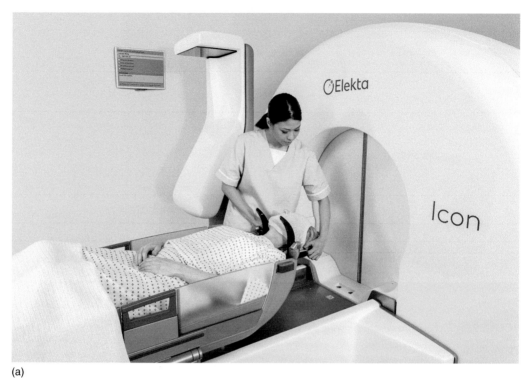

(a)

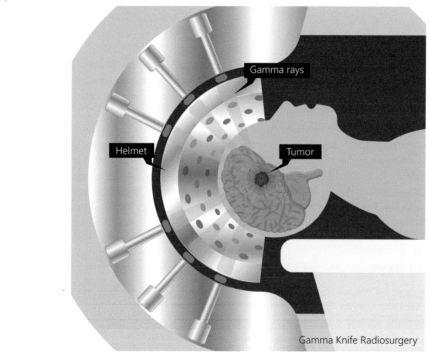

(b)

FIGURE 9.10 (a) Leksell Gamma Knife Icon™. Image supplied courtesy of Elekta Limited. (b) Leksell Gamma Knife. Schematic of hemispherical shell and multiple Co-60 sources used to focus the beam onto a small target volume.

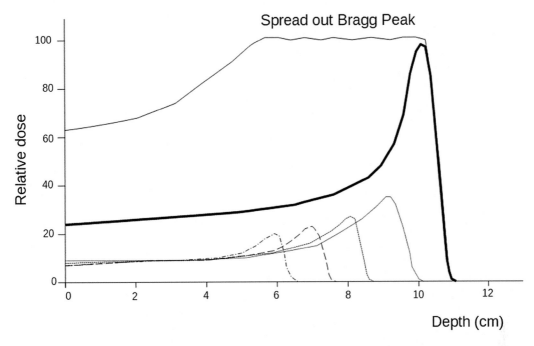

FIGURE 9.11 Individual Bragg peak (bold line) and spread out Bragg peak formed by summing multiple separate peaks.

beyond the target but it must also be spread out in the depth direction to cover the target with an homogenous dose, and also laterally to cover the irregular target shape. Modern proton machines achieve this using pencil beam scanning (PBS) where a magnetic scanner is used to deflect the narrow beam of protons back and forth across the tumour whilst also degrading the energy. This enables dose to be gradually built up in layers into the required distribution accurately conforming to the target shape. The increased conformity achievable with protons means that typically only two or three treatment fields are required where more would be required with photons. Protons are produced by stripping the electron from a hydrogen atom by, for example, ionising hydrogen gas supplied from a gas cylinder. The positively charged proton is then accelerated in a cyclotron to typically 250 MeV, sufficient for most treatment depths. The exit beam from the cyclotron is fed to one or more separate treatment gantries. In the case of multiple gantries, the single proton beam has to be shared and carefully switched according to demand. The proton treatment gantry is much larger than that of a linac but in essence is similar, being mounted isocentrically, having a treatment couch and integrated treatment verification imaging system (Figure 9.12). Proton machines are expensive and limited to a smaller number of specialist centres. Planning for these treatments is performed on a TPS, similar to those used for photons but using an algorithm that models the proton dose distribution. Protons are well suited for treating some complex childhood cancers, as they have the potential to reduce certain side effects such as deafness, loss of intelligence quotient, as well as secondary cancers due to the lower integral dose delivered to normal structures within the body. Protons are also appropriate for treating cases very close to sensitive structures within the brain, head and spine.

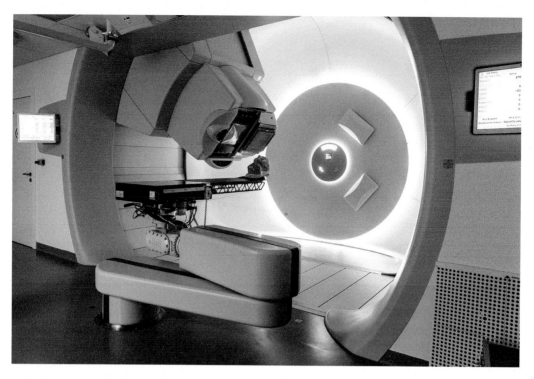

FIGURE 9.12   Proton beam delivery system.

9.3   CLINICAL DOSIMETRY

9.3.1   ABSOLUTE DOSE

Dosimetry, the measurement of radiation dose, attempts to establish a numerical relationship between ionising radiation and the effect that it produces. Such effects may be chemical, physical and biological, but in each case can only occur with the transfer of energy from the radiation to the material being irradiated. Absorbed dose is the term used to describe the amount of energy deposited into the material. The unit of absorbed dose is defined by the International Commission on Radiation Units and Measurements (ICRU) as the energy absorbed (E) per unit mass (m):

$$D = E/m \tag{9.1}$$

In SI units this is joule/kilogram [J/kg] and is given the special name gray [Gy]. The gray is small when compared with other forms of energy dissipation. For a tumour of mass 100 g a total dose of 50 Gy results in a deposited energy of only $50 \times 0.1 = 5J$. This is comparable to the energy required to lift a 0.5 kg mass about 1m against gravity, or the electrical energy delivered to a 1 kW heater in only 5 ms. This latter example indicates that the heating effect of ionising radiation per Gy is very small, with temperature rises typically <0.001°C (0.00024°C per Gy for water). Calorimetry, the method of measuring such small temperature rises, although technically challenging is the only way of directly determining absorbed dose and forms the basis of some national primary radiation standards.

9.3.1.1   Dose as an Indicator of Biological Effect

The energy, from ionising radiation, deposited in a given mass of tissue is just the start of a range of complex processes within the irradiated material. For an individual the biological effects associated with a course of radiotherapy must be related to the energy deposited, although there may be subtle variations from one person to the next because of other patient-specific factors. There is a clear relationship between TCP and absorbed dose so it is accepted that the best way to quantify the delivery of radiotherapy is by prescribing treatment in terms of absorbed dose. Accurate and traceable dose determination is crucial in maintaining consistency between different treatment equipment within a single department, as well as maintaining consistency between equipment in different departments both nationally and internationally. Clinical trials and the adoption of clinical protocols implicitly rely on the consistency of dose measurement.

9.3.1.2 Standards and Traceability

A dosimeter is a device that measures radiation dose. The requirement of such a device is that it exhibits some response (R) to radiation. For example, this may be the charge, in coulombs, liberated by ionised air in an ionisation chamber; or the amount of blackening on a silver halide film exposed to X-rays, determined by the optical density of the film. To determine dose we must have an additional factor, the calibration factor (F) that under the conditions in which the dosimeter was irradiated will convert the measured response into absorbed dose (D):

$$D = F \times R \tag{9.2}$$

where
   D = Measured dose (Gy)
   F = Calibration factor (Gy/unit of response e.g. coulombs)
   R = Response of the detector (unit of response)

For equipment used in the radiotherapy clinic, the calibration factor is routinely determined by comparing the uncalibrated dosimeter with an already calibrated dosimeter. Dosimeters can be placed side by side, or irradiated one after the other in identical positions to achieve this. Assuming that the dose delivered to each dosimeter is the same we arrive at:

$$D_{uncal} = D_{cal} \tag{9.3}$$

$$F_{uncal} \times R_{uncal} = F_{cal} \times R_{cal} \tag{9.4}$$

$$F_{uncal} = F_{cal} \times R_{cal} / R_{uncal} \tag{9.5}$$

where:
   D_{uncal} and D_{cal} = Dose received by the uncalibrated and calibrated radiation dosimeters
   F_{uncal} and F_{cal} = Calibration factors for the uncalibrated and calibrated dosimeters
   R_{uncal} and R_{cal} = Responses of the uncalibrated and calibrated dosimeters

This process of transferring the calibration factor is called an inter-comparison or cross-calibration. In principle, this process is straightforward but must be performed consistently and with care. Dosimetry codes of practice within radiotherapy describe in detail how this should be carried out and careful implementation of this process is one of the key areas of responsibility of a medical physicist working in radiotherapy. At the top of this calibration 'chain' sits a device that cannot be calibrated by comparing with another. This device is called a primary standard and measures (or more correctly realises) the quantity of interest from first principles. In this hierarchy, all devices depend on the accuracy of the primary standard. Figure 9.13 shows this calibration chain from the centrally maintained primary standard down into the hospital clinic. Each vertical arrow indicates that a cross-calibration must take place in order to disseminate the primary standard calibration down the chain. At each of these stages, there is an associated uncertainty introduced. Field instruments are the dosimeters used routinely in the hospital, such as the thimble ionisation chamber.

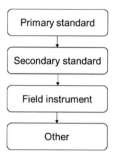

FIGURE 9.13   Calibration chain linking the primary standard to equipment used routinely in the clinic. Each vertical arrow indicates that a cross-calibration is required in order to calibrate the next piece of equipment down the chain.

Other equipment, such as quick dose constancy check devices, are used to confirm daily dose output of the linear accelerators and will be calibrated against the field instruments and these lie on the next step down in this chain. Codes of practice published at national (e.g. IPEM – Institute of Physics and Engineering in Medicine, UK, AAPM – American Association of Physics in Medicine) and international level (IAEA – International Atomic Energy Authority) detail how absorbed dose is to be measured in the hospital and also how the cross-calibration between calibrated and uncalibrated dosimeters is performed. In countries that do not have primary radiation standards, secondary standard dosimetry laboratories (SSDL) are maintained. These laboratories hold secondary standards that have been sent off site and calibrated against a primary standard. Devices are then sent to the local SSDL to be calibrated.

9.3.1.3   Reference Conditions

All radiotherapy treatment machines must be calibrated under well-defined and reproducible conditions. Normally a linear accelerator is calibrated in terms of machine monitor units (MU) where 100 MU set on the machine will deliver 100 cGy (1 Gy) of dose under the standard reference conditions. A typical reference condition would be a 10×10 cm radiation field size defined at a distance of 100 cm between the radiation source and measurement point with the measurement point at the depth of maximum dose on the beam central axis, as shown in Figure 9.14.

9.3.1.4   Practical Measurement in the Clinic

In the hospital, a small thimble-shaped ionisation chamber is routinely used to measure dose. This consists of a 0.1 cm^3–1.0 cm^3 cavity within which ionisation of air occurs within the enclosed volume and is collected between the chamber's axial electrode and its conducting wall (Figure 9.15a). A bias voltage of typically 100–400 V is applied between the axial electrode and conducting wall. The current generated from the ionised air is small, being of the order of 10^{-10}A. This requires a sensitive current measuring device called an electrometer to accurately integrate this signal into a measured charge of typically several nano coulomb. This charge is then multiplied by the appropriate calibration factor and other specific correction factors as defined in the relevant code of practice in order to derive the dose.

9.3.2   Relative Dose

Many measurements performed in radiotherapy are done so in a relative mode, whereby a series of measurements are carried out at different positions within the radiation beam. In these cases, the dose is normalised with respect to one point along the measured curve or plane. For these measurements, we are less concerned with absolute dose but rather the relative distribution of the radiation dose and the difference between multiple measurements with respect to each other.

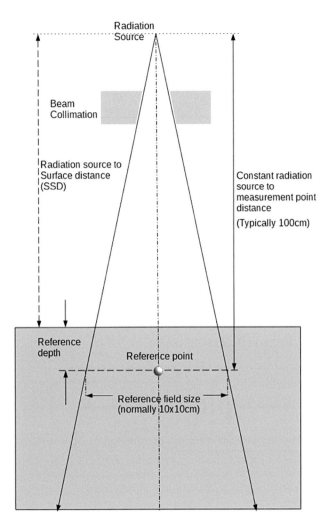

FIGURE 9.14 Typical reference set-up for calibration of an MV X-ray beam from a linear accelerator.

9.3.2.1 Representation of a Radiation Beam – The Isodose Distribution

Water is an ideal measurement surrogate for human tissue and most measurements made to quantify radiation dose distributions are acquired in water. Corrections to account for the different tissue composition in a real patient are applied within the TPS during the treatment planning process. The attenuation of a beam of radiation measured in water can be represented in 2D by an isodose distribution. Rather like the isobars on a weather map joining points of equal pressure, the lines on an isodose distribution join those of equal dose. The distribution is usually displayed in relative mode, normalised to 100% at the maximum dose value on the beam central axis. Figure 9.16 shows a schematic isodose distribution of an MV X-ray beam measured in water.

9.3.2.2 The Beam Central Axis – The Percentage Depth Dose

Plotting the isodose values in direction A–B, along the central axis of the beam from the water surface to the maximum measured depth produces a depth dose curve. When normalised to 100% at the depth of maximum dose, this is called a percentage depth dose curve (PDD). The shape of the

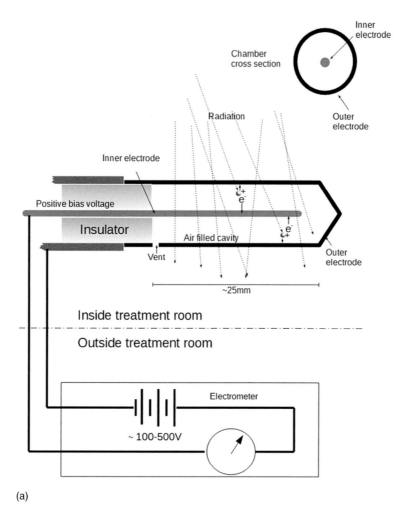

(a)

(b)

FIGURE 9.15 (a) Schematic representation of a thimble ionisation chamber and electrometer. (b) Measurement electrometer which is connected to the ionisation chamber.

(*Continued*)

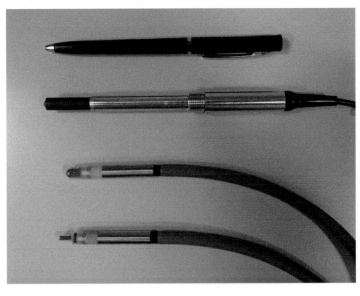

(c)

FIGURE 9.15 (Continued) (c) Thimble type ionisation chambers. The uppermost chamber is used to measure the MV X-ray dose from a linear accelerator. The bottom two small volume chambers are used for X-ray measurements in a water tank. The graphite cap for these two chambers has a protective waterproof cover. The pen at the top of the picture gives an indication of scale.

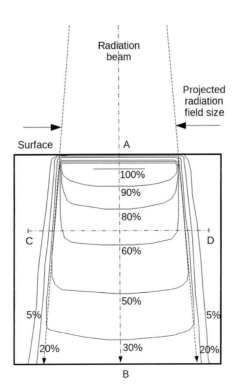

FIGURE 9.16 Representation of an MV X-ray distribution measured in water. The distribution is normalised to 100% at the position of maximum measured dose.

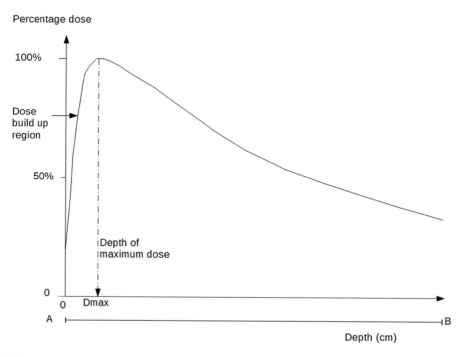

FIGURE 9.17 Representative percentage depth dose curve for an MV photon radiation beam.

PDD curve is a function of the beam energy, field size and radiation source to measurement surface distance (SSD) (Figures 9.17 and 9.18).

$$PDD(d,A,SSD) = 100 \times D(d,A,SSD) / D(dmax,A,SSD) \qquad (9.6)$$

where:

PDD(d,A,SSD) = The percentage depth dose at depth d, for field size A, at radiation source to surface distance SSD.

D(d,A,SSD) = Dose measured at depth d.

D(dmax,A,SSD) = Dose measure at the depth of maximum dose dmax

There are other ways to characterise the attenuation of an MV X-ray beam along the beam central axis such as the use of tissue phantom ratios. In this case, the point of measurement remains at the same distance from the source (i.e. SSD+d = constant, normally the isocentric distance of 100 cm). This is achieved by changing the amount of water above the measurement point so that both the SSD and measurement depth 'd' change but their sum, the distance between radiation source and chamber remains the same. Measurements acquired at different depths are then normalised with respect to the maximum reading (tissue maximum ratio) or other chosen depth, such as 5 cm or 10 cm (tissue phantom ratio). Because the point of measurement remains at the same distance from the source there is no inverse square component as is present in the case of the percentage depth dose curve where the chamber gradually moves further from the source with increasing depth and so the measured dose falls off due to both attenuation in tissue and inverse square fall off. Tissue maximum and tissue phantom ratios are more challenging to measure practically than PDDs and are normally derived from measured PDD data, by in essence removing the inverse square component. They are better suited to calculations involving isocentric treatments [1, 2] where the SSD will vary with beam entry angle but the isocentric distance, normally positioned at the centre of the treated target, is a constant.

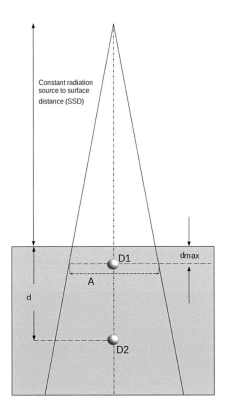

$$PDD(d,A,SSD) = 100 \times D2/D1 \quad (\%)$$

FIGURE 9.18 Measurement set-up for acquiring a percentage depth dose.

9.3.2.3 Percentage Depth Dose and the Build-Up Effect

Of special mention is the build-up effect exhibited by MV photon beams. Absorbed dose is not delivered directly by photons but indirectly by the secondary electrons generated by photon interactions in the tissue, principally forward-directed Compton scattered electrons. As the photon beam enters the patient there are very few electrons present other than those generated in the linear accelerator treatment head or those backscattered from deeper within the patient tissue. As a result, the amount of energy transferred from the electrons to surrounding tissue is small, leading to a relatively small amount of absorbed dose at the surface of the patient. As the photons penetrate further into the tissue and interact to produce electrons there will be a rapid increase in the number of electrons and therefore an increase in the proportion of energy transferred. Electrons have a finite range and deposit a similar amount of energy per unit length of travel as they slow down due to repeated collisions in tissue. This combined effect leads to a gradual build-up in the number of electrons and therefore absorbed dose until a maximum is reached. Beyond this point, the number of new electrons generated is offset by those already produced upstream reaching the end of their range. The depth of build-up is determined by the range of secondary electrons produced, which in turn is proportional to the photon beam energy. The build-up phenomenon results in the beneficial effect of 'skin sparing' in MV external beam radiotherapy. This enables deep-lying tumours to be treated while reducing the severity of treatment-related toxicity to the skin and superficial layers of tissue. At kV photon energies, secondary electrons have a negligible range and it can be assumed that the photon interaction and resultant energy deposited by the secondary electron occurs at the same

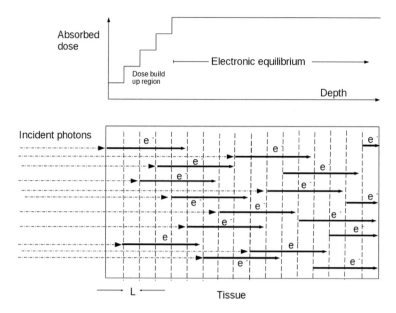

FIGURE 9.19 Simplified representation of the MV build-up effect occurring in successive layers of tissue of thickness 'L' where the mean electron range in tissue is 5 L. As the photon beam is attenuated with increasing depth in tissue the number of electrons generated will correspondingly reduce so the dose will start to fall beyond the depth of dose maximum (rather than remain constant as indicated on this simplified diagram).

place. For this reason, kV PDDs have no build-up with their dose maximum occurring at the skin surface (Figure 9.21). The build-up effect occurring between air and tissue is one example where dose equilibrium is not present. Similar effects occur at all interfaces, such as those between soft tissue and lung or soft tissue and bone, with the magnitude of the build-up or -down effect being related to the energy of the photon beam and the density and composition of materials on either side of the interface. The absence of electron equilibrium is hard to model accurately within a TPS (Figure 9.19).

9.3.2.4 Off-axis Distributions – Beam Profile and Penumbra

Plotting the dose values from the isodose distribution (Figure 9.16) in direction C-D perpendicular to the beam central axis at constant depth 'd' produces a dose profile (Figure 9.20). The dose profile characterises the dose distribution in the off-axis direction i.e. at points lying perpendicularly to the central beam axis, and includes the beam penumbral region where the radiation source is only partially visible from the point of interest. The profile has been renormalised to 100% at the beam central axis. The width and shape of the beam penumbra is determined by geometric considerations such as the physical dimensions of the radiation source, the source to collimation system (within the linear accelerator treatment head) distance and source to measurement distance. In addition, the penumbral width is increased slightly by the effect of laterally scattered electrons generated in water at MV energies which have a greater range than at kV energies. In radiotherapy, the penumbra width is normally defined as the distance between the 80%–20% levels when the profile is normalised to 100% on the beam central axis at the depth of interest. A small penumbra is important to ensure radiation dose is accurately directed to the required target region and not unduly spread out into adjacent normal structures in close proximity. For standard linear accelerators, a penumbra width of approximately ≤5 mm would be expected under typical measurement conditions and radiation focal spot sizes.

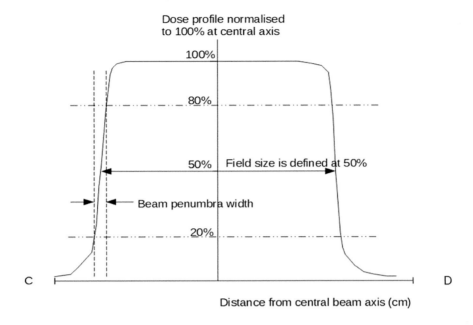

FIGURE 9.20 Beam profile plotted along the axis C–D from Figure 9.16.

9.3.2.5 Attenuation of Beams in Different Tissue Types

Although measurements performed in water act as a good surrogate for interactions in human tissue, they do not accurately represent the attenuation in a 'real patient' where radiation is absorbed differently according to the types, density and shape of tissues traversed. This will change the actual dose distribution received by the patient from that measured in water and must be accounted for in the treatment planning process. At MV energies the predominant radiation interaction in human tissue is via the Compton effect where a photon is scattered by a free electron. The energy absorbed from the photon by this electron is then deposited within the local tissue by direct electron-electron interactions. The probability of a Compton interaction occurring is proportional to electron density (number of electrons per unit mass). The electron density of most tissues are similar and largely determined by the proportion of hydrogen as well as physical density. Hydrogen with only a single proton in its nucleus has twice the electron density per atom than most other elements which typically contain equal numbers of neutrons and protons in their nucleus and therefore half the number of electrons per unit mass. The TPS must in some way determine the different tissue electron densities and apply the appropriate corrections to the radiation beam distribution. Fortunately, the CT scan acquired at pre-treatment simulation is well suited to this. Pixel by pixel information contained in the CT scan is displayed in Hounsfield units (HU) which scale the linear attenuation measured across each pixel to that in water. The HU for each pixel can be converted to the corresponding electron density by a simple look up conversion table within the TPS. The derived electron density map for each CT slice is then used as the basis for subsequent radiation dose calculations.

9.3.3 CHARACTERISTICS OF CLINICAL RADIOTHERAPY BEAMS

The characteristics of a radiation beam are dependent on the type of radiation and energy. External beam radiotherapy is mainly delivered using photons or electrons, with MV photons being by far the predominant treatment of choice. Other modalities, such as proton and heavy-ion treatments, are delivered in a limited number of specialist centres.

9.3.3.1 Kilovoltage X-rays

Clinical kV X-rays generally lie in the range 50–300 kV. Figure 9.21 shows the PDD curves for a range of kV energies. Treatment field size is defined by a range of applicators or cones which are attached to the kV treatment unit. The chosen applicator is brought into contact with the patient's skin surface to cover the required treatment region as well as accurately setting the treatment distance and hence delivered dose rate, which is strongly dependent on the X-ray target to treatment distance (typically ~15–50 cm). The maximum dose is delivered at the skin surface, as described in Section 9.3.2.3, with a rapid fall-off with depth. For this reason, kV units are used to treat the skin and superficial lesions. At kV energies, there is a higher probability of photoelectric interactions, particularly in bone which has a higher effective atomic number compared to soft tissue. This can therefore lead to substantially increased attenuation and hence absorbed dose if the beam traverses bone. The use of kV X-rays has diminished over the years because of the widespread availability of high energy photons and electrons generated by linear accelerators. However, many large hospital departments still maintain a single kV X-ray unit (Figure 9.3).

9.3.3.2 Megavoltage Photons

A typical MV photon isodose, PDD and profile are shown in Figures 9.16, 9.17, and 9.20, respectively. MV photon beams can be delivered to the patient in the form of simple single fields, multiple fields from different directions, to complex arc treatments where the radiation beam is delivered continuously as it is rotated around the patient. The methods and reasons for this are discussed further in the section on treatment planning.

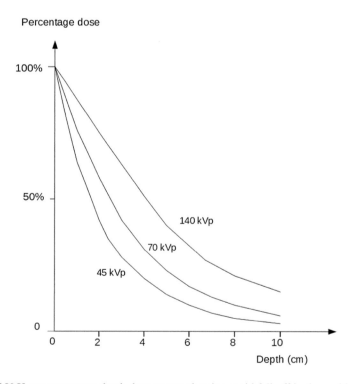

FIGURE 9.21   kV X-ray percentage depth dose curves showing rapid fall-off in dose with increasing depth and maximum dose delivered at the skin surface.

9.3.3.3 Megavoltage Electrons

Unlike photons, electrons are directly ionising, and immediately deposit energy into the irradiated material by multiple coulomb interactions involving electron-electron scattering collisions. As a result, they exhibit a finite range of travel within which all the incident electrons deposit their energy, resulting in a rapid fall-off in dose (Figure 9.22). Electrons are ideally suited to treating skin and superficial lesions of up to several cm in tissue. Most linear accelerators are capable of delivering electrons, covering a typical energy range of 4–20 MeV. Electrons, like kV X-rays, are normally delivered as a simple single-field treatment.

9.3.4 DETERMINATION OF DOSE AT AN ARBITRARY POINT IN A PATIENT

9.3.4.1 Applying Measured Data to Real Situations

There are a huge number of discrete settings available on a linear accelerator, and each of these will alter the radiation dose delivered to the patient. Fortunately, there is only a gradual change in dose when most settings, for example, radiation field size, are adjusted. As a result, a reduced subset of measurements covering the full range of available settings are acquired during the initial machine commissioning process. These measurements are then used as input into the TPS to enable the radiation beams produced by the linear accelerator, under all settings, to be modelled. These data can also be tabulated to enable simple dose calculations to be performed at discrete points, for example, the percentage depth dose curve can be used to calculate the doses at different depths along the beam central axis if the dose delivered at the depth of dose maximum is known.

9.3.4.2 Converting Relative Dose Measurements to Absolute Dose – The Machine Output Factor

The percentage depth dose is normalised to 100% at the depth of dose maximum, but because it is a relative measurement we have lost the link to the absolute dose delivered by the linear accelerator to the patient. This is re-introduced by the use of an output factor that characterises the change in

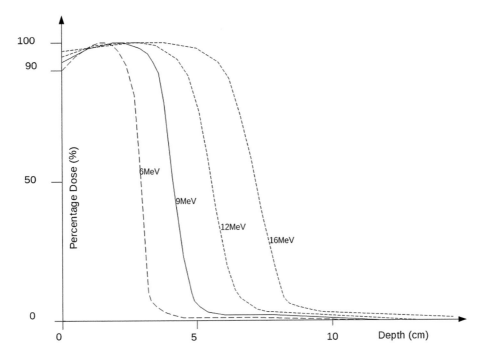

FIGURE 9.22 Representative electron percentage depth dose curves for a range of beam energies.

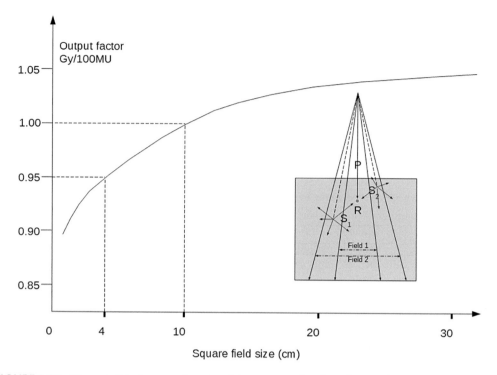

FIGURE 9.23 The output factor curve is a plot of dose measured at the reference depth (R), in this example for a nominal 6 MV beam at a depth of 5 cm with source to phantom surface distance of 95 cm, against field size. It quantifies the change in scattered dose at R with change in irradiated volume as the field size is adjusted.

dose rate (Gy per monitor unit, MU) at the reference depth as the radiation field size is adjusted. For the reference field size (typically 10 × 10 cm) at which the linear accelerator is calibrated, the dose rate is usually set to 1 cGy per MU (1 Gy per 100 MU). As the field size is changed the proportion of scattered radiation reaching the reference point, situated on the beam central axis at the reference depth, will change. Figure 9.23 demonstrates this effect and shows a typical output factor curve. We can now calculate the required linear accelerator MU setting to deliver the dose at a point under simple conditions.

Example of a simple dose calculation:

How many machine monitor units must be set on a 6 MV linear accelerator to deliver a dose of 0.5 Gy at a depth of 5 cm on the beam central axis for a 4 cm × 4 cm field size at a radiation source to surface distance of 95 cm?

From Figure 9.23, the machine output factor, measured at 5 cm depth and 95 cm SSD, for a 10 cm × 10 cm field is 1.00 Gy/100 MU and for a 4 cm × 4 cm field is 0.95 Gy/100 MU

The required monitor units to deliver 0.5 Gy for a 10 cm × 10 cm field = 50 MU (0.5 × 100/1.00)

The required monitors units to deliver 0.5 Gy for a 4 cm × 4 cm field = <u>52.6 MU</u> (0.5 × 100/0.95)

9.3.4.3 Irregular Shaped Fields – The Equivalent Square Concept

It is not practical to measure all possible field sizes and shapes that can be set on a linear accelerator as these can range from simple square and rectangular shapes, to irregular shaped fields defined by the multileaf collimator (MLC). Interpolation of data is required between measured values of quantities, such as the output factor and percentage depth dose data, described above. These are normally measured for a representative range of square field sizes set using the collimators on the linear accelerator. Other field shapes are expressed in terms of an equivalent square (EQS) for interpolation

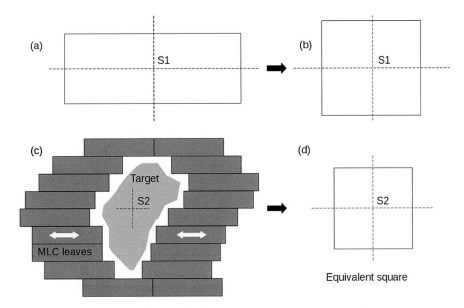

FIGURE 9.24 Demonstration of the equivalent square concept. The equivalent square field (b) has the same central axis scatter component at a given reference depth (S1) as the rectangular shape (a). Likewise the scatter S2 from the irregular field in (c) can be equated to the equivalent square field (d).

purposes. The EQS is defined as that square field size that has the same amount of scattered radiation (more correctly the scatter fluence) at the calculation point as the irregularly shaped field in question. It should be noted that as long as there is a direct line of sight between the point of interest and radiation source then the primary radiation component does not vary with field size or shape. There are a number of ways of calculating the EQS of non-square fields which increase in complexity as the field increases in irregularity. For example, a simple and reasonably accurate manual calculation of EQS for rectangular shapes can be made using Sterling's formula [1] which relates the ratio of the field area to its perimeter (Figure 9.24):

$$EQS = 4\,Area\,/\,Perimeter \tag{9.7}$$

For rectangular fields, this simplifies to

$$EQS = 2WL\,/\,(W+L) \tag{9.8}$$

where:
W = Field width
L = Field length

Methods to calculate equivalent squares for more complex irregular shapes revolve around the calculation of the contributing scatter component. There are several ways to do this with varying levels of complexity and accuracy [1, 2].

9.3.4.4 Corrections for Inhomogeneities

When the radiation beam traverses tissue that is not water it will be subject to a different amount of attenuation (absorption and scatter). If the tissue's electron density is substantially less than water, as in the case of lung or air cavities, the attenuation will be less. When travelling through bone, which is slightly more electron-dense than water, the attenuation will increase. Some form of correction must be applied to account for the effect this will have on the radiation beam. One simple manual way of accounting for different attenuation is by considering the water equivalent path length. This is calculated by correcting the distance traversed across each contributing inhomogeneity according to the ratio of its electron density to that of water. For the case of lung, with an approximate electron density of 0.25 relative to water, the water equivalent path length of 2 cm of lung equates to 0.5 cm of water. Accurate inhomogeneity corrections can only be applied using the TPS. The TPS takes into account not just the change in attenuation of the primary component of radiation, such as the above method on a voxel by voxel basis, but also the lateral extent of any inhomogeneity and the resultant change in the scattered radiation component.

9.3.5 PRINCIPLES OF PATIENT DOSE COMPUTATION

9.3.5.1 Defining the Problem

The calculation of a single point dose on the central axis of a beam under well-defined conditions can be achieved simply using beam data tables and a calculator. Calculating the dose distribution in a patient is much more complex but in essence can be calculated by summing the primary and integrated scattered radiation components at each point in turn (Figure 9.25). Each patient of course differs in physical shape and tissue composition, and the location and size of the target to be treated

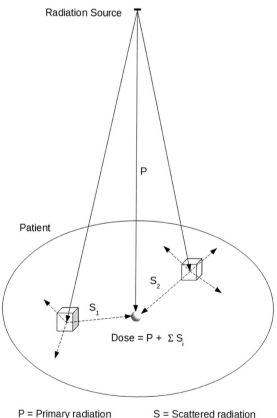

FIGURE 9.25 Primary and scattered dose – the dose at any irradiated point can be calculated by summing the primary (P) and scattered components (S_i) integrated over the irradiated volume.

will be unique to that patient. The resultant complexity means these calculations are performed using dedicated TPS software, running on powerful computers. This software should provide all the functionality required to support the planning process from importing the raw CT images to calculation and evaluation of the final treatment plan. However, central to the process is the treatment planning algorithm integrated within the software. The algorithm must accurately calculate the dose at any point in the patient to determine if a given plan is acceptable, and the computer must be capable of doing this in a short amount of time. There has been considerable development in these algorithms over the years and there are a range of solutions provided by a relatively small number of commercial manufacturers worldwide.

9.3.5.2 Treatment Planning Algorithms

Broadly speaking, modern treatment planning algorithms can be divided into model-based methods or direct methods that simulate radiation transport from first principles. Model-based approaches use a limited set of measured data to configure the model which is then applied to real-life geometries, by applying suitable approximations and corrections. The sophistication of modern model-based algorithms in applying these corrections ensures that for most relevant clinical situations adequate accuracy is obtained. There are situations however where these approximations are not as accurate, for example, at the interfaces between different tissue materials such as lung and air and soft tissue and bone where electronic equilibrium is not present. Simulating radiation transport from first principles can in theory remove all approximations and provide very accurate results within all tissue types as well as the interfaces between these different tissues.

9.3.5.3 Model-based Algorithm

There are many model-based algorithms of increasing complexity, accuracy and computational demands. Kernel-based models are currently in widespread use and will be briefly outlined in this section. The dose delivered to tissue irradiated by MV photons can be considered to take place by a two-stage process. The first stage involves the interaction of the primary beam with the tissue (i.e. the patient), which is then followed by the transport of energy away from the primary interaction site to the rest of the tissue. The first step can be calculated analytically by tracing the primary beam from the radiation source and projecting through the patient, the second stage is calculated using an energy deposition kernel. The energy deposition kernel models how dose, transferred to tissue at the primary radiation interaction site, spreads out within the surrounding tissue.

For the energy deposition kernel, the dose to a point p in tissue arising from primary interactions at a distant point r is:

$$\Psi(r)\frac{\mu}{\rho}(r)K(p-r)dV \tag{9.9}$$

where:

$\Psi(r)$ is the energy fluence at r.

$\frac{\mu}{\rho}(r)$ is the mass attenuation coefficient at r.

$K(p-r)$ is the fraction of the total energy released at r that is deposited at point p.

dV is the small volume element surrounding the point at r.

To calculate the total dose to point p a 3D integration must be performed summing the contributions of the kernels generated from all discrete volume elements dV within the irradiated volume. If the dose energy deposition kernel is assumed to be the same at all primary interaction points (i.e. it is invariant) then a 3D convolution can be performed. In reality, the patient is not homogenous and so the dose deposition kernel will vary according to the local tissue density through which it spreads. More sophisticated approaches take this into account at the expense of increased computation time (Figure 9.26).

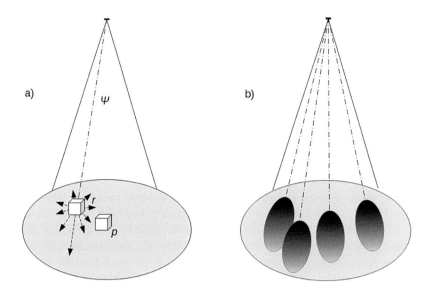

FIGURE 9.26 (a) Dose to a point p in tissue arising from primary interactions at a distant point r is characterised by the dose deposition kernel centred around r. (b) The total dose at each point is determined by a 3D integration of the contribution of all dose deposition kernels (represented by the shaded ellipse) across the irradiated volume, including the one centred at p.

9.3.5.4 Direct Method – Simulation of Radiation Transport Using Monte Carlo-based Approach

Monte Carlo (MC) is a statistical simulation method based on random (stochastic) sampling. MC simulates radiation transport by modelling the paths or tracks of a sufficiently large number of individual particles, typically~10^9 using a random number generated probability distribution governing the individual fundamental interactions involved at each stage of the particle's journey, from the radiation source to the patient. MC methods can accurately compute dose under almost all circumstances, but are computationally intensive because of the huge number of particle journeys that have to be simulated in order to get accurate results. MC was previously in the realm of research, but it is now commercially available in many TPS for both photon and electron beam calculations (Figure 9.27).

9.3.6 DOSIMETRY EQUIPMENT ASSOCIATED WITH COMMISSIONING AND QUALITY CONTROL

9.3.6.1 3D Plotting Tank

The 3D plotting tank, often called the water tank, is an important piece of dose measurement equipment used by a medical physicist to commission, and then provide occasional ongoing checks on the consistency of radiation beams generated from a linear accelerator. The 3D plotting tank is a large clear plastic tank of sufficient size to measure profiles across the largest radiation field size, normally 40 cm × 40 cm, at the machine isocentre. Inside the tank is a system of mechanical support arms that can be motor driven independently in all three orthogonal directions to enable collection of profile and percentage depth dose data. A radiation detector appropriate to the measurement being made can be mounted on the support assembly and controlled remotely via dedicated computer software from outside the treatment room. Measurements are normally made at discrete, selected positions (rather than measured continuously) and the dose integrated over a fixed period of time at each position. The linac dose rate is not constant and may drift by several per cents over the course

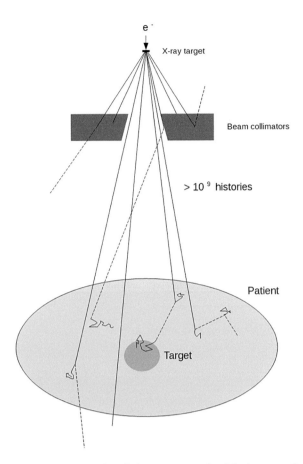

e⁻

X-ray target

Beam collimators

$> 10^9$ histories

Patient

Target

FIGURE 9.27 Monte Carlo simulation of radiation transport. Particle journeys from radiation source to patient are simulated to accurately calculate the dose deposited within the patient.

of a measurement. This would mean that the dose integrated into the measurement chamber could differ even though it was repeating an identical set of readings. To remove this dose rate dependence an additional chamber called the reference chamber is placed at a fixed position in the beam. For a given measurement the reading on this chamber will only change if the dose rate from the machine changes. Dividing the reading from the measurement (signal) chamber with that from the reference chamber removes any dose rate fluctuation effects to produce a stable and reproducible measurement. Measurements are made in relative mode with the software enabling a series of data points to be automatically renormalised to the dose maximum or any other chosen point within the measurement data set. The spacing between successive measurement points is adjusted depending on the expected dose gradient in that region. For example, when acquiring a dose profile the measurement point spacing in the penumbral region will be reduced to ≤1 mm to accurately capture this region where the dose changes rapidly whereas in the central region of the beam which is more uniform it may be ≤5 mm. The plotting tank software also enables the reconstruction of 2D isodoses from a percentage depth dose curve and a series of profiles measured at a range of depths (Figures 9.28 and 9.29).

3D plotting tanks are large and difficult to manoeuvre; they are complex and take a lot of time to set up before they can be used to acquire any measurements. Therefore, they are not appropriate to use for all measurements that may be acquired in the department. Other devices and methods are used routinely to check important beam characteristics on a more regular, routine basis.

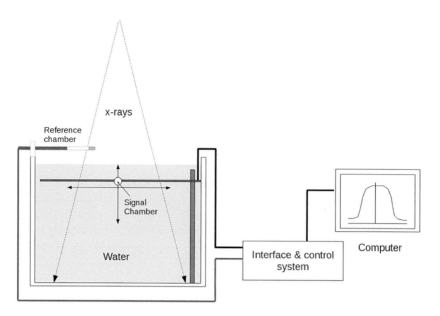

FIGURE 9.28   Schematic of a 3D plotting tank system.

FIGURE 9.29   3D plotting tank being set-up ready for measurement.

9.3.6.2 2D Detector Arrays

2D detector arrays consist of a 2D array of detectors varying in size and resolution depending on the commercial product. Some systems employ diode detectors, whereas other systems employ ionisation chamber detectors. These systems enable rapid acquisition of the beam profiles and can ensure that the beam is being steered correctly onto the X-ray target and producing the required beam uniformity. 2D detectors can also be attached to the linear accelerator treatment head to measure beam profiles at different gantry angles.

9.3.6.3 Other Ancillary Equipment

There is a wide range of ancillary equipment used by medical physicists to perform measurements on linear accelerators, from absolute dose calibration through to routine QC. It is normally the responsibility of the medical physicist to ensure this equipment is calibrated and correctly maintained (Figure 9.30 and Table 9.3).

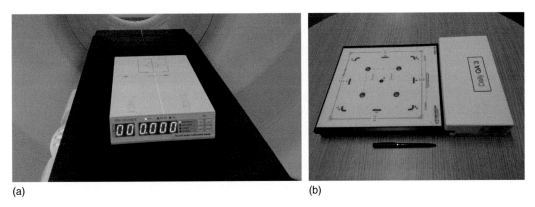

(a) (b)

FIGURE 9.30 QC equipment. (a) Daily dose constancy check device with large clear display visible via a closed-circuit TV system from outside the treatment room (b) Multiple detector device for routine checking of dose constancy and beam uniformity. See also Figure 9.15b for a picture of an electrometer.

TABLE 9.3
Equipment Used for Quality Control

Measurement	Equipment Example	Use
Absolute dose	Electrometer and ionisation chamber	Baseline calibrations, recalibrations and routine checks on dose delivery
Dose constancy check	Large volume ionisation chamber	Quick check of machine dose delivery e.g. each morning when the treatment machine is switched on and daily QC is performed prior to clinical use. Simple to set up and use.
Beam uniformity	2D array of ionisation or semiconductor detectors	Detector array is placed in the path of the beam to ensure the cross-sectional uniformity of radiation is within allowed tolerances. Will be part of a regular QC programme.
Commissioning	3D plotting tank	Measurements of beam profiles and PDD curves at initial machine set-up. Provides a baseline against which subsequent QC measurements are compared. Subsequently used for occasional accurate checks and possible beam adjustments e.g. following major repairs.

9.4 TREATMENT PLANNING

9.4.1 DATA ACQUISITION

The first stage in the treatment planning process is to acquire image data with the patient in the intended treatment position. The purpose of this is to:

- Image the position, extent and shape of the target volume, organs at risk and other relevant structures.
- Acquire sufficient data to enable accurate computation of the dose distribution within the patient.
- Acquire accurate information for appropriate set-up of the patient for treatment delivery.

9.4.2 POSITIONING AND IMMOBILISATION

Patient positioning must be defined, and reproducibly maintained, throughout the complete treatment planning and delivery process. Suitable patient immobilisation is crucial in achieving this for the full treatment course which can extend over several weeks. Immobilisation should help the patient remain as motionless as possible during treatment where the high-dose treatment region could be in close proximity to an organ that is sensitive to radiation. This is particularly important for treatments involving the head and neck and for these treatments customised thermoplastic shells are made to constrain any motion during treatment. To make these individualised shells the thermoplastic material, supplied in flat sheets, is first softened in a hot water bath and then carefully moulded around the patient's anatomy. As the thermoplastic material cools it becomes rigid retaining its moulded shape.

9.4.3 IMAGING FOR TREATMENT PLANNING

CT imaging is at the heart of radiotherapy planning and has been for many years. The patient is aligned, using external lasers in the scanner room, in the most appropriate position for the specific treatment being simulated. Patient setup will be influenced by suitable radiation beam entry positions whilst ensuring the patient is comfortable enough to remain still for an extended period of time. The patient may be lying in a supine position (on their back), or prone position (on their front), with their arms up or down, knees bent, etc. Specific devices might also be used for patient set-up, for example, the use of an angled tilt-board to support the upper body with arms raised up over the head for tangential breast treatments enables the radiation beams to be readily directed towards the breast tissue (Figure 9.31b). Choice of device and level of immobilisation will depend on treatment

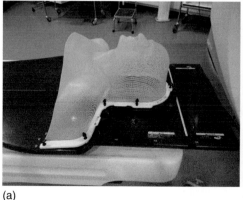
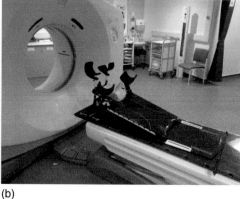

(a)   (b)

FIGURE 9.31 (a) Head and neck immobilisation shell made from a thermoplastic material. (b) Breast board immobilisation system with head rest and adjustable arm supports.

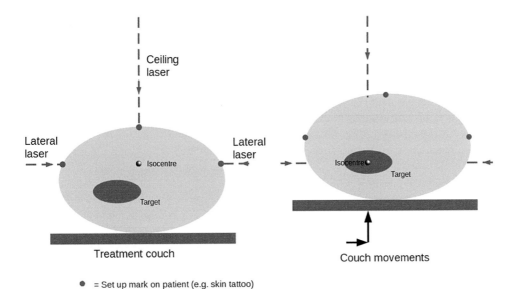

FIGURE 9.32 Example of patient set-up. First, the set-up marks on the patient are aligned to the external room lasers which define the treatment machine isocentre. The couch is then moved by pre-calculated offsets to position the target at the required position for treatment.

site and intent, either palliative or radical. The importance of remaining still during treatment will be explained to the patient as it is much more effective if the patient understands and is compliant with these requirements.

Once the immobilisation and set-up is determined and the patient is comfortable on the couch, the patient is moved such that the external lasers are aligned on appropriate external positions on the patient's body. For the case of abdominal treatments such as the prostate or cervix this will be onto a stable anatomical position on the skin surface. Radio-opaque fiducial markers such as small ball bearings which can be visualised on the CT scan are then taped onto the patient's external anatomy to indicate these laser positions. The patient is then scanned and the CT image assessed, for any unwanted artefacts or movement between setup and scan, and providing the scan is acceptable permanent ink tattoos, in the form of a single point dot, are marked on the patient's skin at the position where the fiducial markers were taped. The CT images are then transferred (exported across the IT network) from the scanner to the TPS. Within the TPS the position of the treatment isocentre, normally placed close to the geometric centre of the target, is determined in relation to the fiducial markers visible on the CT image data. This takes the form of translational offsets in the three orthogonal directions related to the patient, i.e. superior–inferior, left–right or anterior–posterior. On the treatment machine the ink tattoos, representing the original fiducial marker positions, are then set to the external linac lasers so that the patient is in the same starting position as on the CT scanner when the planning scans were first acquired. The patient can then be moved from this known starting position to the correct isocentre position for treatment by moving the couch by the required translational offsets calculated above (Figure 9.32)

9.4.4 Determining the Treatment Target

Accurate definition of the intended treatment target is fundamental to radiotherapy. Location of the gross tumour volume will be derived from one or more sources depending on the site to be treated and may include palpation, biopsy information as well as diagnostic images from CT, MRI and PET scanners. The process of localisation and subsequent treatment of the target introduces a level of uncertainty that will be present along the patient pathway. This uncertainty can be broadly classified as either clinical or geometric and has to be either accounted for in some way, and/or minimised where possible.

9.4.4.1   Clinical Uncertainty

A common problem in medicine is the determination and accurate localisation of any diseased tissue. In surgery, this corresponds to the appropriate extent of tissue to be physically removed from the patient. In radiotherapy, this corresponds to the target volume to be treated with a high dose of ionising radiation to ensure all cancer cells are being targeted. This target volume is normally determined with imaging and therefore is limited by the accuracy and precision of the imaging modality in displaying the extent of diseased tissue. Even with very high image contrast, there will be a layer of microscopic disease spread, which may exist on the periphery of any tumour, that is not visible. This clinical uncertainty is accounted for by adding a margin to the visible tumour. The ICRU provides guidance for a systematic approach to volume definition (ICRU reports 50 and 62 [3, 4]). In this report, the gross tumour volume (GTV) is defined as the gross palpable, visible and demonstrable extent and location of the malignant growth. It then defines the clinical target volume (CTV) as the volume encompassing the GTV but with an additional margin accounting for any microscopic extension of the primary tumour. This additional margin depends on treatment site and can range from zero to several centimetres.

9.4.4.2   Geometric Uncertainty

Geometric uncertainty is more specific to radiotherapy and is introduced along the patient's pathway. Most radiotherapy is delivered in multiple fractions administered over several weeks and there are uncertainties around both patient set-up and treatment delivery over this extended period. As described above, external reference marks are used to align and set up the patient for treatment. Lasers are normally used to align the patient and there are limitations on the accuracy with which this can be achieved on a daily basis. In a similar way, the accuracy with which the linac treatment parameters can be set is subject to tolerances. How accurately can the gantry, collimator or couch angle be set, and what will be the variation in defined radiation field size from day to day. In addition, the internal patient anatomy is subject to change over this time, even though it may appear unchanged outwardly, and although treatment verification imaging may be able to quantify some of this change it will not be able to correct all of this variation. These effects will all contribute to a level of uncertainty in the accuracy of delivered treatment. Once again a margin is used to account for these uncertainties. ICRU report 50 [3] defines the planning target volume (PTV) as the volume fully encompassing the GTV and CTV but with an additional margin accounting for organ motion and the uncertainties inherent in radiotherapy set-up and treatment delivery (Figure 9.33).

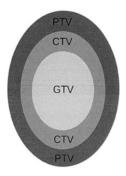

FIGURE 9.33   Margins to allow for uncertainty in radiotherapy. GTV = gross tumour volume, CTV = clinical target volume, PTV = planning target volume.

9.4.5 ORGANS AT RISK

In trying to treat the target we are constrained by the vicinity of sensitive normal (i.e. healthy) tissues defined as the organs at risk (OAR) and in most cases this is the limiting factor on how much and therefore how effective the treatment will be. OARs can be divided into serial and parallel structures which determine how their function is likely to be compromised by excessive radiation. A sub-functional unit is defined as a part of the OAR in which the basic function of the OAR is independently achieved. For serial structures, the sub-functional units are aligned linearly in a serial manner. This means that damage to just one of the sub-functional units will irreparably affect the whole organ. An example of this is the spinal cord, where damage to any section of the cord could lead to radiation myelitis and ultimately paralysis. For parallel structures, the sub-functional units are arranged in a more forgiving pattern with irreparable damage to one sub-functional unit still allowing the remainder of the organ to function albeit in a reduced capacity. The lung is an example of a parallel structure. In terms of the treatment planning process for serial organs, we must never exceed the dose value which will damage any of the sub functional units, termed the dose maximum. For parallel structures, the loss of a certain proportion of the organ volume may be accepted without significantly compromising function. In this case, the absolute volume or proportion of the irradiated organ exceeding a certain dose becomes the important consideration. The TPS should provide software tools to calculate and evaluate different treatment plans according to OAR type. It should also be noted that there are also serial-parallel structures such as the heart which are a combination of the two types (Figure 9.34).

9.4.6 TREATMENT PLANNING PROCESS

9.4.6.1 The Aim of Treatment Planning

Treatment planning determines the way in which a high dose of radiation is to be delivered precisely to the required target volume within a patient. This is achieved using a dedicated

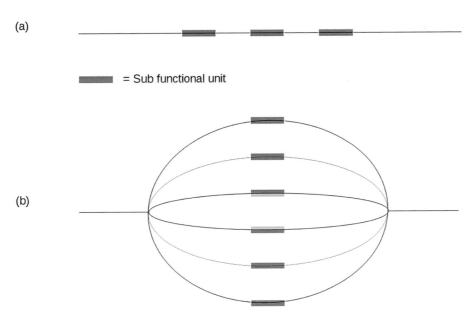

FIGURE 9.34 (a) Serial and (b) parallel structures represented in terms of their sub functional units.

treatment planning computer or system. This planning computer must enable several key functions to be carried out as follows:

- Import of the treatment CT image data acquired with the patient in the treatment position.
- Contouring of the target volume to be treated and OARs to be avoided on all relevant 2D transverse CT images.
- Contouring of the external body contour on all CT slices. This defines the extent of the patient and is required for dose computational purposes by telling the computer where in 3D space the patient is.
- Construction of a 3D patient model from the 2D transverse CT images and defined contours.
- Conversion of the CT pixel by pixel HU to electron density which is required to calculated the radiation dose within the constructed 3D patient model.
- Characterise and position the radiation beams from the linac on to the patient in order to simulate the effect of delivering the radiation to that specific model representing the patient.
- To provide tools to adjust the beam shape and angle of delivery and an algorithm to then calculate the radiation dose inside the patient.
- Adjust the patient plan to meet the required design constraints as required by the oncologist and treatment protocol.
- Evaluate one or more plans for the same patient in order to determine the best plan.
- Export the final approved treatment plan to the treatment delivery machine. The treatment plan describes how to deliver the radiation by defining all necessary settings on the machine. This will include parameters such as gantry angle, field shape, treatment head rotation (collimator) angle, treatment couch position, MU, etc

9.4.6.2   Simple Treatment Planning and the Superposition of Fields

Tumours situated at depth within the body require treatment with photon beams of energy typically 6 or 10 MV. Although less common some linacs do treat at higher MV energies affording greater tissue penetration, which may be advantageous for treatments of larger patients. For successful radiotherapy, the normal objective is to deliver a uniform dose distribution to the defined target, represented by the PTV, while sparing adjacent healthy tissues. Conventionally this is achieved by delivering multiple photon beams from varying angles around the patient's body.

A single photon beam, shown in Figure 9.35a, is not suited to delivering a uniform dose across the target volume. At a given depth and in a direction perpendicular to the beam central axis, the profile is uniform within the central area of the beam; however, in the depth direction the dose decreases with increasing depth across the target as characterised by the shape of the PDD curve. In order to improve the uniformity at depth additional beams can be delivered from different directions. Figure 9.35b shows the effect of two equally weighted photon beams delivered with opposing beam geometry, i.e. one beam entering from above (anteriorly, for a patient lying supine on the treatment couch) and the other beam delivered by rotating the linac treatment head round to a gantry angle of 180° so the beam is directed up through the couch towards the target volume from beneath (posteriorly). The effect of this is a higher relative dose at depth within the body than could have been delivered by a single field with the decreasing contribution from one field being compensated for by the increasing contribution from the opposed field. We now have a dose distribution that is uniform across the target volume. For many simple palliative treatments, this two field 'parallel-opposed' configuration is adequate as it delivers a uniform dose across the irradiated volume and is simple to plan, calculate and deliver. However, as can be seen we are also irradiating regions outside the target i.e. normal tissue, to similar amounts of radiation which is not desirable. Figure 9.35c shows the effect of introducing a further two fields to produce four equally weighted photon beams entering the patient at orthogonal gantry angles of 0°, 90°, 180° and 270° and intersecting at the treatment isocentre positioned at the geometric centre of the target. The effect of this is that the four beams sum their individual dose contributions to produce an even higher relative dose to this target volume. The dose

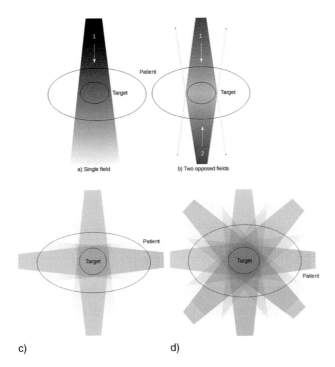

FIGURE 9.35 Superposition of multiple photon beams to improve target volume dose uniformity and reducing dose to adjacent normal tissues. (a) single photon beam (b) parallel-opposed beams (c) four beams and (d) eight beams.

delivered to the patient by the beams as they are attenuated through normal tissue is now shared out between the four beams reducing the high-dose component delivered to normal tissues outside the target volume. The dose delivered to normal tissues is not now the limiting constraint enabling higher doses to be delivered to the target so improving the probability of local tumour control.

The only downside is that the volume of normal tissue irradiated to this lower to intermediate dose is increased. Extending this process you could continue to add more beams from intermediate angles (45°, 135°, 225° and 315°) leading to an even more uniform dose within the central target region and reduced maximum dose in the normal tissue, as shown in Figure 9.35d. Arc or rotational therapy takes this principle to its natural conclusion. In arc therapy, rather than multiple successive static beams delivered as described here, the gantry rotates around the patient delivering radiation continuously as it goes, effectively mimicking an infinite number of static individual beams entering the patient from all angles. At present we have assumed that the target volume and patient is of uniform shape and that the relative dose contribution from each beam is the same. The next step is to consider non-uniform or irregular target shapes.

9.4.6.3 Conforming to Irregular Target Shapes

Conformal radiotherapy is a term meaning measures have been taken to conform the radiation dose distribution to the target volume. This is usually carried out with multiple photon beams, as described above, but with further methods of shaping the beam to the target shape and so reducing the dose to normal tissues. These beams are delivered statically such that each beam delivers its dose to the patient separately at a fixed geometry. Beam geometries are available throughout the full gantry angle range of 360° around the patient and the collimators (or diaphragms) within the linear accelerator treatment head are capable of varying rectangular field sizes. Beam angles and sizes that avoid entering or exiting through an OAR can be chosen. However, the projected shape of an irregular target from any given angle will be different and the photon beam can be further shaped by

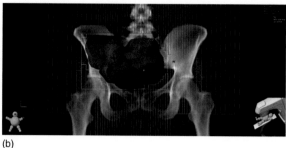

(a)   (b)

FIGURE 9.36   MLC and field shaping. (a) MLC driven to form a complex shape on the linac treatment head. (b) Simulation of MLC positions in the treatment planning system, in this example conforming to the large target volume in red.

the MLC to change the shape from simple rectangles to shapes that conform to this irregular shape, as shown in Figure 9.36.

9.4.6.4   Beam Modulation

As well as shaping the beam, in many cases it is advantageous to also modulate the beam intensity. Beam modulation is the process by which the intensity profile of the incident beam is modified in some way before it enters the patient in order to pro-actively compensate for the subsequent absorption of the beam in the patient which would otherwise lead to a poorer distribution of dose. Wedges are one way of modulating a conformal static beam. Wedges are used to adjust the isodose shape and are typically used to compensate for missing tissue or flatten the isodose distribution in cases where the photon beam enters the patient obliquely. The wedge is either a physical wedge in the path of the photon beam that attenuates the beam through the different thicknesses of the wedge or a dynamic wedge that uses a moving collimator jaw in the treatment head to dynamically modify the delivered distribution into a wedged shape (in 1D). More sophisticated 2D beam intensity modulation techniques are now routinely achieved by adjusting the position of individual MLC leaves forming the basis of IMRT (Figures 9.37–9.39).

Conformal treatment techniques are significantly more complex than that of single photon beams and therefore require a TPS to calculate the resultant dose distribution. To determine suitable beam arrangements, shapes and modulation of the beam, it also requires clear localisation of target volume and OAR. Conformal radiotherapy is 'forward planned'. This describes the process whereby the operator makes specific decisions about the photon beams and calculates the resulting dose distribution essentially by trial and iterative manual adjustment, based on training and experience, until a satisfactory solution is reached. For many treatment sites a common starting point is used in terms of number of beams to use and gantry angles.

In the example shown in Figure 9.40, the dose distribution is achieved by utilising angled beam geometry, beam shaping (through field size and MLC) and beam modulation (through wedges) to achieve a high uniform radiation dose to the target volume (shown in red) whilst reducing the dose to the spinal cord, heart and healthy lung to below a specific dose tolerance for each structure.

9.4.7   Intensity Modulated Radiotherapy

Photon beams used in conformal radiotherapy have a relatively uniform dose distribution across the majority of the field; any modulation of the dose distribution is achieved simply by the patient's external body contour, internal heterogeneous tissues and the use of wedges. IMRT on the other hand utilises non-uniform dose distributions by modulating the intensity of radiation across the field using MLC. IMRT can be delivered either in a step-and-shoot fashion or a dynamic fashion (Figure 9.41). In step-and-shoot IMRT the MLC leaves move to a desired position and then a portion of the beam

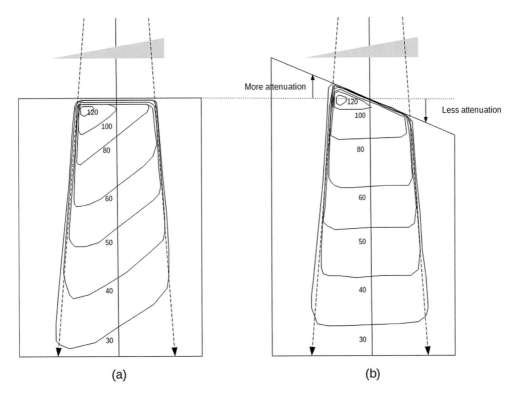

FIGURE 9.37   Beam modulation using wedges. (a) Effect of a wedge on the isodose distribution under standard measurement conditions. (b) Demonstrates the use of a wedge to correct for an oblique entrance surface.

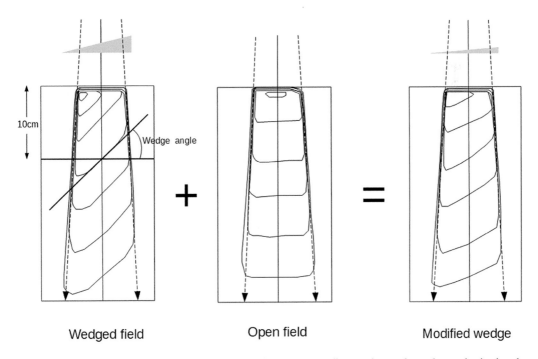

FIGURE 9.38   Generation of wedge fields, method 1. Any intermediate wedge angle can be synthesised as the appropriately weighted sum of wedged and open fields.

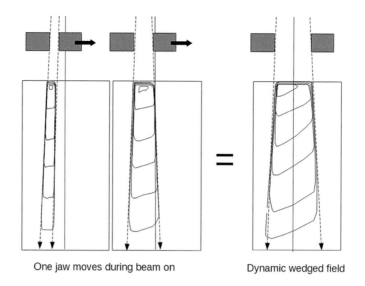

One jaw moves during beam on   Dynamic wedged field

FIGURE 9.39 Generation of wedge fields, method 2. A wedge isodose shape can be produced by moving one collimator jaw across the field when the radiation beam is on. The rate of travel of the jaw determines the wedge angle.

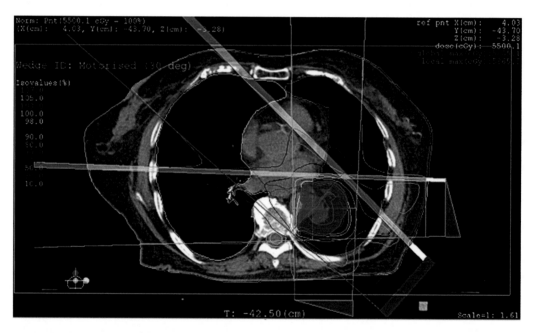

FIGURE 9.40 A conformal radiotherapy treatment plan for a lung target volume.

is delivered before the MLC leaves move on to the next position to deliver another portion of the beam. This means that for each beam (i.e. gantry) angle many differently shaped segments of the radiation beam are delivered to produce a modulated radiation field derived by summing the contributing individual beam segments. In dynamic IMRT the MLC leaves move continuously during treatment delivery achieving similar dose modulations to step-and-shoot IMRT but normally with a faster delivery because the beam is not interrupted between successive discrete segments. With IMRT we are even more capable of shaping dose distributions around target volumes and avoiding

OAR. IMRT is of a much higher complexity than conformal radiotherapy and cannot be forward planned as the number of possible MLC positions are too great and cannot be achieved iteratively in a manual trial and error approach. IMRT is therefore inverse planned, a process whereby the goals (objectives) of a desired dose distribution to the target and limiting doses (constraints) to the organs at risk are stated at the beginning of the planning process. These desired planning requirements are also subject to set penalties which determine relative priorities if compromises need to be made in order to achieve a suitable dose distribution. For example, the dose to one OAR may be set such that it must never be exceeded whereas to another it may be exceeded by a certain amount. The methods and complexities of defining objectives and constraints are subject to the specific algorithm implemented within the TPS (Table 9.4).

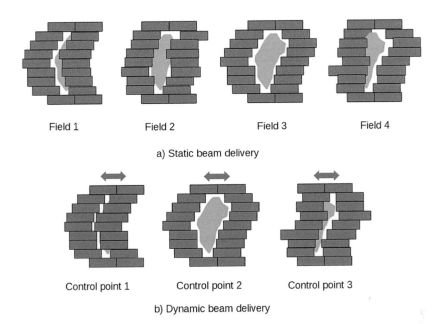

FIGURE 9.41 Use of MLC to modulate the beam. (a) Series of static MLC fields used to build up the required 2D dose. (b) MLCs are dynamically swept across the target whilst the beam is on.

TABLE 9.4
Examples of Inverse Planning Objectives for a Head and Neck Plan

Structure Name		Constraint	Dose [Gy]
PTV	98% of volume	Receives more than	55.8
PTV	95% of volume	Receives more than	57.0
PTV	Median dose	Is	60.0
PTV	Dose maximum	Is less than	63.0
Spinal cord	Maximum dose (to 0.1cc)	Is less than	44
Parotid glands	Mean dose	Is less than	24
Optic nerve	Maximum dose	Is less than	48

In this example the prescribed target dose is 60 Gy to be delivered over 30 treatment fractions.

An example of a possible objective function for the PTV is given in Equation 9.10 where Σ_j represents the summation over all of the voxels (j) within the PTV structure, W is the weighting or penalty factor for the PTV, $D_{j,cal}$ is the calculated dose using the currently defined beam parameters for the jth voxel and $D_{j,px}$ the prescribed dose for the jth voxel:

$$\text{Objective function} = W\Sigma_j \left(D_{j,cal} - D_{j,px}\right)^2 \tag{9.10}$$

The treatment planning software automates the iterative inverse planning process successively adjusting possible beam parameters and recalculating the dose to minimise the overall objective functions for both the PTV and all other included structures. The software must also be sophisticated enough to guard against stopping too soon at an apparent best fit when local minima are found as this may not represent the best possible solution. Modern software will also enable the operator to stop and guide the iteration in a chosen direction.

9.4.8 VOLUMETRIC MODULATED ARC THERAPY

Volumetric modulated arc therapy (VMAT) combines the benefits of both arc therapy and IMRT by delivering the radiation beam as one continuous beam that is being modulated dynamically by moving the MLC whilst also rotating around the patient in an arc. This method delivers the resultant dose distribution without turning the beam off and therefore results in a quicker form of treatment delivery whilst still maintaining the benefits of IMRT (Figure 9.42).

9.4.9 PLAN EVALUATION

Once a treatment plan has been produced it needs to be evaluated to ensure it meets the desired requirements. There are several steps to evaluating a plan and the TPS provides the required tools to achieve this. A basic checklist would begin with reviewing isodose distributions on successive planned CT slices throughout the irradiated volume in order to verify:

- Dose uniformity across the planning target volume (PTV). For conformal radiotherapy, the ICRU 50 [6] recommendation is that this lies in the range 95%–107%, where the 100% dose point is defined at the isocentre which will typically be at the nominal centre of the target.
- Is the whole target covered to adequate dose (e.g. does the 95% isodose contour surround the PTV).
- Assess the doses to each OAR such as the maximum dose to serial structures.

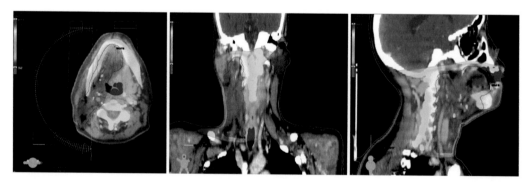

FIGURE 9.42 A VMAT radiotherapy treatment plan for a head and neck target volume. For ease of visualisation the dose distribution is displayed as a colour wash with red representing the high-dose region and blue the lower dose areas. The three displayed sections (transverse, coronal and sagittal) demonstrate the ability of the VMAT plan to conform to the PTV (red line) in 3D.

9.4.9.1 Dose Volume Histograms

A useful tool in evaluating treatment plans is to use dose volume histograms (DVH). The DVH for a given structure is a plot of radiation dose versus the percentage volume of the structure receiving at least that dose. It enables:

- A 3D dose distribution to be represented in simple graphical form.
- An indication of target coverage and doses to sensitive organs.
- Different plans to be evaluated on the same graph and the best one chosen.
- Quick visualisation of the maximum dose to serial organs
- Dose to parallel organs to be assessed by checking the volume of the organ receiving a specific dose, to ensure no constraints are exceeded that could lead to functional damage.

To properly calculate the DVH each structure of interest must be fully delineated on all slices to enable the computer to calculate the doses to all voxels within the structure. To enable this the CT scan range must also be of an appropriate length to include the whole structure. Figure 9.43a shows a representative DVH plot for a target PTV as well as serial and parallel OARs. The typical shape of the target DVH is characterised by most of the volume being covered by a dose close to that prescribed indicated by a steep fall in the curve in this region. For a serial organ, the maximum dose can easily be read off the dose axis where the volume of the structure receiving the given dose or greater reaches 0%. For a parallel organ, the volume (V) of the structure receiving at least a given dose (D_v) is also easily derived in order to assess sub functional damage. These metrics can be quickly compared with the corresponding curves from other plans to assess the relative merits of each plan. Figure 9.43b shows how a DVH is derived. A structures DVH, more correctly called the cumulative DVH, is derived by first calculating the dose to each voxel within that structure and binning (sorting) these doses into discrete intervals. This can be visualised more easily by plotting a differential dose volume histogram which is a histogram of the number of voxels receiving a dose in each of the defined dose intervals. The cumulative DVH is then derived from the differential DVH with, in this example, the total volume of the structure being ten voxels. All 10 voxels receive a dose greater or equal to 10 Gy and so the cumulative DVH is ten voxels (100% volume) at 10 Gy. One voxel receives a dose of between 10 and 15 Gy so that only nine voxels receive a dose of greater than or equal to 15 Gy, etc. It can be seen that the resolution of the curve is a function of the number of bins or dose intervals used in the initial calculation.

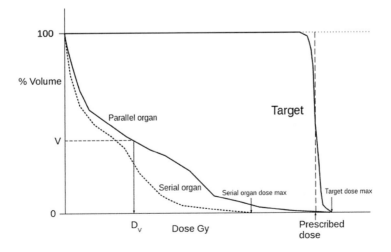

FIGURE 9.43 (a) Cumulative DVH curves for a PTV, serial and parallel structures. Figure 9.43b Derivation of a cumulative DVH

(*Continued*)

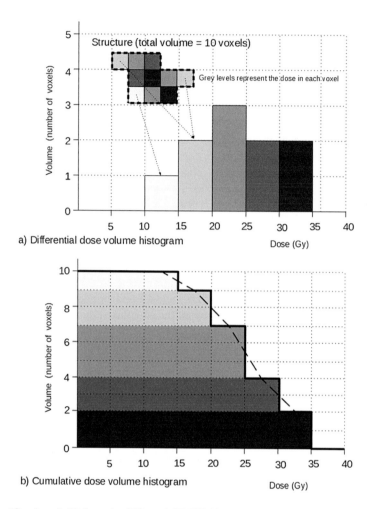

a) Differential dose volume histogram

b) Cumulative dose volume histogram

FIGURE 9.43   (Continued) (b) from the differential DVH (a).

9.4.10 PLAN CHECKING

Once the plan has been evaluated and approved for treatment by the clinician it will be prepared for treatment. This preparation stage will vary depending on individual department processes but it will always be subjected to a series of extensive independent checks. Checks may include such things as the patient demographic data (correct patient), histology results (confirmed diagnosis) all the way through to plan acceptability (e.g. the dose maximum for serial OARs has not been exceeded) and an independent check of the MU settings on the linac required to deliver the pre-scribed dose.

For complex IMRT treatment deliveries, it is normal for a delivery quality assurance check to be performed prior to the first day's treatment. This involves a simulated treatment being delivered on the linac to a measurement device set up on the treatment couch. The dose distribution measured by this device is compared with the expected pre-calculated dose based on a perfect delivery of the intended plan. Further investigation is required if the agreement between the expected and measured doses differ by more than a given tolerance otherwise treatment can commence. Reasons for a dis-crepancy range from incorrect calibration of linac dose delivery, MLC positioning errors, incorrect plan, plan export errors or incorrect plan set-up instructions.

9.5   TREATMENT TECHNIQUES

As previously outlined the fundamental aim of radiotherapy is to deliver an accurate and precise radiation dose to the target volume while minimising the radiation dose to any surrounding and uninvolved normal tissues. Many external beam radiotherapy treatment techniques are available to do this for a variety of treatment sites. The most commonly used treatment modalities are kV X-rays generated by a kilovoltage unit, and MV photons or electrons generated from a linac. Other modalities such as protons, neutrons and heavy ions are limited to more specialist facilities and are not considered further in this section (Table 9.5).

9.5.1   Kilovoltage X-ray Treatment

kV treatments are delivered by kilovoltage units (described in Section 9.2.2) providing X-rays in the energy range 50–300 kV. These units deliver simple, single field treatments to lesions on the patient's skin surface or superficial layers. The lesions are normally clearly visible on the patient's skin surface and therefore the correct field size and treatment margin are easily determined. The higher the X-ray energy, the greater the depth of penetration of the X-rays in tissue, but kV X-rays are limited to penetration depths within the first few centimetres of tissue. kV X-rays are readily attenuated by thin pieces of lead, typically 1 mm per 100 kV will reduce the incident intensity by approximately 99%. Irregular shaped treatment fields can therefore be cut from lead sheets of appropriate thickness and placed onto the skin surface, beneath the chosen applicator, to shape the field.

9.5.2   Megavoltage Photon Treatment

MV photon treatments are delivered by a linac, providing photons in the energy range of typically 4–18 MV (normally 6–10 MV). Treatment methods were introduced in the preceding sections of this chapter and include, in order of increasing complexity, simple single field, parallel-opposed field, conformal, IMRT and VMAT treatment techniques. IMRT and VMAT are now the most commonly used treatment techniques for radical radiotherapy in many centres. VMAT utilises complex technology to deliver radiation dose to the target volume, with beam shaping and beam modulation achieved with dynamic MLC, while the radiation beam is rotated around the patient's body. Accuracy and precision is paramount to achieve tumour control and reduce normal tissue complications, but as already described geometric uncertainty must be accounted for.

TABLE 9.5
Choice of Treatment Type

Treatment Site	Options	Comments
Within skin or superficial layers	kV X-rays or low energy electrons	Simple, single field treatments, with beam shaping using applicators and lead cut-outs.
First few centimetres within the body	Electrons	Simple, single field treatments, with beam shaping using an electron applicator attached to the treatment head and shaped lead or metal alloy inserts.
At depth within the body	MV photons	Multiple field (conformal and IMRT) treatments, beam shaping with MLC and beam modulation with wedges or MLC; or arc (VMAT) treatments, beam shaping and modulation with MLC.
Whole body	MV photons	Patient at extended treatment distance.
Whole body (skin)	Electrons	Multiple field treatments at extended distance.

9.5.2.1 Image-guided Radiotherapy

The pre-treatment CT scan acquired for the purposes of treatment planning may have been acquired up to two weeks before treatment commences. In this time the patient's internal anatomy is likely to have changed, and it can also change between successive fractions of radiotherapy treatment (inter-fractional motion). An example of this would be movement of a prostate target volume due to changes in rectal and bladder filling. There are interventional techniques that can be applied to reduce this inter-fractional motion, such as treating a patient with a full bladder, providing the patient with a pre-treatment enema, etc., but these techniques will only reduce, and not eradicate, inter-fractional motion. A method of ensuring geometric precision of treatment delivery is image-guided radiotherapy (IGRT). IGRT is used to quantify and correct the geometric position of internal anatomy to ensure accurate radiation beam delivery to an inter-fractionally moving target volume. IGRT can be achieved in a variety of ways, primarily using either 2D planar or 3D volumetric X-ray imaging, described further in Section 9.6.2.

9.5.2.2 Respiratory-gated Treatment

As well as inter-fractional motion of organs between successive fractions there is also intra-fractional motion which describes the motion of organs during the actual treatment delivery. One example of this would be the continual movement of a lung target volume due to respiration. A method of ensuring improved geometric precision of treatment delivery for intra-fractional motion is via respiratory-gating. Respiratory-gated treatment techniques interlock the radiation beam so that it can only be delivered at known reproducible points in the breathing cycle. These methods are covered in more detail in Section 9.6.1.2.

9.5.2.3 Stereotactic Radiotherapy Techniques

Stereotactic ablative radiotherapy (SABR) is a specialist technique to deliver very high doses of radiation to often very small target volumes. Typical sites that receive SABR are targets in the brain, lung, liver and spine. Because a very high radiation dose per fraction is delivered to the patient, immobilisation is paramount to achieve target coverage while sparing OAR and minimising uncertainty in treatment delivery. SABR can be delivered through conformal, IMRT or VMAT treatment techniques. An important area of stereotactic radiotherapy is stereotactic radiosurgery which involves single high-dose intra-cranial treatments as described in Section 9.3.5.

9.5.2.4 Adaptive Treatment Techniques

The extended duration of fractionated radiotherapy treatment over a period of several weeks means that the target volume and its surrounding structures may gradually change over time. Planning margins applied to the target volume, and treatment verification imaging will be able to account for and mitigate some of these changes but there may come a point where the target volume and surrounding structures have changed to such an extent that the treatment is compromised, for example by partially missing a target volume or overdosing an OAR that has moved closer to the target volume. These changes can be caused by a change in patient weight, or deformation of internal anatomy due to tumour shrinkage, or other shape changes. In such cases, the treatment can be adapted to account for this change, by using adaptive radiotherapy techniques. One way of performing adaptive radiotherapy to account for these changes would be to repeat the pre-treatment imaging and treatment planning process for the new patient anatomy. This, however, takes considerable time to achieve and the adapted re-plan would not be able to be delivered immediately, meaning treatment would remain compromised until the new treatment plan was ready. An alternative way of adapting radiotherapy which is used for treatment sites that can vary considerably at each treatment, such as the bladder, would be to produce a number of treatment plans for various bladder sizes at the start of radiotherapy treatment and then treat every fraction of radiotherapy with the most appropriate treatment plan for the bladder size during that fraction. The ideal scenario of course would be to rapidly adapt the plan at each fraction based on verification images acquired prior to treatment. This scenario is only just being introduced in radiotherapy, and is currently too time consuming and therefore impractical for most treatment sites. Developments introduced on the latest equipment, including the MR-linac described in Section 9.2.2.3, aim to integrate a daily adaptive workflow as standard.

9.5.3 ELECTRON TREATMENTS

Electron treatments are delivered by linacs (Section 9.2.3) providing MeV electrons in an energy range of typically 4–20 MeV. Electrons, like kV treatment techniques, are normally delivered as single field treatments to lesions on the skin surface up to several centimetres in depth. Radiation beam shaping is achieved by the use of patient-specific custom made lead or metal alloy inserts. The higher the electron energy, the greater the depth of penetration of the electrons in tissue. Electrons exhibit a rapid dose fall-off beyond their useful effective treatment depth which minimises dose to underlying tissue. A tissue equivalent material, known as bolus, can be added on to the patient's skin surface in order to modify the depth of dose deposition in the patient. This can be used to increase the dose at the skin surface and also to modify the depth of penetration to that required (Figure 9.44).

9.5.4 WHOLE BODY TREATMENTS

9.5.4.1 Total Body Photon Irradiation

Total body irradiation (TBI) is a technique for delivering radiotherapy treatment to the entire body; it is primarily used as a treatment for patients in preparation for bone marrow transplantation. Bone marrow transplantation is part of the treatment given for various types of leukaemia, malignant lymphoma and aplastic anaemia. Typically TBI is used to eradicate any remaining tumour cells, not targeted by other treatments such as chemotherapy, and to lower the immune system of the patient to prevent rejection of the transplanted bone marrow. Radiotherapy can be delivered in a variety of ways, the most common treatment technique being to treat at an extended distance with parallel-opposed treatment fields encompassing the entire body. In this setup, the patient is positioned up to 500 cm away from the target with the largest field size set on the linac enabling the resultant

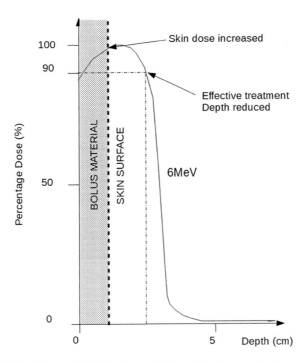

FIGURE 9.44 Use of bolus (tissue mimicking) material to modify the effective treatment depth of electrons.

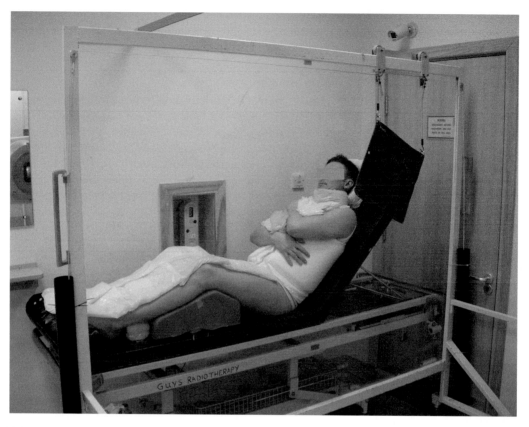

FIGURE 9.45 TBI treatment technique. Patient at an extended distance and treated with two lateral fields. Patient and couch rotated between this (left lateral) field and the right lateral field. Water equivalent bolus material is placed between the legs and around the chest and a thin lead sheet suspended off the screen at the level of the head to improve the overall dose homogeneity across the length of the patient.

divergent field to fully encompass the patient's entire body. There are other treatment methods available including those that utilise matching many static radiation beams along the length of the patient's body or the patient being moved through the radiation beam using an automated couch. The primary goal for all techniques is to irradiate the entire body to a high uniform dose of radiation. Typically the highest photon energy radiation beam available on the linac will be used to maximise beam penetration. For the technique shown in Figure 9.45 a perspex screen is placed in the path of the MV photon beam. This screen generates additional low energy electrons which are absorbed at superficial depths removing the skin-sparing effect and ensuring an homogenous dose throughout the entire body.

9.5.4.2 Total Skin Electron Beam Therapy

Total skin electron beam therapy (TSEBT) is a technique for delivering radiotherapy to the entire patient's skin surface, typically used for patients with the condition of mycosis fungoides (MF), a skin lymphoma. The incidence of MF is quite low and the technique for treating it is involved and therefore this technique is only delivered in a small number of radiotherapy departments. Low energy electrons are the most appropriate treatment modality for TSEBT. Like TBI the aim of TSEBT is to treat the entire patient, but as the electrons are far less penetrating in tissue than photon beams they are ideally suited in treating the skin and superficial layers. Organs at depth in the patient's body are spared by using electrons; however, some shielding is still required to shield sensitive structures

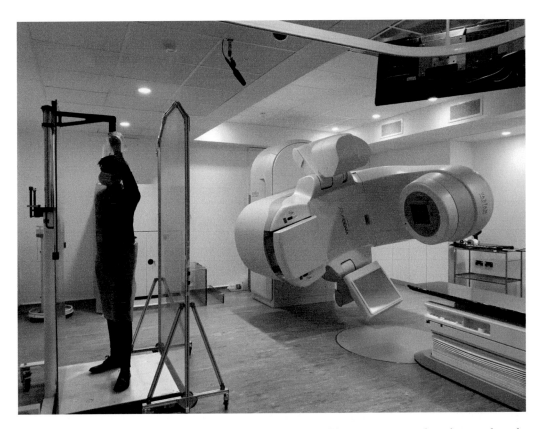

FIGURE 9.46 TSEBT treatment set-up. This technique uses a thin perspex screen degrader to reduce the electron beam energy appropriate to that required for superficial skin treatment. The picture shows one of the six different treatment positions. To ensure complete radiation coverage from foot to head two beams are delivered in each position, one with the treatment head pointing down, as pictured, and the other pointing up. A stand is used to help set up and provide support for the patient during treatment. Actual treatment is delivered with the patient skin uncovered.

on the surface such as the patient's eyes. TSEBT is routinely delivered at an extended treatment distance, like TBI, but a parallel-opposed treatment field arrangement is not suitable to treat the patient's entire skin surface so TSEBT is treated with a multiple field arrangement. The patient typically stands in a range of different treatment positions in front of the linac to try and encompass the entire skin surface, reduce any skin folds and minimise regions of possible overdose due to overlapping beams. Even with this approach not all of the skin can be irradiated and additional top-up fields are required to a range of sites such as the soles of the feet and perineum. Because the treatment is at an extended distance the linac has to be set into a high-dose rate electron mode otherwise the fall of in electron dose at distance would mean treatment times would be too long. Additional care and QC by the physicist is required to ensure that the machine operates correctly in this mode as well as being returned to the normal dose mode following TSEBT (Figure 9.46).

9.6 IMAGING IN EXTERNAL BEAM RADIOTHERAPY

Imaging in radiotherapy can broadly be divided into that required for diagnosis, localisation (treatment planning) and treatment verification purposes. Diagnosis involves the full range of standard imaging modalities such as CT, MR and PET to determine whether disease is present and the extent of the disease spread (staging). Imaging for treatment planning follows positive diagnosis and concerns accurate localisation of the disease to be treated as well as other critical structures to be considered and ideally avoided when planning the treatment. Imaging for treatment verification concerns imaging the patient immediately prior to, or during, treatment delivery to ensure they are in the intended position. Verification imaging in radiotherapy can take many forms that may include 2D planar or 3D volumetric imaging, surface body-mapping techniques, implanted fiducial markers, ultrasound or implanted radio frequency transponders.

9.6.1 Imaging Modalities for Treatment Planning

9.6.1.1 kV CT Simulation

CT is ideally suited to radiotherapy treatment simulation for a number of reasons.

- The kV energies used in CT produce images with adequate soft-tissue differentiation.
- The images have high spatial resolution, contrast to noise resolution and high uniformity, allowing accurate delineation of anatomical treatment planning structures.
- The images have high spatial geometrical accuracy in all dimensions.
- The images can be converted to electron-density distributions required for treatment dose calculation.

A CT simulator is essentially a conventional CT scanner with some additional modifications. These include:

- Wide bore aperture to enable the patient to be set up and imaged in the treatment position with reduced likelihood of collision between the patient or immobilisation equipment and the CT scanner.
- Flat couch top. Conventional CT scanner couches are curved to maximise the imaging field of view. The use of a flat couch top mimics the treatment couch on the linear accelerator and ensures the patient shape and distortion of internal anatomy remains the same. The flat couch top will also have appropriate fixing points to attach patient immobilisation devices in the same relative positions as on the treatment couch.
- External lasers to set up and align the patient accurately and reproducibly against external reference marks. Typically these take the form of a right and left lateral as well as a sagittal laser that also projects a longitudinal line or cross onto the patient from a ceiling mounted position.

Acquired images are exported, normally in the Digital Imaging and Communications in Medicine (DICOM) standard format [5], to the treatment planning system for target and organ delineation by the radiation oncologist. Treatment-site specific CT scanning protocols are designed such that image quality is optimised according to that particular treatment site. The scanning protocol settings will differ, for example, between thoracic and pelvic scanning due to the difference in effective tissue path length through which the kV photons must traverse from kV tube to detector. Variables of use when optimising protocols include kV, mAs, pitch and image reconstruction technique. These settings must be optimised and justified against the imaging dose received by the patient, ensuring doses are kept as low as reasonably practicable (ALARP). The medical physicist must be aware that in some instances large increases in dose may result in small clinically insignificant increases in image quality and therefore not appropriate. This is equally true of cone-beam CT discussed later in the section on treatment verification.

9.6.1.2 Accounting for Motion – 4DCT

Although the patient immobilisation is effective, some treatment sites will be subject to periodic motion. Organs in the thoracic and abdominal region are likely to be affected by respiratory motion. Locally, motion can be either eliminated, reduced or taken into account. Respiratory motion can be eliminated by treating in either end expiration or end inspiration breath-hold. In order to do this, the

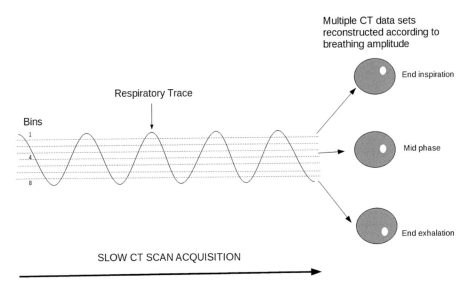

FIGURE 9.47   4DCT scanning process. Multiple 3D CT datasets reconstructed from different points within the breathing cycle.

CT scan must also be acquired in the corresponding breath-hold stage. Alternatively, motion can be quantified by acquiring a 4-dimensional CT (4DCT) scan. This is a slow scan taken over multiple breathing cycles such that the data can be binned (sorted) into discrete parts of the breathing cycle. A respiratory signal is generated from a surrogate for breathing motion, typically using a mouth-piece to measure flow of air to/from the lungs (spirometry) or tracking abdominal/chest movement. This signal is used to separate the CT information into the discrete bins according to the different segments of the breathing cycle from end expiration through to end inspiration and back to end expiration. The data within each binned segment is reconstructed into an image of the patient at that point within the breathing cycle. The resultant 4DCT typically consists of 8–12 separate CT scans each representing a snapshot of the internal anatomy captured at each part of the breathing cycle (Figure 9.47).

Once the motion of the target and surrounding tissues has been acquired in this form a range of planning and treatment options become available. By stepping through the CT scans for each successive phase the full extent of any target motion can be quantified. This can be used to ensure that the CTV-PTV treatment margin described previously is sufficiently large enough to encompass all organ motion during breathing. It should be noted that a conventional single (non-4D) CT scan provides a single snapshot of the patient at a random point within their breathing cycle and from this the full extent of organ motion cannot be fully assessed. Alternatively, the treatment plan can be generated from a single CT scan representing just one discrete phase of the breathing cycle (e.g. end expiration). In such cases, the treatment must also be delivered with the patient's breathing cycle in the corresponding phase as that in which it was planned. In a similar way to the original 4DCT acquisition a breathing surrogate must be used to track the patient's breathing cycle and ensure delivery at the correct moment. This type of treatment delivery is called 'gated' treatment as the radiation is only delivered within an allowed interval where the tumour is at or near to the expected position. If the treatment machine and imaging combination allow, an alternative is to dynamically track the tumour such that the beam is continuously moved in synchrony with any tumour movement. This for example can be achieved with the cyberknife system which continuously images implanted fiducial markers acting as target surrogates. The position of the robotic delivery arm of the cyberknife system is then continuously adjusted to match any movement of the fiducials. Gating or tracking offer the best treatment option in terms of reducing the amount of normal tissue being irradiated

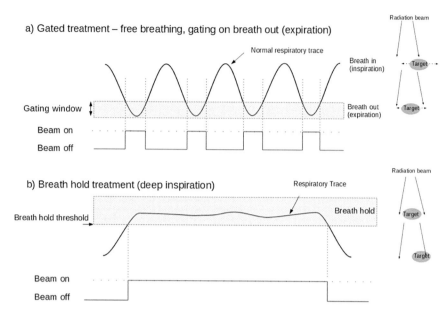

FIGURE 9.48 Methods of treating moving targets. (a) Gated treatment – beam is only turned on when the amplitude of the respiratory trace falls within the defined region. (b) Breath hold treatment – such as deep inspiration breath hold enables the beam to remain on for as long as the patient can maintain breath hold.

but arguably carry an increased risk of geometric miss of the target if there is a mismatch between target motion and resultant positional correction of the delivered radiation beam. Other treatment methods not relying on gating or tracking involve treating the patient in breath hold, for example at deep inspiration breath hold. This relies on the patient holding their breath for an extended period, normally achievable for most patients with suitable coaching. Some form of automatic or manual system must be in place to interrupt the beam as soon as the patient moves from the breath hold position (Figure 9.48).

9.6.1.3   MR Scanning

Although CT currently predominates in radiotherapy, the use of other imaging modalities particularly MR is gradually being introduced. MR offers much improved soft-tissue contrast compared with CT and as such allows increased confidence when determining anatomical boundaries for healthy tissues, for the gross tumour volume and for regions likely to contain microscopic spread of disease. T1- and T2-weighted MR images are useful, for example, in neurological cancer to define the tumour boundaries within the brain. If the patient has had surgery, both pre- and post-operation MR scans can be used. It is often the pre-operative MR that is used to generate the GTV and CTV to ensure that the full site of original disease is irradiated. MR also offers sequences showing functional characteristics of the tissues. Certain imaging sequences such as diffusion-weighted and dynamic contrast-enhanced imaging can highlight the most aggressive or resistant sites of disease. This additional information assists in clinical decision-making and opens up the possibility of more personalised radiotherapy delivery with techniques such as targeted dose escalation where additional boost doses of radiation are applied to specific regions. An issue with MR only simulation is spatial geometric distortion, caused by the perturbation of the magnetic field. This effect tends to increase moving radially away from the centre of the imaging field of view. Another complication is that unlike CT the MR image pixel information has no simple relationship to electron density and therefore cannot be readily used by the treatment planning system to calculate the radiation dose distribution. To increase the accuracy of delineation, MR images can be co-registered with the

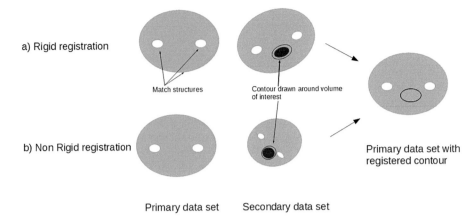

a) Rigid registration

Match structures

Contour drawn around volume
of interest

b) Non Rigid registration

Primary data set with
registered contour

Primary data set   Secondary data set

FIGURE 9.49 CT (primary) and MR (secondary) registration. (a) Rigid registration finds the best match by linear scaling, rotation and translation between data sets. (b) Non-rigid registration also enables non-linear deformation between datasets improving registration accuracy.

planning CT scan so that the defined areas of interest can be precisely transcribed onto the planning CT scan and then used for treatment planning. Rigid registrations permit the secondary dataset, the MR in this case, to be translated and rotated to best match the primary CT dataset. Rigid registration is often compromised because the CT and MR will have been captured at two different time points and often with the patient in a different position on the imaging couch. Head and neck patients, for example, are positioned at simulation planning CT on a flat, hard-top couch in an immobilisation shell that holds the chin up with the neck in an extended position. The neck flexion is therefore different from that on the MR where the patient is typically scanned in a neck-neutral position, on a diagnostic, curved couch. Shoulder positions will often differ due to arm placement, by the side or over the chest. In an effort to correct these external and internal anatomical differences, an alternative to rigid registration is deformable registration which allows non-linear deformation of the secondary dataset (Figure 9.49). Once the CT and MR have been registered and fused, contours drawn on the MR scans are again automatically transferred to the spatially equivalent positions on the CT dataset for treatment planning. MR only treatment simulation however is an active area of development and methods to minimise or correct for distortion and generate pseudo-CT data from MR using various methods including machine-learning approaches are well advanced.

9.6.1.4 PET and PET-CT Imaging

Positron emission tomography (PET) imaging is used to determine regions of increased uptake of the tracer isotope within the tumour and is used in staging how advanced the cancer is prior to treatment, how successful treatment has been in eradicating the tumour and also in determining possible dose escalation regions. Like MR, the PET image pixel information cannot be used directly to calculate radiation dose and PET imaging is not routinely registered with the CT data set. Combined PET-CT scanners enable registered PET and CT images to be acquired on the same machine allowing the possibility of using this CT data set directly for planning.

9.6.2 TREATMENT VERIFICATION IMAGING

9.6.2.1 Purpose

Immediately before treatment, it is necessary to confirm that the patient is set up correctly on the treatment couch and the radiation that is about to be delivered will be directed accurately onto the target as intended. Image-guided radiotherapy relies on accurate verification imaging and this can take several forms depending on equipment type and age. The general requirements of the verification process are as follows:

1. A reference image must be available that represents the intended patient and target position relative to the radiation beam, normally defined by the treatment machine isocentre. This reference image is usually derived from the original CT planning scan.
2. The verification image must contain sufficient information to enable accurate determination of the target position relative to the treatment isocentre or other suitable reference positions.
3. It must be possible to compare the acquired verification image against the reference and determine necessary corrections to be applied to the patient position. This correction is called the set-up error and will consist of small translational and rotational differences.
4. To allow these positional corrections to be applied with accuracy and confidence in order to minimise any set-up error at treatment.

9.6.2.2 2D KV and MV Planar Imaging

2D planar imaging was originally performed using conventional silver halide film which was then developed in film processors. This has now been replaced with high-resolution 2D electronic detector panels mounted onto the gantry of the treatment equipment. For MV imaging the panel is mounted in line with the radiation beam and takes an image of the MV treatment beam exiting the patient. MV images do not provide good tissue contrast as the dominant radiation interaction in soft tissue at this energy is the Compton effect which is proportional to electron density. Most soft tissue has a similar electron density and therefore it is mainly differences in physical density such as between bone, lung, air ways, air and soft tissue that provide useful image contrast. More recently an additional kV X-ray tube and detector panel have been incorporated onto the linac gantry perpendicular to the treatment beam direction (Figure 9.6a and 9.6b). This produces high contrast kV images enabling more accurate determination of soft tissue and subsequent set-up error correction.

For 2D planar imaging, the reference image is computer generated from the planning CT data set and is called a digitally reconstructed radiograph (DRR). At the chosen gantry angle multiple ray line projections of the X-ray source are traced through the 3D CT dataset onto a 2D plane beyond the dataset representing the detector panel position. The ray path of each projection is summed through the traversed voxels, with the attenuation in each voxel represented by its electron density relative to water. This is then appropriately scaled and normalised to produce a grayscale image (Figure 9.50). The limiting resolution of the DRR is determined by the voxel resolution of the CT

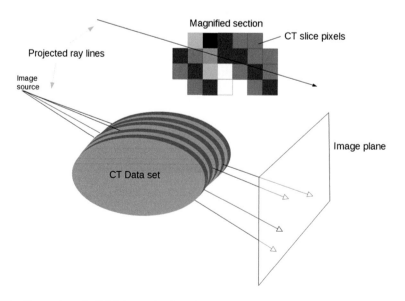

FIGURE 9.50 Generation of a DRR by projection of the X-ray beam through the CT data set.

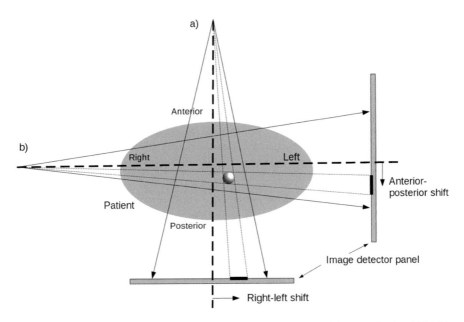

FIGURE 9.51 Orthogonal imaging to resolve set-up corrections in 3D. In this example, the shift of the sphere in each image plane is measured relative to the machine (and imaging system) isocentre position represented by the intersection of the two dashed lines.

dataset. An advantage of DRRs is that the 2D reference image can be constructed at any projection angle through the 3D CT dataset.

The detector panel provides rapid acquisition and display of the verification image and this is matched against the reference image using dedicated software to determine the set-up error and required corrections. The matching process may be automated or manually achieved with the final match confirmed by an experienced operator prior to positional correction. A single 2D planar image only provides information on the set-up error in orthogonal directions in the plane of the image. In order to discern set-up errors in all three orthogonal directions two images must be acquired at different angles, ideally 90° apart (Figure 9.51).

9.6.2.3 3D Volumetric Imaging

As well as providing 2D kV planar images the kV X-ray source and detector panel are able to produce 3D volumetric cone-beam CT (CBCT) images. Multiple 2D kV image projections are acquired as the gantry is slowly rotated once around the patient. The resultant image projections are reconstructed to produce a 3D volumetric image with good resolution, contrast and geometric accuracy in all directions. This 3D data set can be viewed and matched directly slice by slice against the reference planning CT data set. CBCT is the current standard for image guidance in radiotherapy.

9.6.2.4 Non-ionising Radiation Methods

2D planar and CBCT imaging use ionising radiation to image the target either directly or indirectly via surrogates such as small metallic fiducial markers implanted within the tumour or bony anatomy in close proximity to the tumour. Treatment verification methods are possible using non-ionising radiation such as MR imaging, optical body surface mapping (surface guided radiotherapy), ultrasound imaging, tracking of implanted radiofrequency transponders and infrared body marker tracking. Ultrasound and MR image the anatomy directly whereas the other methods indirectly infer position of the tumour based on measurements such as the body surface. MR-linacs are gradually being introduced into clinical use and these machines provide all the benefits of MR imaging, such as superior soft-tissue contrast, to the treatment verification process.

9.6.3   CONCOMITANT IMAGING DOSE

Although radiotherapy delivers a high dose of radiation to the treatment target, it is still important where possible to minimise any additional dose to normal tissue outside the target which may be included within the imaging fields. The majority of treatment verification uses X-ray imaging in the form of 2D planar or 3D kV CBCT and this delivers an additional dose called the concomitant imaging dose to the patient. There is always a balance to ensure that accurate treatment verification is achieved but to minimise any unnecessary additional concomitant imaging dose that will increase the probability of radiation-induced secondary cancers. The key advantage of non-ionising verification imaging such as MR, ultrasound and optical surface mapping techniques is that this is not a constraint.

9.7   REFERENCES AND FURTHER INFORMATION RESOURCES

REFERENCES

[1] *Handbook of Radiotherapy Physics: Theory & Practice*. Editors: P. Mayles, A. Nahum & J.C. Rosenwald. CRC Press, Taylor & Francis, Boca Raton. 2007. (Two volume 2nd edition in press).

[2] *Khan's The Physics of Radiation Therapy*. Editor: J. P. Gibbons. Wolters Kluwer Health; Wolters Kluwer (India) Pvt. Ltd., New Delhi. 2019. 598. 6th edition. ISBN/ISSN 9781496397522.

[3] ICRU (International Commission on Radiation Units and Measurements). *Prescribing, recording and reporting photon beam therapy*. Report No. 50, ICRU, Bethesda, MD. 1993.

[4] ICRU (International Commission on Radiation Units and Measurements). *Prescribing, recording and reporting photon beam therapy* (supplement to ICRU report 50). Report No.62, ICRU, Bethesda, MD. 1999.

[5] Digital Imaging and Communications in Medicine (DICOM) standard. https://www.dicomstandard.org/

FURTHER RESOURCES

In addition to the references above some further references and resources are provided below:

EMITEL, E-Encyclopaedia of Medical Physics and Dictionary. http://www.emitel2.eu/emitwwwsql/encyclopedia.aspx

IPEM (Institute of Physics and Engineering in Medicine, UK), Reports, books and educational material. www.ipem.ac.uk.

BIR (British institute of Radiology), Reports, books and educational material. www.bir.org.uk

ESTRO (European Society for Radiotherapy and Oncology). https://www.estro.org/

AAPM (American Association of Physicists in Medicine). https://www.aapm.org/ Task Group report series (free to access). https://www.aapm.org/pubs/reports/

IAEA (International atomic energy agency, Vienna), Publications, including technical report series. www.iaea.org/publications.

Attix, F. H., 2004, *Introduction to Radiological Physics and Radiation Dosimetry*, ISBN: 9780471011460 |Online ISBN: 9783527617135 |doi:10.1002/9783527617135. Wiley-VCH Verlag GmbH & Co.

Webb, Steve, 2005, *Contemporary IMRT – Developing Physics and Clinical Implementation*, 1st edition, e-book published 2019. CRC Press, Boca Raton. eBook ISBN: 9780429144066. doi:10.1201/9780429144066

ICRU (International Commission on Radiation Units and Measurements), 2010, Prescribing, recording and reporting photon beam therapy – beam intensity modulated radiotherapy (IMRT), Report No.83, Journal of the ICRU, Vol. 10, Oxford University Press, Oxford, UK. doi:10.1093/jicru/10.1.Report83.

ICRU (International Commission on Radiation Units and Measurements), 2004, Prescribing, recording and reporting electron beam therapy, Report No.71, Journal of the ICRU Vol. 4. Oxford University Press, Oxford, UK.

10 Brachytherapy

Mauro Carrara
Fondazione IRCCS Istituto Nazionale dei Tumori, Milano, Italy

Francesco Ziglio
Ospedale Santa Chiara, Trento, Italy

CONTENTS

DOI: 10.1201/9780429155758-10

10.1 INTRODUCTION

Brachytherapy (BT) is a form of radiotherapy that uses one or more sealed radioactive sources placed inside or close to the tumour to be treated. This treatment modality is usually more invasive compared to external beam radiotherapy (EBRT), but in many cases it allows better shaping of dose to the target. Dose gradients across the tumour are steeper than in EBRT partly due to the lower energies but primarily due to the proximity of the source to the irradiated tissues which leads to a rapid fall-off in dose, as the distance from the source increases, due to the inverse square law effect. Even though it can be a labour-intensive technique requiring specific experience and training compared to EBRT, for some cases it has shown its superiority with respect to other currently available radiotherapy treatment modalities [1–4]. Since the radioactive sources are placed in close contact with the site to be treated or are remotely driven into the required location from outside the patient, brachytherapy is only appropriate for some accessible anatomical sites [5]. The main anatomical sites that can potentially be treated are shown in Figure 10.1.

In general, the different categories of brachytherapy applications can be identified as follows:

- *Intracavitary brachytherapy:* The sources are placed into body cavities close to the target volume (e.g. cervix).
- *Interstitial brachytherapy:* The sources are placed directly (or by means of catheters) within the tissue to be irradiated (e.g. prostate).
- *Intraluminal brachytherapy:* The sources are placed by means of catheters in the lumen to be irradiated (e.g. oesophagus).
- *Surface brachytherapy:* The sources are placed over the tissue to be treated (e.g. skin).

Treatments can be also classified with regard to the source dose rate. In particular, as defined by the ICRU [6], *high-dose rate (HDR)* brachytherapy delivers a dose to the dose specification point

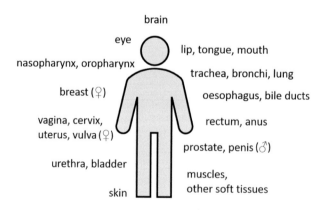

FIGURE 10.1   Anatomical regions that can in principle be treated with brachytherapy.

TABLE 10.1
List of Different Treatment Categories Available in Brachytherapy

Classification	Possible Categories
Type of implant	Intracavitary, interstitial, intraluminal, surface
Dose rate	Low-dose rate (LDR), medium dose rate (MDR), high-dose rate (HDR)
Application duration	Temporary, permanent
Average effective energy	Low energy (LE), high energy (HE)

with a rate that is higher than 12 Gy/h. This value represents a lower limit, and in practice treatments are typically delivered with a dose rate that is more than one order of magnitude higher. In *low-dose rate (LDR)* brachytherapy, the dose rate to the dose specification point is between 0.4 and 2 Gy/h. *Medium dose rate (MDR)* brachytherapy can in principle be defined with a dose rate between 2 and 12 Gy/h, but is not in common use nowadays [7].

With regard to the duration of the source application, both *temporary* and *permanent* brachytherapy treatments exist. In the first case, sources are placed inside the patient for a specified treatment time whereas in the second case, sources are permanently implanted so they completely decay within the patient. HDR brachytherapy applications are always temporary, since they deliver a very high dose in a reduced amount of time; LDR can instead be both.

A further classification can be provided in terms of the average effective energy of the emitted radiation, with an arbitrary threshold defined at 50 keV. Low energy (LE) sources are considered those with an average effective energy below 50 keV [8], high energy sources (HE) higher than 50 keV [9]. LDR treatments are nowadays performed, with a few exceptions, with LE sources, whereas HDR treatments are administered with HE sources (Table 10.1).

This chapter introduces these concepts in more detail, focusing on the different delivery systems and their applications as well as the physical properties of the main radioisotopes in use. A section is dedicated to the important area of radioactive source dosimetry, describing the distinction between HDR and LDR sources. The principles of dose calculation according to the currently recommended formalism are introduced and, finally, the main steps involved in treatment planning are described. References are given throughout the text to provide more reading around each of the described topics.

10.2   DELIVERY SYSTEMS AND APPLICATIONS

10.2.1   BRIEF HISTORICAL BACKGROUND

The history of brachytherapy dates back to the beginning of the 20th century, when the first clinical experiences were documented [10–12]. Up to the 1960s brachytherapy was usually performed by inserting the radioactive source(s) manually into the patient for a pre-calculated amount of time. Ra-226 was the most popular radioactive source at that time, and was manufactured in the form of tubes, needles or small capsules [13, 14]. Even though Ra-226 sources were generally LDR emitters, this manual procedure caused radiation exposure to the clinical staff performing the manual insertion. In particular, the hands of the radiation oncologist performing the implant could receive a very high dose. To minimise exposure from the radioactive sources, the *afterloading* technique was introduced into clinical practice in the 1960s [15–17]. With this technique, empty applicators, catheters or plastic tubes are first inserted or implanted in the patient and subsequently loaded with one or more radioactive sources. In *manual afterloading*, the radioactive sources are fed manually into the previously implanted empty applicators/catheters, substantially reducing exposure of the operating room staff involved in the procedure. In *remote afterloading*, one or more sources are inserted into applicators/catheters by means of a remotely controlled system, further minimising radiation exposure of operators who can remain safely outside the treatment room while the source is driven into position. Since in remote afterloading the radioactive sources are machine driven and no manipulation by the clinical staff in standard conditions is expected, this technology is the one that prevailed with time [16].

Different remote afterloading technologies are available, including machines working with pneumatic handling of multiple LDR sources (i.e. source trains) of similar activity. Thanks to remote afterloading, high-dose rate (HDR) sources could also be introduced, allowing fractionated temporary treatments lasting only a few minutes [17]. Nowadays, the most widespread technology for brachytherapy delivery is the single stepping-source cable-driven HDR afterloader with other afterloading machine technologies now out of production. LDR source afterloaders (e.g. using Cs-137) are no longer in routine clinical use, due to the following advantages of HDR vs LDR afterloading techniques:

- Since HDR treatments are quick, administration of treatments on an outpatient basis is possible (improved patient experience).
- The number of treated patients per week can be increased (higher productivity for the hospital).
- Only one (reusable) source is used to administer the entire treatment (easier for dose calculation and radiation protection purposes).
- By stepping the single source into different discrete dwell positions during treatment the optimised dose distribution can be delivered after implant of the applicator/catheters (increased treatment flexibility for the clinical staff).

The main LDR treatments still routinely delivered are for prostate and ophthalmic eye plaque brachytherapy.

10.2.2   STEPPING-SOURCE **HDR** REMOTE AFTERLOADERS

In single stepping-source HDR remote afterloaders, one sealed radioactive source is welded to the end of a steel cable (Figure 10.2). When the machine is not in use, the source is fully retracted inside a shielded container that protects staff entering the brachytherapy bunker (shielded treatment room) from exposure. In the treatment phase, staff exit the bunker and the source is driven out into the applicator/catheters, inserted within the patient, and after treatment retracted by means of a stepping motor, remotely controlled via the treatment console. Within or in close proximity to the region to

FIGURE 10.2   Picture of a sealed Ir-192 radioactive source welded to the end of the steel drive cable.

be treated, the source steps inside each implanted applicator/catheter in turn, through a series of predefined *dwell positions* with predefined *dwell times* at each of these positions. The source dwell positions and times are calculated in the treatment planning phase and usually transferred to the treatment console via the hospital network.

The standard commercially available HDR radioisotopes are Ir-192 and Co-60, with Ir-192 being the most frequently used worldwide [9]. A more detailed description of the HDR brachytherapy sources and their physical properties is provided in the next section. Typical positional and time accuracies of the currently available machines are $< \pm 1$ mm and $< \pm 1$ s, respectively. Outside the region to be treated, the source moves at a high speed (i.e. several tens of cm per second) to minimise unnecessary irradiation of the patient.

In order not to enter the treatment room and connect the afterloader to each individual catheter separately, the afterloader has a series of different exit channels that allow simultaneous connection of the machine to the different applicator channels and/or catheters implanted in the patient. During treatment, the channels are sequentially selected by the afterloader control system and the source driven down the selected channel in turn. Each channel connection is by means of dedicated flexible *transfer tubes*, and the part of the machine that allows connection of these tubes to the different exit channels is usually called the *indexer*. The machine also contains a *check cable* with a dummy source that is connected to a handling system that is almost identical to that of the radioactive source. Every time a treatment is started, the check cable is run through the same transfer tubes and catheters in order to verify the absence of potential obstructions or connection issues, before delivery of the radioactive source is allowed. A schematic representation of the interior of an Ir-192 afterloader is provided in Figure 10.3.

At the treatment console, audible and visual signals indicate the treatment status and an additional radiation monitor connected to the machine provides an independent check of the source in/out of safe position. The patient in the treatment room is also monitored by means of an audio-visual communication system and treatment can be interrupted at any time in case of emergency.

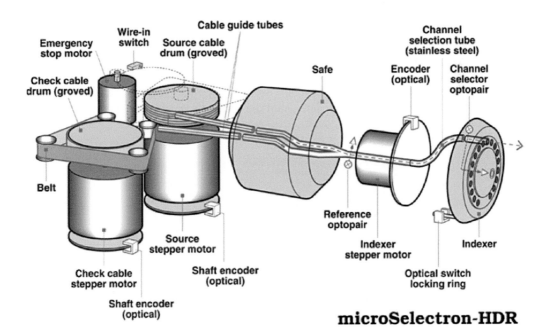

FIGURE 10.3 Schematic representation of the main parts of the microSelectron-HDR Ir-192 afterloader.

(Image courtesy of Elekta.)

Moreover, the system is provided with backup batteries that allow rapid retraction of the source back into the shielded container in case of failure of the mains electrical power supply. Several further interlocks which will stop the source being driven out of the safe position, or automatically retract the source (e.g. door opened, system malfunctioning, high friction torque along the source path, improper connection of transfer tubes) guarantee a safe administration of the treatment with proper radiation protection of patient and staff.

10.2.3 APPLICATORS, NEEDLES AND CATHETERS FOR HDR BRACHYTHERAPY

Applicators, needles and catheters are placed into the tumour or close to it and are used in HDR temporary brachytherapy to contain the radioactive source in position during the treatment. As outlined above these are connected to the afterloader by means of transfer tubes that allow the source to be driven from the afterloader to the programmed dwell positions and back.

Numerous devices are available, with the most appropriate one chosen in relation to the anatomical tumour site and size. Devices should have the following characteristics:

- They should be mechanically stable.
- Their location should be reproducible to allow reliable planning of the source positions.
- Reusable applicators should be easy to sterilise.
- Disposable devices such as catheters and needles should not be expensive.

Hollow needles and catheters are used for interstitial applications [18, 19]. These are implanted into the tissue to be irradiated and then removed following treatment. In some fractionated applications they are left in place until the end of the treatment course and used for more than one irradiation, in other cases a new implant is performed for each single application. Needles are manufactured both in plastic or metal whereas catheters are usually available in plastic, since they have to be partially flexible to adapt to the patient anatomy. Lengths can be adjusted depending on the treated anatomical site. For some applications, for instance, prostate or partial breast implants, templates exist that help place the needles/catheters at predefined positions and with defined separations from each other (Figure 10.4).

In intracavitary brachytherapy, a large variety of applicators with different designs and sizes is available, in particular for gynaecological brachytherapy [20, 21]. Standard vaginal applicators for treatment of the vagina are of a simple design. These consist of cylinders of different diameters and lengths with a catheter hole along the central axis through which the source can be driven (Figure 10.5).

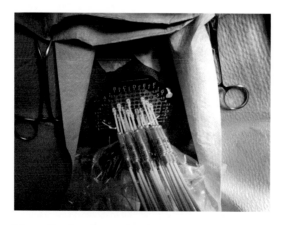

FIGURE 10.4   Example of the application of a template for prostate brachytherapy during an interstitial implant.

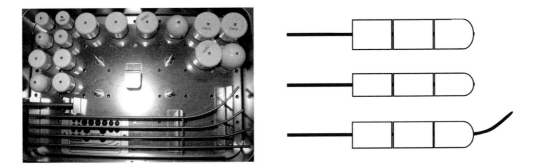

FIGURE 10.5   Vaginal applicator set with a range of different cylinder diameters and central tube shapes available for assembly. On the right, examples of some constructed vaginal applicators are shown.

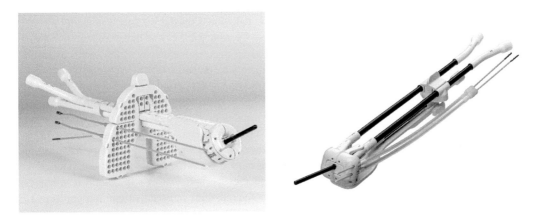

FIGURE 10.6   Examples of two MRI compatible uterovaginal applicators. Both applicators have one intrauterine and two lateral catheters. They also enable interstitial implantation of needles to achieve more tailored dose distributions.

(Images courtesy of Elekta.)

This design enables a symmetrical dose distribution around the central axis. Cylinders with additional peripheral catheters or designed with integrated interior shields are also available to produce more tailored dose distributions, which are not symmetrical around the central axis.

Uterovaginal applicators are designed for the treatment of cervical cancer, with an intrauterine tube of chosen curvature and length inserted into the uterus combined with one ring or two ovoids that are placed into the vaginal fornices. The most modern applicators can be made of materials that are MRI compatible and/or have slots to place additional interstitial needles to allow combined intracavitary/interstitial applications (Figure 10.6) [20–22].

Applicators are available for other treatment sites such as for intraluminal brachytherapy, in particular for the trachea, the oesophagus and the bronchus [23], as well as for the treatment of the skin with superficial brachytherapy.

10.2.4   LDR Applications

LDR brachytherapy implies a much longer treatment time, compared to HDR applications, with the upper limit given by a permanent interstitial implantation of radioactive sources called 'seeds' for the treatment of localised prostate cancer [24, 25]. This is an effective and common treatment, with tens of thousands of patients treated worldwide to date. A multidisciplinary team, typically composed of a radiation oncologist, a urologist and a clinically qualified medical physicist (along with the anaesthetist and theatre team) is required to perform the implantation under general anaesthesia in theatre. The current technique is based on transrectal ultrasound imaging, with the ultrasound probe fixed within a stepper device to the operating table [18, 26]. Transverse ultrasound images are acquired and transferred to a dedicated treatment planning system, where the doctor defines the target volume (the prostate gland) and the organs at risk (urethra and anterior rectal wall), and the clinically qualified medical physicist determines the position of the seeds. The most widely used radionuclides are Iodine-125, Palladium-103 and, more recently, Caesium-131 (Figure 10.7). A more exhaustive description of LDR brachytherapy sources and their physical properties is provided in the next section. The seeds are placed within the prostate gland and in the immediate (few millimetres) surrounding tissue through sharp thin needles, which are passed through the perineum, according to the treatment plan.

A process called post planning is routinely performed after the implantation as part of a comprehensive quality assurance programme that also includes quality control checks on the equipment, software and unused seeds. On a CT (or MR) scan the implanted seeds are identified, relevant structures are outlined, and the actual dose distribution is calculated on the treatment planning system. The dose distribution represented as isodoses (see also Section 9.3.2.1) and dose-volume histograms (Section 9.4.9.1) relates to the integral dose, delivered up to complete decay of the seeds, assuming that the structures and the seed positions were and will be the same during the weeks and months after the procedure. In the rare case of a sub-optimal dose distribution (i.e. under dosed region of the prostate gland), a corrective implant to improve the dosimetric distribution can be performed. It is very important to maintain the quality and accuracy of the implant procedures at a high level since a strong correlation has been found between calculated dosimetric parameters and oncologic outcome [27].

Eye plaque treatment is another clinical application of LDR brachytherapy. Depending on the size and location of the lesion it can be a treatment choice for retinoblastomas, melanoma of the choroid or of the conjunctiva [28]. Conservation of the eye and vision is maintained whenever achievable. Ophthalmic brachytherapy is usually delivered with radioactive gamma-emitting LE iodine-125 seeds temporary loaded onto plaques (Figure 10.8a) or with silver plaques containing a beta-emitting Rh/Ru-106 radioisotope layer (Figure 10.8b). To adapt as much as possible to the site and location of the tumour, seed positions and strengths can be adjusted in the first case, whereas appropriate plaque diameter and shape (circular convex shape with or without notch) can be chosen in the second case from those available in the clinic [29].

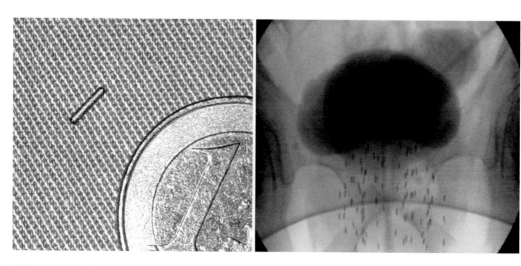

FIGURE 10.7 Example of a seed used for LDR brachytherapy (left image). On the right is a radiological image of a prostate implant showing the numerous implanted seeds in position.

(a) (b)

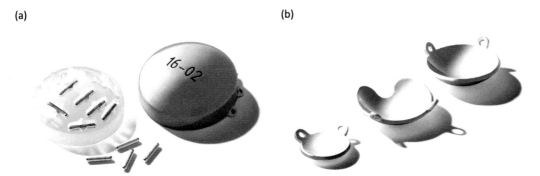

FIGURE 10.8 (a) Example of a LE gamma-emitting eye plaque with seeds, a silicone insert and the gold shield that has to be placed over the seeds after construction; (b) beta-emitting plaques of different dimensions and shapes.

(Images courtesy of Eckert & Ziegler BEBIG GmbH.)

To administer the therapy, the plaque is surgically sutured to the eye sclera over the lesion and left there for the time needed (typically several days) to deliver the required dose.

10.3 RADIOACTIVITY AND DEFINITIONS

Since brachytherapy is based on the emission of photons or particles following radioactive decay of the source, it is important to provide some physical definitions and theoretical descriptions related to radioactivity, which are relevant to clinical practice [30].

Activity A: is the rate of decay of a radionuclide. According to the SI it is measured in decays/s, with 1 decay/s =1 Bq (becquerel). Conversion from Bq to Ci (curie), another unit in widespread use for many decades, is $3.7 \ 10^{10}$ Bq = 1 Ci. The activity A at any time t for a radionuclide varies exponentially with time and can be calculated as follows:

$$A(t) = A(t_0)e^{-\lambda(t-t_0)} \tag{10.1}$$

where $A(t_0)$ is the known activity at a specific time point t_0 and λ is the *decay constant*. The exponential alone results in a value in the range 0–1 for t later than t_0, and can be defined separately as the *decay factor F*

$$F(t) = e^{-\lambda(t-t_0)} \tag{10.2}$$

The decay constant is related to the *half-life $T_{1/2}$* of the source, which is the time required for one-half of the atoms of the radioactive material to decay, according to:

$$\lambda = \frac{\ln 2}{T_{1/2}} \tag{10.3}$$

Combining (10.1) and (10.3), if $t-t_0$ is substituted with $T_{1/2}$, the resulting decay factor is 0.5, as expected, indicating that the activity has halved.

Apparent activity A_{app}: for the sealed sources used in brachytherapy, the resulting radioactive emissions outside the source are filtered by the source shielding and are therefore different from those of the raw radioactive material. To clarify this difference, the apparent activity is defined as the activity at 1 m distance from an unshielded source that would result in the same exposure rate as the filtered source. The apparent activity of a source is important because it is correlated with its rate of emitted radiation (note: one decay doesn't necessarily mean one emitted photon or particle) and therefore with its strength. The dose rate delivered to the patient is therefore related to the apparent activity of the source.

Specific activity A_{specific}: is the activity of a radionuclide per unit of mass

$$A_{\text{specific}}(t) = \frac{A(t)}{m} \tag{10.4}$$

A_{specific} usually has units of Bq g^{-1}, GBq mg^{-1} or Ci kg^{-1}. It is important because it describes the strength of a specific radionuclide source. The higher the specific activity, the smaller is the required mass of radioactive material to obtain the same activity.

Air kerma rate constant Γ_δ: is the air kerma rate $\dot{K}_\delta$ due to photons of energy greater than δ from a point source with activity A, at a distance r in a vacuum, corrected by the inverse square law and normalised by its activity

$$\Gamma_\delta = \frac{r^2 \dot{K}_\delta}{A} \tag{10.5}$$

Γ_δ is characteristic to each radionuclide and is usually measured in Gy s^{-1} m^2 Bq^{-1} or in µGy h^{-1} m^2 MBq^{-1}. Similar to A_{specific}, Γ_δ also describes the strength of a radionuclide to be used in brachytherapy. The higher the air kerma rate constant, the smaller the required activity to obtain the same air kerma rate at a specific distance.

10.4   RADIONUCLIDES FOR HDR BRACHYTHERAPY

An ideal source for use in HDR brachytherapy should be:

- A pure gamma emitter, since beta or alpha particles have short ranges and only deliver unwanted high doses to the small volumes of tissues close to the source, providing no therapeutic benefit.
- A high energy gamma emitter (i.e. a photon-emitting source with average photon energy exceeding 50 keV): energy should be high enough to have an adequate range in the tissues being treated but not too high to cause problems with organs at risk sparing and radioprotection requirements.
- Produced from a stable isotope.
- Have a long half-life: the source should maintain its strength for a long period of time, to allow numerous re-use over a period of time and reduce the frequency of source exchanges.
- Robust: the source is used many times and is subjected to mechanical stress, it should be particularly robust to avoid any possible break with a consequent risk of radioactive contamination.
- Small: a reduced source dimension allows the use of catheters and applicators with adequate dimensions to be inserted/implanted in the patient without being excessively invasive.

Currently, the two radionuclides that are closest to the requirements for an ideal HDR source and that are commercially available are Ir-192 (Iridium) and Co-60 (Cobalt) [31]. A brief description of the physical characteristics of these radionuclides is provided below and summarised in Table 10.2. A detailed description of the recommended dosimetry datasets for each available source model is available in the literature [32–35].

TABLE 10.2

Main Physical Characteristics of the Radionuclides Ir-192 and Co-60 Used for HDR Brachytherapy

Element	Atomic Number (Z)	Mass Number (A)	Half Life $T_{1/2}$ (d)	Approximate Average Photon Energy (MeV)
Cobalt (Co)	27	60	1925	1.25
Iridium (Ir)	77	192	73.8	0.36

10.4.1 IRIDIUM-192

Ir-192 is produced by neutron activation of enriched Ir-191 targets. It has a half-life of 73.8 days which allows it to be used for temporary implants and to be replaced 3–4 times per year. Its most favourable characteristic is its extremely high specific activity, which makes it possible to produce sources of small dimensions with activities of up to 12 Ci. Typically, they have a cylindrical shape of about 1 mm diameter by 3.5 mm in length, including the thin titanium or stainless steel encapsulation material. This is laser welded to the end of a flexible steel cable which is used to accurately drive the source into position (Figure 10.2). The decay spectrum of Ir-192 is very complex, with an energy range of gamma emissions between 0.061 and 1.378 MeV and an average energy of about 0.36 MeV (calculated with a cut-off of $\delta = 10$ keV). Beta emissions, with maximum energy of 0.675 MeV are internally absorbed within the source and by its encapsulation.

10.4.2 COBALT-60

Co-60 is produced by neutron activation of Co-59. It has a half-life of 5.27 years which allows it to be replaced after several years of use. An afterloader source is usually guaranteed for a certain number of transitions (i.e. tens of thousands) where each transition corresponds to an episode where the source is driven out and then back into the safe position. Each of these episodes subjects the source and cable to some stress and therefore a recommended number of transitions is prescribed to ensure integrity and guard against mechanical failure. In a hospital with a consistent workload a source exchange may be required if the maximum number of transitions is reached rather than due to the diminished activity of the source. Since the specific activity of Co-60 is significantly smaller than Ir-192 (i.e. about 8 times), Co-60 sources with dimensions similar to those of Ir-192 can be produced with activities of up to only a few Ci, thus with a reduced strength compared to Ir-192. Of note, even though the specific activity is smaller than Ir-192, the efficacy of this radionuclide in terms of achievable resulting dose rate to the patient is partially increased by its higher air kerma rate constant, which is about three times that of Ir-192. Gamma emissions are mainly at 1.17 MeV and 1.33 MeV energy, with an average value of 1.25 MeV. Beta emissions, with maximum energy of 0.318 MeV are absorbed by the source core and its encapsulation material, as for Ir-192.

10.5 GAMMA-EMITTING RADIONUCLIDES FOR LDR BRACHYTHERAPY

The ideal source for use in LDR brachytherapy has some characteristics that differ from those of the ideal HDR sources. Previously gamma-emitting LDR sources (such as Caesium-137) were also used for temporary gynaecological applications, whereas nowadays gamma-emitting LDR seeds are mainly used for permanent prostate implants [36] or temporary eye-plaque applications. The ideal gamma-emitting LDR source should be:

- A pure gamma emitter: since beta or alpha particles have short ranges and provide high doses only to small volumes of tissues close to the source (as for the ideal HDR source).

- A low energy gamma emitter: since the implants are for longer treatment times, photon energy should be low to avoid the need for demanding radioprotection measures. For LDR prostate implants, high energy photon emitters could in principle be ideal candidates, but in this case the half-life should be short enough to allow the source strength to decay rapidly to a suitable value, with a consequent fast reduction of the radiation exposure outside the patient. This would then enable the patient to return home without having to stay too long in a protected environment in the hospital.
- Small and robust: a reduced dimension allows multiple seeds to be implanted in the patient without being excessively invasive.

A description of the physical characteristics of each available isotope and source model can be found in the literature [32–35]. The source model describes the distribution of radiation around the specific source supplied by a manufacturer and includes the effect of the encapsulation material used to safely seal the source. A brief generic description of these radionuclides is provided below, with relevant physical characteristics given in Table 10.3 for I-125, Pd-103 and Cs-131 [31] (Table 10.3).

10.5.1 Iodine-125

Iodine-125 is produced in a nuclear reactor, through a neutron capture process, irradiating enriched Xenon-124 in gaseous form. It has a half-life of 59.5 days, which makes it suitable for permanent implants. Iodine-125 decays by electronic capture into an excited state of Tellurium-125, that decays to its ground state emitting 35.5 keV gamma rays; characteristic x-rays in the range 27–32 keV are emitted as a result of the electron capture and internal conversion processes.

Seeds used for permanent interstitial prostate brachytherapy are made of a substrate material, on which the liquid Iodine-125 is deposited, and surrounded by a welded titanium capsule, that absorbs the Auger electrons emitted. Typically the seeds are cylindrical, with an external diameter of 0.8 mm and a length of 4.5 mm (Figure 10.7)

10.5.2 Palladium-103

Palladium-103 can be produced in a nuclear reactor, when stable Pd-102 captures a thermal neutron, or in a cyclotron, by bombarding protons onto a rhodium target. It has a half-life of 17 days, which makes it suitable for permanent implants, and it decays via electronic capture with characteristic x-rays in the energy range of 20–23 keV and Auger electrons. Seeds are similar to those containing Iodine-125, with an x-ray marker material, a connecting substrate onto which the radioactive material is absorbed, and an external welded titanium tube to safely encapsulate the source. Palladium-103 was introduced around 1986 for prostate brachytherapy, with the idea that its initial dose rate, around 20 cGy/h at the prescription dose level, three times greater than that of Iodine-125,

---

TABLE 10.3

Main Physical Characteristics for the Radionuclides I-125, Pd-103 and Cs-131 Used in LDR Brachytherapy

Element	Atomic Number (Z)	Mass Number (A)	Half Life $T_{1/2}$ (d)	Approximate Average Photon Energy (keV)
Iodine (I)	53	125	59.5	28
Palladium (Pd)	46	103	17	21
Caesium (Cs)	55	131	9.7	30

might overcome sub-lethal damage repair and cell repopulation that could hinder the efficacy of Iodine-125. However, after three decades of use, no superiority of one radionuclide over the other has been demonstrated in terms of biochemical disease control.

10.5.3 CAESIUM-131

With a shorter half-life (9.7 days) and a higher average energy (30 keV) than Palladium-103, Caesium-131 was introduced into clinical practice for prostate brachytherapy around 2004. The long-term biochemical outcome for patients treated with this radionuclide is comparable to those treated with I-125 and Pd-103.

10.6 BETA-EMITTING RADIONUCLIDES FOR LDR BRACHYTHERAPY

Besides the low energy gamma-emitting I-125 and Pd-103 radionuclides, eye plaque brachytherapy can also be performed in some cases with high energy beta-emitting radionuclides [28, 37, 38]. In particular, their steep dose gradient is optimal for lesions up to 5 mm thick. Rhodium/Ruthenium-106 (Rh/Ru-106) is currently commercially available for this type of treatment.

10.6.1 RHODIUM/RUTHENIUM-106

Ru-106 is a pure beta emitter (E_{max} = 39.4 keV) produced in a nuclear reactor by nuclear fission. With a half-life of 371.5 days, Ru-106 decays to the ground state of Rh-106, which then decays to Pd-106 with a half-life of just 30.1 s (secular equilibrium between the two radioactive elements). In contrast to Ru-106, even though Rh-106 is also a pure beta emitter (E_{max} = 3.5 MeV), it decays to an excited state of Pd-106 which is then followed by emission of de-excitation gamma rays. The photon dose is therefore not negligible, but it is the emitted electrons that are relevant for therapeutic aims, since they can penetrate up to several mm in tissue (falling to around 10% of the prescribed surface dose at 5–6 mm depth).

10.7 SOURCE STRENGTH MEASUREMENT

10.7.1 USE OF DOSIMETRY SYSTEMS WITH TRACEABLE CALIBRATIONS

It is strongly recommended that the intensity of photon-emitting brachytherapy sources is measured by the clinically qualified medical physicist with a dedicated dosimeter, prior to their clinical use. The recommended dosimeter is a well-type ionisation chamber, connected to an electrometer, which together form the *dosimetric system*. The well-type ionisation chamber has a large sensitive volume (typically 200–300 cc) open to the atmosphere producing measurable ionisation currents even for low source activities [39]. In view of its design characteristics this dosimetric system can be used for HE and LE sources and both HDR or LDR. In order to provide an accurate measurement of source strength, it is important that the *dosimetric system* has a calibration coefficient from an appropriate accredited laboratory. Such a laboratory (i.e. Primary Standard Dosimetry Laboratory (PSDL), Secondary Standard Dosimetry Laboratory (SSDL), Accredited Dosimetry Calibration Laboratory (ADCL)) must provide instrument calibration traceable to a primary standard, specific to the radionuclide being measured [40–42]. Historically, source intensity was first defined and measured in terms of *mg-Radium Equivalent*, which was then replaced by apparent activity A_{app} (as previously defined). Nowadays the International Committee for Radiation Units (ICRU) recommends the use of the *Reference Air Kerma Rate* $\dot{K}_{RAKR}$ [30]. In some Countries (e.g. USA and Canada), the Air Kerma Strength S_K is used instead, according to local regulations [32–35]. A brief description of these quantities is provided below:

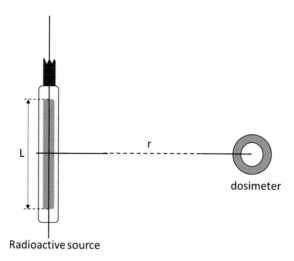

dosimeter

Radioactive source

FIGURE 10.9 Schematic of the experimental set-up for Reference Air Kerma Rate measurement. *L*: length of the active content of the encapsulated source; *r*: distance between the centre of the source and the centre of the dosimeter along the transverse line bisecting the longitudinal axis of the source.

mg-Radium Equivalent: Radium was the first radioisotope used extensively in brachytherapy treatments. This unit was used to allow conversion from the source strength of a given radionuclide to the strength of Radium, providing the equivalent mass of Radium required to produce the same exposure rate at 1 cm. Since Radium is no longer used in clinical practice, this unit is no longer used.

Reference Air Kerma Rate $\dot{K}_{RAKR}$ and *Air Kerma Strength* S_K: These are defined as the air kerma rate *in vacuo* (i.e. a correction is applied for air attenuation and scattering as well as for possible scattering from any nearby objects including walls, floors, and ceilings) due to photons of energy greater than a cut-off value δ, measured along the transverse line bisecting the centre of the longitudinal axis of the source. A schematic of the measurement set-up is provided in Figure 10.9.

The cut-off value δ is chosen to exclude low-energy or contaminant photons that increase the air kerma rate without contributing significantly to the dose at distances greater than 1 mm in tissue. For example, δ is typically 10 keV for HDR brachytherapy sources. Furthermore, the measurement distance r is not uniquely defined but can be any distance that is significantly larger than the maximum linear dimension of the radioactivity distribution (i.e. $r > 10\,L$). The only difference between $\dot{K}_{RAKR}$ and S_K is that the air kerma rate measured at r is corrected for $\dot{K}_{RAKR}$ according to the inverse square law normalised to a reference distance of $r_0 = 1$ m from the source centre, according to:

$$\dot{K}_{RAKR} = \dot{K}_\delta\left(r\right)\left(\frac{r}{r_0}\right)^2 \tag{10.6}$$

Whereas for S_K the air kerma rate is only corrected for the inverse square law:

$$S_K = \dot{K}_\delta\left(r\right)r^2 \tag{10.7}$$

Since $r_0 = 1$ m, the two physical quantities are numerically equal, but differ only in their units, which are typically µGy h⁻¹ and µGy h⁻¹ m² for $\dot{K}_{RAKR}$ and S_K, respectively. The recommended unit for S_K is denoted by the symbol U where 1 U = µGy h⁻¹ m² = cGy h⁻¹ cm².

10.7.2 MEASUREMENT OF THE HDR SOURCE STRENGTH

The measurement of the source strength is performed in the treatment room, with the well-type ionisation chamber connected with an extension cable to the electrometer, which is usually kept outside the bunker. A specific insert is placed in the chamber and connected to the afterloader by means of a dedicated transfer tube (Figure 10.10). From the treatment console, the source is driven towards a specific dwell position inside the insert. This position, called the 'sweet spot', is the source position where the chamber has the highest sensitivity. Several repeat measurements of the ionisation current are performed and converted to a measure of the source intensity using the calibration coefficient provided by the calibration laboratory for the dosimetric system. Since this design of well-type chamber is open to air, the actual mass of air in the chamber and hence the ionisation current produced differs slightly according to the local air temperature and atmospheric pressure. Temperature and pressure are therefore measured and a small correction is applied to the performed measurement to convert to the 'true' air density at which the calibration coefficient is valid. Other influence quantities might also be taken into account to derive the final value. The measured source strength is then compared to the one given in the source certificate provided by the vendor. If data are not in accordance further investigations are required before using the source in clinical practice [43].

FIGURE 10.10 Well-type ionisation chamber and source insert used for strength measurement of the stepping source in HDR brachytherapy.

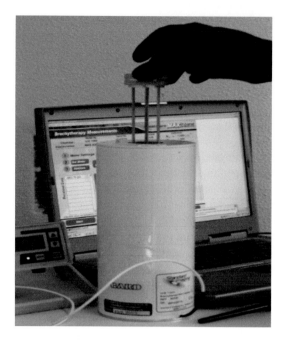

FIGURE 10.11 Strength measurement of LDR brachytherapy seeds with a dedicated well-type ionisation chamber.

10.7.3 MEASUREMENT OF THE LDR SEED STRENGTH

In a prostate brachytherapy procedure the number of seeds implanted is defined according to the treatment planning procedure, and is related to seed source strength, volume of the gland (and, to a lesser extent, by its shape) to be treated and any additional margins required by the clinical situation. In most cases, this number lies between 40 and 100. It is recommended that the source strength of 10% of the seeds are assayed (measured) prior to use, whenever possible. This measurement can be performed a few minutes before implantation, in a sterile environment, where no seed is wasted. Single seeds, or strands consisting of several connected seeds, are inserted into a dedicated holder inside the well-type ionisation chamber connected to the electrometer (Figure 10.11).

The manufacturer provides a calibration certificate, with minimum, maximum and midpoint value of the source strength for the delivered seeds. The midpoint value will be used in the treatment planning system, with the implicit hypothesis that all the seeds have the same source strength. This is an accurate approximation, since the uncertainty in source strength is ±5% within the same batch of seeds. If a discrepancy greater than 3% is measured, further investigation is suggested (i.e. measure another strand, restart the system and repeat the original measurement) [44]. If it is greater than 5%, and 100% of the seeds have been measured, the manufacturer should be contacted in order to find a possible reason (incorrect batch coding or packaging, etc.) so that the appropriate source strength value can be used for planning [42].

10.8 PRINCIPLES OF DOSE DISTRIBUTION CALCULATION

10.8.1 THE AAPM TG-43 ALGORITHM

In 1995, Task Group no. 43 of the American Association of Physicists in Medicine (AAPM) introduced a two-dimensional brachytherapy dosimetry formalism, usually referred in the literature as the TG-43 algorithm. This is currently considered the standard for the calculation of dose distributions around the common designs of cylindrically shaped radioactive sources, both in LDR and HDR brachytherapy [32–35]. The algorithm is valid under the following assumptions:

- Ideal point or line source (i.e. homogeneous and finite linear distribution of radioactivity along the source core).
- Cylindrical symmetry of the source around its longitudinal axis.
- Homogeneous water medium surrounds the source, neglecting patient inhomogeneities as well as any possible inter-source or applicator attenuation effects.
- Finite patient dimensions are ignored.

Even if in some specific clinical applications these approximations might have an impact that is not negligible (e.g. breast implants, some eye-plaque and other superficial treatments, prostate LDR seed implants) [11, 45], for the majority of treated regions in standard conditions and with unshielded applicators, the TG-43 algorithm currently provides the best compromise between calculation accuracy and algorithm sophistication. As shown in Figure 10.12, the chosen coordinate system for the TG-43 algorithm is polar, with the origin placed at the centre of the source and the z-axis oriented along the longitudinal axis of the source. Since the resulting dose distribution is symmetrical, the two-dimensional plane of calculation resulting from the chosen coordinate system can be in any plane running through the z-axis (i.e. any azimuthal angle), providing a three-dimensional dose distribution around the source. The reference point $P(r_0,\theta_0)$ is defined at 1 cm distance from the centre of the source (i.e. $r_0 = 1$ cm) along its transverse bisector (i.e. $\theta_0 = 90°$).

The AAPM TG-43 formalism describes the dose rate (D) at any point (r, θ) around the source according to the general equation:

$$\dot{D}\left(r,\theta\right) = S_K \cdot \Lambda \cdot \frac{G_X\left(r,\theta\right)}{G_X\left(r_0,\theta_0\right)} \cdot g_X\left(r\right) \cdot F\left(r,\theta\right) \tag{10.8}$$

where S_k is the air kerma strength of the source, Λ is the dose rate constant, G_X is the geometry function, g_x is the radial dose function and F is the anisotropy function. The subscript x is substituted with L or P, respectively, depending on whether the ideal line or point source approximation is chosen. Consensus data on the TG-43 parameters for the different source models can be found in the literature [32–35]. This data was either calculated using Monte Carlo simulations or measured in water around the source.

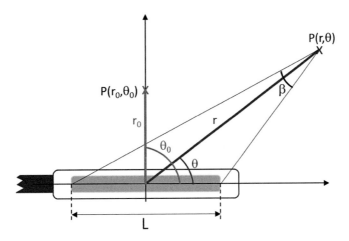

FIGURE 10.12 The polar coordinate system chosen to define the AAPM TG-43 algorithm. The active part of the source is in grey (length L) and is surrounded by its encapsulation.

In general, the first part of the equation:

$$\dot{D}(r_0,\theta_0) = S_K \cdot \Lambda \tag{10.9}$$

gives the dose rate at the reference point $P(r_0, \theta_0)$, and the application of the second part of the equation to $\dot{D}(r_0,\theta_0)$, in equation 10.8, calculates the dose rate to all remaining points $P(r, \theta)$ in water:

$$\dot{D}(r,\theta) = \dot{D}(r_0,\theta_0) \cdot \frac{G_X(r,\theta)}{G_X(r_0,\theta_0)} \cdot g_X(r) \cdot F(r,\theta) \tag{10.10}$$

In the case of multiple dwell positions of the same source (i.e. typically HDR) or multiple sources (i.e. typically LDR), the delivered dose to any point of interest is the superposition of dose distributions to the same point calculated for the source dwelling in each planned dwell position or for each implanted source, respectively.

10.8.1.1   Dose Rate Constant Λ

The dose rate constant Λ is defined as follows:

$$\Lambda = \frac{\dot{D}(r_0,\theta_0)}{S_K} \tag{10.11}$$

and is measured in cGy h^{-1} U^{-1}. Λ is defined for each specific source model of a radioisotope and converts the measured S_K value *in vacuo* at 1 m distance along the axis bisecting the source to the dose rate in water at $P(r_0,\theta_0)$.

Example of the application of Λ: What is the dose delivered in water by an Ir-192 source, see Figure 10.2, with a measured air kerma strength of 40 mGy m^2 h^{-1} at point $P(r_0 = 1$ cm, $\theta_0 = 90°)$ in 1 minute?

Calculation:

Taking $\Lambda = 1.106$ cGy h^{-1} U^{-1} gives:

$$D(r_0,;\theta_0,;1 \text{ min}) = 1.106\,\text{cGy h}^{-1}\text{U}^{-1} * 40000\,\text{cGy cm}^2\text{h}^{-1} * 1 \text{ min} = 442.4\,\text{Gy h}^{-1} * 1/60\,\text{h}$$
$$= 7.37\,\text{Gy}$$

10.8.1.2   Geometry Function G_X

The geometry function $G_X(r,\theta)$ takes into account the inverse square law fall-off in dose corrected for the spatial distribution of radioactivity within the active core of the source when the line-source approximation is considered. Geometry functions for point-source and line-source approximations can be calculated as follows:

$$G_P(r,\theta) = r^{-2} \text{ point source approximation} \tag{10.12}$$

$$G_L(r,\theta) = \begin{cases} \dfrac{\beta}{Lr\sin\theta} & \text{if } \theta \neq 0° \\[2mm] \left(r^2 - \dfrac{L^2}{4}\right)^{-1} & \text{if } \theta = 0° \end{cases} \text{ line source approximation} \tag{10.13}$$

where β is the angle, in radians, subtended by the active length of the source at the point $P(r,\theta)$. $G_X(r,\theta)$ is a pure geometric factor and doesn't consider any interaction of the radiation with the traversed water medium, which is explicitly taken into account in $g_X(r)$. In the TG-43 algorithm, $G_X(r,\theta)$ is further divided by $G_X(r_0,\theta_0)$ to obtain unity at the reference point $P(r_0,\theta_0)$.

10.8.1.3   Radial Dose Function g_X

The radial dose function $g_X(r)$ is defined as follows:

$$g_X\left(r\right)=\frac{\dot{D}\left(r,\theta_0\right)}{\dot{D}\left(r_0,\theta_0\right)}\frac{G_X\left(r_0,\theta_0\right)}{G_X\left(r,\theta_0\right)} \tag{10.14}$$

It takes into account the dose rate change along the transverse axis, due to the interaction of the photons with the traversed water medium (i.e. single scattering, multiple scattering and absorption). The geometrical fall-off effects of the dose are not considered because they are modelled with $G_X(r,\theta)$. g_X is a dimensionless quantity normalised to a value of 1 at r_0. (i.e. $g_X(r_0) = 1$)

An example of the g_L values for the source models of 4 different radionuclides is shown in Figure 10.13. The plots show a much sharper fall-off in g_L for low energy sources (i.e. I-125 and Pd-103) than for the high energy sources (i.e. Ir-192 and Co-60). Energy absorption is much greater at lower energies due to the overwhelming contribution at these energies of the photoelectric effect. Moreover, since the photoelectric effect is very sensitive to small energy variations at these low energies, even small energy differences (i.e. between 21 keV for Pd-103 and 28 keV for I-125) have a significant impact on the fall-off of g_L with distance from the source, with higher energy radiation being less attenuated. At much higher energies such as those of Ir-192 and Co-60, scattering tends to compensate absorption. Even if average energies of the emitted radiation are significantly different, since the photoelectric effect is not predominant at these energies, g_L values remain similar. Figure 10.14 provides an insight into the different contributions to the total radial dose in water from an ideal Ir-192 point-source, calculated by Monte Carlo simulation and multiplied by the square of the radial distance to remove the inverse square law contribution. It shows the dose from the primary photons is exponentially attenuated with depth, whereas the dose contribution from the total scattered radiation component increases with depth. At a depth of approximately 6.5 cm, the contribution to the total dose from scattered photons is higher than the contribution from primary photons.

When scattered, photons lose energy to the surrounding medium so have a lower energy than the primary unscattered photons. It is important to note therefore that the mean energy of the total photon distribution decreases with increasing distance from the Ir-192 source (i.e. the beam is said to 'soften' with depth).

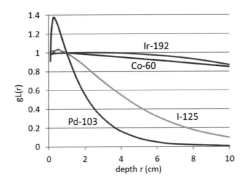

FIGURE 10.13.   Examples of $g_L(r)$ values for different source models for 4 radionuclides.

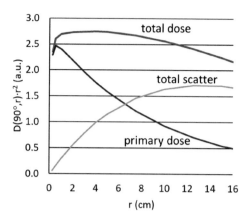

FIGURE 10.14 Simulated contribution to the total radial dose (corrected for the inverse square law) by primary and total scattered photons for an Ir-192 source. Data were taken from 'The CLRP TG-43 Parameter Database for Brachytherapy' by R. E. P. Taylor and D. W. O. Rogers of Carleton University, Canada (https://physics.carleton.ca/clrp/seed_database). Scatter inside the source is not considered in the calculation. Photons escaping the source encapsulation are all considered as primary radiation.

10.8.1.4 The Two-dimensional Anisotropy Function *F*

The anisotropy function $F(r,\vartheta)$ is defined as follows:

$$F\left(r,\theta\right)=\frac{\dot{D}\left(r,\theta\right)}{\dot{D}\left(r,\theta_0\right)}\frac{G_L\left(r,\theta_0\right)}{G_L\left(r,\theta\right)} \tag{10.15}$$

It describes the variation in dose as a function of θ relative to the transverse axis (i.e. $\theta_0 = 90°$) where it is normalised to unity at any radial distance from the source (i.e. $F(r,\theta_0) = 1$). The angular variation is mainly due to interactions of photons produced by the radioisotope within the source itself and with the steel cable to which it is welded, i.e. self-filtration, oblique filtration of primary photons through the cable and source encapsulation and also scattering.

Examples of F values at three different depths for HDR source models of Ir-192 and Co-60 are provided in Figure 10.15. Along the transversal axis (i.e. $\theta_0 = 90°$), F equals 1 as defined for all distances and sources. Approaching the source extremities ($\theta = 0°$ and $\theta = 180°$), F decreases much more for Ir-192 than Co-60 [46]. Since the average radiation energy is lower for Ir-192, radiation interactions within the source material and its encapsulation are more likely, with a consequent decrease in the fluence of primary radiation exiting the source (Figure 10.16).

10.8.1.5 The One-dimensional Anisotropy Function φ

In some cases, for instance, prostate implants with LDR seeds, it is difficult to determine the orientation of each single seed with respect to any point P where the dose is to be calculated. A more accurate equation in the proximity of the seed (distance of $P < 1$ cm) than just approximating each and every seed to a point source, is the simplified line source equation:

$$\dot{D}\left(r\right)=S_K\cdot\Lambda\cdot\frac{G_L\left(r,\theta_0\right)}{G_L\left(r_0,\theta_0\right)}\cdot g_L\left(r\right)\cdot\phi_{an}\left(r\right)\text{valid for any }\theta \tag{10.16}$$

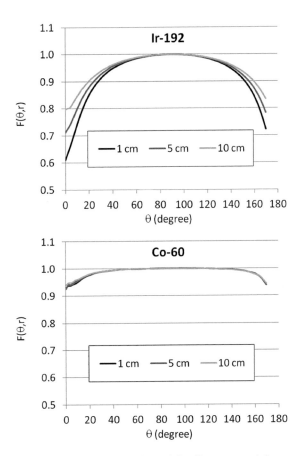

FIGURE 10.15   Examples of $F(r,\theta)$ values for Ir-192 and Co-60 source models.

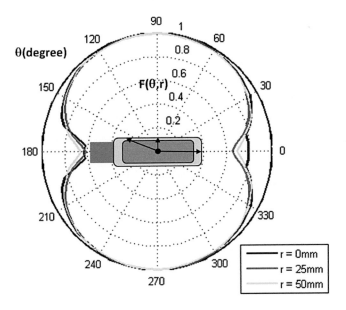

FIGURE 10.16   Anisotropy function of an Ir-192 source in polar coordinates, with a schematic representation of the source and its orientation provided in the middle of the plot. The black arrows represent possible exit paths for a photon emitted from the centre of the source.

φ_{an} is a one-dimensional anisotropy function which is calculated as the average of F over the 4π geometry. In this way, F is approximated by a simpler function which depends only on the distance r of the point P from the centre of the source. Also, G_L is limited only along the source transversal axis θ_0.

10.8.1.6 Comparison of Dose Distributions Arising from Co-60 and Ir-192

At this stage, it might be interesting to compare dose distributions achievable with an Ir-192 and a Co-60 HDR source according to the AAPM TG-43 algorithm (i.e. dose to water in water) in a clinical setting. As an example, a ring applicator with a diameter of 26 mm and with a 40 mm long intrauterine tube was chosen. The achieved percentage isodoses on the para-coronal plane oriented along the length of the intrauterine tube, normalised to the same point at a specific distance from the applicator, are shown in Figure 10.17 for comparison purposes. Despite a significant difference in terms of average gamma energies between Co-60 and Ir-192, the achieved isodoses are very similar. In fact, for these two radionuclides, the impact of the geometry function G in the TG-43 algorithm widely prevails over the small differences between their radial dose and anisotropy functions (see Figure 10.13) at angles not too close to 0° and 180° (see Figure 10.14). The only significant difference in the isodoses is at the top of the applicator and, less pronounced, at the bottom, where the difference between anisotropy functions F of Ir-192 and Co-60 at angles close to 0° and 180° is relevant. In fact, the more energetic the source emission, and therefore the more penetrating the radiation, the closer F will become to unity at all angles.

Although the isodose distributions are well matched, as previously described, there will be differences in treatment times to deliver the same dose due to differences in the physical characteristics of the two sources in terms of air kerma rate constant and specific activity.

10.8.2 Short Overview of Model-based Dose Calculation Algorithms

To overcome some of the limitations of the TG-43 algorithm, new model-based dose calculation algorithms have been introduced in recent years to effectively handle tissue inhomogeneities [11, 45]. In some specific clinical applications (e.g. breast implants, eye-plaque and other superficial

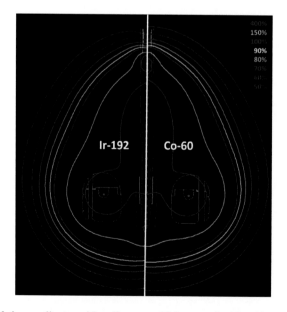

FIGURE 10.17   Virtual ring applicator with a diameter of 26 mm and with a 40 mm length intrauterine tube viewed on the para-coronal plane oriented along the length of the intrauterine tube. Comparison of isodoses achievable with Ir-192 (left side) and Co-60 (right side) according to the AAPM TG-43 algorithm.

treatments) it has been demonstrated that these algorithms can improve dose calculation accuracy. Methods that are commercially available at the present time are based on Monte Carlo simulations, on the solution of the linear Boltzmann transport equation (grid-based Boltzmann solvers) or on point kernel superposition [47–52]. Introduction has been possible because of improvements in computer power and because in brachytherapy the volumes of interest requiring accurate dose calculation are relatively small, and therefore less computer intensive, compared to those in external beam radiotherapy. However, improved dose calculation accuracy is based on additional requirements, such as:

- Quantification of the applicator composition.
- Characterisation of patient tissues and boundaries on the patient images.
- High computer processing power in order to deliver a solution in a feasible time considering the clinical and practical needs.
- More complex commissioning and quality assurance processes.

Currently, the TG-43 algorithm provides the best compromise between calculation accuracy and algorithm sophistication for the majority of treated regions in standard conditions and with unshielded applicators. Even though model-based dose calculation algorithms could improve accuracy in some treatment situations, they are currently not the standard due to their increased requirements in terms of time, cost and complexity. However, in light of their intrinsic flexibility to account for more complex physical situations, these new algorithms are likely to be used more in the future to enable innovative and more sophisticated treatments.

10.9 TREATMENT PLANNING IN BRACHYTHERAPY

10.9.1 INTRODUCTION

As outlined, brachytherapy covers a range of different treatment methods, according to technique and site being treated, and the approach to planning these therefore differs. In general, with stepping sources (i.e. typically HDR) the main objective of treatment planning is to identify the optimal source dwell positions and times to deliver a dose that is as close as possible to that prescribed [11]. This differs from temporary LDR brachytherapy, where sources of usually comparable activities are placed and removed together, with no chance to locally modulate (adjust) the dwell time of a single source. In permanent LDR brachytherapy, sources stay in place for their entire (radioactive) lifetime and the only degrees of freedom are represented by the choice of their positions. Modulation options for the source positions are also different between the different brachytherapy techniques. In Table 10.4, a generic overview of these available options is presented.

TABLE 10.4

Differences between Brachytherapy Modalities in Terms of the Degrees of Freedom for Source Positioning and Dwell Time

Parameter	HDR	LDR Temporary	LDR Permanent
Source dwell times	Modulable, each single stepped position can be different	Modulable, has to be the same for all sources	Not modulable
Source positions	Modulable, with the source constrained within the applicator/catheter	Partially modulable, usually constrained to an applicator	Modulable, sources not constrained to any applicator/catheter

In the following sections, a more detailed description with a few specific clinical examples is provided on the main treatment planning steps in HDR, LDR temporary and LDR permanent brachytherapy. As for so many other examples in medicine, it is important to underline that in brachytherapy several steps lead up to treatment and these constitute a chain of sequential events. For successful treatment, all the links in this chain have to be performed correctly [53]. A problem in just one of these links can in principle cause the treatment to be dangerously sub-optimal or delivered incorrectly.

10.9.2 HDR Brachytherapy Treatment Planning

10.9.2.1 Choice of Treatment Method

Selecting and inserting the right applicator or performing a proper catheter implant is necessary for the subsequent optimal delivery of the treatment [54]. The location of the applicator or catheter in the patient constrains the possible source dwell positions and it cannot in principle be modified after imaging has been performed. The applicator is chosen according to:

- The target to be treated.
- The anatomy of the region to be treated.
- The desired provisional dose distribution.
- The selected imaging modality.

10.9.2.2 3D Imaging

Brachytherapy dose calculation with the TG-43 algorithm does not require information on tissue composition (e.g. electron or mass density) that is provided by CT imaging and can therefore be performed on any clinically suitable image set. Besides CT images, which are considered the standard imaging modality in brachytherapy, MR and ultrasound (US) images can also be used [26, 55, 56]. CT images are very good in discerning tissues with different atomic numbers and are particularly suited for the identification of applicator and catheter positions. However, CT images do not adequately distinguish between different soft tissues of electron density similar to water. In the treatment of cervix cancer, MR images can be acquired since, unlike CT images, they enable identification of possible residual tumour and its development over time. It is important to note that MR-based treatment plans require MR-compatible applicators, which are usually made of titanium or carbon fibre. However, since these materials are inorganic, they are particularly difficult to identify on MR images [21, 57].

In the treatment of prostate cancer, transrectal US images can be used to optimise the insertion of needles or seeds for HDR or LDR brachytherapy, respectively, and to perform dose distribution calculation in real time [58]. The prostate gland, urethra, bladder, rectum and needles or seeds are clearly visible on transrectal US images. An example of an image showing a transversal section of the prostate is shown in Figure 10.18.

10.9.2.3 Target Definition and Organs at Risk Contouring

Prior to treatment planning, it is necessary to identify and define the target to be treated and also the organs at risk, which are the organs around the target that need to be spared as much as possible from radiation. Several digital tools exist to simplify the contouring procedure and make it faster (e.g. interpolation, expansion, automatic recognition). This procedure is the same as for external beam radiotherapy, so whenever technically feasible, the same contouring software can be shared within the radiotherapy department.

10.9.2.4 Applicator and Catheter Reconstruction

In order to define the coordinates for the planned source positions inside the patient, each implanted applicator, catheter or needle has to be reconstructed on the acquired images. As mentioned previously, some imaging modalities are more suited for the identification of applicators (e.g. CT),

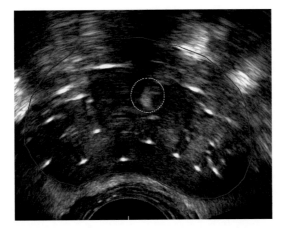

FIGURE 10.18 US transversal section of a prostate (contoured in red) with a series of implanted needles. The urethra, which is an organ at risk, is contoured in yellow.

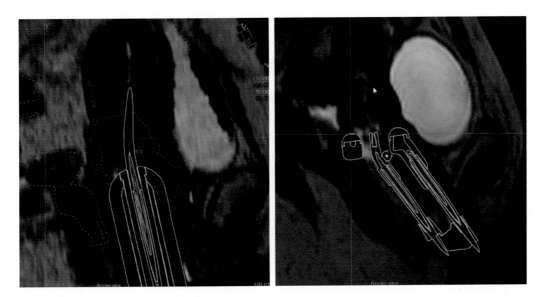

FIGURE 10.19 Examples of MR images with (a) a multichannel (with intrauterine tube) digital applicator and with (b) a cervical digital applicator.

whereas with others this task is more complicated. In MR imaging, for instance, applicators are visible as 'black holes' since they are inorganic. In some specific cases, the availability of a 'digital applicator' available within the planning software which can be manipulated in 3D and superimposed onto the real one on the image dataset, facilitates an accurate applicator position reconstruction [21, 57, 59, 60]. Examples of multichannel (with intrauterine tube) and cervical digital applicators, superimposed onto MR images, are shown in Figure 10.19.

10.9.2.5 3D Dose Distribution Optimisation and Evaluation

The objective of clinical optimisation of a brachytherapy treatment is to deliver the prescribed dose to the target, reducing as much as possible the dose to the surrounding normal tissues including any specific organs at risk. This is achieved by optimising the dwell times of each one of the chosen source positions, which are generally selected by the operator. Among existing optimisation algorithms, some are forward and some are inverse based [61]. The first type considers and optimises only the dose to the target, whereas inverse methods can also take into account the dose to the organs at risk. In general, sample points are randomly distributed over the surface and/or within the volume of the considered target and organs and are used to define and pursue the optimisation objectives by iterative calculation and adjustment of the possible treatment parameters.

To avoid large differences between dwell times of adjacent dwell positions, further conditions can be set in the optimisation algorithms to restrict these differences. Four examples of forward optimisation with the same vaginal cylinder applicator are provided in Figure 10.20. The dose is prescribed 5 mm away from the external surface of the applicator and different restrictions to the dwell time gradient between adjacent source positions are set. The red bars in the histograms indicate the dwell times of the source in the central catheter, from the applicator tip to its bottom position. Allowable dwell time differences are reduced from Figure 10.20 (a) to Figure 10.20 (d), but at the same time the achieved dose distribution is compromised and deviates from that prescribed. Since the number of degrees of freedom is significantly lower (one for every selected source position) than that of an IMRT or VMAT treatment in external beam radiotherapy, optimisation can be calculated essentially in real time (up to a few seconds of computation time). Once the final dwell times for the stepping source in each one of its programmed dwell positions are calculated the resulting dose distributions are reviewed and if satisfactory accepted by the radiation oncologist. The plan will then be second checked by an appropriately authorised person (normally a clinically qualified medical physicist) to ensure that all processes and calculations have been followed correctly. The final approved plan is then sent to the afterloader treatment console ready for treatment delivery.

10.9.3 Temporary LDR Brachytherapy Treatment Planning – Example: Ophthalmic Brachytherapy

In eye-plaque LDR brachytherapy performed with a beta-emitting plaque, treatment optimisation is achieved by selecting the right dimension and shape of the plaque from those available, its correct position over the eye and the time of its removal (i.e. taking into account the known duration of application). Lesion dimensions are usually determined by means of specific ophthalmic imaging techniques, for instance, fundus photography or B-scan ultrasound. An example of imaging and treatment planning for choroidal melanoma is shown in Figure 10.21.

If gamma-emitting seeds are employed, the number and position of seeds within the available slots inside the plaque can furthermore be optimised to achieve the desired dose distribution.

10.9.4 Permanent LDR Brachytherapy Treatment Planning – Example: Prostate Brachytherapy

In prostate LDR brachytherapy, once the seeds are inserted into the prostate, the treatment cannot be optimised further. Optimisation therefore is at the point of selection of the appropriate location for each seed. A post-implant evaluation can however be performed to determine the actual seed density and correct possible cold regions inside the prostate by adding additional seeds. The overall procedure therefore can conceptually be split into two phases, pre-implant planning and post-implant evaluation.

10.9.4.1 Pre-implant Planning

All pre-implant planning processes and treatment delivery is guided by transrectal US images, with the ultrasound equipment connected to the treatment planning system workstation. Sequential transverse images are acquired in the patient's cranio-caudal direction with the aid of a stepper that moves the US probe in discrete steps (not to be confused with the HDR stepping source). The step size usually ranges between 1 and 5 mm and sufficient images are acquired to cover the whole prostate. The reference coordinate system is defined along the transverse plane (x and y) by a template

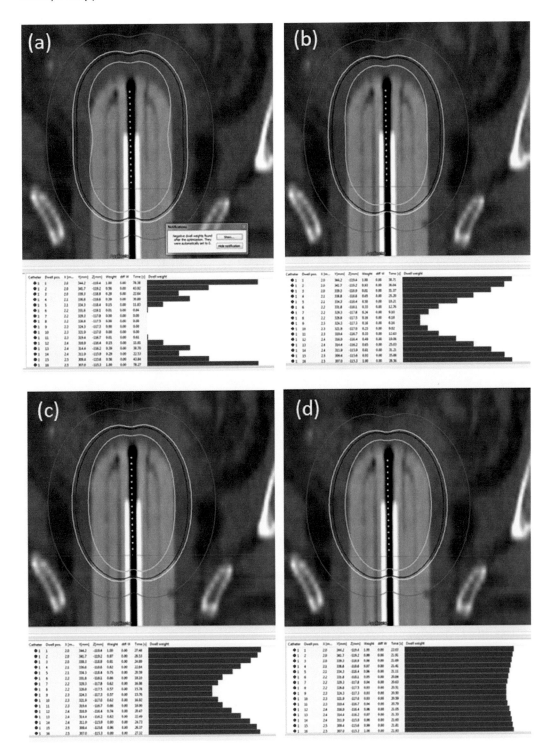

FIGURE 10.20 Four examples of forward optimisation with the same vaginal cylinder case. The dose is prescribed at 5 mm from the surface of the applicator, along the first few centimetres from its tip. The dwell time gradient restriction increases from (a) to (d) and the resulting 100% isodose (red) deviates from that desired.

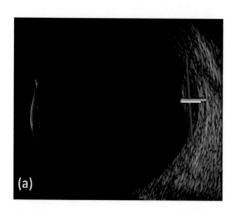
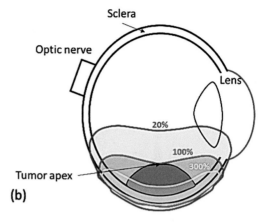

FIGURE 10.21 (a) Ocular ultrasound imaging. The choroidal melanoma is visible over the sclera, with thickness and lateral extension indicated with yellow and red lines, respectively. The green line includes the sclera to define the overall thickness to be treated; (b) Schematic drawing of the human eye with a choroidal melanoma. A few isodoses resulting from the application of an eye-plaque are plotted. The eye-plaque itself is not shown in the image but will be in contact with the sclera and centred over the tumour.

FIGURE 10.22 Template fixed above the stepper and the transrectal ultrasound probe. Two prostate fixation needles (white plastic ends) and one longer needle are visible. A stylet with black markings loaded with seeds is inserted into the needle and is used for their release inside the prostate.

(grid) that is fixed to the stepper, with regularly spaced holes (typically 0.5 cm apart) through which the needles can be inserted. The needles are used to enable the seeds to be fed, through the needles, into their planned position, and then removed. The z-coordinate is defined along the main axis of the US probe (i.e. cranio-caudal direction), with the $z = 0$ coordinate typically placed in line with the base of the prostate. A picture of the template is shown in Figure 10.22.

Once the physician has outlined the target (prostate) and the organs at risk (urethra and anterior rectal wall) on the acquired images, the physicist determines the ideal seed positions, and hence the loading of each needle (Figure 10.23). Seed position optimisation is performed by an inverse and/or forward planned approach, with the seed positions being constrained by the possible needle paths,

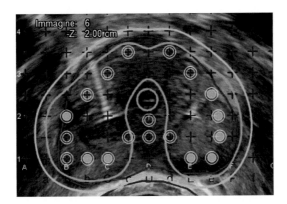

FIGURE 10.23 Example of pre-implant treatment planning. A transrectal ultrasound transverse image is shown, along with the anatomy contours (prostate: red contour; anterior rectal wall: dark blue contour; urethra: small green circle), the template with all possible needle positions (blue crosses) and the planned needle positions (small yellow circles). Computed isodoses representing the total doses delivered are also shown (145 Gy: light blue and 210 Gy: green).

which start from each template hole. Seed positions within the prostate gland are usually planned maintaining a safe distance from the urethra and the rectum so as not to overdose these regions. A set of dosimetric parameters [62] are checked continually during the optimisation process, until an adequate dose distribution for the target and the organs at risk is obtained.

10.9.4.2 Post-implant Evaluation

In the post-implant evaluation (also called post-planning or post-implant dosimetry) the actual seed positions are identified on a set of CT images acquired after implantation. Seed identification is performed either manually or semi-automatically, with a seed finder algorithm.

The dose is then calculated under the hypothesis that the current situation, in terms of prostate morphology and seed position, was and will be the same up to the complete decay of the implanted sources. Isodoses and other dosimetric parameters are used to evaluate the implant quality. An example is shown in Figure 10.24.

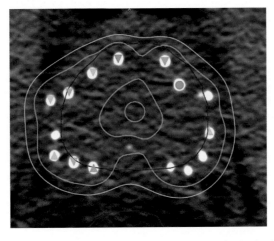

FIGURE 10.24 Example of post-operative evaluation: a CT transverse image is shown, along with anatomy contours (prostate: red contour; anterior rectal wall: dark blue contour; urethra: small green circle), implanted seeds (white spots, with their digital identification shown as green circles or triangles) and calculated isodoses (145 Gy: light blue and 210 Gy: green).

10.10   CONCLUSION

In the field of radiotherapy, brachytherapy is a more labour intensive treatment technique compared to routine external beam radiotherapy treatments. It requires substantial training and an experienced clinical team to be performed successfully, particularly for interstitial and/or intra-operative applications. Working in close collaboration with radiation oncologists and other hospital professionals, clinically qualified medical physicists play an important role in ensuring an accurate treatment is delivered to the patient. A clinically qualified medical physicist will be actively involved with:

- The design of the brachytherapy facility.
- The development of the required specifications for the procurement of sources, treatment equipment, and associated dosimetry systems.
- The quality management of the facility, providing:
  - Acceptance, commissioning and QA of the treatment machine, dosimetry equipment, applicators and of the safety systems (e.g. machine interlocks).
  - Acceptance, commissioning and QA of the treatment planning system.
  - Dosimetry and management of the radioactive source(s) used for treatment.
  - Possible support for patient-specific treatment delivery QA, by means of for instance in-vivo dosimetry measurements.
- Treatment planning and second independent checking.
- Advice on all radiation protection matters associated with the service.

Despite these complexities, the achievable steep dose distributions coupled with the ability to position the radiation source close to the target make brachytherapy an efficient radiotherapy treatment modality for a range of cancers. The role of the clinically qualified medical physicist is crucial in the support and development of an effective brachytherapy service.

REFERENCES

1. Kishan AU, Cook RR, Ciezki JP, Ross AE, Pomerantz MM, Nguyen PL, et al. Radical Prostatectomy, External Beam Radiotherapy, or External Beam Radiotherapy with Brachytherapy Boost and Disease Progression and Mortality in Patients with Gleason Score 9-10 Prostate Cancer. *JAMA* 2018; 319(9):896.
2. Eifel PJ, Yashar CM, Pötter R, Grigsby PW. Curative Radiation Therapy for Locally Advanced Cervical Cancer: Brachytherapy Is NOT Optional. *Int J Radiat Oncol.* 2014; 88(3):537–539.
3. Morton GC, Alrashidi SM. High Dose Rate Brachytherapy in High-Risk Localised Disease – Why Do Anything Else? *Clin Oncol.* 2019; 32:163–169.
4. Holschneider CH, Petereit DG, Chu C, Hsu I-C, Ioffe YJ, Klopp AH, et al. Brachytherapy: A Critical Component of Primary Radiation Therapy for Cervical Cancer. *Brachytherapy.* 2019; 18(2):123–132.
5. Nag S. High Dose Rate Brachytherapy: Its Clinical Applications and Treatment Guidelines. *Technol Cancer Res Treat.* 2004; 3(3):269–287.
6. Chassagne D, Dutreix A, Almond P, Burgers JMV, Busch M, Joslin CA. Dose and Volume Specification for Reporting Intracavitary Therapy in Gynaecology (Report 38). *J Int Comm Radiat Units Meas.* 1985; 20(1).
7. Fischer-Valuck BW, Gay HA, Patel S, Baumann BC, Michalski JM. A Brief Review of Low-Dose Rate (LDR) and High-Dose Rate (HDR) Brachytherapy Boost for High-Risk Prostate. *Front Oncol.* 2019; 9:1378.
8. Yang J, Liang X, Pope C, Li Z. Physics: Low-Energy Brachytherapy Physics. In: Montemaggi P, Trombetta M, Brady LW, editors. *Brachytherapy [Internet].* Cham: Springer International Publishing; 2016 [cited 2020 Jan 30]. pp. 29–39. Available from: http://link.springer.com/10.1007/978-3-319-26791-3_4
9. Perez-Calatayud J, Ballester F, Das RK, DeWerd LA, Ibbott GS, Meigooni AS, et al. Dose Calculation for Photon-emitting Brachytherapy Sources with Average Energy Higher than 50 keV: Report of the AAPM and ESTRO: High-Energy Photon-emitting Brachytherapy Dosimetry. *Med Phys.* 2012; 39(5):2904–2929.

10. Aronowitz JN. Dawn of Prostate Brachytherapy: 1915–1930. *Int J Radiat Oncol.* 2002; 54(3):712–718.

11. Rivard MJ, Venselaar JLM, Beaulieu L. The Evolution of Brachytherapy Treatment Planning: The Evolution of Brachytherapy Treatment Planning. *Med Phys.* 2009; 36(6Part1):2136–2153.

12. Connell PP, Hellman S. Advances in Radiotherapy and Implications for the Next Century: A Historical Perspective. *Cancer Res.* 2009; 69(2):383–392.

13. Kovács G. Modern Head and Neck Brachytherapy: From Radium towards Intensity Modulated Interventional Brachytherapy. *J Contemp Brachytherapy.* 2014; 4:404–416.

14. Dutreix J, Tubiana M, Pierquin B. The Hazy Dawn of Brachytherapy. *Radiother Oncol.* 1998; 49(3):223–232.

15. Holt JG. AAPM Report No. 41: Remote Afterloading Technology. *Med Phys.* 1993; 20(6):1761.

16. Whelan TJ, Aldrich JE, Voruganti SM, Wong OS, Filbee JF, Joseph PJ, et al. A Comparison of Manual and Remote Afterloading in the Treatment of Carcinoma of the Cervix: Increasing Dose Rate with a Flexible Applicator. *Clin Oncol.* 1992; 4(5):294–298.

17. Aronowitz JN. Afterloading: The Technique that Rescued Brachytherapy. *Int J Radiat Oncol.* 2015; 92(3):479–487.

18. Mark R, Vallabhan G, Akins R, Anderson P, Nair M, Neumann T, et al. Interstitial High Dose Rate (HDR) Brachytherapy for Early Stage Prostate Cancer. *Int J Radiat Oncol.* 2005; 63:S304.

19. Dyk PT, Richardson S, Garcia-Ramirez J, Schwarz JK, Grigsby PW. Outpatient High-Dose-Rate Interstitial Brachytherapy in the Treatment of Gynaecologic Malignancies. *Brachytherapy.* 2013; 12:S54.

20. Walter F, Maihöfer C, Schüttrumpf L, Well J, Burges A, Ertl-Wagner B, et al. Combined Intracavitary and Interstitial Brachytherapy of Cervical Cancer Using the Novel Hybrid Applicator Venezia: Clinical Feasibility and Initial Results. *Brachytherapy.* 2018; 17(5):775–781.

21. Hellebust TP, Kirisits C, Berger D, Pérez-Calatayud J, De Brabandere M, De Leeuw A, et al. Recommendations from Gynaecological (GYN) GEC-ESTRO Working Group: Considerations and Pitfalls in Commissioning and Applicator Reconstruction in 3D Image-based Treatment Planning of Cervix Cancer Brachytherapy. *Radiother Oncol.* 2010; 96(2):153–160.

22. Dimopoulos JCA, Petrow P, Tanderup K, Petric P, Berger D, Kirisits C, et al. Recommendations from Gynaecological (GYN) GEC-ESTRO Working Group (IV): Basic Principles and Parameters for MR Imaging within the Frame of Image Based Adaptive Cervix Cancer Brachytherapy. *Radiother Oncol.* 2012; 103(1):113–122.

23. Lettmaier S, Strnad V. Intraluminal Brachytherapy in Oesophageal Cancer: Defining Its Role and Introducing the Technique. *J Contemp Brachytherapy.* 2014; 2:236–241.

24. McLaughlin PW, Narayana V. Progress in Low Dose Rate Brachytherapy for Prostate Cancer. *Semin Radiat Oncol.* 2020; 30(1):39–48.

25. Davis BJ, Horwitz EM, Lee WR, Crook JM, Stock RG, Merrick GS, et al. American Brachytherapy Society Consensus Guidelines for Transrectal Ultrasound-guided Permanent Prostate Brachytherapy. *Brachytherapy.* 2012; 11(1):6–19.

26. Hellebust TP. Place of Modern Imaging in Brachytherapy Planning. *Cancer/Radiothérapie.* 2018; 22(4):326–333.

27. Stock R. Importance of Post-Implant Dosimetry in Permanent Prostate Brachytherapy. *Eur Urol.* 2002; 41(4):434–439.

28. Thomson RM, Furutani KM, Kaulich TW, Mourtada F, Rivard MJ, Soares CG, et al. AAPM Recommendations on Medical Physics Practices for Ocular Plaque Brachytherapy: Report of Task Group 221. *Med Phys.* 2020; 47(5): e92–e124. mp.13996.

29. Simpson ER, Gallie B, Laperrierre N, Beiki-Ardakani A, Kivelä T, Raivio V, et al. The American Brachytherapy Society Consensus Guidelines for Plaque Brachytherapy of Uveal Melanoma and Retinoblastoma. *Brachytherapy.* 2014; 13(1):1–14.

30. ICRU. Prescribing, Recording, and Reporting Brachytherapy for Cancer of the Cervix. *J ICRU.* 2013; 13(1–2):NP.1–NP.

31. Nath R, Rivard MJ, DeWerd LA, Dezarn WA, Thompson Heaton H, Ibbott GS, et al. Guidelines by the AAPM and GEC-ESTRO on the Use of Innovative Brachytherapy Devices and Applications: Report of Task Group 167: Task Group 167: Innovative Brachytherapy Devices and Applications. *Med Phys.* 2016; 43(6Part1):3178–3205.

32. Rivard MJ, Coursey BM, DeWerd LA, Hanson WF, Saiful Huq M, Ibbott GS, et al. Update of AAPM Task Group No. 43 Report: A Revised AAPM Protocol for Brachytherapy Dose Calculations. *Med Phys.* 2004; 31(3):633–674.

33. Rivard MJ, Butler WM, DeWerd LA, Huq MS, Ibbott GS, Meigooni AS, et al. Supplement to the 2004 Update of the AAPM Task Group No. 43 Report: Supplement to AAPM TG-43 Update. *Med Phys.* 2007; 34(6Part1):2187–2205.

34. Nath R, Anderson LL, Luxton G, Weaver KA, Williamson JF, Meigooni AS. Dosimetry of Interstitial Brachytherapy Sources: Recommendations of the AAPM Radiation Therapy Committee Task Group No. 43. *Med Phys.* 1995; 22(2):209–234.

35. Rivard MJ, Ballester F, Butler WM, DeWerd LA, Ibbott GS, Meigooni AS, et al. Supplement 2 for the 2004 Update of the AAPM Task Group No. 43 Report: Joint Recommendations by the AAPM and GEC-ESTRO. *Med Phys.* 2017; 44(9):e297–e338.

36. Aronowitz JN. Don Lawrence and the 'k-capture' Revolution. *Brachytherapy.* 2010; 9(4):373–381.

37. Heilemann G, Fetty L, Dulovits M, Blaickner M, Nesvacil N, Georg D, et al. Treatment Plan Optimization and Robustness of 106 Ru Eye Plaque Brachytherapy Using a Novel Software Tool. *Radiother Oncol.* 2017; 123(1):119–124.

38. Stöckel E, Eichmann M, Flühs D, Sommer H, Biewald E, Bornfeld N, et al. Dose Distributions and Treatment Margins in Ocular Brachytherapy with Ru-106 Eye Plaques. *Ocul Oncol Pathol.* 2018; 122–128.

39. Mukwada G, Neveri G, Alkhatib Z, Waterhouse DK, Ebert M. Commissioning of a Well Type Chamber for HDR and LDR Brachytherapy Applications: A Review of Methodology and Outcomes. *Australas Phys Eng Sci Med.* 2016; 39(1):167–175.

40. Perez-Calatayud J, Ballester F, Carlsson Tedgren Å, Rijnders A, Rivard MJ, Andrássy M, et al. GEC-ESTRO ACROP Recommendations on Calibration and Traceability of LE-LDR Photon-emitting Brachytherapy Sources at the Hospital Level. *Radiother Oncol.* 2019; 135:120–129.

41. Bidmead AM, Sander T, Locks SM, Lee CD, Aird EGA, Nutbrown RF, et al. The IPEM Code of Practice for Determination of the Reference Air Kerma Rate for HDR [192] Ir Brachytherapy Sources Based on the NPL Air Kerma Standard. *Phys Med Biol.* 2010; 55(11):3145–3159.

42. DeWerd LA, Ibbott GS, Meigooni AS, Mitch MG, Rivard MJ, Stump KE, et al. A Dosimetric Uncertainty Analysis for Photon-emitting Brachytherapy Sources: Report of AAPM Task Group No. 138 and GEC-ESTRO: AAPM TG-138 and GEC-ESTRO Brachytherapy Dosimetry Uncertainty Recommendations. *Med Phys.* 2011; 38(2):782–801.

43. Nath R, Anderson LL, Meli JA, Olch AJ, Stitt JA, Williamson JF. Code of Practice for Brachytherapy Physics: Report of the AAPM Radiation Therapy Committee Task Group No. 56. *Med Phys.* 1997; 24(10):1557–1598.

44. Butler WM, Bice WS, DeWerd LA, Hevezi JM, Huq MS, Ibbott GS, et al. Third-party Brachytherapy Source Calibrations and Physicist Responsibilities: Report of the AAPM Low Energy Brachytherapy Source Calibration Working Group: 3rd Party Brachytherapy Source Calibrations and Physicist Responsibilities. *Med Phys.* 2008; 35(9):3860–3865.

45. Beaulieu L, Carlsson Tedgren Å, Carrier J-F, Davis SD, Mourtada F, Rivard MJ, et al. Report of the Task Group 186 on Model-based Dose Calculation Methods in Brachytherapy beyond the TG-43 Formalism: Current Status and Recommendations for Clinical Implementation: TG-186: Model-based Dose Calculation Techniques in Brachytherapy. *Med Phys.* 2012; 39(10):6208–6236.

46. Strohmaier S, Zwierzchowski G. Comparison of 60 Co and 192 Ir Sources in HDR Brachytherapy. *J Contemp Brachytherapy.* 2011; 4:199–208.

47. Terribilini D, Vitzthum V, Volken W, Frei D, Loessl K, van Veelen B, et al. Performance Evaluation of a Collapsed Cone Dose Calculation Algorithm for HDR Ir-192 of APBI Treatments. *Med Phys.* 2017; 44(10):5475–5485.

48. Han DY, Ma Y, Luc B, Wahl M, Hsu I-CJ, Cunha A. Assessment of Volumetric Dose Differences between Calculations Performed with the Advanced Collapsed Cone Engine (ACE) for the Model-Based Dose Calculation Method (TG-186), TG-43, and Monte Carlo. *Brachytherapy.* 2016; 15:S146–S147.

49. Pappas EP, Zoros E, Moutsatsos A, Peppa V, Zourari K, Karaiskos P, et al. On the Experimental Validation of Model-based Dose Calculation Algorithms for [192] Ir HDR Brachytherapy Treatment Planning. *Phys Med Biol.* 2017; 62(10):4160–4182.

50. Iftimia I, Halvorsen PH. Commissioning of the Acuros BV GBBS Algorithm. *Brachytherapy.* 2015; 14:S88.

51. Sinnatamby M, Nagarajan V, Reddy S, Karunanidhi G, Singhavajala V. Dosimetric Comparison of Acuros TM BV with AAPM TG43 Dose Calculation Formalism in Breast Interstitial High-dose-rate Brachytherapy with the Use of Metal Catheters. *J Contemp Brachytherapy.* 2015; 4:273–279.

52. Boman EL, Satherley TWS, Schleich N, Paterson DB, Greig L, Louwe RJW. The Validity of Acuros BV and TG-43 for high-dose-rate Brachytherapy Superficial Mold Treatments. *Brachytherapy.* 2017; 16(6):1280–1288.

53. Kim H, Houser CJ, Kalash R, Maceil CA, Palestra B, Malush D, et al. Workflow and Efficiency in MRI-based high-dose-rate Brachytherapy for Cervical Cancer in a High-volume Brachytherapy Centre. *Brachytherapy.* 2018; 17(5):753–760.

54. Podder TK, Beaulieu L, Caldwell B, Cormack RA, Crass JB, Dicker AP, et al. AAPM and GEC-ESTRO Guidelines for Image-guided Robotic Brachytherapy: Report of Task Group 192: Report of Task Group 192. *Med Phys.* 2014; 41(10):101501.

55. Wang J, Tanderup K, Cunha A, Damato AL, Cohen GN, Kudchadker RJ, et al. Magnetic Resonance Imaging Basics for the Prostate Brachytherapist. *Brachytherapy.* 2017; 16(4):715–727.

56. Nesvacil N, Schmid MP, Pötter R, Kronreif G, Kirisits C. Combining Transrectal Ultrasound and CT for Image-guided Adaptive Brachytherapy of Cervical Cancer: Proof of Concept. *Brachytherapy.* 2016; 15(6):839–844.

57. Richart J, Carmona-Meseguer V, García-Martínez T, Herreros A, Otal A, Pellejero S, et al. Review of Strategies for MRI Based Reconstruction of Endocavitary and Interstitial Applicators in Brachytherapy of Cervical Cancer. *Rep Pract Oncol Radiother.* 2018; 23(6):547–561.

58. Carrara M, Tenconi C, Rossi G, Borroni M, Cerrotta A, Grisotto S, et al. In Vivo Rectal Wall Measurements during HDR Prostate Brachytherapy with MO Skin Dosimeters Integrated on a Trans-rectal US Probe: Comparison with Planned and Reconstructed Doses. *Radiother Oncol.* 2016; 118(1):148–153.

59. Pötter R, Federico M, Sturdza A, Fotina I, Hegazy N, Schmid M, et al. Value of Magnetic Resonance Imaging without or with Applicator in Place for Target Definition in Cervix Cancer Brachytherapy. *Int J Radiat Oncol.* 2016; 94(3):588–597.

60. Hrinivich WT, Morcos M, Viswanathan A, Lee J. Automatic Tandem and Ring Reconstruction Using MRI for Cervical Cancer Brachytherapy. *Med Phys.* 2019; 46(10):4324–4332.

61. Carrara M, Cusumano D, Giandini T, Tenconi C, Mazzarella E, Grisotto S, et al. Comparison of Different Treatment Planning Optimization Methods for Vaginal HDR Brachytherapy with Multichannel Applicators: A Reduction of the High Doses to the Vaginal Mucosa Is Possible. *Phys Med.* 2017; 44:58–65.

62. Nath R, Bice WS, Butler WM, Chen Z, Meigooni AS, Narayana V, et al. AAPM Recommendations on Dose Prescription and Reporting Methods for Permanent Interstitial Brachytherapy for Prostate Cancer: Report of Task Group 137: AAPM TG-137 Report. *Med Phys.* 2009; 36(11):5310–5322.

11 Molecular Radiotherapy

Lidia Strigari

IRCCS Azienda Ospedaliero-Universitaria di Bologna, Bologna, Italy

CONTENTS

DOI: 10.1201/9780429155758-11

11.1 INTRODUCTION

Radionuclide therapy using injected, ingested or implanted radioactivity has been known by a variety of names including isotope treatment, unsealed source therapy, (biologically) targeted radionuclide therapy, internal radiotherapy and, most recently, molecular radiotherapy (MRT). The term 'molecular radiotherapy' is now used to emphasise that treatment is based on the delivery of ionising radiation to tissues through the interaction of a therapeutic agent with targeted molecular sites and receptors. MRT has been used to treat both benign and malignant diseases for many decades. Most early treatments were systemic (whole body) involving oral or intravenous administration of the therapeutic agent and relying on preferential uptake and extended retention by the tumour to provide a high tumour to normal tissue dose ratio. With the increased availability of a wide range of radionuclides and the development of pharmaceuticals able to target specific tumours, the interest and development in MRT has steadily grown. The therapeutic agent is normally called a radiopharmaceutical to reflect the combination of radionuclide and target-specific pharmaceutical. In cancer treatment, MRT often combines the advantage of this target selectivity (like brachytherapy or external beam radiotherapy, EBRT) with that of being systemic, as with chemotherapy, and it may be used as part of a therapeutic strategy with curative intent or for disease control and palliation.

The traditional approach to MRT is to prescribe and administer a fixed amount of radioactivity. More recently, pre-treatment tracer studies have been introduced where a small amount of activity is administered and the uptake in organs and tissues determined via imaging to calculate a more appropriate patient-specific activity required for treatment. This approach also enables the treatment prescription to be calculated and given in terms of absorbed dose to the tumour (or in terms of the limiting dose to an organ at risk). Although this is a substantial improvement over the traditional approach, the accuracy of internal dose calculation is subject to large uncertainties. Conventional internal dosimetry calculations based on standard patient models and tabulated data are now being replaced with patient-specific 3D dose distributions derived from quantitative imaging and multi-modality image registration.

Selective tissue targeting is achieved by a combination of physical and biological mechanisms. Improvements in physical targeting have been achieved by selection of the most appropriate radionuclide. This is based on matching the physical range of emitted particles to the cells within the tumour being treated. Physical targeting has also been improved by more direct routes of administration. Biological targeting has advanced though developments in the carrier pharamaceutical molecules that can provide selective targeting, for example, in the cell nucleus, cytoplasm (thick liquid solution that fills each cell) or on the cell surface membrane receptors. This cell-directed radiotherapy delivers radiation dose directly to the tumour cells with minimal dose to normal tissues.

Examples of MRT treatments currently in routine clinical use include ^{131}I for treatment of thyroid cancer, ^{177}Lu, ^{90}Y or ^{131}I for treating neuroendocrine tumours (NET), ^{90}Y or ^{166}Ho for liver tumours and ^{223}Ra, among others, for palliative therapy of bone metastases (Figure 11.1). MRT employs a range of targeting mechanisms – including metabolic processes, cell surface receptors, extracellular targeting and direct injection into the tumour – to direct the radiopharmaceuticals to the required location. This chapter will introduce several MRT techniques in more detail and outline the practical challenges around internal dosimetry, treatment, radiobiology and radiation protection for patients undergoing this form of radiotherapy.

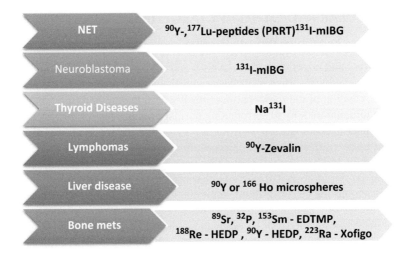

FIGURE 11.1   Examples of disease and the radionuclides and radiopharmaceuticals used to target these sites.

11.2   DELIVERY STRATEGY AND APPLICATIONS

11.2.1   Brief Historical Background

MRT started with the treatment of thyroid disorders using radioactive iodine [131]I. In the USA, in the 1940s, ^{131}I was first used to treat both benign (Graves and multinodular goitre) disease and cancer of the thyroid [1], effectively giving birth to the field of nuclear medicine. The first published application of radioiodine for the treatment of thyroid metastases employed external radioactive counting to estimate the absorbed dose delivered to metastatic deposits and measured biomarkers of response including white cell blood counts [2].

Globally, the use of radioiodine for the management of patients with thyroid disorders is the most well-established MRT treatment. The probability of radiation-induced side effect rates is generally low, and treatment well tolerated.

11.2.2   The Role of the Medical Physicist in the Multidisciplinary Team

MRT should be managed by a multidisciplinary team involving nuclear medicine (NM) physicians, oncologists, radiologists and medical physicists. In the decision to treat with MRT, it is important to include all of these experts in joint discussion with the clinicians managing the patient, normally in the form of attendance at multidisciplinary meetings. Medical physicists provide expert technical guidance to medical staff focusing particularly on the calculated dose to the patient. This offers the opportunity for the clinicians to identify the appropriate treatment strategy. Medical physicists delivering MRT may work primarily in NM (Chapter 6), a specialty known for diagnostic procedures using radiopharmaceuticals targeted to functional aspects of the human physiology (e.g. renal function, bone turnover) or they may be embedded within a radiotherapy team whose primary role is the delivery of EBRT (Chapter 9).

For this reason, it is important that the medical physicist is trained in the subspecialty of MRT either from within the radiotherapy or NM team and included in this multidisciplined workforce as required. The skills required of a medical physicist in delivering high-quality MRT include scientific rigour in calculation, measuring, monitoring and evaluation of the procedures and also the ability to communicate the complexities of the treatment, its side effects and required radioprotection precautions to both staff and patients.

11.3 MOLECULAR RADIOTHERAPY TARGETING

11.3.1 Physical Properties and the Delivery of Radionuclides

Successful delivery and retention within the body rely on the radionuclide being attached to a pharmaceutical or in a chemical format itself that targets the cells of interest. The selective concentration and prolonged retention of the radiopharmaceutical within the target tissues allow high doses to be delivered with minimal toxicity to surrounding normal tissues.

Every radiopharmaceutical used in MRT works on this principle; however, the targeting and mechanism of uptake differs for different types of treatment. The number of radiopharmaceuticals being developed for MRT is increasing and these exhibit a range of different targeting mechanisms. Understanding the site of uptake of the radiopharmaceutical in relation to the cell nucleus is essential when considering the effective therapeutic range of the radionuclide (beta or alpha particles). The mode of uptake of a radionuclide can be broadly classified into intracellular, cell-surface and extracellular mechanisms. Intracellular mechanisms are when a radionuclide is metabolised directly into the cell such as the cytoplasm or DNA within the cell nucleus. For example, ^{131}I –iodide is taken up into the follicular cells of the thyroid, and ^{131}I mIBG is taken up in the cytoplasm and is used for treatment of neuroblastoma. It is also possible to attach radionuclides to the surface membrane of the cell via receptor bindings for hormones, peptides or antibodies. To achieve an effective dose, the tumour cells must provide sufficient binding site opportunities per cell. In addition to this, the radiopharmaceutical must be of a high specific activity (the amount of radioactivity per mass of labelling compound); otherwise, the binding sites will saturate with non-radioactive labelled compound. Extracellular mechanisms are where the radionuclide is brought into close proximity to the tumour cells. This may be by targeting the extracellular osteoid (the bone-like tissue prior to calcification) for bone-seeking agents, by using radiolabelled cells and by more direct routes of administration such as injection into a joint for radiosynovectomy (treatment of inflammatory joint disorders) or arterial injection for ^{90}Y attached to a solid substrate in the form of microspheres for treatment of the liver. The majority of radiopharmaceuticals are in liquid form and administered via an intravenous injection, orally or directly into the organ being treated as for radiosynovectomy.

11.4 MEASUREMENT OF RADIOACTIVITY

11.4.1 Activity Measurement and Traceability to Reference Standards

Radioactive materials are widely used in hospitals for both diagnostic and therapeutic purposes. The levels of activity in radiopharmaceuticals that are administered clinically are governed primarily by the need to balance the effectiveness and the safety of the medical procedure by choosing the minimum radiation dose delivered to the patient needed to achieve the required objective (e.g. diagnostic image quality or therapeutic outcome).

The most commonly used equipment for activity measurement is a well-type gas ionisation chamber (see Figures 10.10 and 10.11 for comparable equipment used in brachytherapy), which normally consists of (a) well-type ionisation chamber, (b) stabilised high voltage supply, (c) electrometer for measuring the small ionisation currents generated within the chamber, (d) processing electronics, (e) display device. The chamber is sealed under pressure and the enclosed gas is usually a noble gas with a high atomic number in order to increase the probability of photon interaction and therefore sensitivity in detecting radiation. The radioactive source is placed within the chamber well and the activity is quantified in terms of the ionisation current generated within the chamber gas and measured by the electrometer. A factor is used to convert the measured current to activity for the given radionuclide. This factor, normally referred to as the calibration factor, is derived by measuring the current generated from a radionuclide of the same type and known activity so that the ratio of true activity per measured current unit can be determined. It is important for MRT that the activity of any radioactive material should be accurately determined prior to administration to a patient.

The need to maintain a high degree of confidence in these measurements requires regular quality control of the instrument carried out in accordance with national and international standards.

Radionuclide calibrators used in medical practice should be traceable to primary measurement standards held at the appropriate National Metrology Institute (NMI). Each NMI confirms the accuracy of its standards by comparisons with the NMIs of other countries ensuring international agreement of measured activity.

The calibration of the instrument (e.g. well chamber) is generally made at the place of manufacture using standard sources with activities that are traceable to a standards laboratory. A certificate should accompany the instrument providing appropriate calibration factors and conditions of operation. A practical method to ensure traceability to the national standard is to calibrate, at least annually, the routine instrument (called a field instrument) against a reference instrument. This can be a 'secondary standard' instrument for which nationally traceable measurements are available for a comprehensive range of radionuclides, geometries and activities. Alternatively, a reliable routine instrument that has been calibrated for the full range of radionuclides for which it is to be a reference could be used.

Several parameters must be assessed to confirm that the radionuclide calibrator maintains its performance requirements over time. A quality control programme generally consists of daily, weekly, monthly and yearly checks and should include tests of precision and accuracy, linearity of activity response (i.e. does the instrument reading change in proportion to the change in activity), test of reproducibility and check of any background signal with no activity present. The sources needed to perform the tests should include a sealed reference source, of known activity, certified to less than +/−5% overall uncertainty. Examples of suitable sources are Co-57, Ba-133, Cs-137, Co-60 to cover the required energy range; of note, Cs-137, Co-60 can be used, without replacement, for tens of years due to their long half-lives. All relevant parameters must be quantified during initial acceptance testing in order to provide baselines against which subsequent measurements can be compared. Less extensive measurements are normally sufficient on an ongoing basis to confirm that there have been no significant changes in performance when compared against the original baselines. However, if the performance changes or there have been major repairs, the detailed measurements required for acceptance testing must be repeated. The IAEA document on Quality Assurance of Radioactivity Measurement in Nuclear Medicine covers all aspects of a QA-programme. More information can be found in the NPL Report 93 [3], IPEM Report No.65 [4] and IAEA Handbook [5]. A general overview of quality control of NM instrumentation is provided in the article by Zanzonico [6].

11.4.2 Imaging Devices for MRT Dosimetry

In nuclear medicine, the imaging of radioactive uptake for diagnostic purposes has led to the development of a broad range of different imaging devices including the intraoperative probe, organ ('thyroid') uptake probe, gamma camera, single photon emission computed tomography (SPECT), positron emission tomography (PET) and combined SPECT/computed tomography (CT), PET/CT or PET/magnetic resonance imaging (MRI) scanners. The two major imaging modalities are gamma camera systems, which detect the spatial distribution of γ rays emitted by a nuclide within the patient, and PET systems, which use the directional correlation between annihilation photons emitted following positron decay to determine the spatial distribution of an injected positron emitting radiopharmaceutical. A more complete description of these imaging modalities is provided in Chapter 6. For MRT, the amount of radioactivity to be administered to guarantee effective therapy needs to be determined prior to treatment. This can be achieved by simulating the treatment by administering a small amount of (known) trace activity and the resultant uptake of this activity in the patient determined using various imaging methods ranging from 2-D planar, 3-D tomographic, SPECT or PET scanner imaging. SPECT can provide quantitative information but the images must be accurately corrected in order to convert recorded counts into activity. Challenges here include the attenuation and scatter occurring within the patient, which can be addressed using several methods [7, 8] including the double or triple-energy window technique (DEW or TEW, respectively).

The basis for the DEW method of scatter correction is that Compton scattered photons lose energy when scattered and therefore contribute more to the lower energy portion of the photopeak than the high energy side. The basis for the TEW method is a correction where the scatter fraction in the photopeak window is estimated by linear interpolation between two adjacent narrow sub-windows. If two or three non-overlapping energy windows are selected, a regression relation can be obtained between the ratio of counts within these windows and the scatter fraction for the counts within the total photopeak window. These methods require that calibration studies be performed to determine suitable regression coefficients for a given system and pair or triplet of energy windows to enable implementation on any modern camera.

A major advance in NM was achieved with the introduction of multimodality imaging systems including SPECT/CT and PET/CT. In these systems, the CT images which are registered in 3D with their corresponding SPECT or PET image, can be used to provide an anatomical context for the functional NM images and allow attenuation correction to improve activity quantification. Other considerations include image reconstruction and registration between the two different imaging modalities where motion between images may impact accuracy, as well as dealing with additional issues such as low spatial resolution, collimator detector response and dead time. Spatial resolution is largely governed by the different lengths and widths of the septa making up the imaging collimator. In general, the longer the septa, the better the resolution but the lower the count rate (sensitivity) for a given amount of radionuclide, due to increased attenuation of the oblique radiation in the walls of the collimating septa. The dead time is the time after each measurement event during which the imaging system is not able to record another event and is a consequence of a finite processing resolution of the imaging system. This results in an increasing loss of counts at higher activities.

11.4.3 QUANTIFICATION OF PATIENT ACTIVITY AND RELATED UNCERTAINTIES

Radiation dose calculations in MRT depend on quantification of the spatial uptake of activity via planar and/or tomographic imaging methods. Both methods have inherent limitations, and the accuracy of activity estimates varies with object size and background radiation levels. Activity quantification with three-dimensional (3D) tomographic imaging, for example, SPECT is theoretically superior to that with 2D planar imaging as problems of organ overlap in 2D may be overcome in 3D along with improved contrast for small regions, leading to more accurate spatial information regarding activity, and thus ultimately radiation dose [9]. Tomographic data is also important in evaluating heterogeneous uptake of activity in organs and resolving issues of underlying or overlying background activity which are obtained using the 2D approach where it is not possible to separate the counts of organs that are partially or fully superimposed over each other in planar images. But many effects complicate accurate SPECT quantification, including the increased effort needed to obtain basic calibration data to convert acquired counts to absolute activity, and to accurately perform corrections for attenuation, scatter and partial volume effects. [10, 11]. The partial volume effect is defined as the loss of apparent activity in small objects or regions because of the limited resolution of the imaging system. If the object dimension to be imaged is less than twice the full width at half maximum (FWHM) resolution of the imaging system, the resultant activity in the object or region (below 20 ml on most imaging systems) is underestimated [12]. The additional contribution to the image from scattered photons is important for radionuclides such as ^{131}I and ^{177}Lu which emit photons of high energy (364 and 208 keV, respectively). If not corrected, this can lead to an overestimate of activity in planar quantification. Many studies have been conducted to validate and improve the quantification of activity uptake with different imaging techniques, but more data is still needed to better characterise limitations regarding object size, background levels, and other variables such as scanner spatial resolution and reconstruction algorithms. Examples of various quantification studies are reported in the literature [13–21]. PET/CT imaging is considered a suitable quantitative modality for activity determination and has been investigated for ^{124}I pre-therapy lesion dosimetry of thyroid tumours and metastases [22].

^{90}Y-PET is of growing interest in the imaging for radioembolisation of liver tumours [23] (injection of radioactive microspheres) and also in peptide receptor radionuclide therapy described later for the treatment of NET. Multimodality ^{90}Y-PET/CT imaging quantification is possible both in phantoms and in patients. Absorbed dose evaluations in clinical applications are strongly related to target activity concentration. PET imaging with time-of-flight (TOF) reconstruction can improve the detectability and quantitative accuracy according to the lesion size [23–25]. The concept of TOF relies on the registration of both annihilation photons and the precise time that each of the two coincident photons is detected. Since the closer photon will arrive at its detector first, the difference in arrival times helps pin down the location of the annihilation event along the line of registration between the two detectors. Emerging algorithms are expected to further improve image quantification for ^{90}Y imaging [26]. More information on these devices, reconstruction methods and quality assurance programmes can be found in the IAEA Handbook [5], a comprehensive volume of Physics in Nuclear Medicine, published in early 2014. In addition, a general overview of quality control of NM instrumentation is provided in the article by Zanzonico [6]. It can be seen that there is still a lot of work for the medical physicist to contribute in this area.

11.4.4 Phantoms

Phantoms are used to both calibrate and verify the activity distributions determined by the imaging system. These come in a range of designs and include point-like sources, homogenous phantoms, spherical sources inserted in cylindrical phantoms (e.g. Jaszczak sphere or NEMA phantom) [12] or patient-like phantoms [27]. Recently 3D-printed phantoms enable more realistic tumour and patient-specific designs to be created and are expected to improve the modelling of the imaging system to accurately localise and quantify activity to feed into image-based dosimetry calculations. This step is necessary to develop personalised patient-specific treatment planning and provide accurate dosimetry.

11.5 PRINCIPLES OF DOSE CALCULATION

At present, MRT is not routinely planned according to the radiation dose to be delivered. Instead, as previously described, most treatments are based on fixed levels of administered activity – sometimes modified according to the patient's weight. But physiology differs from patient to patient affecting how activity is taken up, distributed and retained within the body, and as a result, the actual dose delivered can vary significantly. Thus, the same administered activity might produce a wide range of absorbed doses, both between lesions in one patient and between patients [28]. One major challenge for MRT is the lack of established and adopted procedures for calculating the absorbed dose to a target. One reason why dosimetry-based MRT is not routinely performed is the success of radio-iodine treatments, which have been used to treat thyroid cancer for so long. The high success rate seen when administering a standard fixed activity has led users to surmise that dosimetry is simply not required.

Another confounding issue is that calculation of the absorbed doses from internally delivered radiation is far from straightforward. In EBRT, where delivered dose largely depends upon the incident radiation beam characteristics, the physical anatomy of the patient and their accurate alignment with the beam, dosimetry is highly accurate. However, in MRT, the delivered dose depends upon the biokinetics of the radiopharmaceutical as well as the physiology of the patient. As a result, dosimetry in MRT is patient specific and more challenging to implement. A recent attempt to reinforce the need to implement clinical dosimetry lies in the European Directive 2013/59/Euratom, which mandated that personalised dosimetry-based treatment planning be put in place by February 2018. Despite the complexity of the task, efforts are well underway to implement internal dosimetry, using techniques such as 3D image-based dosimetry.

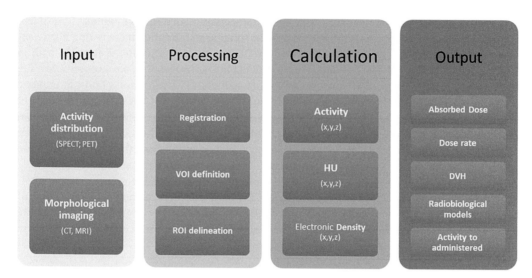

FIGURE 11.2   Flowchart of the absorbed dose calculation process adapted from MIRD Pamphlet number 23 [29]. The abbreviations VOI and ROI refer to the volume (target) and regions (structures or organs) of interest, respectively, that need to be defined. DVH refers to dose–volume histograms which can be used for treatment plan evaluation and are described further in Section 11.6.2.

The current conventional method of calculating absorbed dose delivered internally is known as the medical internal radiation dosimetry (MIRD) scheme. In this approach, the body is considered to be comprised of a series of organs having a significant uptake of radiopharmaceutical with a set of target organs irradiated by these source organs. The MIRD definition of a target is different to EBRT where the target refers to the tumour. In MIRD, the target is any organ or tissue of interest to which the absorbed dose is to be calculated, including normal tissues. The MIRD system has now been in existence for many years and updated guidance and data is issued regularly in the form of pamphlets. In general terms, the dose absorbed by a target from a given radionuclide can be defined as the cumulated activity (which is patient specific and dependent upon biokinetics and uptake distribution) multiplied by a dose factor (a physical characteristic of the radionuclide). The cumulated activity is the total number of decays of the radionuclide and this has to be determined in some way. Dose factors for different radionuclides are published in the MIRD pamphlets. A flowchart outlining the absorbed dose calculation process is given in Figure 11.2, adapted from [29].

11.5.1 Absorbed Dose Calculation

The general equation for absorbed dose calculations is as follows:

$$D = absorbed\ \text{energy}, (\text{E}) / \text{mass} \tag{11.1}$$

The International System of units defines the absorbed dose in joules per kilogram (J/kg) with the special name gray (Gy), where 1 J/kg = 1 Gy. The challenge for MRT dosimetry is that the numerator, E, is not simply derived because it depends on many factors. The absorbed energy in the target needs to be calculated for every source irradiating the target and must incorporate: total number of disintegrations from the source (usually termed source organ), the energy released per disintegration, and the fraction of energy absorbed in the target per disintegration from the source. This last factor is a function of geometry, attenuation of the decay products from the radionuclide between source and target, and absorption properties of the target. The MIRD formalism [30] gives a framework for the calculation of absorbed dose to the target region, from activity in a source region. In addition, MIRD pamphlet No. 21 [31] provides an excellent overview of the current state of MRT dosimetry as well as clarification on the different values and standardised nomenclature recommended. Applying the MIRD methodology and recasting Equation 11.1 in terms of this formulism, the absorbed dose D can be defined as the product between the time-integrated activity and the S value:

$$D = S * \tilde{A} \tag{11.2}$$

The time-integrated activity $\tilde{A}$, named the cumulated activity, equals the total number of decays that take place in a given source region, with units Bq s, while the S-value is the mean absorbed dose per unit of cumulated activity, expressed in Gy /(Bq·s) or as a multiple thereof, for example, in mGy/(MBq s). By dividing Equation 11.2 through by the administered activity, A_0, the ratio of cumulated activity and the administered activity can be represented by the 'residence time' (τ_0) which is the average time that the activity spends in a source region.

$$\frac{D}{A_0} = S * \frac{\tilde{A}}{A_0} = S * \tau_0 \tag{11.3}$$

The absorbed dose per unit activity is named the absorbed dose per cumulated activity (or the absorbed dose per decay). Equation 11.3 allows the separation of aspects related to spatial and temporal behaviour considered in the S factor and the τ_0 calculation, respectively.

A source or a target region can be any defined volume, for example, the whole body, an organ or tissue, a voxel, a cell or even a sub cellular structure. It should also be noted that the source and target can be the same in the case of a self-irradiating tumour. With the source region denoted as r_s and the target region r_t, Equation 11.2 can be written more explicitly as:

$$D(r_t) = S(r_t \leftarrow r_s) * \tilde{A}(r_s) \tag{11.4}$$

$\tilde{A}$ is in effect the area under the curve that describes activity as a function of time in r_s after the administration of the radiopharmaceutical assuming the time of administration $t = 0$. The activity in r_s as a function of time is commonly determined from consecutive quantitative imaging sessions or via direct measurements of the activity on a tissue biopsy, a blood sample or via single probe measurements of the activity in the whole body. Compartmental modelling is a theoretical method that can be used to predict the activity in a source region in which measurements are not feasible. The time-integration period is calculated over infinite time, for example, 0 to ∞ or until a time τ_D, which is considered appropriate for the absorbed dose calculation (e.g. τ_D could be at least 5–10 times the calculated effective half-life of the radionuclide ensuring the remaining number of decays is small in relation to the overall number).

Traditional dose calculation methods have calculated the dose in standard phantoms – human models with known geometries. More recently, with the development of Monte Carlo and dose point kernel techniques, patient-specific dosimetry is beginning to be employed. The S value includes the emitted energy E_i for each nuclear transition i within a given decay, the probability Y_i, of emission of this transition, the absorbed energy fraction ϕ_i and the mass of the target region M(r_t). Δ_i, the equilibrium absorbed-dose constant for each nuclear transition is used to represent the product of the energy emitted E_i and probability of emission Y_i, which equates to the mean energy emitted per given transition per decay within the radionuclide. The absorbed fraction ϕ_i for each transition per decay is defined as the fraction of the energy emitted from the source region that is absorbed in the target region and varies from 0 (zero energy absorbed in r_t) to 1 (all the energy absorbed in r_t). For non-penetrating radiation such as alpha and beta radiation, this will be zero when the source and target are geometrically separated or taken as 1 when $r_t = r_s$.

The S-value depends on the shape, size and mass of the source and target regions, the distance and type of material between the source and the target regions, the type of radiation emitted from the source and the energy of the radiation. The full formalism includes a summation over all the transitions i per decay, as follows:

$$S\left(r_t \leftarrow r_s\right) = \sum_i \frac{\Delta\phi_i\left(r_t \leftarrow r_{s,i}\right)}{M\left(r_t\right)}$$

(11.5)

If the mass of both the source and target regions vary in time, the absorbed fraction will change as a function of time after administration, and a time-dependent version of the internal dosimetry nomenclature must be applied. S values, the mean dose per unit cumulated activity, have been pre-calculated and tabulated for different radionuclides and source-target configurations for both standard adults and children. If necessary, the S values of organs and tissues can be scaled according to patient-specific organ and tissue masses. Electron density scaling is also necessary where it departs from that used in the pre-calculated data (it is generally assumed to be water). More details are summarized in [5].

11.5.2 Time Activity Calculation

To calculate the cumulated activity, the activity as a function of time $A(t)$ must be determined. This can often be described by a sum of exponential functions (generally 1 or more), with A_j the initial activity for the jth exponential, λ_j the biological decay constant corresponding to the jth exponential, λ the decay constant for the radionuclide and t the time after the administration of the radiopharmaceutical. The sum of the j coefficients gives the total activity in the source region at the time of administration of the radiopharmaceutical ($t = 0$):

$$A\left(r_s,t\right) = \sum_j A_j e^{-t\left(\lambda+\lambda_j\right)}$$

(11.6)

The decay constants λ and λ_j equal the natural logarithm of 2 (ln 2 = 0.693) divided by the physical and biological half-life of the radionuclide ($T_{\frac{1}{2},\text{phys}}$ and $T_{1/2,j}$, respectively). The physical and biological half-life can also be combined into an effective half-life $T_{1/2,\text{eff}}$ according to the following equation:

$$\frac{1}{T_{1/2,\text{eff}}} = \frac{1}{T_{1/2,j}} + \frac{1}{T_{1/2,\text{phys}}}$$

(11.7)

The effective half-life is always shorter than both the biological and the physical half-lives alone. The cumulated activity for the relevant time period is commonly calculated as the time integral of an exponential function, but trapezoidal or Riemann integration techniques can also be performed according to the type of time activity behaviour. The number of measurement points are determined carefully considering that both the extrapolation from time zero to the first measurement of activity in the source region and the extrapolation from the last measurement of the activity in the source region to infinity can strongly influence the accuracy in the resultant time-integrated activity.

11.5.3 Basic Assumptions of the MIRD Formalism

Several assumptions are automatically made when the MIRD formalism is applied: activity in the source region is assumed to be uniformly distributed; the atomic composition of the medium is homogeneous; the electron density of the medium is homogeneous; the shape, size and position of the organs are as represented by the human phantoms for which the dose per activity correction factors have been pre-calculated; the mean absorbed dose to the target region is calculated. These assumptions are approximations of reality. The strengths of the MIRD implementation are its simplicity and ease of use. The limitation of these assumptions is that the absorbed dose will vary throughout the calculated volume and maximum or minimum doses to target organs cannot be determined.

11.5.4 Dosimetry on a Voxel Level

The activity present in each image voxel (3D volume element) can be quantified, using SPECT/CT, PET/CT or PET/MRI imaging. Multiple image sequences that display the activity distribution at different points over time after injection may be co-registered to each other to allow for an exponential fit on a voxel-by-voxel basis. A parametric image representing the time-integrated or cumulated activity on a voxel-by-voxel level can thus be calculated. The timing of the image sequences after administration has to be considered carefully taking into account possible exposure of the staff involved in acquiring the images, particularly just after t=0 when the activity is at its greatest. Parametric images that display the biological half-life for each voxel can also be produced by this approach.

The registration of images acquired at different points in time needs to be performed using either rigid or deformable methods as appropriate taking into consideration any organ/tissue movement and changes occurring between each time point. Multimodality imaging such as SPECT/CT and PET/CT facilitates the interpretation of the goodness of any alignment after registration.

Following production of a 3D voxel map of cumulated activity, a suitable method to calculate the absorbed dose distribution within the patient needs to be applied. One approach is where the MIRD schema is defined, at the voxel-level, as a 3D voxel matrix representing the mean absorbed dose to a target voxel per unit activity from a source voxel embedded in an infinite homogeneous medium. Another way of doing this is to use dose point kernels (DPK). A DPK represents the radial distribution of absorbed dose emanating from a point source in an infinite homogeneous propagation medium [32]. At the voxel level a DPK can be represented as a dose voxel kernel (DVK) which is normally generated from Monte Carlo (MC) simulations for the radionuclide of interest. The convolution of a DVK and the activity distribution from an image acquired at a certain time after injection then gives the absorbed dose rate at the voxel level (see section 9.3.5.3 for a similar application in EBRT). This method provides a tool for fast calculation of the absorbed dose rate on a voxel level. The main inaccuracy with this approach is that the convolution relies on the assumed spatial invariance of the DPK and DVK over the volume of calculation. This is valid if the whole body is composed of uniform electron density tissue. In practice, this is not the case and tissue electron density (obtained from the Hounsfield Units [HU] of CT images) varies from point to point. This means that the radial distribution of dose spread of the kernel from each calculation voxel will vary in each direction according to the local tissue densities.

MC simulations can improve the calculation accuracy further by accounting for non-uniform tissue electron density. Registering the activity distribution from a functional image (PET or SPECT) with the tissue electron density distribution from a CT image (see Figure 11.2) provides the required input data to enable a full MC calculation. Although full MC simulations are time consuming, improved computer processing speeds are reducing the calculation time.

11.5.5 Pitfalls and Caveat – Consistency of Reporting

It can be seen that there are several approaches to the calculation of internal doses delivered following injection of a radiopharmaceutical. There are now many studies reporting data on the clinical findings for existing and new diagnostic or therapeutic agents. However, in many of these articles, the description of the methodology applied for dosimetry is lacking or important details are omitted. To be useful to the scientific and clinical community, any dosimetry study should enable accurate reproducibility between centres both nationally and internationally to maintain the safety and the efficacy of the procedure. Methods and information useful for consistent documentation of dosimetry results in individual patient records are discussed further in [33].

11.6 TREATMENT PLANNING SYSTEMS

11.6.1 TARGET DEFINITION AND ORGANS AT RISK CONTOURING

A treatment planning system is a computer-based system running dedicated software to help automate the dose calculation and analysis process. The required input data includes anatomical image data for accurate tumour and organs at risk delineation and some form of parametrised image data necessary for activity quantification. The system should provide an algorithm for the conversion from activity to dose distribution based, for example, on S-values, dose point kernel convolution or MC calculation as outlined in the previous sections. Generally the conversion from activity to dose is accurate with greater uncertainties in dose estimation related to the accurate contouring of the target and organs at risk and on the transfer of these contours from one modality to another, based on rigid or deformable registration, between images taken at different time points and therefore patient and organ position.

Simplified dosimetric approaches might be used for a range of situations, such as when the target cannot be delineated (e.g. bone marrow dosimetry [34]), when the expected effect is related to the mean absorbed dose to the whole body [35–37], when patients' compliance is limited by poor clinical condition for which only measurements with external probes are feasible (e.g. patients in pain with multiple bone metastasis), and when images are obtained with limited counts producing images with poor statistics and resolution (e.g. ^{223}Ra images). In these cases, a simplified approach using tables, calculation spreadsheets, etc. can be used.

11.6.2 DOSE DISTRIBUTION OPTIMISATION AND EVALUATION

The concept of dose–volume histograms (DVHs), extensively used to evaluate tumour and organ dose distributions in EBRT (see 9.4.9), can also be used to display the non-uniformity in the absorbed dose distribution from radionuclide procedures. A differential DVH shows the fraction of the volume that has received a certain absorbed dose as a function of absorbed dose, while a cumulative DVH shows the fraction of the volume that has received at least the absorbed dose given on the x axis (see Figure 11.3).

A truly uniform absorbed dose distribution would produce a differential DVH that shows a single sharp peak indicating that all the voxels within that volume receive the same absorbed dose.

11.6.3 POST-TREATMENT DOSIMETRY

Most MRT still uses fixed activities, but it is recognised that post-treatment dosimetry might be used in order to verify the intended dose prescription and record critical organ doses. In addition, the comparison of pre- and post-treatment dosimetry is expected to validate the robustness of dose effect models thus leading to safer and more effective treatment.

11.7 RADIOPHARMACEUTICAL TARGETING FOR MRT

The number of available and possible radiopharmaceuticals for targeted radiotherapy has increased greatly since the early days. Table 11.1 provides examples of current MRT techniques. The following section describes these techniques in more detail and is divided into beta and alpha emitting radionuclides.

11.7.1 BETA EMITTING RADIONUCLIDES

11.7.1.1 Thyroid Cancer

Based on the British Thyroid Association 2014 [39], the use of radioiodine is only a small part of the overall treatment pathway for a patient with this disease but this guidance illustrates how MRT fits into a complex pathway and the importance of understanding the complete picture from the

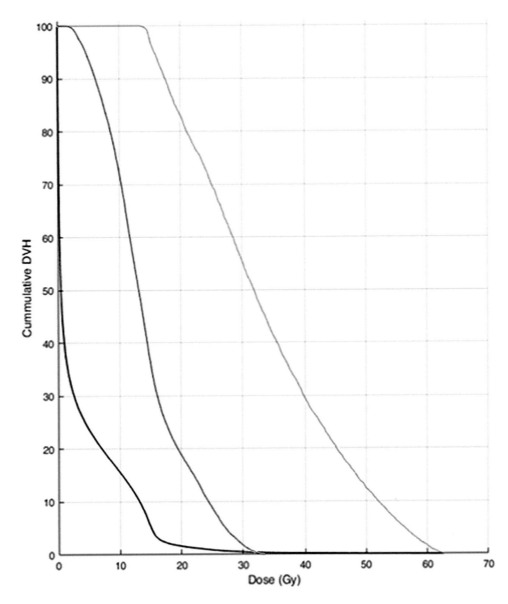

FIGURE 11.3 Examples of cumulative dose–volume histograms (DVHs).

patient perspective. Radioiodine for treatment of thyroid disease is administered orally. It is possible if the patient cannot swallow to administer intravenously but, in such cases, the product must be prepared using aseptic techniques to ensure it remains sterile. The product can be in the form of a capsule where the amount (MBq) of radioactive iodine is pre-loaded into the capsule by the manufacturer. In this case, the only preparation would be to unpack and check the capsule integrity and measure the radioactivity to ensure the patients were getting the radioactivity prescribed by the registered practitioner. Radioiodine can also be delivered as an oral solution requiring the correct amount of radioiodine to be dispensed from a stock vial. Typically a provided stock vial may contain 6000 MBq in a 6 ml solution requiring a small volume of 0.6 ml to be sub-dispensed for a typical 600 MBq treatment for a patient with benign Graves' disease. Dispensing must be done in the correct environment to minimise contamination and ingestion by the operator and is usually

TABLE 11.1

Examples of Current MRT Techniques according to Treated Disease, Radionuclide, Radiopharmaceutical and Method of Post-Treatment Imaging

Treated Disease	Radionuclide/ Radiopharmaceutical Used for Pre-Treatment Imaging	Pre-Treatment Imaging ur Measurement Methods	Radionuclide / Radiopharmaceutical for Therapy	Post-Treatment Imaging or Measurement Methods
Benign thyroid disease	^{131}I NaI, ^{124}I NaI, ^{123}I NaI	Thyroid probe. Gamma camera or SPECT/CT or PET/CT	^{131}I NaI	Thyroid probe. Gamma camera or SPECT/CT
Differentiated thyroid cancer (DTC) with ablative intent and in the case of recurrent disease	^{131}I NaI, ^{124}I NaI, ^{123}I NaI	Gamma camera or SPECT/CT or PET/CTWhole-body probes, blood sampling	^{131}I NaI	Gamma camera or SPECT/CT, Whole-body probes, blood sampling
Neuroblastoma in children and young adults	^{131}I mIBG, ^{124}I mIBG, ^{123}I mIBG	Whole-body probe for bone marrow estimation.Gamma camera or SPECT/CT or PET/CT	^{131}I mIBG	Whole-body probe for bone marrow estimation, Gamma camera or SPECT/CT
Neuroendocrine tumours in adults	^{131}I mIBG, ^{124}I mIBG, ^{123}I mIBG	Gamma camera or SPECT/CT or PET/CT	^{131}I mIBG	Gamma camera or SPECT/CT
Neuroendocrine tumours	Radiolabelled somatostatin analogs	Gamma camera or SPECT/CT	^{177}Lu DOTATATE	Gamma camera or SPECT/CT
Adult neuroendocrine disease	Analogs, as ^{86}Y–DOTATOC^{111}In-DOTATATE	Bremsstrahlung gamma camera or SPECT/CTPET/CT	^{90}Y somatostatin analogs	Bremsstrahlung gamma camera or SPECT/CT or PET/CT
Bone pain palliation	^{99m}Tc-MDP for ^{153}Sm-EDTMP	Gamma camera or SPECT/CT	^{89}SrCl$_2$, ^{153}Sm-EDTMP, ^{186}Re-HEDP, and ^{188}Re-HEDP	Gamma camera or SPECT/CT for ^{153}Sm-EDTMP, ^{186}Re-HEDP and ^{188}Re-HEDP
Treated disease	Radionuclide/ Radiopharmaceutical used for pre-treatment imaging	Pre-treatment imaging or measurement methods	Radionuclide/ Radiopharmaceutical for therapy	Post-treatment imaging or measurement methods
Bone metastases from castration-resistant prostate cancer	^{99m}Tc-MDP	Planar gamma camera imaging	^{223}Ra dichloride	Planar gamma camera imaging
Metastatic castration-resistant prostate cancer	Analog PET ligands	SPECT/CT	^{177}Lu PSMA ligands	SPECT/CT
Liver metastases or primary tumours	^{99m}Tc-MAA or ^{166}Ho-PLLA microspheres	Bremsstrahlung SPECT/CT	^{90}Y microspheres	Bremsstrahlung SPECT/CT, PET/CT or for ^{166}Ho MR
Non-Hodgkin's lymphoma	^{111}In-ibritumomab tiuxetan	SPECT/CT	^{90}Y-ibritumomab tiuxetan	Bremsstrahlung gamma camera or SPECT/CT, PET/CT

(Continued)

TABLE 11.1 (Continued)

Treated Disease	Radionuclide/ Radiopharmaceutical Used for Pre-Treatment Imaging	Pre-Treatment Imaging ur Measurement Methods	Radionuclide / Radiopharma-ceutical for Therapy	Post-Treatment Imaging or Measurement Methods
Radiosynovectomy	^{99m}Tc MDP/HDP/ HEDP and/or ^{99m}Tc-HIG	Planar Gamma camera imaging	^{90}Y, ^{32}P, and ^{186}Re colloid, ^{169}Er citrate	Planar Gamma camera imaging, dicentric chromosomes

Adapted from [38].

performed in a laminar air flow cabinet. To reduce this risk, most centres use capsules for both high (1.1 GBq–7.4 GBq) and low (400–800 MBq) administered activities. However, liquid radioiodine can be more flexible. The use of a gamma camera to perform the activity uptake measurements increases the expense and resources required for the dosimetry calculation but can be advantageous in individual patients because it adds information on the distribution of the activity enabling better correction for background radiation. A camera with a single head for acquisition of anterior views of the thyroid is sufficient in most cases. Although radioactive ^{131}I imaging and therapy are one of the earliest applications of theranostics (where the radionuclide is used for both therapy and diagnosis), and its physical half-life of 8.02 days allows imaging and data acquisition at sequential time points over a period of several days, a number of unresolved clinical questions limit the optimisation of diagnostic techniques and dosimetry protocols. The high energy gamma emissions of 364 keV, as part of the nuclear decay of ^{131}I, result in poor image quality leading to less than optimal evaluation of disease extent. This is due to collimator penetration by the high energy photons combined with the broad energy spectrum of bremsstrahlung radiation. Another factor contributing to poor image quality is the relatively low administered activity (2–5 mCi) for the pre-therapy evaluation. A potential improvement in dosimetry is expected based on the pre-treatment simulation using ^{123}I (gamma emitter) or ^{124}I (PET emitter) as a surrogate. The time activity curve can then be derived as described below for benign thyroid disease treatment.

11.7.1.2   Benign Thyroid Disease

For benign disorders such as thyrotoxicosis and arthritis, radionuclide therapy provides an alternative to surgery or other forms of medical treatment. Radioactive iodine (^{131}I-iodide) therapy for benign thyroid disease is the most established and common MRT. To tailor the therapeutic activity to be administered for radioiodine therapy of benign thyroid diseases such as Graves' disease or hyperthyroidism, pre-therapeutic dosimetry is based on the assessment of the individual ^{131}I kinetics in the target tissue after the administration of a tracer activity. Time schedules for the measurement of the fractional ^{131}I uptake in the diseased tissue are recommended and it is shown how to calculate from these datasets the therapeutic activity necessary to administer a predefined target dose for the subsequent therapy. These measurements are generally performed using a dedicated probe or gamma camera. The first step is to use one of the tracer capsules to evaluate background corrected net count rates, CRp, in a neck phantom that mimics the tissue between the centre of the thyroid target and skin surface. It is also important that the distance between the probe or gamma camera and this patient-like phantom is the same as for the patient measurements [40]. The background count rate for the probe or gamma camera in the absence of any activity sources and for the specific setting of energy window used for imaging is checked each working day. In addition, a check on the measurement constancy for a reproducible geometry is recommended at least once a week using an ^{131}I standard of known activity or a suitable test emitter, e.g. ^{133}Ba. Thereafter, the probe or gamma camera is ready to measure count rates in the target tissue corrected for background to give the net count rate above background.

Thus, the radioactive iodine uptake (RIU) in the thyroid target is calculated as follows:

$$RIU(t) = \frac{CR_T(t)}{CR_P} \tag{11.8}$$

where $CR_T(t)$ is the background corrected net count rates measured in the patient target tissue at time t after the administration and CR_P is the net count rate registered for the tracer activity in the neck-like phantom. If three uptake assessments are planned, measurements at 4 to 6 h, 1 to 2 days and 5 to 8 days after activity administration are recommended. Several equations for fitting experimental data can then be used for activity calculation as reported in [40]. The volume of the target can be estimated by approximating the thyroid lobe or the nodule by an ellipsoid measured using planar imaging and ^{99m}Tc or ultrasound. The mass M is then given by $M = \rho \cdot A \cdot B \cdot C \cdot \pi/6$, where ρ is the physical density and A,B,C are the axes of the ellipsoid. The activity necessary to achieve a specified radiation absorbed dose D in the target mass M is:

$$A = \frac{1}{E} \frac{M * D}{\int_0^\infty RIU(t)\,dt} \tag{11.9}$$

where E, the mean energy deposited in the target tissue per decay of ^{131}I, is 2.808 Gy g/(MBq s), valid for a thyroid with $M = 20$ g and introducing ≤ 5 % error for masses $M \leq 90$ g and with adequate accuracy for most patients. The S factor (i.e., the dose per unit of cumulated activity) is represented by the ratio E/M. The theoretical background and the derivation of the listed equations from compartment models of the iodine kinetics are explained further in ref. [40].

11.7.1.3 Neuroendocrine Tumours

A neuroendrocine tumour (NET) forms from the cells that release hormones (from the endocrine system) into the blood in response to a signal from the nervous system. This may, for example, result in the production of too much hormone which can cause many different symptoms.

Some neuroendocrine cells have proteins (peptide receptors), on their outside surface called somatostatin receptors. The hormone somatostatin attaches to these receptors and causes changes in the cell. For example, they may tell the cell to slow down the production of hormones. A PET scan can reveal if a tumour contains these receptors. Peptide receptor radionuclide therapy (PRRT) is a treatment option for somatostatin expressing (avid) NET using either ^{90}Y or ^{177}Lu labelled to a manufactured form of the hormone somatostatin, called a somatostatin analogue (e.g. dotatate, dotanoc or octreotide). The injected radiopharmaceutical circulates within the bloodstream attaching to the somatostatin receptors on the NET cells from where it enters the cell and kills it from the inside. The physical characteristics of ^{90}Y or ^{177}Lu are significantly different; ^{90}Y undergoes β^- decay with a half-life of 64.1 hours and a decay energy of β(max) 2.28 MeV [4]. ^{177}Lu emits β- particles with a half-life of 6.65 days and with lower energies β(max) of 497 keV (78.6 %), 384 keV (9.1 %) and 176 keV (12.2 %) and low-energy gamma photons, 113 keV (6.6 %) and 208 keV (11 %)]. Generally, ^{90}Y is therefore more penetrating and so effective for larger lesions and ^{177}Lu for smaller lesions (<2cm). The selection of the radionuclide has to be guided by the analysis of clinical condition (tumour size, localisation, and treatment purpose, etc.). Treatment with ^{177}Lu-Dotatate has mild side effects and for patients with metastatic or inoperable tumours is an effective choice. Evidence for the clinical benefit of ^{177}Lu-Dotatate is still under evaluation within the UK. PRRT therapy is administered intravenously over 30–60 minutes. The product may be prepared into a large syringe containing a diluted solution of 20–50 ml and administered using a standard hospital syringe pump to automate delivery and reduce the radiation dose to the hands of staff. The most frequent treatment protocol for ^{177}Lu-Dotatate is currently to administer 7.4 GBq up to four times with a 6- to 10-week interval between each administration. However, protocols delivering cycles of 7.4 GBq until a maximum prescribed absorbed dose to the kidneys and the bone marrow are reached are now being introduced. Although the photon yield is relatively low (see above), the high level of activity administered makes quantitative gamma camera imaging of ^{177}Lu possible. Moreover, it has been demonstrated that patient-specific absorbed doses for ^{177}Lu can be calculated and have a clinical benefit. The absorbed dose limits for the normal tissue and the desirable absorbed dose to the tumour requires further investigation [38].

11.7.1.4 Neuroblastoma

[131]I-metaiodobenzylguanidine (^{131}I-mIBG) has been used in the diagnosis and therapy of neuroblastoma in adult and paediatric patients for many years. Neuroblastoma is a form of cancer that develops in certain types of nerve tissue. To prevent unwanted thyroidal uptake of any free radioactive iodide, that has separated from the mIBG, patients are given stable iodine. This acts as a blocker to minimise uptake of the unwanted radioactive iodine in the thyroid and is administered orally 1–2 days before the planned mIBG treatment and continued for 10–15 days post therapy. Potassium perchlorate is generally used in combination with the stable iodine to facilitate the wash-out of any radioiodine that has still accumulated in the thyroid.

Multiple pre-treatment scans are acquired to determine the predicted level of uptake and retention (cumulated activity) within the patient. The dose to the whole body is used as a surrogate for the bone marrow which is the dose limiting organ. Whole-body dosimetry may be based on repeated measurements obtained using a ceiling mounted Geiger counter, the method of integration of the time activity curve being based on either analytical or empirical approaches. Tumour absorbed doses vary widely and so accurate calculation of lesion doses is expected to provide useful information regarding treatment efficacy. Tumour and normal organ dosimetry can be based on planar scans or multiple planar scans and a SPECT Image. When renal function is impaired because of disease or previous therapy, the radiation doses delivered to some organs (notably to the bone, red marrow [RM] and lungs) might be elevated and need to be considered. More details are reported in ref. [41]. ^{131}I-mIBG is also used to treat NET but is not discussed further in this chapter.

11.7.1.5 Selective Internal Radiation Therapy

Selective internal radiation therapy is a treatment option for patients with primary and secondary liver tumours that cannot be surgically removed. The rationale for this technique relies on the fact that blood supplied to liver lesions is fed mainly by the arterial stream, while normal functional tissue is supplied via the portal vein. Thus, microspheres labelled with an appropriate radionuclide can be directly administered to the liver tumours through the hepatic artery under angiographic guidance and are permanently lodged in microcapillaries, the small blood vessels in and around the tumour. This technique is also called radioembolisation. A pre-treatment planning angiogram is performed where a radio opaque dye is injected through a catheter inserted into the appropriate blood vessel feeding the liver. This dye, visible under x-ray imaging, is then used to confirm the microspheres will take the appropriate route and be located in and around the liver tumour as desired. In the same session, a treatment simulation is then performed, where a radioactive tracer is injected and scintigraphy scanning performed to determine take up of radioactivity in the liver and also that there is no unwanted escape to other parts of the body. Information acquired in the scan can then be used for dosimetric treatment planning (see further below). Toxicity monitoring is of importance as radioembolisation-induced liver disease leading to liver failure can be a severe consequence of standard treatment. At present, ^{90}Y glass, ^{90}Y resin and [166]Ho microspheres are all used to treat primary liver hepatocarcinoma and liver metastases.

TABLE 11.2

Characteristics of Radionuclide and Microsphere Combinations Currently Used for Selective Internal Radiation Therapy (SIRT) for Liver Tumours

Radionuclide (T½ in hours)	90Y (64.1)	90Y (64.1)	166Ho (26.8)
$E_{\beta max}$ in MeV	2.28 (99.9%)	2.28 (99.9%)	1.85 (>90%)
E_y in keV	2× 511 (<0.1%)	2× 511 (<0.1%)	81 keV (6.8%)
Microsphere (MS) material	Glass	Resin	Polylactic acid
Relative embolic effect	Low	High	High
Number of particles	5 million	50 million	30 million
Specific activity (Bq/microsphere)	1.250–2.500	50	330–450
Scout dose	99mTc-MAA	99mTc-MAA	166Ho-MS
Contrast injection during infusion	Possible	Only alternately	Possible
Imaging modality	SPECT or PET	SPECT or PET	SPECT or MRI

For both 90Y glass and resin, simulation scanning is performed with 99mTc-albumin macro aggregate (MAA), as a suitable surrogate, administered under angiographic guidance for quantitative imaging and pre-treatment dosimetry (Figure 11.4). 99mTc-MAA is prepared just before the angio scintigraphic administration. Absorbed dose is calculated using 99mTc-MAA SPECT images. The target mass can be obtained by contouring CT slices to determine its volume in 3D. Assuming an identical 99mTc-MAA and 90Y-microsphere biodistribution, the activity of 90Y in each voxel is directly proportional to the 99mTc counts in the corresponding voxel, allowing a patient-relative camera calibration [39]. The average dose (D) in gray is given by the cumulated activity (A) in GBq, in a target volume of mass M (kg) as follows:

$$D = 49.38 \times A / M \qquad (11.10)$$

where 49.38 is in units of J/GBq.

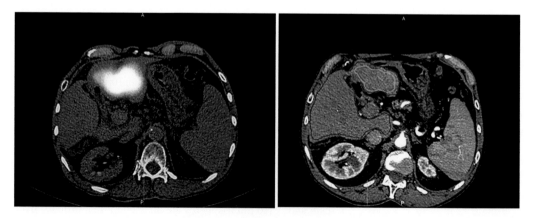

FIGURE 11.4 Example of 99mTc-MAA distribution and liver and target contouring.

For dose calculation, S-values or local kernel-based 3D deposition methods can be applied considering the physical characteristics of ^{90}Y. Because the radiation source and target volume are the same, we define S as S_{self} as it is self-irradiating. In such cases, an absorbed fraction equal to 1 can be used in each voxel and the ratio between the mean deposited beta energy per disintegration and the voxel mass calculated. Dosimetry at the voxel level can be implemented, for example, [62] used a cubic voxel with side d = 4.42 mm, density ρ = 1.03 g/cm^3 and calculated the local deposition coefficient within the voxel as follows:

$$S_{self} = \frac{933\,keV}{d^3 * \rho} = \frac{933 * 1.60 * 10^{-16}\,J}{(0.422cm)^3 * (1.03\,g/cm^3)} = 1.93\,Gy/(GBq \cdot s) \qquad (11.11)$$

where 1 electron volt (eV) has energy = 1.6×10^{-19} J. Assuming a permanent trapping of microspheres in each voxel, the cumulated activity $\tilde{A}_{voxel}$ in each voxel is related to the initial activity by:

$$\tilde{A}_{voxel} = \frac{A_{voxel}}{\lambda_{phys}} = \frac{T_{1/2,phys}}{\ln(2)} * A_{voxel} \qquad (11.12)$$

The convolution of a kernel dose deposition matrix around each voxel source of given activity allows the calculation of the 3D absorbed dose. Post-therapy quantitative imaging can be performed by ^{90}Y bremsstrahlung SPECT or ^{90}Y PET with suitable corrections [42]. The interaction between the high energy electrons from ^{90}Y and tissue can lead to the emission of bremsstrahlung radiation permitting the verification of the post-treatment activity distribution. Moreover, before the advent of tomographic imaging, it was postulated that the decay of ^{90}Y to the 0^+ excited state of ^{90}Zr may result in emission of a positron–electron pair which could be imaged. While the branching ratio for pair-production is small ($\sim 32 \times 10^{-6}$), PET has been successfully used to image ^{90}Y in numerous recent patients and phantom studies. ^{90}Y PET imaging has now been performed on a variety of PET/CT systems, with resolution and contrast superior to bremsstrahlung SPECT. Recently, polylactic acid microspheres labelled with ^{166}Ho have become available for liver treatments. This radionuclide has a 26.8 h half-life, dual beta emission with maximum energy of 1770 keV (49%)–1850 keV (50%) and, importantly, it is paramagnetic and emits gamma photons at 81 keV, with low abundance (6.7%). Gamma photons allow SPECT/CT imaging and dosimetry (some days after therapy to avoid gamma camera saturation) [43], while paramagnetism gives the additional possibility of post-therapy MRI evaluation. ^{166}Ho microspheres have the unique opportunity to be used for simulation (pre-treatment phase) and post-treatments, reducing the uncertainties due to the differences in microspheres observed for ^{90}Y treatment. Several prospective studies are ongoing to assess the efficacy and toxicity of this treatment.

11.7.2 Alpha Emitting Radionuclides

Bone metastases cause pain, fractures and a general decrease in the patient's quality of life and are most commonly treated palliatively. Radium-223 chloride (^{223}RaCl$_2$) in solution is an alpha-pharmaceutical or alpha-particle–emitting nuclide. It has been developed within the past 5 years and is now licensed in some countries to treat bone metastasis in men with prostate cancer. Although treatment is palliative and not curative, it has been shown in clinical trials to prolong life [44]. Whilst EBRT is of value in treating areas of well-localised pain, systemic targeted MRT offers the advantage that multiple sites can be treated simultaneously whilst minimising normal tissue dose. Radium-223-chloride, an alkaline earth metal, mimics calcium and is thus a natural bone-seeking agent being taken into the skeleton when injected into the body. Bone metastases from prostate cancer correspond to areas of increased bone formation and the radium chloride naturally incorporates itself into these tumour cells. The bone mineral hydroxyapatite (extracellular osteoid), which forms

50% of the bone matrix, is its target. Radium-223-chloride forms complexes with hydroxyapatite and subsequently gets incorporated into the bony matrix, targeting tumour cells in close proximity to new bone growth in and around the metastases. Molecular radiotherapy is usually delivered based on a fixed ^{223}Ra activity, sometimes adjusted for the patient's weight. ^{223}Radium has a half-life of 11.4 days, the localised action of alpha emission (with a short path length of 40–100 μm in tissue) helps to preserve the surrounding healthy bone tissue and bone marrow and limits distribution of the agent to soft tissue, thus also minimising the risk of radiation-induced side effects. It has better efficacy and causes minimal toxicity and is well tolerated by patients when compared with beta-emitters (e.g., ^{186}Re-HEDP treatments of bone metastases from prostate cancer).

11.8 TREATMENT REGIME OPTIMISATION

The optimal radiopharmaceutical and regimen to be used is based on the disease course and tumour volume. The calculated dose depends on the used radionuclide, the level of administered activity, the frequency of administration, that is, number of cycles and time between administration. From the viewpoint of the physicist, while the radiopharmaceutical governs the bio-distribution and localisation, the radionuclide is of greater interest, as quantitative imaging and dosimetry are governed by photon energies, physical (and consequently effective) half-lives and particle emission ranges. In EBRT, the use of fractionation allows better sparing of organs at risk than a single irradiation while maintaining the same tumour control. For MRT, a similar daily fractionation scheme is not possible although sequential administrations have been empirically tested to minimise toxicity. Such interval treatments can allow recovery of platelets or stem cell transplants as in the treatment of children with metastatic neuroblastoma [45]. However, this extended interval could permit tumour cell repopulation as well as a reduction of uptake during therapy likely due to the features of surviving repopulating cells, an aspect that complicates the situation. These factors can be taken into consideration using radiobiological models, to predict patient outcome and optimise future treatment schedules [46].

11.9 DOSE EFFECT RELATIONSHIP

To derive dose effect relationships, both the calculation of the absorbed doses to organs or the whole-body and the effects which might impact outcomes for the patient need to be identified.

Dose–response curves can also be derived from fixed activity outcome trials if dosimetry can be assessed. Tissue response depends on many factors, including cumulative (total) radiation dose delivered to the tissues, dose penetration and the radiosensitivity of the targeted cells. Optimal treatment is achieved when the radiation dose is entirely absorbed by the target tissue although in practice this is almost never the case because of biological turnover and uptake of the radiopharmaceutical in other areas. Numerous examples of threshold doses for efficacy and toxicity are reported, for example, in [56, 57] and described in the following section.

11.9.1 RADIOBIOLOGY AND DOSE EFFECT MODELS

Radiobiology concerns the study of the effect of irradiation on different endpoints as a function of dose (Chapter 4). In vitro, the fraction of surviving cells after irradiation allows determination of the relative biological effectiveness (RBE) which quantifies the ratio of absorbed dose required to achieve a given biological response versus a reference radiation (typically a beta emitter or EBRT radiation). However, the RBE varies as a function of absorbed dose and therefore a single RBE value is limited in its utility because it cannot be used to predict response over the wide range of absorbed doses experienced in clinical practice. This requires the need for standardised predictive bioeffect modelling to incorporate the different fractionation schemes and dose rates encountered for both MRT and EBRT [47].

Recently, the impact of radiobiology has been enhanced by the application of therapies with alpha emitters (e.g. Alpha-Radioimmunotherapy, another type of MRT) for which radiobiology through in vitro and in vivo studies will permit a more effective understanding of the RBE of alpha particles measuring the increasing effect in terms of killing of tumour cells or normal tissues.

The linear-quadratic model (LQ) for cell survival adapted originally in EBRT and brachytherapy can also be applied to therapy using radionuclides. [48]. This permits the calculation of a biological equivalent dose (BED) that enables different treatment regimens to be compared in terms of their biological effectiveness. It should be noted that this will differ dependent on the type of tissue or organ being considered. In continuous therapy such as MRT, the repair process of sub-lethal damage takes place during the radiation dose delivery and, therefore, a more general formalism is required. Assuming an exponentially decreasing dose rate and a complete decay of the source, Dale [49] demonstrated that the BED function is given by the following expression:

$$BED = D * \left[1 + \frac{G * D}{(\alpha / \beta)} \right] \tag{11.13}$$

where G is the Lea–Catcheside factor [50], D is the absorbed dose and α and β are the parameters of the LQ model [48] which vary according to tissue type. For a mono exponential clearance of activity:

$$BED = D * \left[1 + \frac{D * \lambda_{eff}}{(\alpha / \beta)(\mu + \lambda_{eff})} \right] \tag{11.14}$$

where D is the absorbed dose, μ is the exponential repair rate constant that quantifies the rate of sub-lethal damage repair and λ_{eff} is the effective clearance rate constant given by the sum of the physical decay and the biological clearance rate constants [51].

11.9.2 EXAMPLE: RED MARROW DOSIMETRY

For most non-myeloablative radionuclide therapy, RM is the first dose-limiting organ. The standard calculation of mean absorbed dose to RM is generally described as the sum of the self-absorbed dose in the RM and the absorbed dose from activity in the remainder of the body (RB). Using Equation 11.2, the dose to the red marrow D_{RM} can be written as:

$$D_{RM} = A_0 \left(\tau_{RM} * S_{RM \leftarrow RM} + \tau_{RB} * S_{RM \leftarrow RB} \right) \tag{11.15}$$

where τ_{RM} and τ_{RB} are the residence times of activity in the RM and remainder of the body and $S_{RM \leftarrow RM}$ and $S_{RM \leftarrow RB}$ are the mean doses to the RM per cumulated activity from the RM itself and remainder of the body, respectively. These values can be determined with the aid of computer programs such as OLINDA/EXM [52] with additional information on calculation methodology in [61].

11.9.3 EXAMPLE: KIDNEY DOSIMETRY

Radiation dose to the kidneys is of clinical significance for peptide receptor radionuclide therapy because of the high uptake of the peptides after glomerular filtration and retention of the radionuclide.

According to the multiregion model described in MIRD Pamphlet 19 [53] for kidneys, the mean dose to tissues of the renal cortex, D_{cort}, is given by:

$$D_{cort} = A_0 \left(\tau_{cort} \cdot S_{cort \leftarrow cort} + \tau_{med} \cdot S_{cort \leftarrow med} \right) \tag{11.16}$$

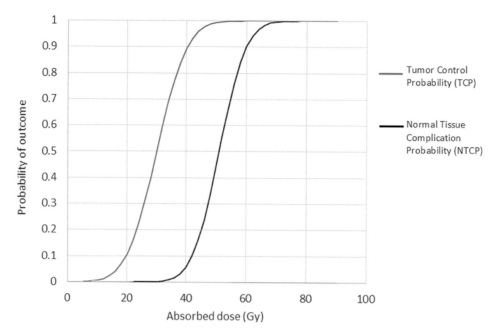

FIGURE 11.5   Example of TCP/NTCP curves.

where τ_{cort} and τ_{med} are the mean residence times for activity in the renal cortex and medullary papillae and $S_{cort\leftarrow cort}$ and $S_{cort\leftarrow med}$ the mean dose to the cortex per cumulated activity from the cortex and medullary, respectively.

The BED for a monoexponential clearance can be calculated using Equation (11.14) assuming the exponential repair rate $\mu = 0.24$/h and the α/β ratio $= 2.5$ Gy for the kidneys, using the value of λ_{eff} calculated from images with/without radiometric measurements.

11.9.4   TUMOUR CONTROL AND NORMAL TISSUE COMPLICATION PROBABILITY

Based on outcome data collected from patients undergoing MRT, two types of curve can be derived, the tumour control probability (TCP) and the normal tissue complication probability (NTCP) to describe the dose–effect relationship. The classical expressions of TCP and NTCP and their combination are based on models for the surviving fraction of cancer cells after radiation treatment developed for EBRT.

The TCP/NTCP curves (Figure 11.5) present a sigmoidal-like behaviour of effect against delivered dose which allows the estimation of the absorbed dose needed to improve the tumour control rate or maintain the toxicity of healthy tissues under a given rate, respectively. Strigari et al. [54] derived the first TCP/NTCP model for radioembolisation based on ^{90}Y resin microspheres. A clear correlation between absorbed dose and response has also been reported for the treatment of pancreatic neuroendocrine tumours. Similar to EBRT, these models and dose constraints can be adopted in the treatment planning strategy for MRT [55] enabling the most effective tumour control while minimising adverse side effects to healthy tissues.

11.9.5   SECONDARY TUMOURS AND RADIATION-INDUCED SIDE EFFECTS

When cure is feasible, the long-term consequences of radionuclide therapy (e.g. fertility disorders, leukaemia or other secondary cancers) compare favourably with the risks associated with and accepted for chemotherapy and radiotherapy. Several trials are producing clear evidence that radionuclide therapy can prolong patient survival; however, the optimal dosage to reduce the secondary cancer risk is still under evaluation. In addition, the adoption of specific dose constraints should help to reduce the rates of other long-term issues such as on the cardiovascular system as well as neurodegenerative effects, etc. This aspect is considered of particular importance for long-term surviving patients (e.g., children and young adults) where the use of dose constraints to limit organ doses is under investigation, for example, to the heart and its sub-volumes such as the left ventricle or left anterior descending artery. These parameters are expected to be of potential relevance for MRT when significant cardiac uptake is present [56].

Currently, the potential impact of internal dosimetry has not yet been fully explored and there is still a lot of work to be done. This requires ongoing development of accurate imaging of accumulated activity and organ delineation, improved dose calculation algorithms, standardised calculation and reporting methodology and requires prospective and retrospective clinical studies.

11.10 RADIOPROTECTION

The decision to admit a patient into hospital for MRT is based on the risk to the public from the radiation used for treatment as well as safely managing the clinical condition. Assessment of the risk from a patient must ensure that the exposure to the public remains less than the permitted dose-limit of 1 mSv per annum. In most cases, this is usually constrained to 0.3 mSv to allow for the possibility of multiple exposures to a member of the public from different sources that they may come into contact with over the course of a year, but there are circumstances where 1 mSv is appropriate. The decision on the exact level to use for the given situation will usually be made by the medical physics expert (MPE), an appropriately experienced medical physicist.

For thyroid treatments in the US, patients are allowed home if the measured effective dose rate is <70 µSv per hour at 1 m This is slightly less restrictive than other nations but is a compromise based on cost-effectiveness and efficiency [57–59].

There are two risks to the public to consider, the risk from the radiation dose rate emanating from the patient and the risk from potential radioactive contamination. These two aspects are considered separately, and for each type of MRT, each risk will vary and one may dominate. For example, in treatments with ^{131}I radioiodine, due to the high energy gamma emission of this isotope, there is always a radiation dose rate issue to consider.

Monitoring patients during inpatient treatment helps predict the day of patient discharge guaranteeing safe conditions for potential third-party radiation exposure such as to family members, co-workers and other members of the public [60]. Advice to be adopted by patients following treatment is provided in the form of clear written instructions. It is vital that patients have time prior to treatment to digest this information to ensure they make an informed decision when consenting to have treatment. All decisions on post-therapy restrictions must be sensible and based on sound evidence of radiation risk. For established MRT such as radioiodine treatments, there are established guidelines for admitting patients to hospital for treatment or treating patients as outpatients and allowing them home straight away. One role of the MPE is to understand the scientific basis for this guidance in order to decide what is required in the case of new types of treatments or to assess treatment options for unusual situations. An example would be whether to allow a patient treated with >800MBq ^{131}I to leave hospital earlier than normal due to exceptional circumstances (e.g. poor clinical condition such as an elderly patient with dementia) bearing in mind that radioiodine is predominately excreted in urine for 48 hours post treatment which brings into consideration the issue of potential radioactive contamination.

11.10.1 RADIOPROTECTION STRATEGIES FOR UNUSUAL SCENARIOS

Unusual scenarios will be discussed first with local colleagues including, where relevant, those involved with the management of emergency scenarios (e.g. potential heart attack, hypertension episode). Advice may also be sought from colleagues at other centres performing MRT. In the rare circumstances that a patient dies within a short time frame from an event that could not be predicted, that is, road traffic accident, unexplained cardiac arrest, etc., a pre-existing risk assessment should have already been prepared and available outlining the risks and to put in place practical aspects to reduce the risk for other staff, workers and members of the public potentially close to the patient, for example, in mortuary refrigerators or involved in burial arrangements. The risk assessment should include information on approximate dose-rates after treatment plus other issues to consider, for example, to be aware of particular organs that concentrate the radioactivity and advise on ways to reduce risk such as allowing the radioactivity to decay over a few days may be possible for short lived isotopes.

11.11 CONCLUSIONS

MRT is an exciting and evolving field offering bespoke radiotherapy targeting solutions for a wide range of treatment sites. The accurate calculation of internal dose distributions is subject to more uncertainties than in EBRT being strongly dependent on patient-specific factors such as how the radiopharmaceutical is taken up, distributed and retained within the tumour and other tissues. The radiobiological effects of these treatments are quite different from EBRT with the radiation delivered over short distances from beta and alpha emitting radionuclides. MRT has an exciting future and the medical physicist will play a key role in its ongoing development.

REFERENCES

1. Hertz B Saul Hertz (1905–1950) discovers the medical uses of radioactive iodine: The first targeted cancer therapy thyroid cancer. *Adv Diagnosis and Therapy.* 2016.
2. Seidlin SM, Marinelli LD and Oshry E. Radioactive iodine therapy; effect on functioning metastases of adenocarcinoma of the thyroid. *J Am Med Assoc.* 1946; 132:838–847.
3. Protocol for Establishing and Maintaining the Calibration of Medical Radionuclide Calibrators and Their Quality Control, Measurement and Good Practice Guide No. 93, National Physical Laboratory, 2006. https://www.npl.co.uk/special-pages/guides/establishing-maintaining-calibration-radionuclide.pdf
4. Quality Standards in Nuclear Medicine, IPEM Report 65, https://www.ipem.ac.uk/
5. Bailey DL, Humm JL, Todd-Pokropek A, van Aswegen A. (Ed.). (2014). Nuclear Medicine Physics: A Handbook for Teachers and Students Endorsed by: American Association of Physicists in Medicine (AAPM), Asia–Oceania Federation of Organizations for Medical Physics (AFOMP), Australasian College of Physical Scientists and Engineers in Medicine (ACPSEM), European Federation of Organisations for Medical Physics (EFOMP), Federation of African Medical Physics Organisations (FAMPO), World Federation of Nuclear Medicine and Biology (WFNMB). International Atomic Energy Agency (IAEA): IAEA.
6. Zanzonico P. Routine quality control of clinical nuclear medicine instrumentation: A brief review. *J Nucl Med.* 2008 Jul;49(7):1114–1131. doi: 10.2967/jnumed.107.050203. Review. PMID: 18587088.
7. Yanch JC, Flower MA, Webb S. A comparison of deconvolution and windowed subtraction techniques for scatter compensation in SPECT. *IEEE Trans Med Imaging.* 1988;7(1):13–20. doi: 10.1109/42.3924. PMID: 18230449.
8. Ogawa K, Harata Y, Ichihara T, Kubo A, Hashimoto S. A practical method for position-dependent Compton-scatter correction in single photon emission CT. *IEEE Trans Med Imaging.* 1991;10(3):408–412. doi: 10.1109/42.97591. PMID: 18222843.
9. Siegel JA, Thomas SR, Stubbs JB, Stabin MG, Hays MT, Koral KF, Robertson JS, Howell RW, Wessels BW, Fisher DR, Weber DA, Brill AB. MIRD Pamphlet No 16 –Techniques for quantitative radiopharmaceutical biodistribution data acquisition and analysis for use in human radiation dose estimates. *J Nucl Med.* 1999;40(suppl):37S–61S. [PubMed: 10025848]
10. King M, Farncombe T. An overview of attenuation and scatter correction of planar and SPECT data for dosimetry studies. *Cancer Biother Radiopharm* 2003;18:181–190. [PubMed: 12804043]

11. Dewaraja YK, Wilderman SJ, Ljungberg M, Koral KF, Zasadny K, Kaminiski MS. Accurate dosimetry in I-131 radionuclide therapy using patient- specific, 3-dimensional methods for SPECT reconstruction and absorbed dose calculation. *J Nucl Med.* 2005;46:840–849. [PubMed: 15872359]

12. Hoffman EJ, Huang SC, Phelps ME. Quantitation in positron emission computed tomography: 1. Effect of object size. *J Comput Assist Tomogr.* 1979 Jun;3(3):299–308. doi: 10.1097/00004728-197906000-00001. PMID: 438372

13. Brown DW, Kirch DL, Trow RS, LeFree M, Steele PP. Quantification of the radionuclide image. *Semin Nucl Med.* 1973;3(4):311–325. doi:10.1016/s0001-2998(73)80025-x

14. Willowson KP, Tapner M, QUEST Investigator Team, Bailey DL. A multicentre comparison of quantitative (90)Y PET/CT for dosimetric purposes after radioembolisation with resin microspheres: The QUEST Phantom study. *Eur J Nucl Med Mol Imaging.* 2015 Jul;42(8):1202–1222. doi: 10.1007/s00259-015-3059-9. Epub 2015 May 13. PMID: 25967868; PMCID: PMC4480824.

15. D'Arienzo M, Cozzella ML, Fazio A, De Felice P, Iaccarino G, D'Andrea M, Ungania S, Cazzato M, Schmidt K, Kimiaei S, Strigari L. Quantitative 177Lu SPECT imaging using advanced correction algorithms in non-reference geometry. *Phys Med.* 2016 Dec;32(12):1745–1752. doi:10.1016/j.ejmp.2016.09.014. Epub 2016 Sep 28. PMID: 27692753.

16. D'Arienzo M, Cazzato M, Cozzella ML, Cox M, D'Andrea M, Fazio A, Fenwick A, Iaccarino G, Johansson L, Strigari L, Ungania S, De Felice P. Gamma camera calibration and validation for quantitative SPECT imaging with (177)Lu. *Appl Radiat Isot.* 2016 Jun;112:156–164. doi:10.1016/j.apradiso.2016.03.007. Epub 2016 Mar 16. PMID: 27064195.

17. Mezzenga E, D'Errico V, D'Arienzo M, Strigari L, Panagiota K, Matteucci F, Severi S, Paganelli G, Fenwick A, Bianchini D, Marcocci F, Sarnelli A. Quantitative accuracy of 177Lu SPECT imaging for molecular radiotherapy. *PLoS One.* 2017 Aug 14;12(8):e0182888. doi: 10.1371/journal.pone.0182888. PMID: 28806773; PMCID: PMC5564164.

18. D'Arienzo M, Pimpinella M, Capogni M, De Coste V, Filippi L, Spezi E, Patterson N, Mariotti F, Ferrari P, Chiaramida P, Tapner M, Fischer A, Paulus T, Pani R, Iaccarino G, D'Andrea M, Strigari L, Bagni O. Phantom validation of quantitative Y-90 PET/CT-based dosimetry in liver radioembolisation. *EJNMMI Res.* 2017 Nov 28;7(1):94. doi: 10.1186/s13550-017-0341-9. PMID: 29185067; PMCID: PMC5705539.

19. Sjögreen K, Ljungberg M, Wingårdh K, Minarik D, Strand SE. The LundADose method for planar image activity quantification and absorbed-dose assessment in radionuclide therapy. *Cancer Biother Radiopharm.* 2005;20(1):92–97. doi:10.1089/cbr.2005.20.92

20. Ramonaheng K, van Staden JA, du Raan H. The effect of tumour geometry on the quantification accuracy of planar ^{123}I phantom images. *Phys Med.* 2016;32(10):1344–1351. doi:10.1016/j.ejmp.2016.03.016

21. Pereira JM, Stabin MG, Lima FR, Guimarães MI, Forrester JW. Image quantification for radiation dose calculations – Limitations and uncertainties. *Health Phys.* 2010;99(5):688–701. doi:10.1097/HP.0b013e3181e28cdb

22. Jentzen W, Freudenberg L, Bockisch A. Quantitative imaging of (124)I with PET/CT in pretherapy lesion dosimetry. Effects impairing image quantification and their corrections. *Q J Nucl Med Mol Imaging.* 2011 Feb;55(1):21–43. PMID: 21386783

23. Martí-Climent JM, Prieto E, Elosúa C, Rodríguez-Fraile M, Domínguez-Prado I, Vigil C, García-Velloso MJ, Arbizu J, Peñuelas I, Richter JA. PET optimization for improved assessment and accurate quantification of 90Y-microsphere biodistribution after radioembolisation. *Med Phys.* 2014 Sep;41(9):092503. doi: 10.1118/1.4892383. PMID: 25186412

24. Carlier T, Eugène T, Bodet-Milin C, et al. Assessment of acquisition protocols for routine imaging of Y-90 using PET/CT. *EJNMMI Res.* 2013;3(1):11. Published 2013 Feb 17. doi:10.1186/2191-219X-3-11

25. Fabbri C, Bartolomei M, Mattone V, et al. (90)Y-PET/CT imaging quantification for dosimetry in peptide receptor radionuclide therapy: Analysis and corrections of the impairing factors. *Cancer Biother Radiopharm.* 2015;30(5):200–210. doi:10.1089/cbr.2015.1819

26. Scott NP, McGowan DR. Optimising quantitative ^{90}Y PET imaging: An investigation into the effects of scan length and Bayesian penalised likelihood reconstruction. *EJNMMI Res.* 2019;9:40. https://doi.org/10.1186/s13550-019-0512-y

27. Gear JI, Cummings C, Craig AJ, et al. Abdo-Man: A 3D-printed anthropomorphic phantom for validating quantitative SIRT. *EJNMMI Phys.* 2016;3(1):17. doi:10.1186/s40658-016-0151-6

28. Denis-Bacelar AM, Chittenden SJ, Dearnaley DP, et al. Phase I/II trials of [186]Re-HEDP in metastatic castration-resistant prostate cancer: Post-hoc analysis of the impact of administered activity and dosimetry on survival. *Eur J Nucl Med Mol Imaging.* 2017;44(4):620–629. doi:10.1007/s00259-016-3543-x

29. Dewaraja YK, Frey EC, Sgouros G, et al. MIRD pamphlet No. 23: Quantitative SPECT for patient-specific 3-dimensional dosimetry in internal radionuclide therapy. *J Nucl Med.* 2012;53(8):1310–1325. doi:10.2967/jnumed.111.100123

30. Loeevinger R, Berman M. A schema for absorbed-dose calculations for biologically-distributed radionuclides. *J Nucl Med.* 1968;9–14.

31. Bolch WE, Eckerman KF, Sgouros G, Thomas SR. MIRD pamphlet No. 21: A generalized schema for radiopharmaceutical dosimetry--standardisation of nomenclature. *J Nucl Med.* 2009;50(3):477–484. doi:10.2967/jnumed.108.056036

32. Bardiès M, Kwok C and Sgouros G. Chapter Dose point kernels for radionuclide dosimetry. In *Therapeutic Applications of Monte Carlo Calculations in Nuclear Medicine* ed H Zaidid and G Sgouros (Bristol: IOP Publishing).

33. Lassmann M, Chiesa C, Flux G, Bardiès M; EANM Dosimetry Committee. EANM Dosimetry Committee guidance document: good practice of clinical dosimetry reporting. *Eur J Nucl Med Mol Imaging.* 2011;38(1):192–200. doi:10.1007/s00259-010-1549-3

34. Lassmann M, Hänscheid H, Chiesa C, et al. EANM Dosimetry Committee series on standard operational procedures for pre-therapeutic dosimetry I: blood and bone marrow dosimetry in differentiated thyroid cancer therapy. *Eur J Nucl Med Mol Imaging.* 2008;35(7):1405–1412. doi:10.1007/s00259-008-0761-x

35. Violet J, Jackson P, Ferdinandus J, et al. Dosimetry of ^{177}Lu-PSMA-617 in metastatic castration-resistant prostate cancer: Correlations between pretherapeutic imaging and whole-body tumour dosimetry with treatment outcomes. *J Nucl Med.* 2019;60(4):517–523. doi:10.2967/jnumed.118.219352

36. Buckley SE, Chittenden SJ, Saran FH, Meller ST, Flux GD. Whole-body dosimetry for individualized treatment planning of 131I-MIBG radionuclide therapy for neuroblastoma. *J Nucl Med.* 2009;50(9):1518–1524. doi:10.2967/jnumed.109.064469

37. Buffa FM, Flux GD, Guy MJ, et al. A model-based method for the prediction of whole-body absorbed dose and bone marrow toxicity for 186Re-HEDP treatment of skeletal metastases from prostate cancer. *Eur J Nucl Med Mol Imaging.* 2003;30(8):1114–1124. doi:10.1007/s00259-003-1197-y

38. Stokke C, Gabiña PM, Solný P, et al. Dosimetry-based treatment planning for molecular radiotherapy: A summary of the 2017 report from the Internal Dosimetry Task Force. *EJNMMI Phys.* 2017;4(1):27. Published 2017 Nov 21. doi:10.1186/s40658-017-0194-3

39. Perros P, Boelaert K, Colley S, et al. Guidelines for the management of thyroid cancer. *Clin Endocrinol (Oxf).* 2014;81(Suppl 1):1–122. doi:10.1111/cen.12515

40. Hänscheid H, Canzi C, Eschner W, et al. EANM Dosimetry Committee series on standard operational procedures for pre-therapeutic dosimetry II. Dosimetry prior to radioiodine therapy of benign thyroid diseases. *Eur J Nucl Med Mol Imaging.* 2013;40(7):1126–1134. doi:10.1007/s00259-013-2387-x

41. Giammarile F, Chiti A, Lassmann M, Brans B, Flux G; EANM. EANM procedure guidelines for 131I-meta-iodobenzylguanidine (131I-mIBG) therapy. *Eur J Nucl Med Mol Imaging.* 2008;35(5):1039–1047. doi:10.1007/s00259-008-0715-3

42. Pasciak AS, Bourgeois AC, McKinney JM, et al. Radioembolisation and the dynamic role of (90)Y PET/CT. *Front Oncol.* 2014;4:38. Published 2014 Feb 27. doi:10.3389/fonc.2014.00038

43. Elschot M, Nijsen JF, Dam AJ, de Jong HW. Quantitative evaluation of scintillation camera imaging characteristics of isotopes used in liver radioembolisation. *PLoS One.* 2011;6(11):e26174.

44. Hoskin P, Sartor O, O'Sullivan JM, et al. Efficacy and safety of radium-223 dichloride in patients with castration-resistant prostate cancer and symptomatic bone metastases, with or without previous docetaxel use: A prespecified subgroup analysis from the randomised, double-blind, phase 3 ALSYMPCA trial. *Lancet Oncol.* 2014;15(12):1397–1406. doi:10.1016/S1470-2045(14)70474-7

45. Gaze MN, Chang YC, Flux GD, Mairs RJ, Saran FH and Meller ST. Feasibility of dosimetry-based high-dose 131I-meta-iodobenzylguanidine with topotecan as a radiosensitizer in children with metastatic neuroblastoma. *Cancer Biother Radiopharm* 2005;20:195–199.

46. Sarnelli A, Negrini M, D'Errico V, Bianchini D, Strigari L, Mezzenga E, Menghi E, Marcocci F and Benassi M. Monte Carlo based calibration of an air monitoring system for gamma and beta+ radiation applied radiation and isotopes: Including data, instrumentation and methods for use in agriculture, industry and medicine. 2015;105:273–277.

47. Hobbs RF, Howell RW, Song H, Baechler S, Sgouros G. Redefining relative biological effectiveness in the context of the EQDX formalism: Implications for alpha-particle emitter therapy. *Radiat Res.* 2014 Jan;181(1):90–98. doi: 10.1667/RR13483.1.

48. Fowler JF. The linear-quadratic formula and progress in fractionated radiotherapy. *Br J Radiol.* 1989;62:679–694.

49. Dale RG. The application of the linear-quadratic dose-effect equation to fractionated and protracted radiotherapy. *Br J Radiol.* 1985;58:515–528.

50. Lea DE, Catcheside DG. The mechanism of the induction by radiation of chromosome aberrations in Tradescantia. *J Genet.* 1942; 44: 216–245.

51. Loevinger R, Budinger TF, Watson EE. *MIRD Primer for Absorbed Dose Calculations.* New York: The Society of Nuclear Medicine, Inc, 1991. revised edition.

52. Stabin MG, Sparks RB, Crowe E. OLINDA/EXM: The second-generation personal computer software for internal dose assessment in nuclear medicine. *J Nucl Med.* 2005;46:1023–1027. [PubMed: 15937315]

53. Bouchet LG, Bolch WE, Blanco HP, Wessels BW, Siegel JA, Rajon DA, Clairand I, Sgouros G. MIRD Pamphlet No 19: Absorbed fractions and radionuclide S values for six age-dependent multiregion models of the kidney. *J Nucl Med.* 2003;44:1113–1147. [PubMed: 12843230]

54. Strigari L, Sciuto R, Rea S, Carpanese L, Pizzi G, Soriani A, Iaccarino G, Benassi M, Ettorre GM, Maini CL. Efficacy and toxicity related to treatment of hepatocellular carcinoma with 90Y-SIR spheres: Radiobiologic considerations. *J Nucl Med.* 2010 Sep;51(9):1377–1385. doi: 10.2967/jnumed.110.075861. Epub 2010 Aug 18. PMID: 20720056

55. Strigari L, Konijnenberg M, Chiesa C, Bardies M, Du Y, Gleisner KS, Lassmann M, Flux G. The evidence base for the use of internal dosimetry in the clinical practice of molecular radiotherapy. *Eur J Nucl Med Mol Imaging.* 2014 Oct;41(10):1976–1988. doi: 10.1007/s00259-014-2824-5. Epub 2014 Jun 11. PubMed PMID: 24915892

56. Moyade VS. The heart matters: A review of incidental cardiac uptake on Ga-68 DOTA peptide PET-CT scans. *Nucl Med Commun.* 2019 Oct;40(10):1081–1085. doi: 10.1097/MNM.0000000000001064

57. Woodings S. Radiation protection recommendations for I-131 thyrotoxicosis, thyroid cancer and phaeochromocytoma patients. *Australas Phys Eng Sci Med.* 2004;27:118–128.

58. Ross DS, Burch HB, Cooper DS, et al. American thyroid association guidelines for diagnosis and management of hyperthyroidism and other causes of thyrotoxicosis [published correction appears in Thyroid. 2017 Nov;27(11):1462]. *Thyroid.* 2016;26(10):1343–1421. doi:10.1089/thy.2016.0229

59. American Thyroid Association Taskforce On Radioiodine Safety, Sisson JC, Freitas J, McDougall IR, Dauer LT, Hurley JR, Brierley JD, Edinboro CH, Rosenthal D, Thomas MJ, Wexler JA, Asamoah E, Avram AM, Milas M, Greenlee C. Radiation safety in the treatment of patients with thyroid diseases by radioiodine 131I: practice recommendations of the American Thyroid Association. *Thyroid* 2011;21:335–346.

60. D'Alessio D, Giliberti C, Benassi M, Strigari L. Potential third-party radiation exposure from patients undergoing therapy with 131I for thyroid cancer or metastases. *Health Phys.* 2015;108(3):319–325. doi:10.1097/HP.0000000000000210

61. Sgouros G. Bone marrow dosimetry for radioimmunotherapy: Theoretical considerations. *J Nucl Med.* 1993;34:689–694. [PubMed: 8455089]

62. Chiesa C, Mira M, Maccauro M, et al. Radioembolisation of hepatocarcinoma with (90)Y glass microspheres: development of an individualized treatment planning strategy based on dosimetry and radiobiology. *Eur J Nucl Med Mol Imaging.* 2015;42(11):1718–1738. doi:10.1007/s00259-015-3068-8

12 Optical and Laser Techniques

Elizabeth Benson
King's College Hospital NHS Foundation Trust, London, UK

Fiammetta Fedele
Guy's and St Thomas' NHS Foundation Trust, London, UK

CONTENTS

DOI: 10.1201/9780429155758-12

12.1 OPTICAL RADIATION IN MEDICINE

The use of optical radiation in medicine is widespread and extends to many different specialist areas. Optical radiation applied for medical treatment and diagnosis is termed 'artificial optical radiation' and its use in the workplace in the United Kingdom is governed by The Control of Artificial Optical Radiation at Work Regulations, 2010. Within these regulations, artificial optical radiation is defined as 'any electromagnetic radiation in the wavelength range between 100 nm and 1 mm which is emitted by non-natural sources'. That is, artificial optical radiation is electromagnetic radiation with wavelengths within the infra-red, visible or ultra-violet region of the electromagnetic spectrum.

The Control of Artificial Optical Radiation at Work Regulations, 2010 were implemented to transpose European Directive 2006/25/EC into UK law. In the Directive and its associated European Commission non-binding guide to implementation, wavelength ranges for infra-red (IR), visible and ultra-violet (UV) radiation are defined and it is these ranges that will be applied here, as seen in Figure 12.1. Further sub-divisions are defined within these wavelength ranges as described below.

12.1.1 Infra-red Radiation (1 mm–780 nm)

Infra-red radiation is further sub-divided into three wavelength ranges with different properties and applications as shown in Table 12.1.

12.1.2 Visible Radiation (380–780 nm)

All wavelengths of visible radiation may pass to the retina of the eye, as this is part of the mechanism by which we are able to see. Therefore, the whole of the visible spectrum is part of the retinal hazard region. Specific wavelengths of visible radiation correspond to different colours of visible light; these wavelengths are absorbed well by different materials within the skin and are therefore useful for different applications, as shown in Table 12.2.

12.1.3 Ultra-violet Radiation (400 nm–100 nm)

UV radiation has the highest energy of the optical radiations and this is key to its medical application. UV wavelengths are absorbed in the front structures of the eye and do not reach the retina. As with infra-red radiation, ultra-violet radiation is sub-divided into three wavelength ranges. These are described in Table 12.3.

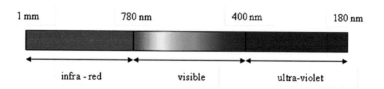

FIGURE 12.1 Artificial optical radiation wavelength ranges.

TABLE 12.1
Infra-Red Wavelength Ranges, Properties and Example Applications

Wavelength	Properties	Applications
Near infra-red (IRA) 1400 – 780nm	Penetrates to the back of the eye and retina, forms part of the 'retinal hazard region' 1400 – 400 nm.	Diode lasers • ophthalmology Nd:YAG lasers (1064 nm) • ophthalmology Low-level laser therapy (c. 900 nm) • physiotherapy Thermal cameras • thermometry • gait laboratories
Mid infra-red (IRB) 3000 – 1400 nm	Energy rapidly absorbed in water therefore useful for cutting tissue.	Holmium:YAG lasers (2100 nm) • urology, gynaecology, ear, nose and throat (ENT), gastroenterology
Far infra-red (IRC) 1 mm – 3000 nm	Energy rapidly absorbed in water therefore useful for cutting tissue.	Carbon dioxide lasers (10600 nm) • gynaecology, ENT, dental

TABLE 12.2
Commonly Used Visible Wavelengths, Properties and Applications

Wavelength	Properties	Applications
620–780 nm (red)	Absorbed well by melanin in the skin, green pigments and porphyrins.	Alexandrite laser (755 nm) • hair removal Ruby laser (694 nm) • ophthalmology, hair removal, tattoo removal Diode lasers (635 nm) • photodynamic therapy
570–590 nm (yellow)	Absorbed well by haemoglobin and coloured pigments in the skin.	Diode lasers (570–590 nm) • ophthalmology, tattoo removal
490–570 nm (green)	Absorbed well by melanin, haemoglobin and red pigments.	KTP:YAG lasers (532 nm) • ophthalmology, urology, treatment of vascular lesions, tattoo removal
400–490 nm (blue)	Absorbed well by porphyrins.	Diode lasers (405–420 nm) • photodynamic therapy, tattoo removal Blue light LED sources (430–540 nm) • neonatal phototherapy (peak at 460 nm) • dental curing (peak 460 nm) Filtered halogen and xenon sources • dental curing (peak between 460 and 480 nm)

TABLE 12.3
Ultra-Violet Wavelength Ranges, Properties and Example Applications

Wavelength	Properties	Applications
UVA 315–400 nm	Absorbed in the lens of the eye.	Excimer laser (XeF – 350 nm) • laser eye surgery, neurosurgery UVA fluorescent tubes (broadband, peak at 365 nm) • phototherapy (dermatology) • infection control • pest control (catering facilities) • genetics laboratories UVA1 filtered discharge lamps (340–400 nm) • genetics laboratories
UVB 280–315 nm	Absorbed by the cornea.	Excimer laser (XeCl – 308 nm) • laser eye surgery, neurosurgery Narrowband fluorescent tubes (TL01 311 nm) • phototherapy Broadband fluorescent tubes • phototherapy • laboratories
UVC 280–100 nm	Absorbed at the surface of the cornea.	Excimer laser (KrF – 248 nm, ArF – 193 nm) • laser eye surgery, neurosurgery Narrowband UVC fluorescent tubes (250–260 nm) • germicidal

TABLE 12.4
Effect of Localised Temperature Rise on Body Tissue

Temp. (°C)	Cell Effect
60–80	Collagen is denatured causing tissue contraction and coagulation
> 100	Cellular water boils and makes a phase transition to steam, expansion causes explosive rupture of cell walls
250–400	Tissue burns and becomes carbonized
500	Carbonised tissue burns and evaporates

12.1.4 COHERENT AND NON-COHERENT ARTIFICIAL OPTICAL RADIATION SOURCES

Artificial optical radiation may be further divided into coherent and non-coherent radiation. For artificial optical radiation emitted from a coherent source, all photons within the radiation beam will have the same frequency, wavelength and direction and move in phase with each other. This is not true for non-coherent radiation, which is generally broadband radiation formed of photons of different energies. The type of radiation, coherent or non-coherent, is linked to the mechanism of excitation of light emission.

Lasers are coherent sources of artificial optical radiation. All other sources of artificial optical radiation are non-coherent; therefore, these are also termed non-laser sources. Exposure to low levels of non-coherent artificial optical radiation is an everyday occurrence from sources such as computer screens and general lighting. Normal exposure to low levels of laser radiation is possible but less commonplace, for example, from laser pointers and CD or DVD players. When applying both types of artificial optical radiation source medically, the wavelength and dose are controlled to ensure that diagnosis and treatment are carried out safely.

12.1.5   OPTICAL RADIATION TISSUE INTERACTIONS

When the tissue interactions of artificial optical radiation are considered, they are categorized into four different effects: photo-thermal effects, photo-mechanical effects, photo-chemical effects and photo-ablative effects. These effects are used to predict how optical radiation may be applied for diagnosis and treatment, for example, the use of photo-thermal effects for coagulation in ophthalmology, the use of photo-mechanical effects for lithotripsy, the use of photo-chemical effects for UV treatment of skin conditions and the use of photo-ablative effects for laser eye surgery. The applications of these effects are discussed in more detail in Sections 12.2.1 and 12.3.1.

While these effects are applied for medical diagnosis and treatment, they are also the effects that may cause damage to the eye or skin when using artificial optical radiation. The difference between an effect causing a benefit and an effect causing an injury is the exposure that the individual receives. There are artificial optical radiation exposure limits which are set out in Annexes I and II of Directive 1006/25/EU; these are discussed in more detail in Section 12.4.

12.1.6   PHOTO-THERMAL EFFECTS

The thermal effects of artificial optical radiation are the effects that are most likely to cause damage. These effects may be considered with regard to two processes: coagulation and vaporization. Either of these effects may occur, dependent on the temperature to which the tissue is heated.

Coagulation occurs at temperatures between 60 and 80 °C; at this temperature, collagen fibres within the tissue shrink and coagulation occurs. Vaporisation occurs at temperatures greater than 100 °C when water within tissue cells boils and vaporises to become steam. The volume expansion associated with steam production leads to explosive rupture of the cell walls and the removal of tissue in this way is called vaporisation. Any further increase in temperature leads to further vaporisation until all cellular water is removed. Following this, carbonisation and burning of tissue will follow at 250 – 300 °C and 500 °C, respectively.

12.1.7   PHOTO-MECHANICAL EFFECTS

Photo-mechanical effects are caused by short (< 10 μs), high-power exposure to optical radiation. This type of exposure may result in two outcomes:

a) the fast local rate of change of temperature causes a high pressure gradient and associated mechanical waves or 'shock waves' which cause damage to cells.
b) high intensity electric fields are induced causing ionisation and leading to the production of localised plasma. This plasma expands, forming a shock-wave which exerts mechanical forces on the surrounding tissue, causing ablation and fracturing (photodisruption).

12.1.8   PHOTO-CHEMICAL EFFECTS

The body relies upon photo-chemical interactions to carry out several of its important natural functions; these include vision, formation of vitamin D and melanin production in tanning, for example. The chemical changes and interactions caused by the action of optical radiation may be used to diagnose and treat medical conditions, for example, the diagnosis of skin conditions by identifying bacteria that fluoresce, applying blue light to oxidise bilirubin and treat jaundice and applying visible light to activate cancer treatment drugs in photodynamic therapy. Exposure to optical radiation may also cause damage by photo-chemical effects, for example, sunburn. This damage may be cumulative, for example, the formation of cataracts from UV exposure over time or skin ageing.

12.1.9 PHOTO-ABLATIVE EFFECTS

Photo-ablative effects are only observed when using lasers and not when using non-laser sources of artificial optical radiation. For photo-ablative effects to occur, the energy of the laser beam must be great enough to break molecular bonds within the target material. Following the breaking of bonds, the atoms dissociate or vaporize with no damage to adjacent material. Photo-ablation may be used to precisely remove tissue or bone, depending upon the absorption of the laser wavelength in the material to be ablated. Due to the precision of the laser ablation process, this effect is applied in laser eye surgery.

12.2 LASERS

The basic components of all lasers are the same: an energy source and the lasing material or active medium which may be a solid, a liquid or a gas. The lasing material is contained within the laser cavity which has mirrored ends and acts as a resonant cavity for the amplification of optical radiation. A basic laser construction is shown in Figure 12.2. The acronym LASER stands for Light Amplification by Simulated Emission of Radiation, and this describes the process by which a laser beam is formed within the lasing medium and the laser cavity.

In basic laser theory, the energy source provides energy to the lasing medium. This energy excites atoms within the lasing material and raises them from ground level (A) to a higher energy level (A*). Following excitation of the atoms within the lasing material, the generation of a laser beam depends upon three processes (Figure 12.3).

1) **Spontaneous emission** – some excited atoms will spontaneously lose the energy they have gained and drop back down to ground level, A. When this happens, the atom loses energy and emits a photon with energy A* – A, a 'trigger' photon (Figure 12.3b). The value of A* – A depends on the individual energy levels of particular lasing materials and can be used to calculate the frequency of the emitted photon as in Equation 12.1.

$$A^* - A = hf \qquad (12.1)$$

where h is Planck's constant and f is the frequency of the photon (Hz)
This can in turn be used to calculate the wavelength of the photon by applying Equation 12.2.

$$c = f\lambda \qquad (12.2)$$

where c is the speed of light and λ is the wavelength of the photon (m)

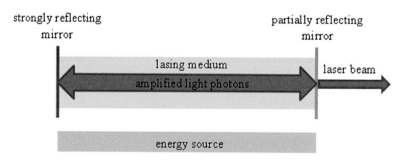

FIGURE 12.2 Laser construction.

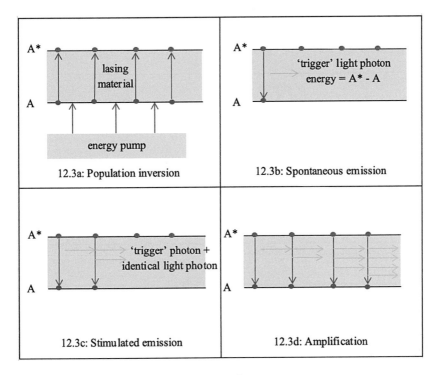

FIGURE 12.3   Laser processes in a two-level laser pumping system.

2) **Stimulated emission** – if a trigger photon encounters an excited atom, it will force the excited atom to emit a photon that is identical, that is, with the same wavelength and phase. The light photons emitted are reflected by the mirrors at either end of the laser cavity. They continue to pass through the lasing material and cause further stimulated emission and amplification of the original trigger photons (Figures 12.3c and 12.3d).

3) **Population inversion** – the majority of the lasing material atoms are excited rather than in the ground state. Therefore, there is an increased probability that a trigger photon will encounter an excited atom and amplification is maximized (Figure 12.3a).

When sufficient light photons have been generated to produce a laser beam with a certain power, the photons are released via the partially reflecting mirror and a beam shutter. In general, a laser beam has the following characteristics and it is these that set the laser apart from other sources of non-coherent optical radiation.

a) **Monochromatic** – the beam is made up of photons with a single wavelength or a small number of discrete wavelengths.

b) **Low divergence** – the photons in the beam spread out from the axis of the main beam very slowly and so the beam maintains its power or energy per unit area characteristics over long distances.

c) **Coherence** – all photons in the laser beam have the same frequency and wavelength and are in phase with each other.

There are of course exceptions and not all lasers conform to these beam characteristics.

The beam is delivered to its point of application via a beam delivery device, often an optical fibre incorporated into a device such as a slit lamp, colposcope or endoscope. Articulated arms with mirror systems are also commonly used for CO_2 laser delivery. The delivery devices associated with different laser applications are given in Table 12.5.

TABLE 12.5

Common Medical Lasers and Their Characteristics by Type of Lasing Medium

Laser	Lasing Medium Type	Wavelength (nm)	Characteristics
CO2		10600	Wavelength highly absorbed by intracellular water – applied in removal of the tissue and sealing of blood vessels. Wavelength does not penetrate deeply into the tissue (c. 100–300 μm); therefore, damage to deeper tissue is minimal. Delivered by a mirror system in an articulated arm as the wavelength is readily absorbed by most standard optical fibres.
Argon	Gas	488, 514	The blue/green wavelengths of the argon laser are well absorbed in haemoglobin and melanin. These wavelengths pass through water and can therefore be used in aqueous environments such as the bladder.
Excimer: Argon Fluoride (ArF) Krypton Fluoride (KrF) Xenon Chloride (XeCl) Xenon Fluoride (XeF)		193 248 308 353	Extremely precise photoablation of the tissue with no damage to surrounding tissue.
Nd:YAG (Neodymium doped: yttrium aluminium garnet) KTP:YAG (potassium-titanyl phosphate doped:YAG) Frequency-doubled Nd:YAG		1064, 532	The Nd:YAG 1064 nm wavelength is transmitted through water and can be used with a wide range of delivery devices. Photo-thermal tissue ablation is not as successful as for other lasers such as CO_2 or KTP:YAG and the beam energy is scattered and well absorbed by proteins in the tissue so greater damage to deeper tissue is caused. The second harmonic of an Nd:YAG beam may be isolated by directing the 1064 nm beam through a potassium-titanyl phosphate (KTP) crystal to achieve a 532 nm beam. This process is termed frequency doubling and lasers using this process are referred to as KTP:YAG or frequency-doubled Nd:YAG lasers. The 532 nm frequency-doubled beam is absorbed well in haemoglobin and melanin and transmitted through water. It is more efficient than the Nd:YAG for photo-thermal coagulation and vaporization of the tissue. Lasers may be capable of operating at both 1064 and 532 wavelengths. The beam is usually delivered using optical fibre.
Ho:YAG or THC:YAG (thulium, holmium doped: YAG)		2100	Strongly absorbed by water, used for photo-thermal cutting and ablation of the bone and cartilage. Output is pulsed and can be Q-switched for use in lithotripsy.
Er:YAG (Erbium doped:YAG)		2940	Strongly absorbed by water, more suited to superficial applications and less common than other YAG lasers.
Alexandrite	Solid state	755	Well absorbed by deoxygenated haemoglobin, longer visible wavelength and long pulse width allows deeper penetration and treatment of larger vessels.
Ruby		694.3	The first laser to be demonstrated. Had wide applications in several medical areas but has now been mostly replaced by more efficient lasing media with similar wavelengths.
Diode lasers	Semi-conductor	400–450 600–900 1100–1600	Generally compact and therefore easily transportable. Can be transmitted via conventional optical fibres.
Pulsed dye	Dye	300–1800 1100–1600	Visible wavelengths longer than 532 nm allow deeper penetration.

12.2.1 Types of Laser and Applications

The properties of a certain laser depend on several different laser characteristics including the laser's pumping method, lasing material and output modes. These properties indicate the applications for which a specific laser may be successfully used.

12.2.1.1 Pumping Methods

In Section 12.2, the concept of population inversion was introduced. The energy to achieve population inversion is provided by the laser energy pump in a process referred to as 'pumping'. Different lasing materials are pumped in different ways: gas lasers tend to use electrical current, while lasers with a solid state or liquid lasing material tend to be optically pumped, for example, using a flashlamp.

The length of time before spontaneous emission occurs for an excited lasing material atom gives an indication of how easy it is to achieve population inversion in that particular material. In the two-level pumping system described in Section 12.2, population inversion cannot be achieved as spontaneous emission occurs very quickly, resulting in a constant transfer of atoms between the ground and excited energy levels. For some materials, it is possible to make use of an energy level between the ground and excited states. This allows the atoms in the material to remain in a higher energy state for a longer period of time so that population inversion is more easily achieved. Where possible, two energy levels between the excited and ground state may be used for a four-level pumping system. This system further prolongs the time for which lasing material atoms are in an excited state and the arrangement is used in YAG (Yttrium Aluminium Garnet) lasers. See Figure 12.4 for a representation of the different laser pumping schemes.

12.2.1.2 Lasing Material

As described in Section 12.2, different lasing materials have different energy levels; therefore, it is the lasing material that determines the energy of the light photons emitted within the laser cavity. According to the electromagnetic spectrum, light photons with a certain energy and frequency will have a certain wavelength and so the lasing material also defines the wavelength emitted by a certain laser. Indeed, lasers tend to be referred to according to their lasing material. Tables 12.5 and 12.6 indicate some common medical lasing materials, their wavelengths and common applications.

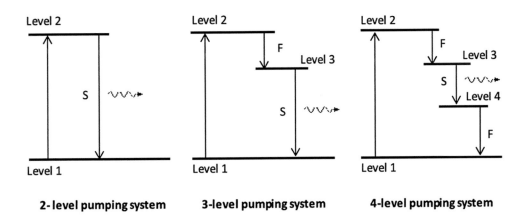

F – fast transition, no light photon emitted

S – slow transition, light photon emitted

FIGURE 12.4 Laser pumping systems.

TABLE 12.6
Common Medical Applications of Lasers

Ophthalmology	
Lasers	**Delivery Devices**
Nd:YAG, KTP:YAG, Diode (810 nm, 577 nm), Argon	**Optical fibre with slit lamp or laser indirect ophthalmoscope (LIO)**

Iridotomy – a small hole is made on the rim of the iris to allow fluid to flow between the back and the front of the eye and therefore reduce pressure in the back of the eye that may result from glaucoma.

Vitrectomy – removal of some or all of the vitreous humour from the eye. The vitreous is the gel-like substance that fills the back of the eye. Used to treat vitreous floaters and prior to retinal detachment treatment.

Photocoagulation – laser photo-thermal effects used to 'weld' the retina back into place when treating retinal detachment. Also used to seal unhealthy blood vessels in the eye and to treat wet age-related macular degeneration.

Capsulotomy – after cataract surgery, the lens posterior cavity may thicken (posterior lens capsule opacification) making it hard for light to pass through to the retina. A laser is used to make a hole in the lens posterior capsule so that the light can pass.

Excimer	**Precisely constrained automated eye tracking system**

Laser eye surgery – laser ablation is used to precisely shape the cornea and improve vision.

Urogynaecology	
Er:YAG, CO^2	**Handpiece with articulated arm**

Stress incontinence – Laser photo-thermal effects used to remodel collagen in endopelvic fascia and pelvic floor tissue.

Urology	
Nd:YAG, KTP:YAG, Ho:YAG, Tm:YAG	**Optical fibre with endoscope**

Lithotripsy – Q-switched pulses are used to photo-mechanically destroy urinary tract stones.

Benign prostatic hyperplasia (BPH) – Laser photo-thermal effects used to remove and coagulate the tissue from around the prostate.

Surgical lasers	
CO^2	**Articulated arm with mirror system or specialised optical fibre**

Skin and genital lesions – photo-thermal removal of skin lesions.

Colposcopy – photo-thermal removal of abnormal cervical cells.

Otolaryngology or ear, nose and throat (ENT) – removal of the tissue, for example, tumours, growths and warts.

Dental – removal of the excess skin in the mouth.

Lipomas – removal of benign, fatty tumours from the spinal cord.

Gynaecology – treatment of lesions in the lower genital tract, treatment of cervical disease

Nd:YAG, KTP:YAG	**Optical fibre with fetoscope for TTTS, endoscope for otolaryngology and laparoscope or hysteroscope with sapphire or quartz tip for gynaecology.**

Treatment of twin-to-twin transfusion syndrome (TTTS) – photo-thermal coagulation and ablation of placental blood vessels to prevent transfer of blood from one twin to another.

Otolaryngology – as for CO_2.

Gynaecology – as for CO_2 and removal of the tissue in the uterus (laparoscopy – see argon laser).

Ho:YAG	**Optical fibre**

Transmyocardial laser revascularization (TMLR) – revascularization of the myocardium based on diversion of arterial blood flow. Laser is used to create myocardial channels through which blood can flow to ischaemic regions.

(Continued)

TABLE 12.6 (Continued)

Surgical lasers cont.

Lasers	Delivery Devices
Diode (1470 nm, 810 nm)	Optical fibre

Endovenous laser vein treatment (EVLT) – photo-thermal coagulation and ablation of varicose veins.

Endometrial laser intrauterine thermotherapy (ELITT) – removal of tissue in the uterus.

Argon	Optical fibre with laparoscope

Laparoscopy – keyhole surgery in the abdomen and pelvis: the visible wavelength passes through water so that the laser can be used in aqueous environments.

Excimer (308 nm)	Optical fibre

Cardiology – precise laser ablation to remove coronary and peripheral plaques.

Neurology

Diode	Optical fibre

Neurophotonics – use of near infra-red laser beams to investigate the characteristics of brain tissue.
Optical coherence tomography (OCT) – technique analogous to ultrasound but using light instead of sound waves. Light reflections are used to image brain functional activity.

Excimer	Optical fibre

Brain bypass surgery – precise laser ablation used to weld blood vessels, bypassing damaged vessels and revascularizing areas of the brain.

Dermatology and aesthetics

Nd:YAG, KTP:YAG, Alexandrite, pulsed dye, ruby	Optical fibre with various laser handpieces

Vascular lesions – photo-thermal treatment of port wine stains, dilated capillaries, telangiectasias, rosacea and venous lakes as well as other vascular anomalies.

Scar therapy – photo-thermal treatment of tissue scarring and keloids.

Pigmented skin lesions – photo-mechanical treatment of various pigmented lesions, e.g. 'café au Lait', freckle reduction, etc using Q-switched mode.

Tattoo removal – Q-switched pulses are used to break down dye pigments in the surface skin layers. The dye fragments are then removed by the body.

Hair removal – photo-thermal treatment of the hair follicles to remove hair and reduce regrowth. Different lasers have different effects depending on type of hair and required longevity of effect. Diode lasers at 800 nm may also be used for this application.

CO_2, Nd:YAG, Er:YAG	Optical fibre

Skin resurfacing – Applied for treatment of atrophic scarring, e.g. acne scarring and scarring from chicken pox.

Photodynamic therapy (PDT)

Pulsed dye, diode	Optical fibre

Actinic keratosis (AK) and basal cell carcinoma (BCC) – A photosensitized drug is administered to the patient and is activated by exposure to the appropriate wavelength. In the presence of oxygen, free radicals are formed which kill the problem cells. Non-coherent light sources are also used effectively for PDT.

Physiotherapy

Diode	Optical fibre

Low-Level Laser Therapy (LLLT) – Use of laser light at various wavelengths to stimulate wound healing and reduce inflammation.

TABLE 12.7

Common Non-Laser Sources in Healthcare

Application	Non-laser Source	Type	Wavelength (nm)	Characteristics
General/Examination lighting	Illuminator/desk lamp Standard/compact fluorescent General lighting LED	Incandescent Gas discharge low pressure LED	350–1000 nm	These sources are designed to produce bright but not too intense visible light. Standard and compact fluorescent sources might produce residual UV mercury lines at 313 nm and 365 nm. In typical condition of use, emissions are not a hazard for people with average sensitivity to optical radiation.
Genetics/Disinfection	UVC fluorescent light source	Gas discharge low pressure	250-260 nm Peak 254nm	This source is highly erythematic and exposures of just few seconds can induce harm.
Genetics/Sample transilluminators	UVA and UVB fluorescent lamps	Gas discharge low pressure	280-400 nm	These sources can provoke erythema for short exposures (~30 min).
Aesthetic/ dermatology/ physiotherapy	Intense pulsed light (IPL) / intense light source (ILS) / intense continuous light system (ICL)	Xenon/ krypton arc (flash) lamp	400–1400 nm	Broad spectrum source is filtered to produce wavelengths suitable for the relevant procedure. Light is typically delivered to the skin via a handpiece and the skin should be cooled during treatment. Often used in beauty therapy applications with safety considerations similar to lasers (see MHRA guidance).
Pest Control Used to keep harmful insects away from food	Broadband UVA	Gas discharge low pressure fluorescent lamps	315–400 nm Peak 360 nm few visible lines	This source is generally used in canteens as its wavelengths attract pests.
UV Phototherapy For the treatment of skin diseases such as psoriasis, eczema and vitiligo	Broadband UVB	Gas discharge low pressure fluorescent lamps	280–400 nm Peak 312 nm Few visible lines	This UVB light can penetrate through the stratum corneum of both the eye and skin, and through the epidermis of the skin (0.05 mm), and it has a high erythematic effect for exposures of a few seconds.
	Narrow band UVB (TL01)		311–313 nm Peak 312 nm Few visible lines	This narrow band UVB light peaks at the wavelength that is most efficient in treating skin diseases such as psoriasis. It is less erythematic than lower wavelength UV.
	UVA or PUVA		315-400 nm Peak 354 nm Few visible lines	This broadband UVA source is about 100 times less erythematic than UVB sources, but it can penetrate deeper into the skin dermis (1-2 mm) and interact with the basal cell layers and can also reach the eye lens. It is generally used with photosensitizers.
	UVA1	Filtered high pressure Gas discharge	340-400 nm Peak 360-370nm Visible lines	Broadband UVA that penetrates deeply in the dermis but is less erythematic than standard UVA.

(Continued)

TABLE 12.7 (Continued)

Application	Non-laser Source	Type	Wavelength (nm)	Characteristics
Blue light photography For the treatment of neonatal jaundice and Crigler-Najjar syndrome, affecting the metabolism of bilirubin	Blue-light	Fluorescent tube Blue LED Filtered halogen lamps	400–600 nm Peak 500 nm	Blue light can penetrate deep into the skin and is used to break down bilirubin. Retinal cells are particularly sensitive to light in this region and can be seriously and irreparably damaged by it.
Photodynamic therapy For the activation of drugs that treat skin cancer	Red light	Filtered xenon source LED	615–640 nm Peak 630 nm	The intense red light can penetrate deep in the skin to subcutaneous tissue (3-4 mm) and is used to photo-activate drugs that treat skin cancer
Operating theatre lights	Bright visible light	Halogen sources	380–1000 nm	Very bright sources, emitting primarily in the visible, and potentially a blue light hazard. The natural aversion response of the eye usually prevents them causing any blue light damage.
		Metal-halide LED		
Phototesting	Monochromators	Filtered xenon lamps	Selectable narrowband emissions in the 290–700 nm range	The biological effects of the light will depend on the central wavelength of the emission.
	Solar simulators	Filtered xenon lamps	290 nm	A bright source with a significant UVB and blue light component, that can cause both erythema and ocular damage.

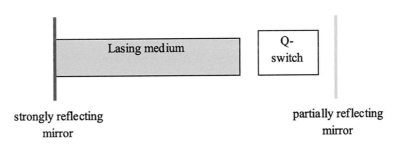

FIGURE 12.5 Q-switching.

12.2.1.3 Laser Output Modes

Lasers may be operated in different modes which can influence the application for which they are used. These are described in the following sections and illustrated in Figure 12.6.

12.2.1.3.1 Continuous Mode

The simplest mode in which a laser may operate is continuous mode, that is, the laser beam is continuously emitted and applied via the delivery device. The time for which the beam is applied is therefore determined by the operator using the exposure switch. The power (W) of continuous beam laser outputs is measured to indicate the energy deposited by the beam per second. When the area over which the beam acts is taken into account, the irradiance (Wm^{-2}), or power deposited per unit area of the beam, can also be calculated. This is used to indicate the potential a particular laser beam and delivery device has to treat or to cause damage.

12.2.1.3.2   Pulsed Mode

Lasers are commonly operated in pulsed mode and indeed it is often possible to use the same laser in either continuous or pulsed mode. In pulsed mode, the laser has a defined pulse length and therefore pulse frequency (number of pulses per second). Pulse lengths seen in medical applications are generally in the order of milliseconds (ms) with pulse repetition frequencies in the order of kilohertz (kHz).

As the pulse length of a particular laser pulse is known when using pulsed mode, the output of pulsed laser beams is measured in joules (J). Therefore, the power deposited in a specific time is indicated. When the area over which the beam is acting is taken into account, the radiant exposure of the beam (Jm^{-2}) may be calculated to indicate the treatment or damage potential of the beam.

12.2.1.3.3   Q-Switched Mode

Q-switched mode is a specialist laser mode used in certain applications. A Q-switch is an electro-optical component which stops optical radiation when activated and transmits optical radiation when deactivated. Such a device is placed between the mirrors at either end of the laser cavity (Figure 12.5). Initially, while the lasing material is pumped, the Q-switch is activated and lasing cannot occur as the optical path through the laser cavity is blocked. Pumping energy is stored in the upper energy level, allowing for maximum population inversion. When the Q-switch is deactivated, the energy stored in the upper energy level is released to give a high-energy 'giant pulse' with duration in the order of nanoseconds (ns). Q-switching is used to photo-mechanically break-up material such as kidney stones or tattoo dye pigments.

12.2.1.4   Laser Applications

Lasers are applied for many different and varied purposes in medicine. Some of the most common fields of application and associated procedures are shown in Table 12.6. This is not an exhaustive list; laser use in medicine is constantly evolving, and therefore, there are many emerging laser applications and also some outmoded laser applications which may not be included here.

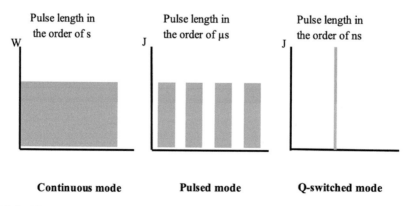

FIGURE 12.6   Illustration of pulse lengths for different laser modes.

12.3 NON-LASER SOURCES

Non-laser optical sources generate broadband spontaneous emissions from electrically excited materials, which can be metals (incandescent sources), gases (gas discharge sources) or semiconductor junctions (light-emitting-diodes).

12.3.1 INCANDESCENT SOURCES

Incandescent sources exploit the black body effect, by means of which a body heated at high temperatures radiates energy. They use a metallic filament, generally tungsten, heated by the passage of an electric current. The increase in temperature excites the electrons in the filament to higher energy states, and energy is emitted in the form of photons once the electrons return to their stable state. The filament is enclosed in a glass or quartz bulb filled with an inert gas (such as argon) that protects the filament from oxidation.

The radiant power (L) at a specific wavelength (λ) from a perfect black body is only a function of temperature and follows Planck's law:

$$L(\lambda) = \frac{A}{\lambda^5 \left(e^{\frac{B}{\lambda T}} - 1 \right)}$$ (12.3)

where c is the speed of light, $A = 2hc^2$, $B = hcK_B$, h is Planck's constant, K_B is Boltzmann's constant, T is the temperature in Kelvin.

The wavelength at which the emission is maximum also depends on the temperature according to Wien's law:

$$\lambda_{\max} = \frac{b}{T}$$ (12.4)

where $b = 2898$ mK.

The higher the temperature, the higher the contribution at shorter wavelengths (Figure 12.7).

Incandescent source filaments reach a temperature of 2000–3000 K, mostly emit in the infra-red (90%) and are not as efficient as other sources at producing emissions in the UV or visible range.

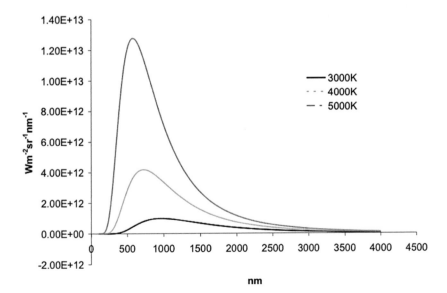

FIGURE 12.7   Spectral emissions of black bodies at different temperatures.

12.3.2 GAS DISCHARGE SOURCES

Gas discharge sources are made of tubes filled with a mixture of mercury and an inert gas such as argon or xenon. An electric field is applied across the tube, which will accelerate electrons and favour ionisation of more mercury particles and excitation of electrons to higher states.

As the electrons recombine, they will emit photons at characteristic wavelengths (also known as mercury lines), primarily 254 nm (Figure 12.8).

The tubes often have a phosphor coating applied to their internal walls, which when excited by the 254 nm radiation will emit a fluorescent spectrum dependent on the phosphor itself. Fluorescent tubes contain mercury at low pressures (a few atmospheres, i.e. few hundred thousand pascal).

High pressure mercury lamps (10–100 atm), also called 'arc lamps' because a discharge can also occur between the two electrodes outside of the glass envelope, have a broader emission spectrum (Figure 12.9).

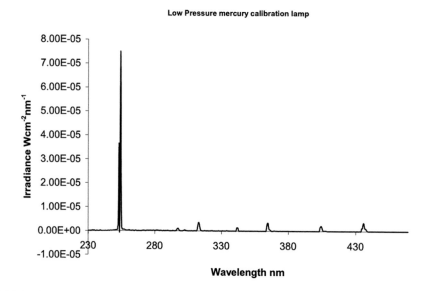

FIGURE 12.8   Spectral emissions of a low-pressure mercury lamp.

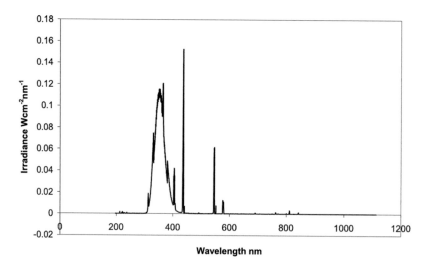

FIGURE 12.9   Example of ultraviolet fluorescent light spectrum.

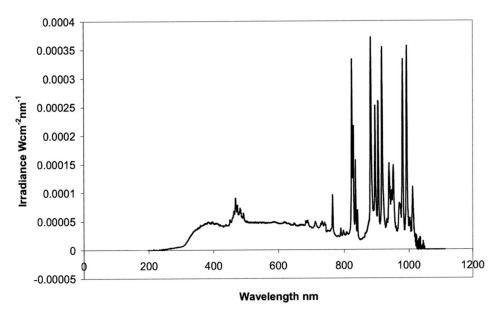

FIGURE 12.10 Spectral emissions of a xenon lamp.

Metal halide lamps and xenon arc lamps are particular classes of high-pressure lamp. In metal halide lamps, an element with a low excitation threshold is added to the mixture, such as an alkaline metal. This results in additional output, especially in the UV range. Xenon arc lamps, by contrast, do not contain mercury, just pure xenon. They require even higher powers and reach higher temperatures but produce a continuum spectrum from the UV to the infra-red (Figure 12.10). These lamps have a potential risk of explosion.

Xenon and krypton flash lamps are a particular kind of arc lamp. These lamps are used in intense pulsed light sources (IPLs) to produce a broadband optical output (400–1400nm). This broadband source is then filtered to achieve the wavelengths required for the relevant application.

12.3.3 LIGHT EMITTING DIODE (LED) SOURCES

Light emitting diode (LED) sources are based on solid state diodes, which are chips of semiconducting material doped with impurities to create p-n junctions. These junctions have an excess of holes on one side (p-side) and of electrons on the other (n-side). At equilibrium, an energy gap (E_g) separates the electron conduction band from the holes valence band (Figure 12.11), and if an electron and a hole recombine, this energy is emitted as a photon of wavelength

$$\lambda = \frac{E_g c}{h} \tag{12.5}$$

where h is Planck's constant and c is the speed of light.

The likelihood of recombination is maximum for electron/hole pairs separated by the minimum energy gap E_g but is also possible at other energy levels.

LED emissions follow a typical Gaussian distribution with a theoretical bandwidth dependent on the junction temperature (T in kelvin):

$$\Delta\lambda = 1.8\,K_B T \tag{12.6}$$

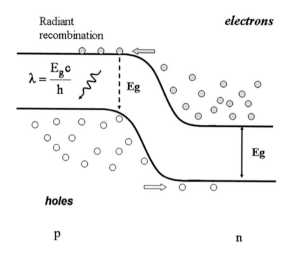

FIGURE 12.11   LED junction schematic.

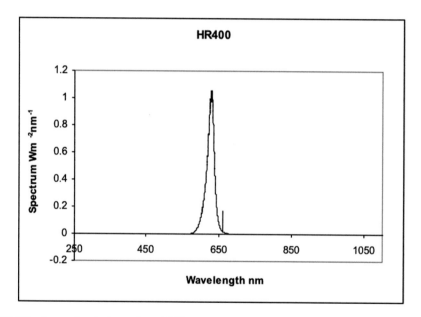

FIGURE 12.12   Spectral emissions of a red LED source.

and a maximum wavelength of emissions equal to

$$\lambda_{max} = \frac{1}{ch}\left(E_g + \frac{K_B T}{2}\right)$$ (12.7)

where h is Planck's constant, K_B is Boltzmann's constant and c is the speed of light.

A junction at 25 °C, with E_g = 25 eV will, for example, produce radiation in the red part of the spectrum with λ_{max} = 490 nm and a theoretical bandwidth of 9 nm.

12.3.4   NON-LASER SOURCES IN HEALTHCARE

Non-laser optical sources are used widely in healthcare. Table 12.7 lists common sources and their characteristics. The most common examples are examination lights and general lighting, followed by operating theatre lights, pest control sources and genetics laboratory instrumentation. Therapy sources are generally found in dermatology, neonatal units and ophthalmology, whilst curing lights are common in dentistry. Some highly specialized lights are employed in phototesting (3–4 units in the whole UK) and research units.

12.4 OPTICAL RADIATION PROTECTION

The hazards of artificial optical radiation have been investigated by the International Commission on Non-Ionising Radiation Protection (ICNIRP), the international body that advises on non-ionising radiation safety. ICNIRP originally published guidelines on exposure limits for incoherent visible and infra-red (1997) and UV radiation (1996), and for laser radiation of wavelengths between 1000 μm and 180 nm (1996). These limits have been updated and reviewed since they were first published. In 2000, a separate set of guidelines was published regarding limits of exposure to laser wavelengths in the retinal hazard region (400 nm–1.4 μm).

The introduction of the EU Directive on health and safety requirements for workers exposed to artificial optical radiation in 2006 (2006/25/EC) saw the adoption of the ICNIRP limits into European law. The exposure limit values (ELVs) for exposure to non-laser and laser sources, set out in Annexes I and II, respectively, of this Directive, have been directly adopted from the guidelines recommended by ICNIRP. These were formally implemented into UK law, under the Health and Safety at Work etc. Act, by the Control of Artificial Optical Radiation at Work Regulations, 2010.

When considering the hazards associated with optical radiation sources, it is important to consider several factors. Initially, the organs at risk of exposure must be identified, which for optical radiation are the eye and the skin, and the effects of the radiation beam interaction with these organs must be considered. Secondly, any hazards that are not associated with the optical radiation beam itself are considered: for example, electrical hazards associated with high-voltage power supplies and the fire hazard associated with the use of some lasers.

When considering damage to the eye and skin, artificial optical radiation may cause damage by any of four mechanisms. These mechanisms are the same as the tissue interactions described in Section 12.1 which govern the application of artificial optical radiation, that is, photo-thermal, photo-mechanical, photo-chemical and photo-ablative effects. The difference between a tissue interaction being a damage mechanism rather than an effect is the dose administered. A treatment dose is administered to provide a desired effect, an increased dose may cause damage, an undesired effect.

12.4.1 Hazards of Optical Radiation

The hazards associated with the optical radiation beam itself are highly dependent on the wavelength of the beam. This determines how far into the eye or skin optical radiation is able to penetrate and therefore which structures may be vulnerable.

12.4.1.1 Ultra-Violet Radiation Hazards

12.4.1.1.1 Eye

UVA wavelengths are absorbed at the lens causing photo-chemical effects such as the formation of cataracts. These effects are cumulative and therefore more significant for longer exposures. UVB wavelengths are mainly absorbed in the cornea and conjunctiva and also cause photo-chemical damage such as photokeratitis (snow blindness or welder's flash), a painful but temporary effect. For shorter exposures to UVA and UVB, heat damage is the more prominent hazard. However, for UVC wavelengths, photo-chemical effects dominate even for short exposures. UVC wavelengths are absorbed in the cornea and the damage threshold for photo-chemical effects is very low. Light wavelengths shorter than around 200 nm are heavily absorbed in air and are not considered here.

12.4.1.1.2   Skin

UV wavelengths are absorbed in the outer layer of dead skin cells (stratum corneum, 8–20 μm); therefore, the possibility of thermal damage is low. However, thermal damage may occur for higher exposures, by thermal induction or by vaporization or removal of the outer skin, particularly if the skin is repeatedly exposed. Prolonged and repeated exposure to UV wavelengths can cause photo-chemical damage such as erythema or sunburn. The long-term photo-chemical effect of UV exposure is damage to collagen and DNA within the skin cells causing skin aging and skin cancer.

12.4.1.2   Visible Radiation Hazards

12.4.1.2.1   Eye

Optical radiation with a wavelength in the visible spectrum is able to reach the retina. As the radiation travels through the cornea and lens towards the retina, it is focused. Therefore, a beam with a certain energy per unit area at entry to the eye may have an energy per unit area several orders of magnitude higher when it reaches the retina. Visible light reaching the retina is absorbed in the surface layer, the epithelium, and may cause retinal burns due to very high, localised temperature rises. As the incident light wavelength increases towards the infra-red spectrum, light is absorbed in the layers below the epithelium such as the choroid. Short duration (< 10 s) exposures tend to produce thermal effects. Longer duration (> 10 s) exposures cause photo-chemical effects such as photoretinitis (blue-light injury) and lesions. The damage associated with these photo-chemical effects is cumulative, resulting with the possibility of reduction in sensitivity of photosensitive cells and changes to night and colour vision.

12.4.1.2.2   Skin

Visible optical radiation can cause damage to the skin by photo-thermal mechanisms. The longer the wavelength of the optical radiation, the further it is able to penetrate into the skin. Wavelengths that travel deeper into the skin dissipate their energy and are also scattered by structures within the skin so that their energy is further diffused. This results in a higher injury threshold for longer wavelengths.

12.4.1.3   Infra-red Radiation Hazards

12.4.1.3.1   Eye

Optical radiation in the near infra-red range forms part of the retinal hazard region (400–1400 nm). Optical radiation beams of these wavelengths are able to travel through the cornea, lens and vitreous fluid and are focused onto the retina. Due to their longer wavelength, near-infra-red wavelengths pass to the choroid layer of the retina and are associated with the same effects as those described for visible radiation.

Optical radiation in the mid-infra-red range is absorbed in the lens of the eye. The main hazard associated with this is cataract formation caused by photo-chemical effects. The formation of cataracts may not be immediate but the effect may be cumulative and is associated with long term exposures.

Far-infra-red optical radiation wavelengths are entirely absorbed in the front of the eye. The cornea is the part of the eye most at risk when considering wavelengths longer than 2000 nm. The damage caused by exposure depends upon the wavelength of the incident optical radiation beam as this governs the distance the beam can travel into the cornea. If the beam is absorbed in the outer cellular layer (corneal epithelium), natural replacement of epithelial cells will heal any photo-thermal lesions and there will be no permanent damage. However, thermal damage to deeper layers of the cornea may cause scarring and corneal opacity.

12.4.1.3.2 Skin

Damage to the skin from near-infra-red radiation is caused by photo-thermal effects, as described for visible radiation skin effects. The longer wavelengths of the near-infra-red spectrum penetrate deeper into the skin to the epidermis (50–100 µm) and dermis (1–4 mm). The longer wavelength of near-infra-red radiation means that its energy is dissipated and scattered as it travels through the skin and therefore the threshold for skin injury from exposure to infra-red radiation is higher than for visible wavelengths.

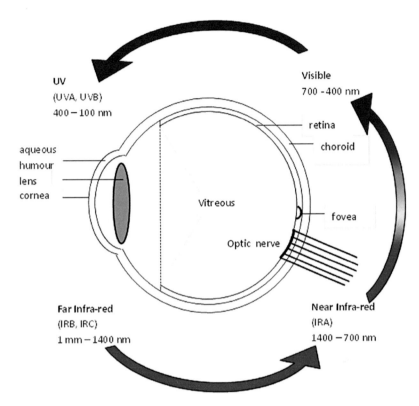

FIGURE 12.13 Anatomy of the eye showing structures that optical radiation of different wavelengths is able to reach.

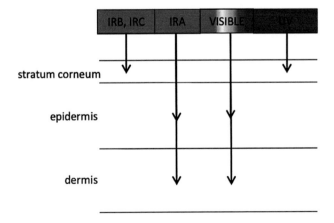

FIGURE 12.14 Structure of the skin in relation to penetration by optical radiation of different wavelengths.

As for UV, far-infra-red wavelengths are absorbed in the stratum corneum. These wavelengths are unlikely to cause thermal damage.

12.4.2 Laser Safety

Alongside EU Directive 2006/25/EU and the Control of Artificial Optical Radiation at Work Regulations, IEC standards in the 60825 'Safety of laser products' series describe the requirements for the safe use and manufacture of laser products. The parts of this series of particular relevance for the use of lasers in medicine are as follows.

- IEC TR 60825-1: 2014 – Safety of laser products – Part 1: Equipment classification and requirements. This part of the standard describes the process of laser classification.
- IEC TR 60825-8: 2006 – Safety of laser products – Part 8: Guidelines for the safe use of laser beams on humans. Describes the safety measures associated with the application of lasers in healthcare.
- IEC TR 60825-14: 2004 – Safety of laser products – Part 14: A user's guide. Describes the practical implementation of the measures described in other parts of the standard.
- IEC TR 60825-5: 2019 – Safety of laser products – Part 5: Manufacturer's checklist for IEC 60825-1. Describes the safety requirements, equipment and test methods required by the manufacturer for implementation of IEC TR 60825.

These standards, along with the UK Medicines and Healthcare Products Regulatory Agency (MHRA), document 'Lasers, intense light source systems and LEDs – guidance for safe use in medical, surgical, dental and aesthetic practices' and European Commission 'Non-binding Guide to Artificial Optical Radiation' provide practical advice regarding the implementation of laser safety legislation.

12.4.2.1 Laser Classification

Compliance with the relevant parts of IEC TR 60825 enables the laser product to meet applicable legal requirements. Detailed within the standard is the recommended process of laser classification, determination of the accessible emission limit and requirements for product labelling. The accessible emission limit (AEL) takes into account the laser output and access to the laser beam and is the maximum exposure a user has access to when using a laser under normal conditions. This is usually stated in watts for continuous wave lasers and joules for pulsed lasers.

The AEL for the treatment beam and aiming beam must be indicated on the laser as shown in the standard. As a general rule, a higher AEL indicates a higher laser class. The seven laser classes and their associated hazards are described in Table 12.8. In accordance with EU Directive 2006/25/EU, all Class 3B and Class 4 lasers in use in the workplace must be adequately risk assessed.

Taking the ELV for the eye (given in Annex II of Directive 2006/25/EU) for a particular laser and the AEL for that laser, the nominal ocular hazard distance (NOHD) may be calculated. The NOHD indicates the distance away from the laser beam where the ELV ceases to be exceeded. That is, where the AEL is equal to the ELV. At distances greater than the NOHD, the ELV for the laser will not be exceeded and the beam is safe for viewing. The NOHD can be calculated using Equation 12.8 where all quantities are expressed in SI base units.

$$Nominal\ ocular\ hazard\ distance = \frac{\left[\sqrt{\frac{4.Power}{\pi.ELV}} - initial\ beam\ diameter\right]}{Divergence, d} \quad (12.8)$$

TABLE 12.8

Laser Classes, Description of Associated Hazards and Example Devices

Class	Description of Associated Hazard	Example Device
Class 1 (embedded)	Laser completely enclosed. May be hazards if interlocks are over-ridden.	Laser in a CD or DVD player or laser printer.
Class 1	Very low-level power.	
Class 1C	Very low-level power designed for contact application, e.g. to skin. Skin maximum permissible exposure may be exceeded during application, and therefore, there is a risk to the skin; however, ocular hazard must be prevented by engineering means.	Home use hair removal device.
Class 1M	Safe under conditions of normal use, may be hazardous when using viewing optics.	A disconnected fibre optic communication system.
Class 2	Safe for short exposures – eye is protected by natural aversion response.	
Class 2M	Safe for short exposures – may be hazardous when using viewing optics.	Bar code scanner.
Class 3R	Low risk of injury for normal use, may be dangerous if used improperly by untrained individuals.	High-power laser pointers.
Class 3B	Direct viewing is hazardous.	Physiotherapy lasers for LLLT.
Class 4	Risk of injury to the eye and skin. Associated fire hazard.	Surgical lasers.

M – magnifying optical viewing instruments

R – reduced/relaxed requirements for manufacturer and user, e.g. no key switch

B – historical

C – contact

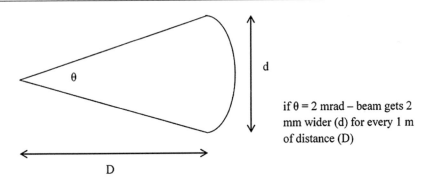

if θ = 2 mrad – beam gets 2 mm wider (d) for every 1 m of distance (D)

FIGURE 12.15   Laser beam divergence.

Divergence, expressed in radians, is a measure of how the laser beam spreads out along its path and can be calculated as shown in Figure 12.15 and Equation 12.9.

$$d = D\theta \qquad (12.9)$$

Equation 12.8 demonstrates that the NOHD depends strongly upon the individual properties of each laser such as power, spot size, divergence and wavelength – which is taken into account when determining the ELV. The NOHD is used to define the controlled area for a specific laser; for surgical lasers, this is generally taken to be the room in which the laser is used as the NOHD often

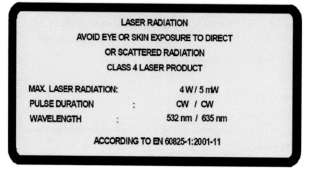

FIGURE 12.16   Examples of labels for Class 3R and Class 4 lasers and laser apertures.

exceeds the boundaries of the treatment room. For some surgical lasers, for example Ho:YAG, the NOHD is typically shorter. However, this is always dependent upon each individual laser's properties and should always be verified before a smaller controlled area is defined.

The manufacturer is required to provide AEL values and accompanying information that allows for the calculation of ELV, NOHD as well as personal protective equipment (PPE) requirements for a particular laser. Some of this more detailed information may be found in the laser user manual. However, it is a recommendation of IEC TR 60825-1 that lasers be labelled with a laser warning sign, that the laser aperture be labelled and that information relevant to the specific class of laser is given. It is recommended that Class 3B and Class 4 lasers are labelled as shown in Figure 12.16 with a warning of the hazards, the organs at risk, the nature of the laser radiation (i.e. visible or invisible), the laser class, the laser wavelength and the name and publication date of the standard to which the laser has been classified.

12.4.2.2   Non-Beam Hazards

The hazards of Class 3B and Class 4 laser exposure to the eye and Class 4 laser exposure to the skin are well known and safety measures are put in place to limit exposure. Therefore, laser injury, particularly to the eye, is unlikely. In practice, there are several other non-beam hazards associated with laser use which are more likely to cause serious injury.

12.4.2.2.1   Fire and Electrical

If laser treatments are carried out in areas where anaesthetic gases are used, such as operating theatres, patient drapes, oxygen and heat from the laser beam form a fire triangle. Therefore, there is a high risk of fire and even of explosion in these situations. This risk can be reduced by avoiding the build-up of flammable gases under drapes, keeping drapes surrounding the treatment area damp and ensuring that all endotracheal tubes are laser resistant to prevent damage from the laser beam.

Mis-use or lack of maintenance of laser units may also result in fire. Due to high electrical power demands, some lasers may require a dedicated, high-voltage power supply. The use of the incorrect power supply may lead to equipment malfunction and even electrocution – a hazard for both the operator and the patient. Electrical malfunctions caused by use of incompatible accessories or improper maintenance may also result in electrocution or present a fire hazard. Due to the high power demands of laser units, they become hot when in prolonged use; if the unit is not cleaned regularly, dust and debris in the filters may ignite.

12.4.2.2.2   Laser Plume

The debris created when tissue is evaporated, vaporized or ablated using a laser may contain harmful bacteria and viruses, blood fragments, hair, cellular material along with fumes associated with smoke which may also be toxic or carginogenic. If the laser application is internal, then this plume is normally considered to be contained within the patient, but this should be considered for each separate laser application. For external laser applications, the use of a smoke evacuator which is capable of removing particles 0.01 µm in size is recommended, along with face masks for users and for the patient.

12.4.2.2.3   Chemical

Organic dyes used in pulsed dye lasers may be toxic, corrosive and carcinogenic. Under normal conditions of operation, exposure to these chemicals does not occur. However, exposure may occur when servicing pulsed dye lasers or in case of accidental damage. The use of the organic dyes associated with pulsed dye lasers is controlled and monitored under Control of Substances Hazardous to Health legislation. Where pulsed dye lasers are in use, their users should be aware of the associated risks.

12.4.2.2.4   Optical (Beam Delivery)

As discussed in Section 12.2.1, several different devices are used to deliver the laser beam to the point of application. There are some risks associated with the use of these devices.

- optical fibre – fibre snap and damage to the end of the fibre during use – re-usable fibres should be checked prior to use.
- articulated arm – aiming beam and treatment beam misalignment – check prior to laser use.
- slit lamp (ophthalmology) – damage to filters protecting the ophthalmologist's eye – check prior to use.

12.4.2.2.5   Mechanical (Manual Handling)

Laser units may be heavy and difficult to move. The Management of Health and Safety at Work Regulations 1999 should be observed at all times when handling heavy equipment.

12.4.2.3   The Laser Safety Hierarchy

EU Directive 2006/25/EC requires that a risk assessment be carried out for all Class 3B and Class 4 lasers in use in the workplace. In the UK, the Health and Safety Executive has published general guidance regarding how risk assessment should be carried out. This guidance identifies the '5 Steps to Risk Assessment' and includes consideration of how risks may be reduced once they have been identified. The Health and Safety Executive recommends asking the following questions:

- Can I get rid of the hazard altogether?
- If not, how can I control the risks so that harm is unlikely?

Practically, it is recommended that the following possibilities are considered:

- trying a less risky option;
- preventing access to the hazards;
- organising your work to reduce exposure to the hazard;
- providing welfare facilities such as first aid and washing facilities;
- involving and consulting with workers;
- issuing protective equipment.

The MHRA laser guidance document organizes these considerations into a hierarchy of laser controls (Figure 12.17). Engineering controls may allow the risk to be removed or significantly reduced; therefore, it is recommended that these are applied first. Engineering controls should be followed by administrative controls which aim to change the manner in which people work using methods such as signage and training. Only if all risks cannot be reduced to as low as reasonably practicable using engineering and administrative controls, should the use of PPE be considered as a method of reducing risk.

12.4.2.4   Engineering Controls

Engineering controls applied to laser systems are most effective when considered and implemented at the design phase, as part of a risk assessment. However, in most cases, a laser is introduced into an existing environment and so this may not be possible. Therefore, retrospective engineering controls must be applied.

12.4.2.4.1   Enclosing the System and Room Design

Completely enclosing the laser system would be the most effective way to protect users from the laser beam and this method is used with great success in industry. However, this approach is often not practical or possible in healthcare. Even though completely enclosing the laser is generally not possible, careful consideration of the laser room configuration is straightforward and effective and can have similarly successful results. For example, consider the ophthalmology laser room shown in Figure 12.18(a). Rearrangement of the room so that the treatment beam is not aiming at the door, as in Figure 12.18 (b), significantly decreases the risk from the laser beam to anyone entering the room.

12.4.2.4.2   Environment

When considering room set-up, the equipment and furnishings within the treatment room should also be considered. The reflection of a Class 4 laser beam may cause damage to the eye or skin and represent a fire hazard. Therefore, any non-essential reflective items such as mirrors should be covered or removed and essential items such as surgical trolleys, surgeon's panels, taps, curtains,

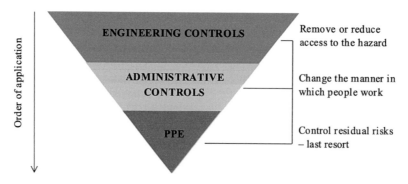

FIGURE 12.17   The laser safety hierarchy.

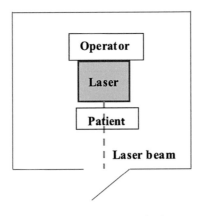 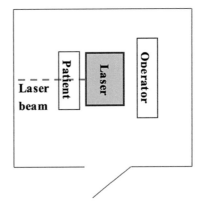

(a) Beam pointing towards door   **(b)** Beam pointing away from door

FIGURE 12.18   Laser room layout.

chair cushions, etc. should have low reflection finishes and be made of fire-retardant materials to mitigate associated risks. The position of any windows and vision panels should be considered, along with the properties of the laser, to ensure that suitable specialist laser blinds are fitted where required.

Signage is a simple and effective method of controlling access to potentially hazardous areas. As for ionising radiation, there are several different types of laser signage available including plain, illuminated and interlocked. Signage should meet the requirements of IEC TR 60825-14:2004 (Figure 12.19). The type of signage installed will depend on the individual requirements of the specific laser environment and application. Signage is of particular importance in environments such as operating theatres where it may not be possible to lock access doors.

To reduce the fire hazard associated with the use of Class 4 lasers, non-essential flammable gases should be removed from the laser environment. Suitable fire extinguishers should be readily available for use should a laser fire occur.

12.4.2.5 Administrative Controls

Particularly in the healthcare environment, the use of engineering controls alone cannot reduce the risks associated with laser use to an acceptable level. Therefore, administrative controls are also applied. Administrative controls are a series of laser management measures that are put in place to ensure that lasers are used by correctly trained people and according to correct procedures, therefore minimising the risk to members of staff and patients.

12.4.2.5.1 Personnel

For the laser safety system within an organization to be effective, it requires management. This should be documented in the organisation's 'Laser and Artificial Optical Radiation Safety Policy', or similar. Within this document, certain laser safety roles are defined in order to facilitate compliance with EU and UK legislation. These roles include the Laser Protection Adviser (LPA), Laser Protection Supervisor (LPS) or Laser Safety Officer (LSO), and authorized users.

An LPA may be internal or external to an organization and has expertise in matters of lasers and laser safety; therefore, they are able to assist the employer in fulfilling statutory and regulatory requirements. The MHRA laser guidance document recommends that it is good practice for all NHS and private laser providers to consult an LPA with regard to Class 4 and Class 3B laser systems. For private providers, the requirements of the Local Authority determine whether the appointment of an LPA is required.

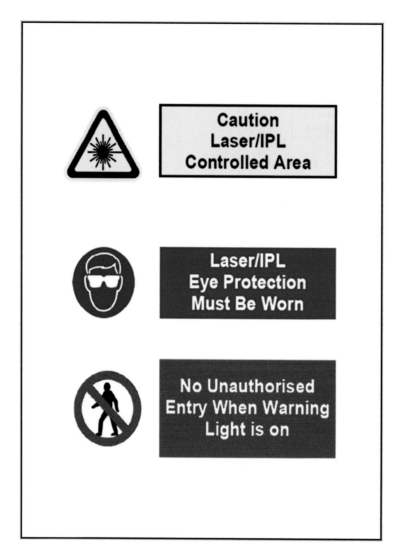

FIGURE 12.19   An example of British Standard laser signage.

The appointed LPA oversees laser safety within all departments of an organization. The appointment of a Laser Protection Supervisor (LPS) or Laser Safety Officer (LSO) is recommended at a departmental level to assist the LPA in local matters of laser safety. It is recommended that the LPS or LSO should be suitably trained and formally appointed.

Individuals who are authorized to use lasers are referred to as 'authorised users'. There may be more than one authorised user per laser. It is recommended that an authorisation mechanism and register of authorised users be in place for each laser.

12.4.2.5.2   Documentation

The organisational laser and artificial optical radiation policy has been mentioned previously and should be in place in all organisations using Class 3B and 4 lasers. This document describes the arrangements for the safe use of lasers within the organization and therefore defines the required documentation for compliance with artificial optical radiation regulations and guidance.

In accordance with legislation, all Class 3B and Class 4 lasers must have a prior risk assessment in place before being put into use. A risk assessment may be carried out by the LPA or LPS and should consider the risks associated with the hazards of laser use and the measures that should be put in place to reduce those risks to as low as reasonably practicable.

Prior risk assessment is generally used to define the laser controlled area and to compile Local Rules for the controlled area. A Local Rules document should be in place for every laser within an organization and describes the normal and emergency operating procedures to be followed during laser use. All members of staff involved in laser use should have read the Local Rules and should be asked to sign to say that they have done so. The MHRA guidance document suggests that the Local Rules should contain the following information.

- Details of the management structure
- Contact details for laser safety role holders (LPA, LPS, lead user)
- Register of Authorised Users
- Laser key keeping arrangements
- Description of the Controlled Area
- Description of the nature of the hazard associated with the equipment
- Details of controlled and safe access to the equipment area
- Training requirements for equipment users
- Responsibilities of the equipment user
- Methods of safe working, including equipment layout
- Simple pre-use safety checks and instructions
- Requirements for PPE
- Prevention of use by unauthorised persons
- Adverse event and equipment fault procedures and logs
- Use of loan or demonstration equipment
- Procedures relating to temporary staff
- Procedures relating to visiting engineers

As for all medical procedures, laser treatments and use should be described by a protocol and operating instructions. This is a method of standardising care and ensuring that defined procedures and treatment parameters are applied. All authorized users should be aware of the treatment protocol to which they are working to ensure similar outcomes for all patients. The laser operating instructions are usually found within the user manual.

12.4.2.5.3   Equipment

Laser equipment should be managed as part of an organisation's Equipment Management Policy. With regard to this, there are several phases within a laser's lifecycle that should be considered.

- Procurement: what equipment should be purchased and what actions and testing are required before the equipment can be put into use? A risk assessment may be carried out here to ensure that all actions are considered.
- Acceptance (pre-use checks) and commissioning: acceptance testing checks that the equipment has been received as specified, without any faults. Commissioning tests are used to set a baseline for future tests. Following successful completion of acceptance and commissioning tests, the equipment may be put into use.
- Normal service: once the equipment is in service, a record should be kept of any planned or unplanned maintenance including faults, upgrades and modifications.
- QA testing: daily or weekly tests, or even tests before every use, should be carried out, often by users, to ensure the continuing safety of the laser. QA tests may also be carried out less frequently, for example, annually or bi-annually by a specialist.
- Routine servicing: servicing by a qualified service agent at regular intervals, for example, every 6–12 months.

- Audit: regular, often annual, assessment to ensure that the required standards for safe laser and IPL use are being maintained within the department.
- Decommissioning: procedures regarding removal or disposal of equipment at the end of its life. Tests may be carried out, depending on whether the equipment is to be sold or disposed of. Dye lasers in particular may require specialist consideration due to the toxic dye.

The MHRA guidance document details recommended tests and the stages at which they should be carried out. It may be that the tests described require specialist test equipment. If such equipment is not available, it may be possible to request that a service engineer carry out the tests at installation and provide the results. It is not mandatory to carry out all of the tests described in the guidance document, the tests that are carried out will depend upon the organisation's equipment and artificial optical radiation management policies.

Loan laser equipment is often brought into the healthcare environment, sometimes as a trial prior to buying similar equipment, or as a demonstration. Before loan equipment can be put into use, a prior risk assessment must be carried out. Indemnity forms for such equipment should also be in place if it is to be used without the supervision of a company representative. The equipment management policy may also require that acceptance and commissioning tests be carried out on the equipment.

12.5.2.5.4 Training

All those involved in laser use should be suitably trained. The MHRA guidance document suggests three levels of laser training. Table 12.9 shows the different roles and the recommended training for each role. The MHRA also recommends syllabus content for core of knowledge and laser safety awareness courses, as do The British Medical Laser Association, Institute of Physics and Engineering in Medicine, Institute of Physics and Society for Radiological Protection.

12.4.2.6 Personal Protective Equipment

PPE should be the last resort with regard to reducing the risk associated with laser use to as low as reasonably practicable. Therefore, it should be demonstrated in the prior risk assessment that all possible engineering and administrative controls have been applied, before the use of PPE has been considered. The reduction of risk to an acceptable level in healthcare without the use of PPE is often not possible and therefore PPE is used in the healthcare sector when using lasers.

Laser safety glasses and eye shields are the most commonly used laser PPE in the healthcare environment and they are used by patients and by staff. There are many different types of eye protection available. It is important to realise that eye protection must be chosen based on its specification and the specification of the laser.

TABLE 12.9
Laser Roles and Required Training

Training		Provided by	Authorised User	Laser Assistant	Other Healthcare Staff
Equipment training		Laser supplier	✓	✓	X
Procedural training		Laser supplier/training course/ clinical lead	✓	X	X
Safety training	Core of knowledge		✓	✓	X
	Laser safety awareness course	LPA	X	X	✓

The optical density of the eyewear reduces the incident beam to a beam that is safe for viewing after it has passed through the lenses of the eyewear. Different types of eyewear have different optical densities for different laser wavelengths and for lasers operating in different modes. The protection a pair of glasses offers for different wavelengths and modes is printed onto the lenses or frame of the glasses. Protection for continuous wave operation is denoted by D, pulsed operation by I and Q-switched by R. The manufacturer's mark and a CE mark are also on the lens. Lenses tend to be different colours to provide protection at different wavelengths.

Taking all this into account, laser glasses are usually not transferrable between different lasers and even the upgrade of a laser may mean that laser glasses of a different optical density are required. The optical density of the lens required for protection from certain lasers can be calculated, although description of the method is outside the scope of this book. The LPA should always be consulted with regard to the purchase and use of laser safety glasses.

Laser safety glasses used in healthcare are manufactured and tested according to BS EN 207:2009 'Personal eye protection equipment. Filters and eye-protectors against laser radiation (laser eye protection)'. Laser glasses used in laser adjustment work must be manufactured and tested according to BS EN 208:2009 'Personal eye-protection. Eye protectors for adjustment work on lasers and laser systems (laser adjustment eye-protectors)'.

12.4.3   Non-Laser Source Protection

12.4.3.1   Regulations

The key regulations in the UK for non-laser optical sources are the same as for other hazardous equipment (including lasers); the Health and Safety at Work etc. Act (1974) (HSWA) and the Personal Protective Equipment at Work Regulations (1992). These regulations lay a duty on the employer to provide staff with adequate training and protective equipment and in turn make each employee responsible for operating the equipment safely for themselves and other persons who might be affected, such as patients or members of the public. As highlighted in Section 12.2.1.2, these common regulations have more recently been supplemented by specific regulations for optical equipment: the Control of Optical Radiation at Work Regulations (2010) (CAOR), which regulate occupational exposure and refer to the European Artificial Optical Radiation Directive (AORD) 2006. Taking into account the different nature of laser and non-laser sources, the Directive has separate sections and exposure limits for coherent and non-coherent sources, which also require different assessment procedures. The remainder of this chapter will focus on the exposure limits and assessment procedures relevant to non-laser sources and conclude with a sample assessment.

12.4.3.2   Exposure Limits and Action Spectra

Exposure limits for non-laser optical radiation have been established by the International Commission on Non-Ionising Radiation Protection (ICNIRP) and adopted in the AORD. The limits are based on numerous empirical observations, laboratory experiments and epidemiology studies and take into account specific optical spectral ranges.

There are two limits in the ultraviolet region (180–400 nm), but up to 13 limits for visible and infra-red optical radiation (400 nm – 1 mm). This is because the visible and infra-red emissions mainly affect the retina of the eye, for which the size of the retinal image and the length of exposure determine the maximum level of radiation which can be absorbed with no permanent effects. However, surveys of optical sources in healthcare have shown that only four of these limits (known as a, b, d and g), apply to these sources. This is essentially because non-lasers sources found in healthcare are primarily extended sources, in terms of how they are seen by the eye, and are used for reasonably prolonged times by staff during the day. These four ELVs are shown in Table 12.10.

The first two limits, the ultraviolet exposure limits (a and b), take into account superficial interactions with the body and are specified in terms of weighted radiant exposures. The radiant exposure

TABLE 12.10

Exposure Limits That Apply to Typical Non-Laser Sources in Healthcare, *t* Indicates Exposure Times in s, and α the Angle Subtended to the Source in Sr.

Limit Index	Hazard	Physical Quantity	Spectral Range, nm	Action Spectrum (see Figure 12.14)	ELV
a	Actinic hazard (H_{eff})	Effective radiant exposure	180–400 nm	$S(\lambda)$	30 Jm^{-2} daily value over 8 hours
b	Near-UV hazard (H_{UVA})	Radiant exposure	315–400 nm	UVA box function	104 Jm^{-2} daily value over 8 hours
d	Retinal blue-light hazard (L_B) t > 10000s	Effective Radiance	300–700nm	$B(\lambda)$	100 Wm^{-2}sr^{-1} for t > 10000s For α ≥ 11 mrad
g	Retinal thermal hazard (L_R) t > 10s	Effective radiance	380-1400 nm	$R(\lambda)$	2.8 107/C$_\alpha$ Wm^{-2}sr^{-1} for t > 10s where C$_\alpha$=α for 1.7 ≤ α ≤100 mrad C$_\alpha$ = 100 for α> 100 mrad

is defined as the energy received by a unitary surface, measured in Jm^{-2}. The weights applied are specific to the relevant limit and represent the variation in biological hazard with wavelength (i.e. energy and penetration properties). These weights are tabulated in the AORD as dimensionless action spectra (Figure 12.14) with values ranging between 0 and 1 as a function of wavelength. The most hazardous wavelengths have the highest weights, close to 1. Action spectra have been established empirically through experiments using monochromatic sources or via epidemiological studies.

The first UV limit (a, 30 Jm^{-2}), is also known as the actinic hazard limit because the UV wavelengths have often been referred to as 'actinic radiation', able to cause significant adverse health effects. The limit takes into account possible erythematic or carcinogenic effects for the skin and the cornea of the eye; it uses an UV action spectrum $S(\lambda)$ (Figure 12.14) which combines observation of genetic modifications linked to non-melanoma cancer and shorter term erythemal effects.

The second limit (b, 10^4 Jm^{-2}), known as the near UV hazard limit, is specific for UVA radiation (315–400 nm) and relates in particular to possible photo-chemical damage to the lens (formation of cataracts) observed with exposure to natural or artificial UVA sources. At present, it is still not known if this risk varies with wavelength within the UVA range, so the action spectrum is set equal to 1 between 315 and 400 nm and zero otherwise (Figure 12.14).

The hazard associated with non-laser visible and infra-red light is production of either photo-chemical or thermal damage. As explained previously, these effects depend on the way the beam enters the eye, and so in these wavelength ranges limits are expressed as weighted radiances, rather than irradiances. The radiance is the energy emitted by a source, across a specific field of view, per unit volume, and is measured in Wm^{-2} sr^{-1}. Thus, the estimation of visible and infra-red hazards requires knowledge of the field of view, expressed as the angle subtended to the source (ω, in radians) and the solid angle of view (Ω, in steradians). These can be approximated as follows.

$$\omega = \frac{d}{L} \tag{12.10}$$

$$\Omega = \frac{d^2}{A} \tag{12.11}$$

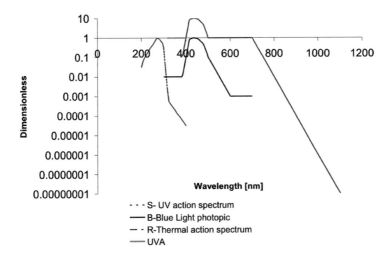

FIGURE 12.20   ICNIRP action spectra for optical radiation.

where d is the minimum distance from the source, L is the diameter of a circular source or average of the dimensions of a non-circular source and A is the area of the source.

The blue light limit (d, 100 Wm^{-2}sr^{-1}) relates primarily to the photo-chemical effect of light in the blue light region (around 450 nm) on the retinal cells and is weighted according to the blue light action spectrum B(λ) (Figure 12.20) which peaks in that region.

The thermal hazard limit (g, 2.8 10^{7}C$^{-\alpha}$Wm^{-2}sr^{-1}) addresses the thermal damage that can be induced in the same cells by both blue light and infra-red A radiation and is weighted using the thermal hazard action spectrum R(λ) (Figure 12.20). This action spectrum shows an anomaly, in that it has a weight higher than one and peaking at 10. This has been corrected by ICNIRP and the limits have been adjusted accordingly, but the revision has not been translated into the regulations at the time of publication of this book. Assessment done using the higher weighting is of course more restrictive and so errs on the side of safety.

Before discussing the equipment used to measure exposure to non-laser sources, it is helpful to discuss the rationale for risk assessment of these sources and hence to understand when such measurements are necessary.

12.4.3.3   Risk Assessment Rationale for Non-Laser Sources

As for laser sources, risk assessment of these sources needs to follow the five-step process recommended by the Health and Safety Executive.

1. *Identify the hazard.* This means gathering information on the emission spectrum of the source and on output (irradiance) levels. This might be done via available documentation or direct measurements.
2. *Decide who might be harmed and how.* This involves considering the use of the source, who might be exposed and for how long.
3. *Evaluate the risk and decide on precautions.* This means using the information gathered to estimate possible exposures during current work practice and the likelihood of these exposures occurring. It is then necessary to decide what controls and protective equipment, if any, are required to eliminate the risk or mitigate it to an acceptable level.
4. *Record your findings and implement corrective actions.*
5. *Review your assessment and update if necessary.* Assessment will need to be reviewed periodically to check whether a change of practice or equipment has altered the risk.

In some cases, information available via the manufacturer or from publications such as the EU Non-binding Guide to Optical Risk Assessment might be enough to establish the extent of the hazard arising from a source. Sources that have been developed specifically for photobiology, or in recent years since the publication of the Directive, are likely to have already been classified by the vendor according to the IEC Standards for Photobiological Sources (IEC 62471:BS EN 62471 2008) as lying in a specific risk group. This helps identify how hazardous the source is without any further assessment. Indeed, the Non-Binding Guide recommends starting an assessment with a risk group estimation. This risk grouping is analogue to the risk classification for lasers, and it is described below.

12.4.3.4 Risk Grouping

The risk group of a photobiological source can be estimated by measuring the output of the source at the minimum distance of exposure. This distance is usually 20 cm for hand-held sources or 1 m or more (according to use) for other sources.

There are four risk groups.

- *Exempt*. Long and short distance exposures are safe.
- *Risk Group 1*. Can be harmful for long exposures at short distances.
- *Risk Group 2*. Can be harmful for brief exposure at short distance.
- *Risk Group 3*. Can cause harm for short exposures (of the order of seconds) at medium to long distances (>1 m).

Table 12.11 reports the specific criteria used for grouping. A source is placed in a risk group when it poses at least one of the listed hazards at an exposure time less than the maximum indicated in the table. It must of course be taken into account that non-laser sources tend to be broadband and often pose multiple optical hazards at the same time; therefore, more than one exposure limit may apply. For example, if a source emits only UV radiation, with the UV actinic exposure being negligible and the maximum exposure time (MET) to near UV above 1000 s, the source will be classified as exempt. Pest control or UV hand hygiene boxes are examples of exempt sources using predominantly UVA bulbs. Conversely, if the MET for a UV source, with respect to the near UV limit b, is less than 1000 s, but greater than 300 s, the source will be classified as belonging to risk group 1 even if it does not exceed any other limit.

Risk Group 3 sources, which in healthcare include for example ultraviolet phototherapy units and IPLs will always require a specific risk assessment and control measures.

Risk Group 2 sources will become a hazard only if exposure is prolonged; therefore, they may require assessment according to the local procedures.

TABLE 12.11
Non-Laser Lamp Risk Group Classification

Risk Group	UV Hazard (E_{Eff}) Max Exposure Time [s]	Near UV Hazard (E_{UVA}) Max Exposure Time [s]	Retinal Blue Light (L_B) Max Exposure Time [s]	Retinal Thermal Hazard (L_R) Max Exposure Time [s]
Exempt	30000	1000	10000	10
1	10000	300	100	10
2	1000	100	0.25	0.25
3	<1000	<100	<0.25	<0.25

Risk Group 1 and exempt sources will only pose a risk in very specific limited circumstances. For example, hand-held examination UV sources are only hazardous if used inappropriately.

12.4.3.5 Measurement Instrumentation

Often it is possible to gather enough information about a source from either datasheets or published data to make an assessment of the exposure levels and associated risks. When this is not possible, the source will need to be tested to estimate its output. Testing equipment, spectroradiometers and radiometers, is based on the use of photodiodes and thermocouples. Spectroradiometers are used to gather spectral information from optical sources. They measure the light output, and specifically the irradiance of a source, as a function of wavelength. Two kinds of spectroradiometers can be used for this task: double monochromators and single monochromators.

Double monochromator spectroradiometers exploit a rotating diffraction grating or a prism to extract a narrowband component around a selected central wavelength and reconstruct a whole spectrum by mechanically changing the wavelength section at each step. In single monochromator charged coupled device (CCD) spectroradiometers, on the other hand, light diffracted by a fixed grating is focused on an array of CCD elements. Double monochromator systems demonstrate better rejection of stray light (i.e. light from outside the band measured), whilst CCDs have the advantage of being faster and more portable and are very useful for environmental measurements. However, calibrated broadband radiometers are less expensive and more widely used. These instruments directly estimate weighted irradiances of sources with known spectral output, such as phototherapy sources, and can be useful in making environmental assessment of new phototherapy units.

In summary, to evaluate the risk arising from a non-laser source, information is needed regarding both the spectral region of emission and the output levels of the source. This is illustrated in the following practical example concerning the assessment of sources used for optical therapy.

12.4.3.6 Practical Assessment of Optical Therapy Sources

In terms of radiation protection, therapy sources need separate assessments for patients, staff and members of the public. Patients need exposure of all or part of the body for therapeutic benefit. The efficacy of non-laser therapeutic sources depends on the accuracy of the dose the patient receives, so it is good practice to perform regular quality control and output checks on treatment units. In addition, treatment protocols need to be designed in such a way as to avoid any unnecessary exposure to parts of the body other than the part to be treated. The most common types of therapy sources are UV phototherapy cabins, neonatal blue light therapy units for the treatment of jaundice, and red lights used in photodynamic therapy (which uses light to activate cancer treatment drugs). In the case of staff and members of the public, exposure needs to be completely avoided or kept below the AORD exposure limits.

The dosimetry of treatment units is based on the use of radiometers responsive over the specific spectral regions of interest, with calibration traceable to national standards. Figure 12.21 illustrates the two dosimetry methods used for UV phototherapy cabins.

- *Direct method* (Figure 12.21a), whereby a member of staff wearing full protective equipment enters the cabin and takes a series of measurements from different exposure points.
- *Indirect method* (Figure 12.21b), which is based on the use of tripods and estimations of shielding factors that compensate for the tripods not having the same cabin occupancy as a person.

As the cabin irradiance is non uniform, varying with height and across panels, estimation of patient dose is based on the average of a set of measurements which can range from a minimum of 4 (one per panel) to 24 or more.

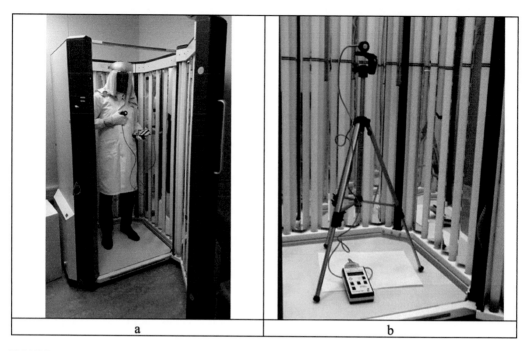

a   b

FIGURE 12.21   UV Cabin dosimetry methods: (a) direct method, (b) indirect method.

BIBLIOGRAPHY

A Non-Binding Guide to the Artificial Optical Radiation Directive 2006/25/EC, Radiation Protection Division, Health Protection Agency.

Bertolotti, Mario. 1983. *Masers and Lasers, An Historical Approach*. Adam Hilger Ltd, Bristol. ISBN 0-85274-536-2

Bogdan Allemann, I., & Goldberg, D. J. (eds). 2011. *Basics in Dermatological Laser Applications*. Published online Karger AG Basel. ISBN: 978-3-8055-9788-3. eISBN: 978-3-8055-9789-0

Brown, Ronald. 1968. *Lasers, Tools of Modern Technology*. Aldus Books, London.

BS EN 207 (2009). Personal eye-protection equipment. Filters and eye-protectors against laser radiation (laser eye-protectors).

BS EN 208 (2009). Personal eye-protection. Eye protectors for adjustment work on lasers and laser systems (laser adjustment eye-protectors).

BS EN 62471 (2008). Photobiological safety of lamps and lamp systems.

Carruth, J. A. S., & McKenzie, A. L. 1986. *Medical Lasers Science and Clinical Practice*. Adam Hilger Ltd.

Coleman, A., Fedele, F., Khazova, M., Freeman, P., & Sarkany, R. (2010). A survey of the optical hazards associated with hospital light sources with reference to the control of artificial optical radiation at work regulations 2010. *J. Radiat. Prot. Res.* 30(3), 469.

Coleman, A. J., Aneju, G., Freeman, P. & Fedele, F. (2020). *Thermal Radiation Exchange Model of Whole-Body UV Phototherapy*. Biomedical Physics & Engineering Express. https://iopscience.iop.org/article/10.1088/2057-1976/abac1d/pdf

Council Directive 2006/25/EC on the minimum health and safety requirements regarding the exposure of workers to risks arising from physical agents (artificial optical radiation) (19th individual Directive within the meaning of Article 16(1) of Directive 89/391/EEC) 2006, OJ L 114.

Health and Safety Executive, Control of Artificial Optical Radiation at Work Regulations 2010, S.I no. 1140. http://www.legislation.gov.uk/uksi/2010/1140/pdfs/uksi_20101140_en.pdf

Health and Safety Executive, Health and Safety at Work etc. Act, 1974, c. 37 http://www.legislation.gov.uk/ukpga/1974/37/contents

Health and Safety Executive, Risk Assessment, A brief guide to controlling risks in the workplace, INDG163(Rev4) (2014). http://www.hse.gov.uk/pubns/indg163.htm

Health and Safety Executive, The control of Substances Hazardous to Health Regulations (2002). Approved Code of Practice, Crown Copyright 2013, ISBN 978 0 7176 6582 2 http://www.hse.gov.uk/pubns/priced/l5.pdf

Henderson, A. Roy, 1997. *A Guide to Laser Safety*. Chapman & Hall.

ICNIRP2000. Revision of the guidelines on limits of exposure to laser radiation of wavelengths between 400 nm and 1.4 μm. *Health Phys.* 79(4), 431–440.

ICNIRP 2004. Guidelines on limits of exposure to ultraviolet radiation of wavelengths between 180 nm and 400 nm (Incoherent Optical Radiation). *Health Phys.* 87(2), 171–186.

ICNIRP 2013. Guidelines on limits of exposure to incoherent visible and infrared radiation. *Health Phys.* 105(1), 74–91.

ICNIRP 2013. Guidelines on limits of exposure to laser radiation of wavelengths between 180 nm and 1,000 μm. *Health Phys.* 105(3), 271–295.

IEC TR 60825-1:2014, Safety of laser products – Part 1: Equipment classification and requirements.

IEC TR 60825-14:2004, Safety of laser products – Part 14: A user's guide.

IEC TR 60825-5:2019, Safety of laser products – Part 5: Manufacturer's checklist for IEC 60825-1

IEC TR 60825-8:2006, Safety of laser products – Part 8: Guidelines for the safe use of laser beams on humans.

Kitsinelis, S., & Kitsinelis, S. (2015). *Light Sources: Basics of Lighting Technologies and Applications*. CRC Press.

Li, T. H., & Luo, G. Y. (2015). Design of light source of agricultural UVALED pest control lamp in food production. *Adv. J. Food Sci. Technol.* 9(1), 36–39.

Medicines and Healthcare products Regulatory Agency, Lasers, intense light source systems and LEDs – guidance for safe use in medical, surgical, dental and aesthetic practices, Crown copyright, September 2015.

NICE 2006. Interventional procedure overview of intrauterine laser ablation of placental vessels for the treatment of twin to twin transfusion syndrome, IPG198.

Nouri, Keyvan 2011. *Lasers in Dermatology and Medicine*. Springer-Verlag London Ltd. ISBN 978-0-85729-280-3.

Petersen, R. C., & Sliney, D. H. (1986). Toward the development of laser safety standards for fibre-optic communication systems. *Appl. Opt.* 25(7), 1038–1047.

Tulleken, C. A., Verdaasdonk, R. M., Berendsen, W., & Mali, W. P. 1993. Use of the excimer laser in high-flow bypass surgery of the brain. *J. Neurosurg.* 78(3), 477–480.

Yu, Peter P., & Cardona, M. 2010. *Fundamentals of Semiconductors: Physics and Materials Properties*. Springer, London.

13 Ionising Radiation Protection

Cornelius Lewis
King's College Hospital NHS Foundation Trust, London, UK

Jim Thurston
Dorset County Hospital, NHS Foundation Trust, Dorchester, UK

CONTENTS

DOI: 10.1201/9780429155758-13

13.1　RISKS OF IONISING RADIATION

The risks of ionising radiation have been discussed extensively in Chapter 4. From a radiation protection perspective, the principal issues are as follows:

- Any dose of ionising radiation can cause harm.
- Even very small doses of ionising radiation may produce cell mutations which can ultimately lead to cancers. Latent periods before a cancer develops may be years or tens of years. These are called stochastic effects.
- Above certain dose thresholds, ionising radiation will cause organ or whole system damage within a relatively short period of time, days to months. These are called deterministic effects.

As knowledge of the effects of ionising radiation has increased, systems and practices to facilitate protection from harmful effects have been developed. The main agency responsible for developing a framework for protection is the International Commission on Radiological Protection (ICRP) which was established in 1928. Much of the work of ICRP is undertaken by Task Groups which focus on specific issues and are usually time limited. There are also four permanent committees, reporting to the main Commission, which are as follows: Radiation Effects, Doses from Radiation Exposure, Radiological Protection in Medicine and Application of the Commission's Recommendations.

13.2　PRINCIPLES OF RADIATION PROTECTION – JUSTIFICATION, OPTIMISATION AND LIMITATION

The principles of radiation protection developed by ICRP are designed to address the issues listed above.

The first principle is justification. Any dose of ionising radiation may cause harm to individuals, and so before use in a practice (e.g. radiography, radiotherapy, laboratory research, nuclear power generation, etc), the benefit must be shown to be greater than the risk.

If a practice is justified, there is a need to ensure that the amount of ionising radiation used is kept at the lowest level required to produce the desired benefit thus reducing the risk of stochastic effects. This is the principle of optimisation, usually referred to by the acronym, ALARA – as low as reasonably achievable (also, in some regulations, ALARP – as low as reasonably practicable). Optimisation should not be confused with minimisation. For example, in radiographic imaging, a low dose may produce an image but the quality of that image may not be sufficient to deliver an effective clinical diagnosis.

The final principle is limitation. Doses of ionising radiation should be kept below the levels which are known to cause deterministic effects or produce significant risks of stochastic effects.

13.3　A FRAMEWORK FOR IONISING RADIATION PROTECTION

The basic framework for radiation protection developed by ICRP proposes that there are various sources of ionising radiation (naturally radioactive, artificial sources and electrically generated) which, through a series of pathways or networks, may expose individuals (Figure 13.1). Exposures to ionising radiation fall into three categories: existing exposures, planned exposures and emergency or accidental exposures.

Existing exposures are otherwise known as natural background radiation including cosmic radiation, solar radiation and naturally occurring terrestrial sources, for example, radon gas and heavy radioactive elements contained in granite and other rocks.

Planned exposures are those which arise from 'practices'. These may include nuclear power generation, industrial sources, research and medical uses. Conversely, emergency or accidental exposures are those which arise from practices (including conflict) but not in a controlled manner.

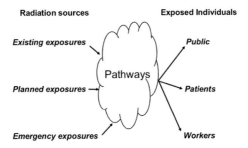

FIGURE 13.1 ICRP framework.

Individuals who may be exposed are also considered in three separate categories: members of the public, radiation workers and patients. Groups or populations of individuals may be exposed to a single source of radiation (source-related exposures) or a single individual may be exposed to a variety of separate sources (individual-related exposures).

The principle of justification is source-related. Any source that is used as part of a practice must be shown to produce a benefit that is greater than the risk it entails.

Optimisation is also source-related. For every source used in a practice, the ALARA principle must be employed. The specific exposure situation should be evaluated and procedures adopted to ensure exposures from the source are optimised. This may involve the application of dose constraints or, in patients, dose reference levels which, if exceeded, would trigger further actions.

Limitation (of dose) is individual-related for planned exposure situations because the individuals concerned may be exposed to a number of different sources. This involves the application of internationally agreed limits. Dose limits for individuals are designated as either stochastic or deterministic. Stochastic limits are those established to manage the potential cancer risk from whole body effective doses whilst deterministic limits are used to ensure that doses to particular organs (specifically skin and eyes) do not exceed the levels which would produce immediate effects.

The widely quoted risk of future malignancies (stochastic risk) following radiation exposure to adults is $5 \times 10^{-2} \mathrm{Sv}^{-1}$. This implies that for every 100 individuals who receive 1 Sv of radiation, there will be five cancer fatalities. Doses of ionising radiation for workers and members of the public are usually of the order of mSv and so the risk would be five cancer fatalities for every 100,000 exposed individuals. To facilitate any planned exposure, a certain level of risk must be tolerated.

Risks across a whole range of industries are subdivided into three categories: acceptable, tolerable and unacceptable (Figure 13.2). The tolerable region is generally accepted as between 1 in 1,000 and 1 in 100,000 for workers. For radiation exposure, assuming a risk of $5 \times 10^{-2} \mathrm{Sv}^{-1}$, this translates to a whole body tolerable dose limit of 20 mSv for radiation workers. The tolerable region for members of the public is formally set at a factor of 10 lower although, for practical and precautionary reasons, the tolerable limit is set at 1 mSv. These limits do not include any dose the worker or member of the public may receive from natural background radiation.

Deterministic limits for individual organs are set well below the threshold doses at which deterministic effects are known to occur. Currently, these are 20 mSv (equivalent dose) for the eye and 500 mSv for the skin, hands and feet.

Exposures from natural radiation sources cannot be controlled by the application of these principles. For natural sources (existing exposures), other procedures are invoked. A particular example is radon gas concentrations in buildings. Where radon gas concentrations exceed specific limits, measurements are made and actions taken to reduce the concentration, for example, equipping buildings with pumps to extract the gas from the building.

Similarly, the principles cannot be applied fully to emergency and accidental exposures. In these situations, actions are taken to avert exposures, for example, by imposing exclusion zones such as those implemented following the Chernobyl nuclear accident.

FIGURE 13.2 Risk stratification.

13.4 RADIATION PROTECTION IN A MEDICAL CONTEXT

Exposures to ionising radiation in medicine are considered to be planned exposures (practices) to which the principles of radiation protection apply in whole or in part.

In the case of medical radiation workers, justification is implicit because the exposure is given for the benefit of the patient – and, of course, the worker also benefits from receiving a salary. The focus for workers is on optimisation and limitation. Reliance on optimisation and limitation is also the focus for members of the public who may be exposed to radiation from medical procedures.

In contrast, limitation does not apply to patients because the effectiveness of the treatment or diagnosis might be reduced, potentially decreasing any benefit. Thus, for patients, the focus is on justification and optimisation.

13.5 RADIATION PROTECTION FOR HEALTHCARE WORKERS

The use of radiation in hospitals and associated healthcare facilities is highly controlled and regulated. National governments produce legislation to effect control on the use of radiation in hospitals, and whilst specific national requirements vary, there is a commonality of approach based on the recommendations of ICRP (ICRP 2007).

Healthcare establishments must normally be authorised by the relevant national authority (e.g. the UK Health and Safety Executive) before they are permitted to use ionising radiation. Once authorised, establishments are required to undertake risk assessments and subsequently implement actions to ensure that risks are adequately controlled.

13.5.1 Risk Assessment

Prior to commencing any new work with ionising radiation, an assessment must be carried out to identify the risks involved. The assessment is used to determine the policy and procedures to ensure that the work can be undertaken in accordance with good radiation safety practice and in compliance with the relevant regulations. The principal requirements of a risk assessment are to evaluate the radiation hazards, identify those who may be exposed to the hazards and determine the protection measures required to provide adequate controls. Risk assessments must be documented and re-assessed on reasonable timescales or when changes occur which may affect the level of risk.

Major issues to consider in a risk assessment include the following:

- The nature of the radiation source involved;
- Radiation dose rate to which individuals may be exposed;
- The possibility of surface or airborne radioactive contamination;
- Access to areas where dose rates or contamination levels may be significant;

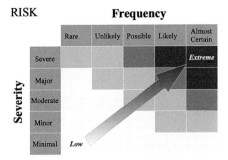

FIGURE 13.3 Risk matrix.

- Engineering control measures to reduce the risk and level of exposure;
- The availability and adequacy of working procedures;
- The availability of personal protective equipment (PPE);
- The consequences of possible failures of control measures.

Risks identified during an assessment are usually graded in terms of frequency with which the risk might occur and the severity of the hazard if it does. The overall risk is the combination of these two factors and usually represented qualitatively or numerically on a 5 × 5 matrix (Figure 13.3).

Following a risk assessment, actions are taken to ensure that appropriate measures for optimisation and dose limitation are implemented. Risks assessed as high or extreme are obviously given a higher priority for action.

13.5.2 Actions to Achieve ALARA

There is a hierarchy of measures to achieve ALARA. These range from actions to prevent exposure to those designed to ensure that exposures do not breach dose limits. These include, in order of application, engineering controls and warning devices, systems of work, PPE and the use of dose constraints.

13.5.2.1 Engineering Controls and Warning Devices

These measures can be implemented to ensure that sources or devices can be used safely. One of the most common engineering controls is the use of interlocks which ensure a device cannot be energised or a source exposed unless the interlock is closed. Interlocks are commonly used in radiotherapy bunkers. Staff leaving the bunker must close a door or gate which will enable power to be supplied to the device or source mechanisms. When an interlock is closed warning signs will illuminate and sometimes there will be an accompanying acoustic warning. If the gate or door is opened during an exposure, the power will automatically shut off or the source will be returned to its shielded container thus terminating the exposure.

Radiotherapy bunkers and similar environments will also be equipped with one or more panic buttons to ensure that any member of staff inadvertently remaining in the bunker can terminate the exposure.

Warning signs are widely used in radiation environments. These will display the international warning sign for radiation, the trefoil, and will usually describe the type of hazard and risk, for example, x-rays, contamination (Figure 13.4). At the entrance to almost every diagnostic x-ray room, an illuminated warning sign is displayed. This normally provides a two-stage warning with one panel, displaying the radiation trefoil, illuminated if power is supplied to the x-ray device and a second, normally red, panel illuminated to warn of an exposure in progress. The exposure warning panel will usually include a 'do not enter' instruction (Figure 13.5).

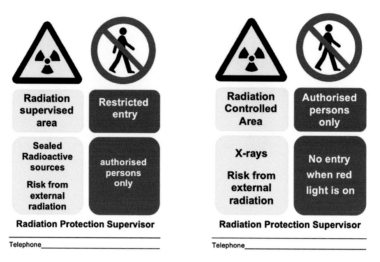

FIGURE 13.4 Radiation warning signs.

FIGURE 13.5 X-ray room warning sign.

13.5.2.2 Systems of Work

The main purpose of Systems of Work is to describe the key arrangements and instructions for restricting exposure in a particular area. The level of detail provided will vary according to the complexity of the radiation work involved.

A System of Work will describe the area in which the work is to be undertaken and provide working instructions relevant to that particular area to ensure the work can be undertaken safely. These instructions would normally indicate the categories of staff who are permitted access to the area, access arrangements and the requirements for personal dosimetry or protective equipment. Details of contingency arrangements in the event of foreseeable accidents should also be provided.

Employers of staff who are required to work in radiation environments must also develop a management system to ensure safe working practices (see below) which will include nominated individuals with specific responsibilities for radiation safety. Systems of Work will identify the responsible persons and provide contact details to ensure any concerns can be raised or problems reported.

13.5.2.3 Time, Distance and Shielding

Instructions for working safely with ionising radiation will emphasise the importance of time, distance and shielding.

13.5.2.3.1 Time

Self-evidently, the less time an individual is exposed to a source, the lower the resulting exposure. A particular environment in which this can be particularly effective is a Nuclear Medicine Department where sealed and unsealed sources are regularly used. Adequate preparation before dispensing a radiopharmaceutical or administering to a patient will reduce the time of exposure. Once a patient is injected, they will become a radiation source and so any member of staff involved in their care must be conscious of the time they spend with the patient.

13.5.2.3.2 Distance

Distance can be very effective in reducing exposure. The relationship between distance and exposure for a point source follows an inverse square relationship, exposure $\propto$ 1/distance^2. Increasing the distance by a factor of 2 reduces the exposure by a factor of 4, tripling the distance reduces exposure by a factor of 9, and so on (Figure 13.6).

Many sources will not be point sources but the inverse square law may still be applied as a reasonable estimate if the distance from the source is at least $10X$ where 'X' is the longest dimension of the source.

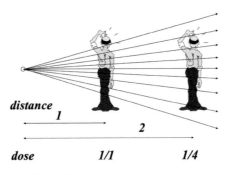

FIGURE 13.6 Distance/exposure relationship.

13.5.2.4   Shielding

The use of materials to shield sources of ionising radiation is essential in achieving ALARA. Common materials used for shielding in medical practices are lead, concrete and perspex. The choice of shielding material depends on the type and energy of the radiation.

Lead is the shielding material of choice for x- and gamma-ray sources below an energy of up to around 150 keV. At these energies, attenuation is principally through the photoelectric effect which increases in proportion to the cube of the atomic number (Z^3). Above these energies, Compton scatter gradually becomes the predominant process in which attenuation is dependent on electron density.

Lead is used extensively in the shielding of x-ray rooms – in walls, doors and protective shielding for operators. The x-ray tube has a lead shield to ensure that emissions are restricted to the exit port and lead collimators are used to define the x-ray field.

In Nuclear Medicine departments, vials of radiopharmaceuticals are shielded in lead pots. Syringe shields, usually made from tungsten because of its lower toxicity, are used to protect staff administering radiopharmaceutical injections. Basic gamma camera rooms are not usually shielded because the dose rates from injected patients tend to be quite low. However, if the camera incorporates a CT system (SPECT/CT), shielding must be provided as it would be for CT installations. Shielding is essential in PET or PET/CT facilities because the dose rates are significantly higher with or without the CT component. Whilst the photoelectric effect is not the principal attenuation process for PET, lead is normally used to provide the barrier. The shielding for PET rooms typically requires lead thicknesses of several centimetres in contrast to x-ray rooms where lead thicknesses are usually of the order of a few millimetres.

In radiotherapy rooms which employ x- and gamma-ray energies between 5 and 25 MeV, lead does not provide any advantage over other materials. Barriers in radiotherapy rooms predominantly use concrete. Barrier thicknesses in radiotherapy installations are typically of the order of 1–2.5 metres of high density concrete.

13.5.2.5   Personal Protective Equipment

In some situations, staff working in radiation environments cannot be protected by engineering controls or Systems of Work and so PPE is provided. The most common reason for using PPE in medicine is for interventional procedures, for example in cardiology or interventional radiology when staff have to be present in the room with the patient during the exposure.

Areas of the body for which PPE may be provided include the trunk, thyroid, eyes and hands. PPE is principally designed to protect staff from scattered radiation. Lead aprons protect the trunk from the neck down to a level approximately 10 cm above the knee. They come in a variety of forms although the most common is the wrap-around apron which provides shielding front and back. Other designs include front-only protection and two-piece (vest and skirt) aprons. The shielding provided in a lead apron will be either 0.25 mm or 0.35 mm lead equivalent, the latter suggested for use with primary x-ray energies in excess of 100 kV. Lead aprons if appropriately worn are capable of reducing exposure to the wearer by between 60% and 90% depending on the x-ray energy.

The material within the apron is a lead-rubber compound although 'light lead' aprons are also available to reduce fatigue in the wearer. These use composites of tin, antimony or tungsten and are designed to have x-ray absorption edges at appropriate energies. There is some debate as to the effectiveness of 'light lead' aprons and, specifically, the range of x-ray energies over which they provide suitable levels of protection.

The thyroid gland is recognised as a particularly radiosensitive organ. Staff who work close to patients during radiological investigations are usually provided with a thyroid collar which would normally have a lead equivalent thickness of 0.5 mm which should reduce exposures by up to 90%. Thyroid collars are worn around the neck and outside the accompanying lead apron.

The eye, and specifically the lens of the eye, is also radiosensitive. In contrast to the thyroid where protection is required against stochastic effects, the lens of the eye may suffer a deterministic effect leading to cataract formation. Investigations summarised in the 2007 recommendations of the ICRP (ICRP, 2007) concluded that cataracts could result from much lower exposures than previously assumed which prompted the annual limit for exposure to the eye of 150 mSv per annum being reduced to 20 mSv (equivalent dose). Eye protection is provided using spectacles or goggles with lead equivalent thickness of up to 0.75 mm. Prescription spectacles are also available.

Protection for hands can be provided with either gauntlets or gloves usually impregnated with a heavy metal oxide, although rarely lead. Gauntlets, which will provide greater levels of protection, are large and bulky which makes them unsuitable for handling small objects. Impregnated latex gloves which are sterile and disposable are also available for undertaking clinical procedures although many clinicians prefer not to use them because they reduce dexterity and tactile sensitivity.

13.5.2.6 Dose Constraints

An important element in achieving ALARA is the use of dose constraints. As the term implies, these are dose levels which should not be exceeded in normal practice. Constraints are set prospectively at some fraction of the dose limit. If a constraint is exceeded during a specific period of time, an investigation should be undertaken to determine causal factors and the individual or individuals concerned may, if considered appropriate, be removed from any further radiation work for that period.

13.5.3 DOSE LIMITATION

As discussed above, stochastic and deterministic dose limits for radiation workers and members of the public have been recommended by the ICRP. To ensure that dose limits are not exceeded, radiation workers may be monitored. Members of the public are rarely monitored because strict controls and optimisation processes, including environmental monitoring, prevent them from being exposed to levels which may lead to limits being exceeded.

13.5.3.1 Radiation Workers

Staff working with radiation must receive suitable training. In some countries, radiation workers are sub-divided into two categories, classified and non-classified (sometimes referred to as Category A and Category B). Classified radiation workers are defined as those who may exceed 3/10ths of a dose limit and therefore must be monitored and have periodic health checks. Non-classified workers are not required to be monitored (or to have health checks), but often are to ensure their doses are not approaching the level at which they need to be classified.

The majority of staff working in healthcare are non-classified. Specific groups who do need to be classified include interventional radiologists and cardiologists (particularly for eye exposure) and nuclear medicine/PET staff, particularly radiopharmacy workers (for exposures to extremities).

13.5.3.2 Personnel Monitoring

Personnel monitors can be provided for the whole body, eyes and extremities (Figure 13.7). Most monitors are passive devices (giving a retrospective measure of radiation dose received) but electronic monitors, which provide an active, real-time display of radiation dose, are sometimes used in specific environments.

Whole body monitors, often referred to as film badges, are available for measuring exposures from X, gamma and beta radiations. Some monitors use photographic film, of the type used in dentistry, but more commonly thermoluminescent detectors are used. The badge holder has various 'windows' with different levels of filtration to enable some characterisation of the radiation source. Whole body monitors should be worn on the torso at chest or waist level and underneath any PPE that is being worn. Staff performing interventional studies may also wear whole body monitors at neck level to indicate thyroid dose.

Eye dose monitoring is a particular concern for clinical staff performing interventional procedures on patients, for example angioplasty, and a variety of dosimeters and techniques have been

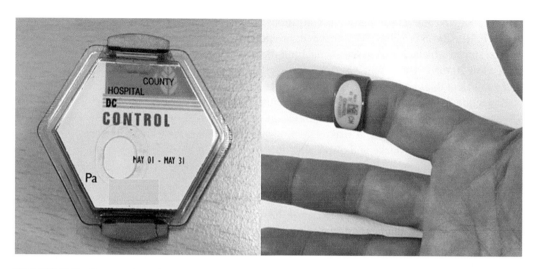

FIGURE 13.7   Personnel monitors.

developed more recently to address this. The key issue is measuring the dose as close to the lens of the eye as possible. The simplest and, arguably, most accurate method is to use an adjustable headband with Velcro fasteners incorporating a thermoluminescent dosimeter. Ideally the dosimeter should be placed above the eye which is closest to the radiation source. Some clinicians find the wearing of headbands uncomfortable and, as an alternative, thermoluminescent dosimeters have been designed which can clip to the side of spectacles. A third technique is to record the dose using a whole body monitor worn at neck level and applying a correction factor for distance usually of around 0.75 although this is not formally accepted (Martin, 2011).

Staff handling unsealed source, for example, administering radiopharmaceuticals in Nuclear Medicine, and possibly those performing interventional procedures, should be monitored for doses to the extremities (hands). Two techniques are available and used depending on the circumstance of the exposure. Clinicians involved in interventional procedures will receive exposures to the hand as a whole. In this situation, rings incorporating a thermoluminescent dosimeter can be used. However, staff involved in the administration of radiopharmaceuticals and similar activities are more likely to receive exposures to fingers, particularly the tips of the fingers, which can vary considerably depending upon the proximity of the source. Rings can be used but a correction factor would normally be applied as the dose to the finger-tip could easily be several times the dose recorded at the base where the ring is situated. Finger stalls are used preferably in this situation. These are pliable plastic sleeves worn over the finger incorporating a thermoluminescent dosimeter at the tip.

13.5.3.3   Monitoring Periods

The time for which a personnel monitor is worn (the wear period) will depend on the particular circumstances of the exposure. Where the risk of excessive exposures is assessed to be low, for example, a radiographer working in a general x-ray department, a whole body monitor can be worn for up to three months. Staff who are required to wear eye or extremity monitors are more likely to receive higher exposures and the wear period would normally not exceed one month. Finger stalls are particularly fragile and can be exchanged as often as every two weeks or sooner if they become damaged.

If a higher than usual exposure is received, an investigation of the circumstances must be undertaken and any causal issues addressed. If an incident occurs or for other reasons a higher than usual exposure is anticipated, the dosimeter(s) should be sent immediately for processing and suitable actions taken when the readings are returned.

13.5.3.4 Contamination Monitoring

The most significant hazard arising from radioactive contamination is ingestion. Personal monitors are of limited value in areas where unsealed radioactive sources are used because localised areas of contamination may not be detected. Reliance on the use of personal monitors in such areas could be counterproductive because they may lead to a false sense of security.

In areas where unsealed sources are used (nuclear medicine departments, radiopharmacies, laboratories), routine contamination monitoring is essential. This should be performed with instruments designed to detect emissions from the radioactive sources being used. Geiger counters will detect most gamma radiations and if the probe is equipped with a thin window can also detect beta radiation. However, they have relatively low sensitivity and, particularly for beta and low energy gamma radiation, a thin window proportional counter is more effective. Scintillation detectors may also be used although less effective in detecting beta and low energy gamma radiation.

Staff should be encouraged to monitor themselves after each episode of potential contact, for example, after a patient is injected with a radiopharmaceutical or has been scanned with a gamma camera. Whilst there is potential for any part of the body to become contaminated, hands are particularly vulnerable. Surfaces may also become contaminated following accidental releases and should also be monitored frequently.

In a nuclear medicine department, the optimal regimen would be to have wall-mounted hand monitors, or even combination hand and foot monitors, at the exit of every room where unsealed sources are used. Surfaces, particularly in injection rooms and patient toilets, should be monitored at least daily.

If a member of staff or surface is found to be contaminated, immediate action should be taken to establish the severity and extent of contamination. Suitable actions, including the careful removal of contaminated clothing and cleaning of surfaces, must then be taken to decontaminate the area using previously agreed Systems of Work which staff have been trained to apply. Particular attention should be given to assessing the dose to any staff member involved including the possibility of internal contamination.

13.5.4 Management of Radiation Protection

The use of radiation in workplaces is highly regulated because of the significant risks involved. As with all health and safety issues, the employer (e.g. Chief Executive Officer) must take responsibility. The employer should establish a policy which will outline the scope of any work being undertaken and the employees to which it applies. It will also outline a framework for managing risks to staff, patients and members of the public to ensure compliance with relevant legislation. This should incorporate the appointment of experts and clarification of their responsibilities, the means of communicating policy objectives, arrangements for protection (e.g. risk assessments, Systems of Work, personal monitoring and PPE) and training.

The number of expert advisers will depend on the scope of work being undertake but will certainly include experts to advise on arrangements for the safety of staff and members of the public – the Radiation Protection Expert (also referred to as the Radiation Protection Adviser) and one to advise on patient protection – the Medical Physics Expert. If the employer uses radioactive substances, a Radioactive Waste Adviser may also be appointed. In each department or area in which radiation is used, a supervisor who has received specific training in radiation protection issues would normally be appointed. Their duties would include ensuring that local protocols and Systems of Work are adhered to, administering personal motoring for staff in the area, inspecting PPE and liaising with management and experts over any matters of concern.

Communication of radiation safety requirements is clearly essential and this is normally achieved through one or more committees (e.g. Radiation Protection Committee, Medical Exposures Committee) with suitable representation from relevant stakeholders.

One of the most important requirements in any system for radiation protection is staff training. Agencies responsible for health and safety consistently report that an absence or lack of suitable training is a contributory factor in accidental or excessive exposures. Training should be available for all staff, both directly involved in work with radiation or simply working in an environment where radiation exposure is a risk, including portering and domestic staff. The level and depth of training must be tailored to meet the needs and ability of individuals.

13.6 RADIATION PROTECTION FOR PATIENTS

Until the last quarter of the 20th century, radiation protection in medicine focused on staff and members of the public. This was based on the premise that patients received radiation as part of diagnosis or treatment and was therefore implicitly justified and, of course, international dose limits could not apply. Optimisation of doses to patients was not considered an issue.

A study published by the UK National Radiological Protection Board (NRPB, 1986) demonstrated that optimisation was a significant problem. The study compared effective doses for a variety of procedures across 20 hospitals in the UK. It showed that doses for the same procedure could vary considerably. For example, the study reported that the minimum effective dose delivered for an x-ray of the abdomen was 0.12 mSv and the maximum, 9.94 mSv (a factor of over 80 times greater). Of the 12 studies compared, the difference between minimum and maximum effective doses was between 10 and 100 except for chest x-rays where the factor was over 400.

In 1990, the report of a joint working party between the UK British Institute of Radiology and NRPB was published (NRPB, 1990) which summarised

> Earlier studies by RCR on the effective use of diagnostic x-rays, and data from recent surveys of radiology practice by NRPB, enabled a rough estimate to be made of the extent of clinically unhelpful, repeated and unoptimised x-ray examinations in the UK. This indicated that a reduction of nearly one-half in the current collective dose to the population from medical x-rays might be possible without detriment to patient care.

Studies such as these persuaded regulatory bodies that action was required to improve justification and optimisation. The first such action, which preceded the publication of these studies, came from the European Union in 1984 in the form of a Directive (later repealed and currently incorporated in the Basic Safety Standards [EU 2013/59]) which required member states to establish basic protection measures for patients undergoing diagnosis or treatment using ionising radiation.

13.6.1 JUSTIFICATION OF MEDICAL EXPOSURES

An assessment that the benefit from radiation exposure exceeds the potential risk is required for all exposures. This applies not only to patients undergoing diagnosis or treatment but also to those undergoing non-medical imaging procedures which will include those on normal/healthy volunteers in research studies, health screening applications (including for immigration), occupational/public health studies and medico-legal exposures.

With the exception of medico-legal exposures, the benefit from research, health screening and occupational studies may not be to the particular individual involved but to society as a whole. New techniques may be developed, cancers may be detected and treated at an early stage and disease threats to populations may be revealed, for example, the detection of tuberculosis. Medico-legal procedures are undertaken to establish and confirm the detriment to specific individuals following incidents. A particular example might be radiological examinations undertaken following road accidents to influence the level of compensation to victims. In essence, the benefit would be financial.

The justification of a medical exposure should involve a number of considerations:

- The specific objectives and potential diagnostic or therapeutic benefits of the exposure;
- The characteristics of the individual involved;
- The detriment the exposure may cause;
- The efficacy, benefits and risk of alternative techniques having the same objective but involving less or no exposure to ionising radiation.

There are in the region of 2.5 million diagnostic procedures performed in the UK National Health Service every month (NHS England, 2018) and each should be individually justified. Specifically, there is a need to establish what information will be gained by undertaking the procedure and how the result could subsequently affect the patient's clinical management.

Age is an important factor to consider in radiation exposures. Whilst the average detriment for stochastic effects is quoted as 5×10^{-2} Sv^{-1}, there is a considerable variation with age. Below 10 years of age, the risk may be as high as 15×10^{-2} Sv^{-1}, whilst over the age of 70, it may be 1×10^{-2} Sv^{-1} or lower (HPA, 2011) and justification of exposures must take this into account. A study (Hall and Brenner, 2008) hypothesised that a CT scan of the abdomen in young children, below the age of 10, could result in a lifetime cancer risk exceeding 10%. Thus, whilst justification is necessary for all exposures, it is particularly important for children and young adults.

Procedures carried out on females of child-bearing age, often assumed to be women between the ages of 12 and 55, is a further issue in justification. Exposures in pregnancy are permissible but the potential detriment to the foetus is an additional factor to consider during the process of justification (see Section 7.3). This is particularly important when the exposure is to the whole body, as in nuclear medicine studies, or directly to the abdomen.

The radiological basis for determining stochastic radiation risks, the linear no-threshold hypothesis, assumes that risk is cumulative. This is a precautionary approach and the implication for justification is that exposures should not necessarily be considered in isolation if a specific exposure is linked to others. For example, patients being investigated for a particular condition may require a series of exposures. Similarly, patients being treated for chronic conditions or followed up to determine their response to treatment (e.g. chemotherapy patients) may need repeated exposures over a lengthy time period.

Another aspect of justification is the consideration of other techniques which do not require exposure to ionising radiation. An obvious alternative is ultrasound which is particularly useful for imaging soft tissue structures. With the continuous improvement of ultrasound technology, an increasing number of diagnostic studies can be undertaken with this modality. A particular example is the use of ultrasound to evaluate lung disease or impairment in an intensive care setting which, previously, would have required multiple chest x-rays or CT scans.

CT scans are associated with higher patient doses and so adequate justification is imperative. MRI is sometimes an alternative to CT scanning and can often be used to provide equivalent diagnostic information, particularly in soft tissues. However, there are factors other than radiation dose to consider when deciding if MRI could be used in place of CT. MRI is an expensive modality and so there is limited provision. In addition, MRI scanning is more time consuming than CT. Selecting MRI over CT purely on the basis of radiation exposure could place further pressure on MRI capacity with the unintended consequence of producing significant delays in diagnosis leading to poorer patient outcomes.

13.6.2 Patient Referral

Requests for diagnostic procedures involving ionising radiation are often made by clinicians with little expertise in radiation risk. Whilst there are management processes in place to avoid unjustified exposures (see below), it is useful to prevent or reduce the number of such requests being made.

Guidelines have been established in most developed countries to help referrers make the best, safest and most suitable use of valuable and limited resources (e.g. ESR, 2014). Although principally intended for use by clinicians in primary and emergency care, they are applicable in all healthcare settings and provide advice on the most appropriate imaging investigation for a range of clinical indications.

Whilst guidelines can reduce the number of inappropriate requests made to imaging departments, the most essential requirement is for requesting clinicians to provide sufficient information for a radiologist or other expert to determine if the request is justified. For example, a request simply stating 'Chest X-ray' would never be acceptable.

13.6.3   Optimisation of Medical Exposures

With the increasing sophistication of medical radiation technology, there are a growing number of dose optimisation features available, some examples of which are described below.

To use these and other features of any technology, it is essential that staff have received suitable training. Regular users (e.g. radiographers and nuclear medicine technicians) receive comprehensive initial training which is updated as new technologies become available. A problem can arise with 'casual' users, in particular clinical staff who perform interventional procedures. The traditional methodology for instructing such users, 'see one, do one, teach one', is hardly suitable for modern technologies, and many countries now require a more thorough and formalised training system.

13.6.3.1   Optimisation Technologies in Planar X-Ray Systems

13.6.3.1.1   Filtration

X-ray filtration is an established means of reducing unnecessary exposure by removing low energy x-rays from the beam which would be fully absorbed and not contribute to image formation (see Chapter 5). All x-ray systems include inherent filtration, totalling around 2.5–3.0 mm of aluminium. In recent years, there has been interest in using additional filtration to further optimise the x-ray beam in specific applications. The use of thin copper filters to reduce dose in paediatric applications is becoming more widely used, although still the subject of debate (Jones *et al.*, 2015).

13.6.3.1.2   Automatic Exposure Control (AEC)

AECs are also an established feature of x-ray systems. They comprise a transmission ionisation chamber placed immediately in front of the imaging receptor to terminate the exposure when sufficient dose has been delivered to the film or digital detector to form an acceptable image of diagnostic quality. In fluoroscopy, an AEC (or automatic brightness control – ABC) adjusts the x-ray current and voltage to ensure the fluoroscopic image is of adequate quality.

The correct functioning of the AEC is critically important in digital imaging systems. In older x-ray systems using film, a mal-functioning AEC was immediately apparent because processed film could be seen to be either over- or under-exposure (too dark or too light). Digital systems are designed to compensate for different levels of exposure by post-processing the image and can therefore tolerate large variations in exposure. An under-exposed digital image will produce a readable although noisy image but an over-exposed image may not be apparent to the operator. Quality assurance of the AEC in a digital system is essential.

13.6.3.2   Optimisation Technologies in Fluoroscopy

13.6.3.2.1   Pulsed Fluoroscopy

Fluoroscopy is used extensively for interventional studies producing a real-time image enabling the operator to adjust their intervention during the course of the procedure. Fluoroscopic images are produced using much lower x-ray tube currents than in static imaging to reduce exposure to both patient and operator and as a result are much noisier. The ability of the human visual system to smooth signal variations reduces the impact of noise in the image.

Despite the lower instantaneous dose levels, lengthy interventional procedures can lead to very high overall exposures. A technique used to reduce exposure levels is to replace continuous operation of the x-ray tube with pulsed operation. Typically there will be between 5 and 30 pulses per second, each of the order of a few milliseconds in duration. The pulse frequency is determined by the specific application with imaging of rapidly moving structures requiring higher frequencies. The operator is generally unaware of any flicker in the image.

Instantaneous dose rates per pulse are higher than for continuous fluoroscopy; nevertheless, the reduction in overall exposure can be dramatic. For example, a relatively high pulse frequency of 15 pulses per second can achieve a 25% reduction in overall exposure for the same signal to noise ratio in the image.

13.6.3.2.2 Last Image Hold

In any fluoroscopic procedure, the radiologist or physician will not be viewing the image continuously. In older systems, the fluoroscopic image is only displayed when the x-ray tube is energised which prevents the clinician assessing the current progress of the intervention before proceeding to the next step.

Modern fluoroscopic systems using digital technology provide a 'last image hold' function in which the last image is 'frozen' on the display after the exposure has been terminated. A recent development of this facility is electronic collimation which allows the operator to adjust field dimensions without exposing the patient.

13.6.3.3 Optimisation Technologies in CT Scanning

13.6.3.3.1 Automatic Exposure Control

In older generations of CT scanners in which there was no AEC, the x-ray tube current was fixed for the entire duration of the exposure. Given that the cross-sectional body shape for the major part of the body (thorax and abdomen) is oval, the CT scanner would deliver a much higher dose in an anterior-posterior orientation than in a lateral orientation and this is clearly sub-optimal.

This was initially addressed through the use of longitudinal modulation in which attenuation along the length of the body was assessed from the projection view (scout view or topogram) used to establish the correct positioning for the subsequent scan. As the scan was in progress, the x-ray tube current was adjusted to compensate for the attenuation previously determined. This was particularly valuable for reducing exposure to the thorax where the presence of the lung tissue significantly reduces attenuation.

Further developments led to the introduction of rotational modulation in which attenuation information assessed from a single rotation was used to adjust the x-ray tube current for the subsequent rotation. The latest generation of CT scanners combines both longitudinal and rotational modulation.

13.6.3.3.2 Iterative Reconstruction

Until recently, the standard method for reconstructing CT images from the raw attenuation data was a process referred to as filtered back projection (see Chapter 5). More recently, as computational power has increased, a new technique has been developed, iterative reconstruction (also see Chapter 5). In this technique, the raw attenuation data from the scanner is used to create an initial image using the standard technique of back projection. The initial image is then forward projected and correlated with the original raw data from the scanner and corrections applied based on the computed differences. After correction, the revised raw data is again back projected. Following a series of iterations, the differences between the computed forward projections and the original raw attenuation data become minimal, and once a pre-set level is reached, the image is displayed.

Iterative reconstruction produces images which are less noisy than those from filtered back projection. Because image noise is inversely related to radiation dose, the implication is that images produced using iterative reconstruction require less radiation dose. It has been claimed that dose reductions in excess of 50% can be achieved using this technique.

13.6.3.3.3 Cardiac Scanning

The use of CT scanning to diagnose and evaluate coronary artery disease is becoming more routine because it is a less invasive alternative to coronary angiography. Diagnostic images of the heart can only be of diagnostic value if the heart motion can be frozen otherwise there would be significant blurring in the image. This is achieved using cardiac gating in which an ECG signal is acquired at the same time as the CT scan and used to select a particular segment of heart motion to reconstruct the image.

Gating the cardiac CT image can be performed retrospectively and prospectively. In retrospective gating, the cardiac image is acquired continuously and the ECG signal used once the scan has been completed to select and reconstruct the relevant section of the cardiac cycle. Because the data is acquired continuously, any section of the cardiac cycle can be reconstructed. This has the advantage of being able to produce cine sequences of cardiac motion but also the considerable disadvantage of requiring high radiation doses to produce the data required.

With prospective gating, the ECG signal is used to trigger the scanner to produce x-rays and acquire images only during the part of the cardiac cycle of interest. This results in dose reduction of a factor of 2 or greater. There are limitations with this technique. It cannot be used to create cine sequences. More importantly, it requires a relatively slow heart rate, less than 65 beats per minute, often achieved by administering beta blockers prior to the scan to slow down the heart rate.

13.6.3.4 Other Issues in Patient Dose Optimisation

13.6.3.4.1 Patient Shielding

The use of patient shielding has been common practice for many years. It is applied when organs which are particularly radiosensitive (e.g. gonads, breast, thyroid) lie close to the imaging field.

In planar x-ray imaging, the most common device is the gonad shield which is a rigid, T-shaped piece of lead enclosed in plastic. It can be placed on the surface of the patient, for example, during abdomen imaging, to protect the ovaries in females or the gonads in males. Flexible pieces of lead rubber can also be employed to protect the abdomen during chest imaging. Some dentists will require patients, particularly children receiving orthopantomograms or cone beam CT studies (CBCT), to wear lead aprons or thyroid collars to avoid scattered exposure to the thyroid and other organs.

In CT imaging which generally delivers higher patient doses, there is concern over doses to the eyes during brain scanning and to the breast from chest scanning. Because CT images are acquired by rotating the x-ray beam around the region of interest, shielding becomes a more complex problem. The solution adopted has been to use shields which incorporate materials such as bismuth. This allows more transmission of the x-ray beam than lead, and thus sufficient information to permit image reconstruction, but still provides for some dose reduction to structures lying immediately beneath the shield.

The use of patient shielding has long been a cause for debate. It is argued that correct positioning of the imaging field to avoid sensitive structures is, in most cases, sufficient and much of the scattered dose to sensitive organs will be due to internal scatter. In CT, the use of bismuth shielding can increase noise in the image thus degrading image quality and possibly also increase overall patient dose because tube currents may increase to penetrate any shielding device.

A recent comprehensive review of patient shielding was published by the UK British Institute of Radiology (BIR, 2020). One of the driving forces for the review was the considerable variation in the application of patient shielding. The review concluded that in the overwhelming majority of cases patient shielding provided no significant benefit in terms of dose reduction to radiosensitive organs. A particular concern addressed by the review was the situation in which shielding was specifically requested by patients. The review recommended that operators (radiographers) should have received adequate training to be able to explain why shielding was not required.

13.6.3.4.2 Equipment Selection, Commissioning and Quality Assurance (QA)

Devices utilising ionising radiation in radiology, nuclear medicine or radiotherapy are amongst the most expensive medical technologies, ranging from around £100,000 for the most basic x-ray system to several million for a state-of-the-art CT scanner or External Beam Radiotherapy system. Careful specification and option appraisal is required to ensure the most appropriate device is purchased. Specification should include both clinical and technical considerations, including features associated with patient dose optimisation, and involve relevant experts, including Medical Physicists.

After purchase and installation, the device should be commissioned which involves thorough testing to ensure that it meets the quoted specifications. Once the device enters service, it should receive routine QA assessments at regular intervals to ensure that it continues to deliver to specification. The frequency of QA is variable although the ideal is for operators to perform simple, system-wide, tests on a routine (daily) basis to indicate if there are potential problems that require more detailed and lengthy investigation by technical experts (IPEM, 2005).

13.6.3.4.3 Image Viewing

Correct interpretation of images is clearly essential in both justification and optimisation. Although most radiological images are now viewed on a monitor screen, film is still occasionally used. In both cases, the viewing conditions are important. A radiological image displays variations in contrast and detail (sharpness). Images of dense structures, such as bone, produce significant contrast variation, but for soft tissue images, the variation is more subtle. Even in high contrast images, there may be small but important details which might not be observed in poor viewing conditions.

Images should be viewed in subdued lighting whether this is in a reporting area or a fluoroscopy room. Monitor screens should be correctly adjusted for brightness and contrast and should be part of a QA programme. It is also important to avoid glare and screen reflections. Screens should be regularly cleaned with an anti-static cloth to avoid the build-up of dust.

With the increasing volume of image reporting required, fatigue amongst reporters is also a potential problem, and there have been a number of reports of fatigue affecting diagnostic accuracy (Stec et al., 2018). One of the ways of potentially supporting radiologists is the use of Machine Learning/Artificial Intelligence (AI). A number of systems have been developed for identifying abnormalities and these have mainly been in areas where there are large volumes of routine studies (e.g. chest x-rays and mammography). There is also growing interest in the use of AI in the scanning modalities, CT and MRI, which produce large amounts of data from a single study.

There is no likelihood of AI replacing radiologists, at least for the foreseeable future, but it could soon be employed for pre-screening images to prioritise reporting or for comparing current and previous images of a particular patient (Hosny et al., 2018).

13.6.3.4.4 Dose Area Product (DAP), Dose Length Product (DLP) and Diagnostic Reference Level (DRL)

One of the most important developments for patient dose optimisation in diagnostic radiology has been the introduction of devices to provide real-time or immediate, post-exposure indication of patient doses.

In planar x-ray and fluoroscopy, a transmission ionisation chamber is attached to the end of the x-ray tube and collimator assembly. The ionisation current from the chamber is a measure of the total exposure over the area of the x-ray field – hence the name DAP meter. The unique advantage of such a device is that it is independent of distance between the x-ray tube and the patient because

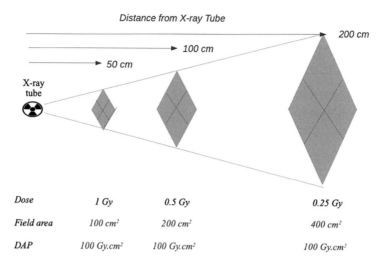

FIGURE 13.8 Principle of dose area product meter.

as the exposure decreases in accordance with the inverse square law the exposed area increases as the square of the distance from the x-ray focus (see Figure 13.8). The dose-area product is expressed in units of Gy.cm^2 (or similar multiples) and can be calibrated to determine entrance air kerma from which patient doses can be calculated.

A similar quantity used in CT scanning is DLP which is product of the standardised CT dose index (CTDI$_{vol}$) and the length of the scan, usually expressed in units of mGy.cm. It is not a measure of patient dose but using a series of factors determined for different sections of the body (head, chest, abdomen) DLP can be converted into effective dose, hence a measure of risk.

Because of their accessibility, measurements of DAP and DLP are particularly valuable in patient dose optimisation. Surveys of radiological and CT investigations at both national and international levels have established expected DAP and DLP values, referred to as DRLs, for most routine radiological procedures in standard-sized patients (e.g. Gov.UK, 2018). Healthcare establishments are advised, and in many countries legally obliged, to set local DRLs based on published national values and undertake routine auditing against them to ensure their local practice remains optimised.

DRLs are also available for Nuclear Medicine procedures and expressed in terms of administered activity (Gov.UK, 2020).

13.6.3.4.5   Use of Equipment

The correct use of equipment and, in particular, the appropriate use of dose optimisation features is essential in ensuring patients are not exposed to unnecessary radiation doses. The emphasis is on adequate training, specifically on the more complex systems. Radiography staff who routinely use the equipment will almost certainly have received comprehensive training but there can be a problem with students and agency staff. Appropriate supervision and local induction training is essential.

An additional element in ensuring dose (and image) optimisation is the use of protocols which have been developed and tested by expert users. Protocols should include a specification of the number of views required, the features that must be included in the imaged field and any positional requirements. They should also indicate the equipment settings to be used and give an indication of the expected dose from the procedure (e.g. DAP, DRL).

13.6.3.5   Patient Dose Optimisation in Radiotherapy

The fundamental issue in radiation therapy is to ensure that the prescribed radiation dose is delivered to the target tissue volume whilst minimising the dose to surrounding tissues (concomitant dose).

The introduction of techniques such as intensity modulated radiotherapy (IMRT) and tomotherapy has significantly improved dose delivery (see Chapter 9).

In addition, radiotherapy is increasingly being combined with imaging (portal imaging, CT and MRI) to refine dose delivery (*also* Chapter 9). It is currently possible to monitor changes in tumour volume during a course of radiotherapy and make suitable adjustments to the treatment.

A particular issue is the treatment of tumours in organs subject to movement during treatment, for example, the movement of lungs and other organs during respiration. Techniques have been developed which utilise gating to modulate the radiotherapy beam during respiration (Giraud *et al.*, 2011) and have reported up to 20% dose reduction to healthy tissues. A disadvantage of gating is that it will extend treatment times.

13.6.3.6 Failures in Dose Optimisation

Reporting of incidents which lead to 'significant accidental and unintended exposures' is encouraged and often legally required in most countries. This applies to both imaging and radiotherapy. The majority of such incidents are due to human error and often involve the failure to follow protocols or reflect a lack of training.

In the UK, annual reports are produced by the regulatory body (Care Quality Commission) summarising trends and highlighting good practice or areas for improvement. In the 2018/19 annual report (CQC, 2019), a total of 1009 incidents were reported. Of these, 796 were in diagnostic radiology, 75 in nuclear medicine and 138 in radiotherapy.

The most common incident reported in diagnostic radiology (53%) was an exposure to the wrong patient because either the radiology department had failed to correctly identify the patient prior to exposure or the wrong patient had been referred initially. Other incidents included setting the wrong exposure (9%) and imaging the wrong part of the anatomy (6%). Equipment failures amounted to a mere 2% of the total.

Identification failures were also common in Nuclear Medicine reports (40%) although errors in dose administration, the wrong radiopharmaceutical or the wrong administered activity, were slightly higher (44%). Of the remaining reports, 6% related to a failure to ascertain that a patient receiving nuclear medicine therapy was pregnant leading to an unintended exposure to the foetus.

In radiotherapy, almost half the incidents reported (43%) were associated with imaging, either at the stage of treatment planning or verification imaging during treatment. The majority of the remaining incidents (34%) were errors in accurately selecting the treatment volume.

13.6.4 Management of Patient Exposures

The control and management of radiation exposures to patients is a subset of general health and safety regulation in most countries. As such, the Chief Executive Officer of the organisation must assume overall responsibility. Their duties will encompass the delegation of specific tasks to adequately trained individuals, the provision of procedures and protocols to control exposures and the development of systems for quality assurance and clinical audit. They must also ensure that mechanisms are in place to report incidents and maintain training records.

Other individuals with specific responsibilities include referrers, practitioners and those with responsibility for the practical aspects of exposures, sometimes termed operators. The referrer, who will be a healthcare professional, is the individual who requests an exposure. Their principal responsibility is to ensure that sufficient information is available on the request form to allow justification of the subsequent exposure. Justification is the responsibility of the practitioner, for example, a dentist, radiologist, nuclear medicine physician or radiotherapist.

Once a referral has been justified, the practitioner will authorise the exposure to proceed. The volume of medical exposures that are requested means that practitioners cannot reasonably be expected to examine the request form for every single exposure. An alternative is for practitioners

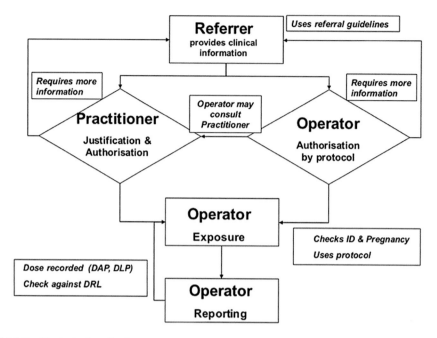

FIGURE 13.9 Control of medical exposure.

to prepare written justification guidelines which, if met, allow the operator to take responsibility for authorisation.

After authorisation has been given, the responsibility for optimising the exposure passes to the person responsible for the practical delivery of the exposure (operator) who may be a radiographer, nuclear medicine technologist, dental technologist or any other individual who has received suitable training. Practitioners, if involved in the practical aspects of delivering the exposure, would also become operators.

Operators are responsible for delivering exposures in accordance with written protocols. As well as controlling the equipment the operator is also responsible for other practical issues, for example, ensuring the correct patient is being exposed, determining pregnancy status, monitoring exposures against local DRLs and subsequently recording this information. Quality assurance and maintenance of equipment are also operator responsibilities because actions taken during QA or maintenance may affect doses delivered by equipment. This requires that medical physicists and some maintenance engineers need to be recognised as operators. The evaluation of a clinical exposure is another operator responsibility because the results may lead to further patient exposure.

An additional appointment is that of Medical Physics Expert (MPE). The MPE is required to have sufficient knowledge, experience and training to be able to provide advice on the radiation physics of medical exposures.

The flowchart shown in Figure 13.9 illustrates the steps entailed in delivering a medical exposure and the involvement of the various duty holders.

13.7 EXPOSURES IN PREGNANCY

13.7.1 RADIATION EFFECTS IN PRE-NATAL DEVELOPMENT

Ionising radiation is particularly damaging during cell division and so more rapidly dividing cells have greater radiosensitivity. Cells in the mammalian foetus are more rapidly dividing and so specific attention is given to any exposure which may involve irradiation of the foetus.

Radiation exposure may affect the foetus in one of four ways, dependent on gestational stage (ICRP, 2003):

- Foetal death
- Organ/tissue malformations
- Mental retardation
- Cancer in childhood (before the age of 15)

Foetal death is believed to occur in the very early period after conception, up to 3 weeks gestation, before the embryo has implanted in the uterus. There is no evidence in humans for this effect but the results of animal experiments suggest that it is a deterministic effect, implying a dose threshold, which will not occur below foetal doses of 100 mGy.

There is also no human evidence to quantify the effects of radiation in producing malformations. Animal data indicates that it is also a deterministic effect with a threshold dose in the region of 100–500 mGy and occurs predominantly during the period of organogenesis, between 8 and 16 weeks after conception.

Limited data exists to quantify radiation effects in brain development, principally from the follow-up children exposed as embryos when the atomic bombs were dropped on Hiroshima and Nagasaki. Whilst the data is scarce, and thus subject to significant error, the effect is expressed as a reduction in IQ and, as a consequence, an increase in the number of children with severe mental retardation. The effects are thought to be at their most severe between 8 and 15 weeks of embryonic development but may also occur at a lower rate between 16 and 25 weeks. Mental retardation is considered to be a deterministic effect with estimates of a dose threshold between 300 and 700 mGy.

The only pre-natal radiation effect thought to be stochastic in nature is the risk of inducing a cancer that progresses to diagnosis in childhood (i.e. by age 15). There are a number of studies which have produced data on the cancer risks in children. In addition to studies of the bomb survivors, one of the most influential was the Oxford Survey of Childhood Cancers (Stewart *et al.*, 1958, Bithell *et al.*, 2018) which is described as the largest case-control study of childhood cancer ever undertaken. The risk of childhood cancer, to age 15, is believed to be approximately $6 \times 10^{-2}.Gy^{-1}$. The lifetime risk, over all ages, would be higher by a factor of at least 2.

13.7.2 Radiation Workers in Healthcare

For the purposes of occupational radiation protection, the foetus is considered to be a member of the public and so the dose limit to the abdomen of medical radiation workers is restricted to 1mSv (over the natural background level) during the term of the pregnancy. This is not an issue for the majority of workers because the average annual dose, even for those who have classified status, is less than 0.5 mSv (HSE, 2017).

The important issue for workers who are or may be pregnant is to alert their employer as soon as possible. This will prompt a risk assessment and ensure that working practices and dose readings are more closely monitored. In the majority of cases, over 98% (BIR, 2009), pregnant staff will not need to alter their duties in any way. A small number of those working in specific areas (e.g. interventional radiology, nuclear medicine therapy) may need to adopt additional measures to reduce their potential exposures.

Staff working in environments where unsealed radioactive sources are used (e.g. Nuclear Medicine Departments) face the additional risk of ingestion from contamination from dealing with very ill patients and accidental releases. Pregnant staff may be advised to avoid specific aspects of their work where there may be increased risks of contamination. Another potential issue for pregnant staff is the use of PPE, specifically lead aprons. As the pregnancy advances into the third trimester, it may become more difficult to ensure that the lead apron fits correctly, although at that stage of pregnancy it is probably the weight of the apron rather than the fit that is more problematic.

A minority of staff, despite their training and previous experience, may be excessively concerned about adverse effects due to radiation exposure during pregnancy and may request an alteration to their duties. Most employers would treat such requests sympathetically although they are not obliged to do so.

13.7.3 PATIENT EXPOSURES

All medical radiation exposures must be justified and when the patient also happens to be pregnant that justification should include a consideration of potential harm to the foetus when assessing benefit. Based on the evidence of radiation damage to the foetus, it is only the stochastic risk of childhood cancer that is of concern, at least for diagnostic procedures. Radiation therapy is very rarely given during pregnancy but not absolutely ruled out (Kal and Struikmans, 2005).

The most important issue is to ascertain if a patient is pregnant immediately prior to the exposure taking place and rigorous procedures have been developed to establish this (HPA, 2009). Females of child bearing age, normally taken to be between 12 and 55, who require an exposure in which the primary beam may irradiate an area between the diaphragm and knees (excluding the arms) or those referred for a nuclear medicine procedure should be asked if they are or may be pregnant. In most cases, this is a relatively straightforward, although confidential, enquiry but procedures need to take into account how to deal with minors (under age 16), unconscious patients or those with language difficulties.

If pregnancy is definitely established and this was not previously known, justification for the procedure needs to be reconfirmed with the practitioner (radiologist, nuclear medicine physician) and/or referring clinician. The critical deciding factor is dose and therefore risk to the foetus with an arbitrary distinction between low and high dose/risk set at 10mGy. At this foetal dose, the risk of childhood cancer is believed to be of the order of 1 in 1000 which is approximately half the natural risk (1 in 500). Other factors which would need to be taken into consideration are any potential benefit to the foetus of making a diagnosis or the additional risk to the mother and foetus of delaying a procedure until later in pregnancy or post-partum. Patients who have been referred for procedures involving high foetal doses will often be booked to attend during the first 10 days of their menstrual cycle when pregnancy is highly unlikely.

Inadvertent exposures during pregnancy are not uncommon. Sometimes a patient will not know or may even deny they are pregnant. If a patient has been irradiated inadvertently during pregnancy, an investigation must be undertaken to determine the foetal dose and correct any procedural failures. The patient must also be informed and provided with a careful and sensitive explanation of the additional risk to her unborn child. Risks to the foetus from inadvertent exposure even at high doses would rarely justify termination of pregnancy.

REFERENCES

BIR, 2009. *Pregnancy and Work in Diagnostic Imaging Departments.* https://www.rcr.ac.uk/system/files/publication/field_publication_files/Pregnancy_Work_Diagnostic_Imaging_2nd.pdf

BIR, 2020. *Guidance on Using Shielding on Patients for Diagnostic Radiology Applications.* https://www.bir.org.uk/media/414334/final_patient_shielding_guidance.pdf

Bithell, J. F., G. J. Draper, T. Sorohan, and C. A. Stiller, 2018. Childhood Cancer Research in Oxford I: The Oxford Survey of Childhood Cancers. *Br. J. Cancer* 119, 756–762.

CQC, 2019. *IR(ME)R Annual Report 2018/19.* https://www.cqc.org.uk/sites/default/files/IRMER_annual_report_2018_2019_final.pdf

ESR, 2014. *IRefer: Making the Best Use of Clinical Radiology.* http://www.eurosafeimaging.org/wp/wp-content/uploads/2014/02/RCR_EuroSafe-Imaging-Poster.pdf

EU, 2013. Council Directive 2013/59 *laying down basic safety standards for protection against the dangers arising from exposure to ionising radiation,* 17-01-2014; OJ L 13. 1–73.

Giraud, P., E. Morvan, L. Claude, F. Mornex, C. Le Pechoux, J.-M. Bachaud, P. Boisselier, V. Beckendorf, M. Morelle, and M.-O. Carrère, 2011. Respiratory Gating Technique for Optimisation of Lung Cancer Radiotherapy. *J. Thorac. Oncol.* 6, 2058–2068.

Gov.UK, 2018. *National Diagnostic Reference Levels.* https://www.gov.uk/government/publications/diagnostic-radiology-national-diagnostic-reference-levels-ndrls/national-diagnostic-reference-levels-ndrls

Gov.UK, 2020. *Notes for Guidance on the Clinical Administration of Radiopharmaceuticals and Use of Sealed Radioactive Sources.* https://www.gov.uk/government/publications/arsac-notes-for-guidance

Hall, E. J., and D. J. Brenner, 2008. Cancer Risks from Diagnostic Radiology. *Br. J. Radiol.* 81, 362–378.

Hosny, A., C. Parmar, J. Quackenbush, L. H. Schwartz, and H. J. W. L. Aerts, 2018. Artifical Intelligence in Radiology. *Nat. Rev. Cancer* 18, 500–510.

HPA, 2009. *Protection of Pregnant Patients During Diagnostic Medical Exposures to Ionising Radiation.* Documents of the Health Protection Agency. RCE-9. ISBN 978-0-85951-635-8.

HPA, 2011. *Radiation Risks from Medical X-Ray Examinations as a Function of the Age and Sex of the Patient.* HPA-CRCE-028.

HSE, 2017. *Occupational Exposure to Ionising Radiation.* https://www.hse.gov.uk/statistics/ionising-radiation/cidi.pdf

ICRP, 2003. *Biological Effects after Prenatal Irradiation (Embryo and Fetus).* ICRP Publication 90. *Ann. ICRP 33 (1-2).*

ICRP, 2007. *The 2007 Recommendations of the International Commission on Radiological Protection.* ICRP Publication 103. *Ann. ICRP 37 (2-4).*

IPEM, 2005. *Recommended Standards for the Routine Performance Testing of Diagnostic X-Ray Systems.* IPEM Report 91, ISBN 1 903613 24 8.

Jones, A., C. Ansell, C. Jerrom, and I. D. Honey, 2015. Optimization of Image Quality and Patient Dose in Radiographs of Paediatric Extremities Using Direct Digital Radiography. *Br. J. Radiol.* 88(1050), 20140660.

Kal, H. B., and H. Struikmans, 2005. Radiotherapy During Pregnancy: Fact and Fiction. *Lancet Oncol.* doi:10.1016/S1470-2045(05)70169-8.

Martin, C. J., 2011. Personal Dosimetry for Interventional Operators: When and How Should Monitoring Be Done? *Br. J. Radiol.* 84, 639–648.

NHS England, 2018. *Diagnostic Imaging Dataset: Annual Statistical Release 2017/18.* https://www.england.nhs.uk/statistics/wp-content/uploads/sites/2/2018/11/Annual-Statistical-Release-2017-18-PDF-1.6MB-1.pdf

NRPB, 1986. *A National Survey of Doses to Patients Undergoing a Selection of Routine X-Ray Examinations in English Hospitals.* NRPB R-200, London, HMSO.

NRPB, 1990. *Patient Dose Reduction in Diagnostic Radiology–Report by the Royal College of Radiologists and the National Radiological Protection Board.* Documents of the NRPB, Vol. 1 No. 3, ISBN 0 859513 27 0.

Stec, N., D. Arje, A. R. Moody, and E. A. Krupinski, 2018. A Systematic Review of Fatigue in Radiology: Is It a Problem? *Am. J. Roentgenol.* 210, 799–806.

Stewart, A., J. Webb, and G. W. Kneale, 1958. A Survey of Childhood Malignancies. *Br. Med. J.* 1, 1495–1508.

14 Image Processing

King's College London, London, UK

CONTENTS

14.1 INTRODUCTION

Digital images are commonplace in modern medicine and diagnostic and treatment planning decisions are increasingly being informed by such images. Computerised image processing techniques are playing a growing role in manipulating these images in order to aid in the clinical decision-making process. This chapter gives an overview of some of the key techniques that can be used to process digital medical images.

The term *image processing* can be defined as the analysis and manipulation of a digital image, either to improve its quality or to extract features or information from the image. Image processing can be divided into a number of fields, including:

- *Image restoration*: the process of taking a corrupted or noisy image and estimating the clean, original image.
- *Image filtering*: the process of producing a new (filtered) image from an original image; the intensities of the filtered image typically represent some function of the original intensities.
- *Image enhancement*: a subclass of image filtering techniques that process digital images so that they look better to an observer.
- *Image registration*: the process of transforming an image so that its content aligns with that of another reference image.
- *Image segmentation*: the process of partitioning an image into different regions.

In this chapter, we will focus on two of these fields: image filtering (including some image enhancement) and image segmentation. For further details of digital image processing in general (i.e. not specific to medical imaging), the reader is referred to (Gonzales & Woods, 2008).

Throughout this section, the example images used to illustrate the techniques presented will be 2D, but remember that in medical imaging many images are 3D and it is fairly straightforward to extend most of these techniques to 3D.

DOI: 10.1201/9780429155758-14

14.1.1 Image Filtering

As stated above, image filtering techniques produce a new, filtered, version of an image by processing the intensities of the original image in some way. Image filtering techniques can be broken down into *spatial domain operations* (i.e. those that process the original image) and *frequency domain operations* (i.e. those that perform the operation on a Fourier transform of the image). Often the same, or a similar, operation can be performed in both ways.

14.1.1.1 Spatial Domain Filtering

Spatial domain operations can be further broken down into *point-processing* and *mask-processing* operations. Both produce a new version of the original image; the difference lies in which pixel/voxel intensities are used to determine the new intensity at any given point. For point-processing operations, the new intensity at image coordinates (x, y) is determined solely from the original intensity at the same coordinates, that is,

$$g(x,y) = T\big(f(x,y)\big) \tag{14.1}$$

where $f(x, y)$ represents the original image, $g(x, y)$ is the new image and T is an *intensity mapping function*.

Figure 14.1 illustrates the concept of an intensity mapping function. The function computes a new intensity for any given original intensity. In Figure 14.1, the mapping is defined by a piecewise linear function, and the function has the effect of enhancing the intensities in the middle of the intensity range, in this case the distinction between grey and white brain matter in an MR image of the head. Therefore, this operation can be seen as an example of image enhancement. To form an intensity mapping function, in principle any function can be defined, so long as there is a single unique output intensity for any given input intensity. For instance, Figure 14.2 shows another example. This time a CT scout image is processed using a technique known as *gamma correction* (or a *power law transformation*). Gamma correction operations are defined by the general equation (Equation 14.2).

$$g(x,y) = c \cdot f(x,y)^{\gamma} \tag{14.2}$$

where γ is the gamma value and c is a constant. Depending on the value of γ, gamma correction can have different effects. If $\gamma < 1$ (as in Figure 14.2), the effect is to enhance the darker parts of the

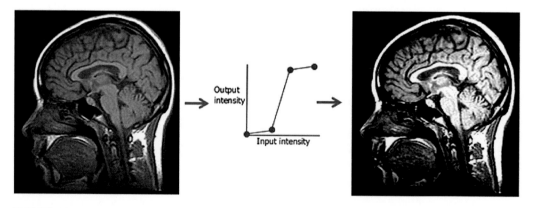

FIGURE 14.1   Piecewise linear intensity mapping applied to a 2D MR slice of the head.

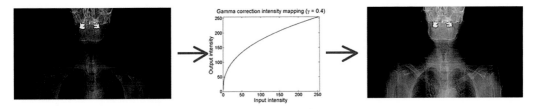

FIGURE 14.2 Gamma correction of a CT scout image using a gamma value of 0.4.

image, in this case making visible the features around the shoulders which were previously hard to visualise. Conversely, if $\gamma > 1$, the effect is to enhance the brighter parts of the image.

Both of the point-processing image filtering operations illustrated above have the effect of enhancing a particular part of the intensity range, for example, the middle (Figure 14.1) or the lower end (Figure 14.2). However, there may be occasions in which it is not as clear as this which part(s) of the intensity range we wish to enhance. In such cases, the *histogram equalisation* technique can be applied. A histogram is a summary of the distribution of intensities in an image. For example, Figure 14.3a–b shows the CT scout image and its associated histogram. The horizontal axis of the histogram represents the intensity and the vertical axis represents a count of how many image pixels have each intensity. Histogram equalisation aims to transform the image intensities so that the histogram of the resulting image is uniform, or 'equalised'. The discrete formulation of the histogram equalisation algorithm works as follows. We first denote the histogram of the original (unfiltered) image by

$$h\left(r_k\right) = n_k, \ k = 0 \ldots L - 1 \tag{14.3}$$

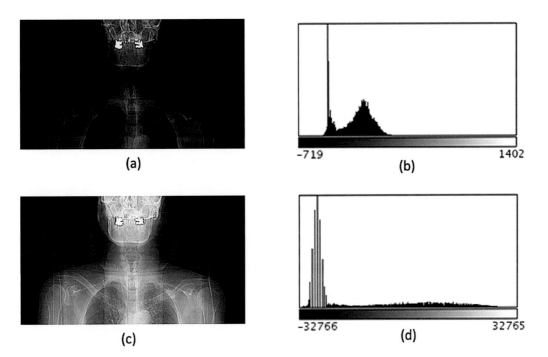

FIGURE 14.3 Histogram equalisation: (a) CT scout image; (b) histogram of (a); (c) histogram equalised version of (a); (d) histogram of (c).

in which r_k represent the L different intensity values possible in the image (normalised to between 0 and 1) and n_k are the counts of pixels with each of these intensities. We then define probabilities of occurrence for each intensity value,

$$p(r_k) = \frac{n_k}{n} \qquad (14.4)$$

where n is the total number of pixels in the image. Then, the intensity mapping, T, is defined as

$$s_k = T(r_k) = \sum_{j=0}^{k} p(r_k), \ k = 0 \ldots L-1 \qquad (14.5)$$

where s_k represents the new, histogram equalised, intensities. These new intensities are also in the range 0 to 1 so they can be multiplied by $L - 1$ in order to recover the original intensity range. An intuitive explanation of histogram equalisation is that it replaces each (normalised) intensity with the cumulative sum of the probabilities of all intensities up to and including itself. Figure 14.3c shows the result of applying histogram equalisation to the CT scout image in Figure 14.3a, and Figure 14.3d shows the histogram of the resulting image. Note the difference in the limits of the x-axis in Figure 14.3b and Figure 14.3d: the original image actually used a very small portion of the available intensity range compared to the equalised image.

An extension of standard point-processing operations can be made so that the new intensity depends on the original intensities at the same location in *multiple* original images, that is,

$$g(x,y) = A\big(f_1(x,y), \ f_2(x,y)\big) \qquad (14.6)$$

where A is a function of multiple (in this case, two) original images. Such operations are known as *arithmetic* or *logical image operations* (depending on whether A is an arithmetic or logical/Boolean operation). A common example of an arithmetic image operation is *digital subtraction angiography* (DSA), illustrated in Figure 14.4. DSA is commonly performed in interventional radiology in order to allow improved visualisation of blood vessels. In DSA, a *mask* fluoroscopy image is taken of the region of interest, followed by a *live* image of the same region after the injection of an intravenous contrast agent. The subtraction of the mask from the live image produces the DSA image, which highlights the vessels.

The alternative to point-processing operations is *mask processing* operations. With mask processing operations, the new intensity at a given pixel depends not only on the original intensity at that pixel, but also on the original intensities at a set of nearby (or *neighbouring*) pixels. This gives

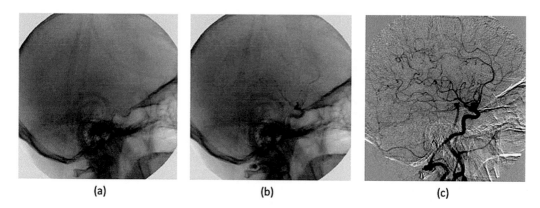

(a)   (b)   (c)

FIGURE 14.4 Digital subtraction angiography: (a) mask image; (b) live image; (c) subtraction image.

mask processing techniques more flexibility in defining the processing operation and permits some interesting effects to be produced.

There are a number of types of mask processing operation. One of the simplest is the family of *rank filters*, in which all pixels' intensities within the neighbourhood are ordered (i.e. ranked), and then one of them is chosen as the new intensity. Precisely which one is chosen determines the type of rank filter, for example, the highest valued one (the *max filter*), the lowest valued one (the *min filter*) or the middle one (the *median filter*). The median filter is the most commonly used rank filter and is often applied to reduce the effects of noise on images. For example, Figure 14.5 shows the use of the median filter for speckle reduction in ultrasound imaging.

Another family of mask processing techniques is *convolution filters*. With convolution, a new image intensity is computed as the weighted sum of the original intensities within the neighbourhood. That is,

$$g(x,y) = \sum_{s=-a}^{a} \sum_{t=-b}^{b} w(s,t) f(x+s, \ y+t) \tag{14.7}$$

where $w(s, t)$ is an array of filter coefficients that determine the precise nature of the convolution operation (Gonzales & Woods, 2008). Some examples of the effects of different convolution filters are shown in Figure 14.6.

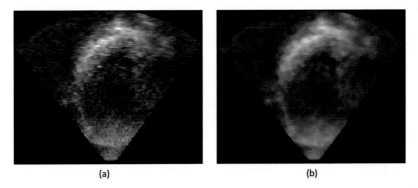

(a)   (b)

FIGURE 14.5   Median filtering for ultrasound speckle reduction: (a) cardiac ultrasound image; (b) median filtered version.

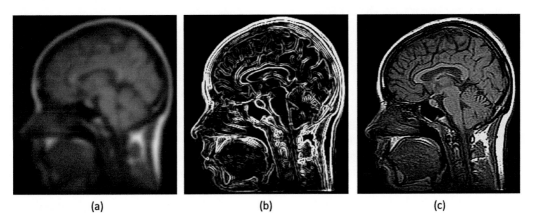

(a)   (b)   (c)

FIGURE 14.6   Mask processing operations on the MRI brain image: (a) Guassian smoothing; (b) Sobel edge detection; (c) unsharp masking.

Figure 14.6a shows the result of applying the following convolution filter to the original MR image shown in Figure 14.1,

$$w = \frac{1}{16} \times \begin{pmatrix} 1 & 2 & 1 \\ 2 & 4 & 2 \\ 1 & 2 & 1 \end{pmatrix} \tag{14.8}$$

This filter is known as a *Gaussian smoothing filter* because its coefficients approximately follow a Gaussian (or *normal*) density function centred on the middle of the filter. In this case, the filter was actually applied several times to emphasise the smoothing effect, that is, the output of the first filtering was used as the input to the second filtering, and so on.

Figure 14.6b shows the result of *edge detection* on the same original image. Edge detection aims to highlight areas of the image in which the intensity changes quickly, that is, the first spatial derivative of the image is high. The image shown in Figure 14.6b was produced by applying two separate filters, specifically

$$w = \begin{pmatrix} -1 & -2 & -1 \\ 0 & 0 & 0 \\ 1 & 2 & 1 \end{pmatrix} \tag{14.9}$$

and its transpose. The results of the two filtering operations were combined by taking the pixel-wise square root of the sum of the squares of the two filtered image intensities. These two filters are known as the *Sobel edge detection* filters and are commonly used in medical image processing to highlight tissue boundaries.

Finally, Figure 14.6c shows an example of *edge enhancement*. Edge enhancement aims to preserve the original image content but to enhance (or sharpen) the edges. One way of doing this is known as *unsharp masking*, which involves subtracting a scaled, smoothed version of an image from itself. As such, it represents a series of image processing operations (both mask processing and point-processing), but it can also be represented as a single convolution filter.

14.1.1.2 Frequency Domain Filtering

As well as performing image filtering operations in the spatial domain (i.e. the techniques described above), similar operations can be performed in the frequency domain. Frequency domain representations of 1D signals are based on the fact that any signal can be expressed as a weighted sum of sine waves of different frequencies and phases. For 2D signals (i.e. images), we also need to consider the direction of the sine waves. The *2D Fourier transform* converts an image into its frequency domain representation, which specifies the relative contributions of the sine waves of different frequencies, phases and directions needed to reconstruct the original image. An example of a frequency domain representation of a 2D CT slice of the head can be seen in Figure 14.7a–b. The magnitudes (i.e. intensities) shown in Figure 14.7b represent the relative contributions of the sine waves. The centre of this image represents the 'DC component' of the image, which is the contribution of a sine wave of zero frequency (this is the average intensity in the image). The further away from the centre we go, the higher the frequency of the sine wave, and the direction of the sine wave is indicated by the angle of each point to the centre. Phases are not shown in this figure.

The frequency domain representation of an image can be used to facilitate image filtering operations. The basic idea can be seen in Figure 14.7, which illustrates the process of *lowpass filtering* (or smoothing), which is a similar operation to the spatial domain Gaussian smoothing filter discussed above and shown in Figure 14.6a. Following the arrows, first, we compute the Fourier transform (Figure 14.7b) of the original image (Figure 14.7a). Next, we process the Fourier transform in some way (Figure 14.7c). In this case, we have kept the central part of the frequency domain image (i.e. the low frequency sine waves) and set to zero the contributions of all of the higher frequency

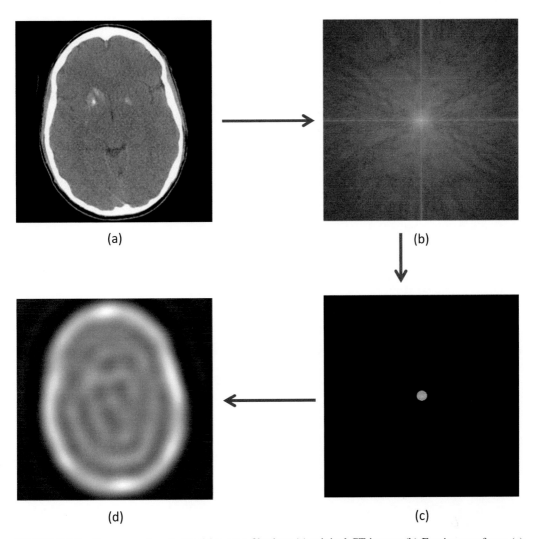

(a)

(b)

(d)

(c)

FIGURE 14.7 Frequency domain ideal lowpass filtering: (a) original CT image; (b) Fourier transform; (c) edited Fourier transform; (d) inverse Fourier transform.

sine waves. In other words, the low frequencies are *passing* and the high frequencies are being blocked – hence, the name *lowpass filter* for a smoothing operation such as this. This processing of the frequency domain image is implemented using a *masking* operation: a binary image (or *mask*) containing only 1s and 0s, with the same dimensions as the frequency domain image, is multiplied point-wise with the frequency domain image to produce a masked version. This masked frequency domain image is then used as the input to an inverse Fourier transform to produce the filtered image. We can see from Figure 14.7d that the result is a smoothed version of the original.

Similarly, we can also perform edge detection, or *highpass filtering*, in the frequency domain. This is illustrated in Figure 14.8a–d and essentially follows the same steps as the lowpass filtering operation, with the exception that the mask allows high frequencies to pass and blocks low frequencies. The result is an edge detection of the original image.

However, now look closely at the lowpass and highpass filtered images shown in Figure 14.7d and Figure 14.8d. In addition to the processed images we would expect, we can also see some artefacts. Specifically, there are 'rings' in the vicinity of sharp edges. These are known as *ringing artefacts* and it can be instructive to look a bit more closely into how and why they occur.

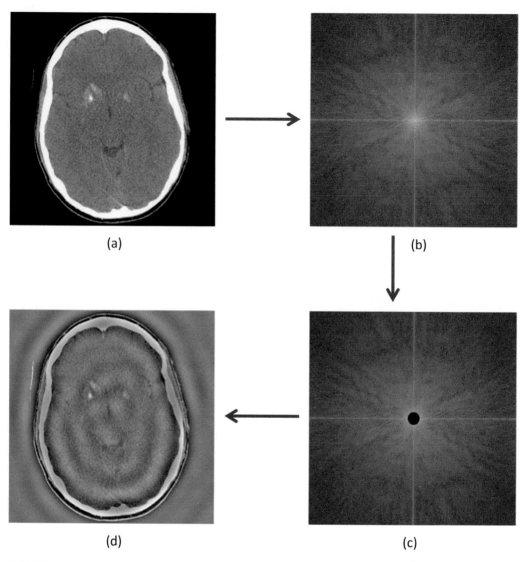

FIGURE 14.8 Frequency domain ideal highpass filtering: (a) original CT image; (b) Fourier transform; (c) edited Fourier transform; (d) inverse Fourier transform.

To understand the presence of ringing artefacts, we need to introduce the *convolution theorem*. The convolution theorem states equivalence between *convolution* in the spatial domain (i.e. the original image) and *masking* (i.e. point-wise multiplication) in the frequency domain. Formally, we can write:

$$f(x,y) \otimes h(x,y) \Leftrightarrow F(u,v)H(u,v) \tag{14.10}$$

where $f(x, y)$ is the spatial domain image, $h(x, y)$ is the convolution filter, $\otimes$ represents the convolution operation, $F(u, v)$ is the Fourier transform of $f(x, y)$ and $H(u, v)$ is the frequency domain mask. The $\Leftrightarrow$ symbol indicates that the two terms represent a Fourier transform pair, that is,

$$FT\left[f(x,y) \otimes h(x,y) = F(u,v)H(u,v)\right] \tag{14.11}$$

and

$$f(x,y) \otimes h(x,y) = FT^{-1}\left[F(u,v)H(u,v)\right] \qquad (14.12)$$

where *FT* represents the Fourier transform operation.

Put simply, the convolution theorem states that if we do a masking operation on the Fourier transform of an image, and then do an inverse Fourier transform on the result, it has the same result as convolving the original image using a convolution filter. *This is true if the convolution filter used is the inverse Fourier transform of the mask used in the frequency domain*, that is, $H(u, v)$ is the Fourier transform of $h(x, y)$.

Now, returning to the lowpass filtered image shown in Figure 14.7d, it is clear that we could get the same result by convolving the spatial domain image using a convolution filter that is formed by taking the inverse Fourier transform of the circular mask, $H(u, v)$. Figure 14.9 illustrates what this looks like. Figure 14.9a shows the circular frequency domain mask, Figure 14.9b shows its inverse Fourier transform and Figure 14.9c shows a 1D profile through this 2D convolution filter. The filter shown in Figure 14.9b is called a *sinc* function, and it is evident from Figure 14.9c that it features oscillations away from its centre. It is these oscillations that cause ringing artefacts.

The masking operations illustrated in Figure 14.7 and Figure 14.8 are known as *ideal lowpass* and *highpass filters*. The name refers to the 'ideal' frequency response of the filters, that is, they feature a sharp cut-off between passed and blocked frequencies.

The problem of ringing artefacts with ideal lowpass and highpass filters can be mitigated by using a mask that has a smoother transition between passed and blocked frequencies. For example, Figure 14.10a and Figure 14.10c show the result of applying *Butterworth* lowpass and highpass filters to the same CT head slice. Note the lack of ringing artefacts. Mathematically, the frequency domain masks for the lowpass and highpass Butterworth filters are defined as

$$H(u,v) = \frac{1}{1+\left(D(u,v)/D_0\right)^{2n}} \qquad (14.13)$$

and

$$H(u,v) = \frac{1}{1+\left(D_0/D(u,v)\right)^{2n}} \qquad (14.14)$$

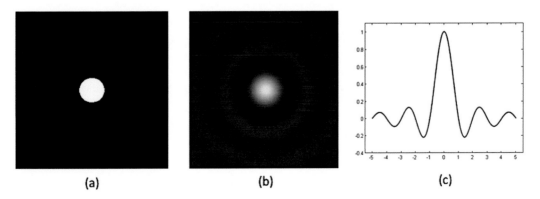

(a)   (b)   (c)

FIGURE 14.9   The cause of ringing artefacts: (a) an ideal lowpass filter; (b) the inverse Fourier transform of (a) which is a sinc function; (c) a 1-D plot of the sinc function.

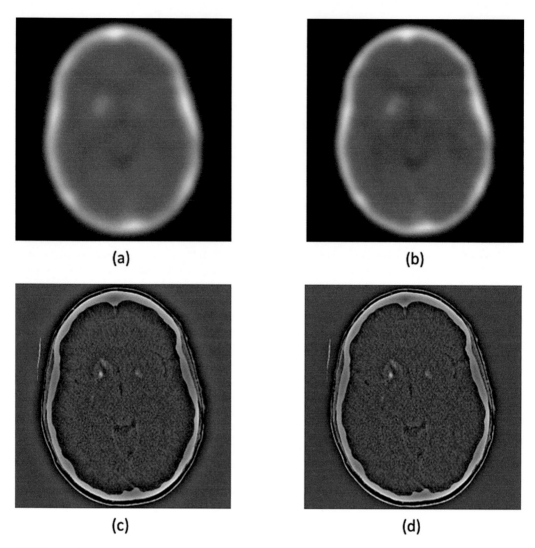

FIGURE 14.10 Butterworth and Guassian filters: (a) Butterworth lowpass filter of CT image; (b) Guassian lowpass filter; (c) Butterworth highpass filter; (d) Guassian highpass filter.

respectively, where $D(u, v)$ represents the distance from the centre of the Fourier transform, D_0 is the frequency 'cut-off' and n is the order of the filter (the results shown in Figure 14.10 were generated using a second order filter, i.e. $n = 2$). These equations result in similar looking masks to the 'ideal' mask shown in Figure 14.9a but with a smoother transition.

An alternative to the Butterworth filters is to use a Gaussian function to define the transition in the masks, i.e.

$$H(u,v) = e^{-D(u,v)^2 / 2D_0^2} \tag{14.15}$$

for the lowpass filter and

$$H(u,v) = 1 - e^{-D(u,v)^2 / 2D_0^2} \tag{14.16}$$

for the highpass filter. These filters also feature a smooth transition at the 'cut-off' frequency of D_0, and the results of applying them to the CT head slice can be seen in Figure 14.10b and Figure 14.10d. Once again, note the lack of ringing artefacts.

14.1.2 Segmentation

The second image processing operation we will look at is *segmentation*. Segmentation can be defined as the process of (manually or automatically) partitioning an image into a number of different regions. Segmentation is normally followed by another operation such as visualisation or quantitative analysis. For example, we may want to partition a medical image into different organs and/or types of tissue and then make quantitative measurements such as the volume of a tumour.

Segmentation algorithms can be manual, semi-automatic or fully automatic. There are a variety of free or commercial segmentation packages (e.g. ITKSnap[1]) for performing manual segmentation, but this approach tends to be time-consuming and operator-dependent. Fully automatic segmentation is a challenging task and an active research topic. Good reviews can be found in (Iglesias & Sabuncu, 2015), (Heimann & Meinzer, 2009) and (Smistad, Falch, Bozorgi, Elster, & Lindseth, 2015). In this section, we will focus on semi-automatic techniques, which involve some automatic processing but also some operator interaction. We will illustrate a simple semi-automatic segmentation pipeline on a medical case study.

14.1.2.1 Case Study: Semi-Automatic Segmentation of the Left Ventricle from MR

This case study illustrates a simple pipeline for segmenting the left ventricle (LV) from a 2D axial cardiac MR slice. Our input image is shown in Figure 14.11a, in which the LV is the large bright

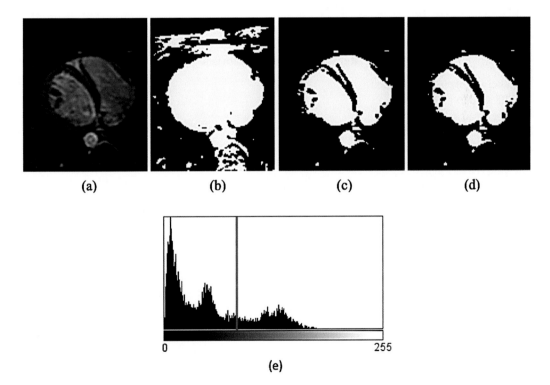

FIGURE 14.11   Thresholding of an MRI image of the left ventricle: (a) original image; (b)–(d) binary images produced using different thresholds; (e) image histogram of (a) showing in red the threshold used to produce (d).

region on the left of the image, the right ventricle (RV) is the large bright region on the right and the aorta is the smaller circular region below the LV. One motivation for performing such an operation might be to compute and track intra-ventricular volume over the cardiac cycle with a view to diagnosing or characterising heart disease.

We will start off with *thresholding*. Thresholding is a simple operation involving creating a binary image in which pixels have an intensity of 1 if the intensity in the original image was greater than a given threshold, and 0 otherwise. This is illustrated in Figure 14.11. Figure 14.11b-d show the results of applying the thresholding operation with three different (manually chosen) thresholds, and we can see that the results are quite different. This shows the importance of choosing an appropriate threshold, and this is, to a certain extent, a subjective choice. However, analysis of a histogram of the image intensities can sometimes be informative, as is shown in Figure 14.11e. The red line in this histogram shows the threshold used to produce the binary image in Figure 14.11d, and it can be seen to be approximately equidistant between two peaks in the histogram, which represent the bright blood pool and the slightly darker ventricular myocardium (i.e. muscle) in the image.

Although a simple operation, thresholding is often used as part of a semi-automatic segmentation pipeline, although on its own it is rarely sufficient to produce satisfactory results. This can be seen in Figure 14.11d, in which the thresholded image still contains the RV, aorta and some image noise or parts of other blood vessels.

One effective way of eliminating these unwanted regions is to use *region growing*. The basic idea of region growing is to manually identify one or several pixels that are known to be 'inside' the region we are interested in (these are called *seed points*), and then to use an automatic algorithm to 'grow' these seeds into a partitioned region. The growing will continue so long as the image content of the new pixels is similar to the content at the seed(s). Pseudocode for the region growing algorithm is shown below.

Region growing pseudocode

Label all pixels as undefined
Add seed to pixel queue
Label seed as 'inside'
While pixel queue is not empty
        Remove point from head of pixel queue
        For each neighbour of the removed point
                If neighbour undefined, and neighbour has similar intensity to seed
                        Add neighbour to pixel queue
                        Label neighbour as 'inside'
                End if
        End for
End while

As can be seen, the region growing algorithm relies on two key concepts: *similarity* and *neighbourhood*. Similarity typically involves some kind of intensity comparison, and when applying the algorithm to a binary image as we are it normally means the intensities should be the same, that is, both 1, since we will always choose a seed that is 'inside' the region. The concept of neighourhood,

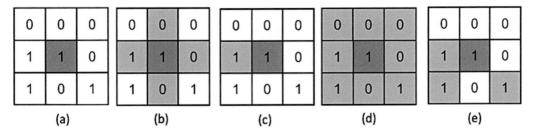

FIGURE 14.12   Neighbours and adjacency: (a) highlighted centre pixel in a simple binary image; (b) 4-neighbours of the centre pixel; (c) 4-adjacent pixels; (d) 8-neighbours; (e) 8-adjacent pixels.

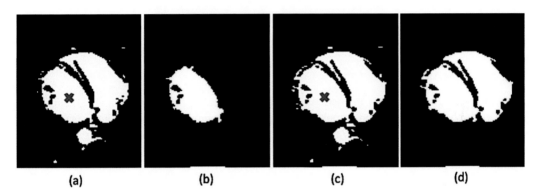

FIGURE 14.13   Region growing: (a) thresholded MRI image indicating in red the position of a seed point: (b) binary image after region growing; (c)-(d) the same process but using a different threshold value.

and the related concept of *adjacency*, is illustrated in Figure 14.12. As can be seen, we can actually define two different types of neighbourhood in 2D images: 4-neighbours or 8-neighbours, depending on whether or not we include diagonally neighbouring pixels. *Adjacent* pixels are those that are neighbours but also similar, and the region growing algorithm will iteratively grow the region into these adjacent pixels.

The results of applying the region growing algorithm to our cardiac MR slice are shown in Figure 14.13. The segmentations shown in Figure 14.13b and Figure 14.13d were generated using 8-neighbours and the seed point is shown as a red cross. The only difference between the two was the binary input images, which were produced using slightly different thresholds. These thresholded images are shown in Figure 14.13a and Figure 14.13c. The segmentation shown in Figure 14.13b is the 'correct' one, in that it contains only the LV. That shown in Figure 14.13d, on the other hand, has 'leaked' into the RV, because the lower threshold caused a 'bridge' of pixels with value 1 between the two ventricles. This again illustrates the importance of choosing an appropriate threshold, and some trial and error may be required.

If repeated on a slice-by-slice basis, or using a 3D version of the region growing algorithm, the simple process outlined above can result in good segmentations. As an illustration, Figure 14.14 shows a 3D rendering of the LV from the full 3D MR volume from which the 2D slice shown in Figure 14.11a was extracted. In more challenging cases, other operations or manual interactions may be added to the segmentation pipeline.

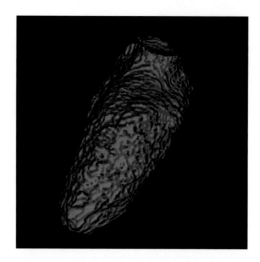

FIGURE 14.14   Surface rendering of a 3D left ventricle segmentation.

14.2   SUMMARY

Image processing concerns the manipulation of digital images, typically with a view to enhancing the image for better visualisation or making quantitative measurements. We have introduced a range of image filtering operations, which can be performed either in the spatial domain (i.e. on the original image) or the frequency domain (on a Fourier transform of the image). The convolution theorem is an important piece of theory that establishes equivalence between these two approaches. We have also seen how a simple semi-automatic image segmentation pipeline can work using thresholding and region growing. In reality, many segmentation problems are more challenging than this and a more complex series of operations may be required to produce satisfactory segmentations.

ACKNOWLEDGEMENTS

The MR image used in Figure 14.1 is in the public domain. All image processing operations were performed using the ImageJ software (Abramoff, Magalhaes, & Ram, 2004). The CT scout image used in Figure 14.2 is from the ImageJ sample images. The original CT image used in Figure 14.7, Figure 14.8 and Figure 14.10 was reused from (Yacoub, 2013) under the Creative Commons Attribution license.

NOTE

1   www.itksnap.org

BIBLIOGRAPHY

Abramoff, M. D., Magalhaes, P. J., & Ram, S. J. (2004). Image processing with image J. *Biophotonics International, 11*(7), 36–42.

Gonzales, R. C., & Woods, R. E. (2008). *Digital Image Processing.* Prentice Hall. https://www.pearson.com/us/higher-education/program/Gonzalez-Digital-Image-Processing-4th-Edition/PGM241219.html.

Heimann, T., & Meinzer, H.-P. (2009). Statistical shape models for 3D medical image segmentation: A review. *Medical Image Analysis, 13*(4), 543–563.

Iglesias, J. E., & Sabuncu, M. R. (2015). Multi-atlas segmentation of biomedical images: A survey. *Medical Image Analysis, 24*(1), 205–219.

Smistad, E., Falch, T. L., Bozorgi, M., Elster, A. C., & Lindseth, F. (2015). Medical image segmentation on GPUs – A comprehensive review. *Medical Image Analysis, 20*(1), 1–18.

Yacoub, H. A. (2013). Abnormal magnetic resonance imaging and hemichorea associated with non-ketotic hyperglycemia. *Journal of Neurology Research, 3*(5), 146–149.

15 Emerging Techniques

Michele Avanzo
IRCCS Centro di Riferimento Oncologico, Aviano, Italy

Tony Greener
Guy's and St Thomas' NHS Foundation Trust, London, UK

Luigi Rigon
Università degli studi di Trieste, Trieste, Italy

Slavik Tabakov
King's College London, London, UK

CONTENTS

DOI: 10.1201/9780429155758-15

15.1   INTRODUCTION

The development of medical physics is closely linked with that of contemporary medicine, in particular medical imaging, radiotherapy and related specialities. In 2011, the International Labour Organization (ILO, Geneva) decided to include both 'medical physicist' and 'biomedical engineer' in the International Standard Classification of Occupations (ISCO) under the professional categories 'Science and Engineering Professionals' and 'Health Professionals' (Smith and Nusslin, 2013). This remarkable achievement of the International Union for Physical and Engineering Sciences in Medicine (IUPESM), International Organisation for Medical Physics (IOMP) and International Federation for Medical and Biological Engineering (IFMBE) benefits the current 30,000 strong workforce of medical physicists globally and recognises the contribution that physicists and engineers make to healthcare.

The exceptional inter-professional collaboration between physics and medicine has delivered three Nobel Prizes in the past 50 years.

1977 – R. Guillemin, A. Schally, R. Yalow ("for the development of radioimmunoassays of peptide hormones").

1979 – A. M. Cormack, G. Hounsfield ("for the development of computer assisted tomography").

2003 – P. Mansfield, P. Lauterbur ("for their discoveries concerning magnetic resonance imaging").

The speed at which medical physics is developing is astonishing. New methods are developed and introduced into clinical practice all the time, in some cases revolutionising medicine. For example, new imaging methods developed over the 20-year period between 1988 and 2009 include the following.

1988 – Electronic portal imaging devices (EPID) in radiotherapy

1990 – Helical computed tomography

1993 – Functional MRI

1995 – Flat panel detectors for diagnostic radiology

1996 – Very high-frequency (VHF) digital ultrasound

1998 – Multislice computed tomography

1999 – Digital mammography

2000 – SPECT/CT hybrid imaging systems

2001 – PET/CT hybrid imaging systems

2002 – 3T MRI

2007 – Digital tomosynthesis

2009 – PET/MR hybrid imaging systems

Most of these medical imaging devices are discussed in this book. In this chapter, brief descriptions of three recent developments are given as examples of emerging techniques with significant potential: phase contrast imaging, radiomics and ultra-high dose rate radiotherapy (FLASH-RT).

15.2   PHASE CONTRAST IMAGING

15.2.1   INTRODUCTION

A typical way to describe the process of image formation in X-ray radiology is to represent the X-ray beam as a stream of particles, namely photons, which impinge on the sample. The latter can absorb or scatter these particles, attenuating the stream to a different extent in different regions of the sample, thereby creating contrast. Thus, for instance, bones show a strong contrast in X-ray radiology due to their higher attenuation of X-rays compared to soft tissues.

However, X-rays are electromagnetic waves, with a wavelength in the range 0.01–0.1 nm, about ten thousand times smaller than visible light. The wave nature of X-rays is typically ignored in diagnostic X-ray imaging, because the particle model provides a simple and effective description of image formation. This section will provide an introduction to novel imaging techniques that take advantage of the wave characteristics of X-rays. These techniques are commonly referred to as X-ray phase contrast imaging.

The fundamental characteristics of the interaction between X-rays and an imaged object using a wave model are summarized as follows. Let us suppose that X-rays are represented by plane waves travelling along the z direction, in a reference system where xy represents the object plane. As a consequence of the interaction with the object, the X-ray wave is modulated by the so-called object transmission function $T(x,y)$ (Rigon, 2014). $T(x,y)$ is a complex-valued function that can be written as:

$$T(x,y) = A(x,y)e^{i\phi(x,y)} \tag{15.1}$$

where

$$A(x,y) = e^{\frac{-2\pi}{\lambda}\int \beta(x,y,z)dz} \tag{15.2}$$

represents the wave amplitude reduction (λ being the X-ray wavelength) and

$$\phi(x,y) = \frac{-2\pi}{\lambda}\int \delta(x,y,z)dz \tag{15.3}$$

represents the phase shift introduced by the object. In these expressions, $\beta(x,y,z)$ and $\delta(x,y,z)$, respectively, are the imaginary part and the real part decrement of the refractive index $n(x,y,z) = 1 - \delta(x,y,z) + i\beta(x,y,z)$, which characterises the object, and the integrations are carried out over the object along the z direction.

Equation (15.2) accounts for an intensity reduction of $A^2(x,y) = e^{\frac{-4\pi}{\lambda}\int \beta(x,y,z)dz}$, which corresponds to the well-known Beer-Lambert law:

$$I_{out}/I_0 = e^{-\int \mu(x,y,z)dz} \tag{15.4}$$

where I_0 is the intensity of the beam impinging on the object plane, I_{out} is the intensity of the outgoing beam and $\mu(x,y,z) = 4\pi\beta(x,y,z)/\lambda$ is the linear attenuation coefficient of the object. This intensity reduction, that is, X-ray attenuation, is the physical process that generates contrast in conventional X-ray imaging.

Conversely, the phase shift $\phi(x,y)$ in Equation 15.3, plays no role in conventional X-ray imaging. In fact, X-ray detectors are not sensitive to the phase of the radiation but only to its intensity. However, with suitable techniques, often (but not always) requiring coherent X-ray sources, the phase shift can generate a detectable intensity modulation. In other words, the phase shift $\phi(x,y)$ serves as a supplementary source of contrast (in addition to attenuation), which is the basis of phase contrast X-ray imaging.

Before reviewing the different phase contrast techniques, it is worth underlining some common characteristics resulting from the definition of phase shift $\phi(x,y)$ in Equation 15.3.

- The phase shift $\phi(x,y)$ is proportional to the real part of the refractive index decrement δ, while absorption is related to the imaginary part β. In biomedical imaging, at typical X-ray energies and for soft tissues, δ is much larger than β. For instance, in mammography δ is in the order of $10^{-6} - 10^{-7}$, while β is typically $10^{-8} - 10^{-10}$ (Lewis, 2004). This suggests that, in principle, phase effects can be more significant than absorption effects, and that it may be possible to visualize soft tissue details that are hardly detectable in conventional radiology.

- Phase contrast imaging does not require X-ray absorption, so phase contrast can be obtained with no radiation dose delivered to the object. In practice, it is impossible to completely exclude absorption. However, phase contrast imaging can be performed at very low dose levels, compared to conventional x-ray imaging.
- Low-dose phase-contrast images can be obtained using X-ray energies that are higher than usual. Both δ and β decrease with increasing energy, that is, with decreasing wavelength, but δ is roughly proportional to λ^2 while β is proportional to λ^3. The faster decrease of β at high energy allows us to choose an X-ray energy where β is small enough to deliver low dose, while δ is large enough to permit good phase contrast.
- When interacting with the object, X-rays undergo refraction and scattering through very small angles. The deviation angle α between the incoming and outgoing X-ray waves can readily be shown to be proportional to the gradient of the phase shift, measured in the object plane xy, that is, $\alpha \cong \dfrac{2\pi}{\lambda} \left| \nabla_{xy}\phi(x,y) \right|$. In biomedical samples, these deviations are generally of the order of 1-10 μrad, too small to be relevant in conventional X-ray imaging.
- As in all imaging techniques, contrast stems from spatial variation of a given physical quantity in the object. In the particular case of phase contrast imaging, not only the phase shift $\phi(x,y)$ but also its gradient, $\nabla_{xy}\phi(x,y)$, and its Laplacian, $\nabla^2_{xy}\phi(x,y)$, are often key quantities. Thus, phase contrast can be particularly strong at the edges of the object or at interfaces within it, where the gradient and the Laplacian typically take high values.

15.2.2 Coherent X-Ray Sources

A simple approach to coherence takes into account temporal (or longitudinal) coherence, which relates to the energy spectrum of the imaging system (the narrower the spectrum, the higher the temporal coherence), and spatial (or lateral) coherence, which relates to the source size and the geometric characteristics of the system (the smaller the source size and the larger the source-to-sample distance, the higher the spatial coherence). Thus, a monochromatic point-like source set at large distance defines an ideally coherent imaging system. The degree of coherence of a real source depends on how well these characteristics are approximated.

While coherent X-ray sources are not necessary for some phase contrast techniques, they are fundamental for others. Most phase contrast techniques have been introduced, or at least developed, taking advantage of highly coherent X-ray sources. Unfortunately, conventional X-ray tubes have rather poor temporal coherence (broad energy spectrum) and spatial coherence (small source-to-sample distance). Microfocus X-ray tubes have good spatial coherence, but at the cost of a low beam intensity. Conversely, modern synchrotron radiation X-ray sources have a high degree of coherence and provide very intense beams. Therefore, they are ideal sources for most phase contrast techniques. However, synchrotrons are large-scale and expensive facilities so, recently, more compact and less costly X-ray sources have been proposed. Examples include liquid-metal-jet (Hertz et al., 2014) and inverse Compton scattering (Carroll et al., 2003) sources. Liquid-metal-jet sources are microfocus tubes where the solid-metal anode is replaced by a continuously flowing liquid metal jet. This makes it possible to overcome the typical power limit of conventional X-ray sources, that is, the anode heat load, and to obtain intense and spatially coherent X-ray beams. In inverse Compton scattering sources, a high-power laser pulse interacts with a relativistic electron beam producing photons in the mega-electron-volt (MeV) energy range. Attractive features of this approach include high intensity, tunable energy, relatively narrow bandwidth and small-angle cone-beam geometry.

15.2.3 X-Ray Phase Contrast Imaging Techniques

X-ray phase contrast imaging is still under development, and arguably the field has not reached its full potential yet. Different phase contrast imaging techniques have been introduced in recent decades, and some of these have already found important applications in medical imaging. A brief review is given below.

15.2.3.1 Propagation-Based Phase-Contrast Imaging

Propagation-Based Phase-Contrast Imaging (PPCI) is very easy to implement, provided that the X-ray source has a suitably high degree of spatial coherence: all that is needed is to choose appropriate values for the source-to-sample and sample-to-detector distances. In particular the latter, called the propagation distance, must be sufficient to allow the X-ray wave emerging from the object to evolve before impinging on the detector (Figure 15.1a). This results in edge enhancement, proportional to $\nabla^2_{xy}\phi(x,y)$, as long as the detector has sufficiently high spatial resolution not to miss this effect.

15.2.3.2 Analyser-Based Imaging

Analyser-Based Imaging (ABI) usually requires a monochromatic and highly collimated X-ray beam. ABI makes use of an analyser crystal, placed between the sample and the detector, which 'analyses' the X-rays emerging from the object. The analyser crystal acts as a narrow angular bandpass filter, modulating the intensity of the X-rays reaching the detector according to the deviations they have experienced traversing the object. When set at a given angle (the Bragg angle) with respect to the X-ray beam, the analyser satisfies the diffraction condition, whereby it acts as an X-ray 'mirror', which efficiently reflects the primary beam towards the detector (Figure 15.1b). However, as the crystal is detuned from the Bragg angle (typically by 10-100 µrad), the reflectivity rapidly drops and vanishes. This principle defines the reflectivity curve, or rocking curve, $R(\theta)$ of the crystal as a function of the detuning angle. The tiny width of the rocking curve makes ABI an ideal tool to highlight X-ray deviations. Thereby, ABI highlights refraction, ultra-small-angle scattering (in the microradian range), and extinction (i.e. the rejection of small-angle scattering, in the milliradian range); these effects, alongside attenuation, contribute to the overall contrast.

The precise nature of the contrast depends on the detuning angle. When the angle is zero (i.e. when the analyser is set at the peak of the rocking curve), undeviated X-rays have the highest probability of reaching the detector, while deviated photons are rejected or attenuated. On the other hand, with suitable detuning angles, photons subjected to refraction or ultra-small-angle scattering have the highest probability of reaching the detector, while undeviated X-rays are rejected or attenuated. In particular, with an almost complete rejection of the direct beam, ABI allows us to perform dark-field imaging, whereby only photons refracted or scattered at ultra-small angles appear in the image. The different effects contributing to contrast in ABI can be exploited by collecting images at different

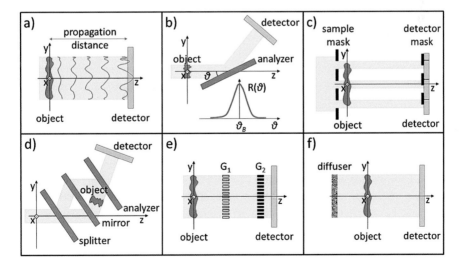

FIGURE 15.1 Synoptic sketch of six different X-ray phase contrast imaging techniques: a) Propagation-Based Phase-Contrast Imaging; b) Analyser-Based Imaging; c) Edge Illumination X-ray Imaging; d) Interferometry; e) Grating Interferometry; f) Speckle Imaging. In each case, the X-ray beam (in yellow) is assumed to be parallel and is directed from left to right, along the z axis.

detuning angles: in this way, parametric images depicting apparent absorption (i.e. attenuation and extinction), refraction and ultra-small-angle scattering can be obtained (Diemoz et al., 2018).

15.2.3.3 Edge Illumination X-Ray Imaging

The principle of edge illumination is to create a very narrow beam which hits the sensitive region of the detector at its edge. Therefore, any small deviation introduced by the object, which moves the narrow beam inside (or outside) the sensitive region of the detector, will cause an increase (or decrease) in the recorded intensity. Thus, phase contrast resulting from very small X-ray deviations is obtained in a straightforward and practical way. This imaging technique has strong similarities with ABI, in its mathematical model and in the images that it can produce (Diemoz et al., 2018). However, unlike ABI, it does not need a monochromatic beam and can be implemented using a conventional X-ray tube. A typical set-up implementing the edge illumination principle requires two highly-absorbing masks with slits, or apertures. One of these 'coded apertures' is placed before the object, to create an array of tiny beamlets, and one in front of the detector, so that each beamlet hits the edge between a sensitive and an insensitive region (Figure 15.1c). Moreover, by shifting the first (sample) mask with respect to the second (detector) mask, different fractions of each sensitive region of the detector can be illuminated. This is somewhat analogous to ABI, when the analyser crystal is moved to different positions along the rocking curve.

15.2.3.4 Interferometry

Interferometric techniques used in optical imaging can be translated into the X-ray realm. For instance, an X-ray Mach-Zehnder interferometer can be built by means of three parallel crystals through which the beam passes (Laue geometry): the first (splitter) splits the beam into two beams traveling in different directions, the second (mirror) reflects both beams towards the third crystal (analyser), where they recombine creating an interference pattern. The object is introduced in one of the interferometer arms, altering the interference pattern (Figure 15.1d). In principle, this scheme allows for very precise measurement of the phase shift $\phi(x, y)$, thus revealing subtle phase contrast. However, this technique has two important drawbacks. The first is that, like ABI, it requires a high degree of coherence. Secondly, due to the very small wavelength of X-rays compared to visible light, the crystal system must be very precisely aligned, which hinders applications where a reasonably large field of view (i.e. several square centimetres) is desired.

15.2.3.5 Grating Interferometry

Another interferometric approach, which relaxes the coherence and mechanical requirements of crystal interferometry, is grating (or Talbot) interferometry, whereby two gratings, usually called G_1 and G_2, are placed between the sample and the detector. G_1, the phase grating, creates a periodic phase shift pattern (with a period of few microns) which creates strong interference fringes at a given distance. G_2, the analyser grating, features periodic slits in a highly absorbing material, with the period matching that of the interference pattern (Figure 15.1e). Therefore, scanning G_2 with respect to G_1 (phase stepping) allows measurement of the interference pattern on the detector. Of course, the interference pattern is altered by the phase shift introduced by the object, and, as in ABI, absorption, refraction, and ultra-small-angle scattering contrast can be obtained.

In principle, Talbot interferometry requires a high degree of spatial coherence, However, the introduction of a third grating, G_0, close to the source, allows the same principle to be implemented with spatially incoherent sources (Talbot-Lau interferometer). G_0, an absorbing mask with transmitting slits, creates an array of individually coherent, but mutually incoherent, sources, which, under suitable geometric conditions, contribute constructively to image formation.

15.2.3.6 Speckle Imaging

Speckle imaging is implemented by introducing an additional element in the X-ray beam, namely a diffuser that generates a strong speckle pattern (Figure 15.1f). Common sandpaper or biological filter membranes with micron sized structures are typical examples (Zdora, 2018). While this speckle pattern can overwhelm the X-ray image of the object, modulations introduced in the speckle pattern by the object reveal its properties, in terms of refraction, transmission and ultra-small angle scattering. Thereby, parametric phase contrast images can be obtained. X-ray speckle-based imaging does not need high degrees of spatial or temporal coherence and requires a very simple set-up (basically a foil of sandpaper). Therefore, it can be readily implemented using conventional X-ray sources and, despite its recent introduction, has the potential to increase the range of applications of phase contract imaging.

15.2.4 Applications

Despite the differences in requirements and implementations, all the techniques outlined above share the ability to highlight phase contrast, i.e. to reveal the contrast hidden in the phase shift term $\phi(x,y)$ (Equation 15.3). When the technique does not reveal the phase shift directly (but rather its gradient or its Laplacian), $\phi(x,y)$ can be obtained with suitable post-processing algorithms (phase retrieval). Most phase contrast techniques allow refraction and small- or ultra-small-angle scattering to be highlighted, alongside attenuation, and allow dark field imaging. Moreover, phase contrast techniques are not limited to planar imaging, but can be implemented as forms of tomography, and can in principle take advantage of suitable contrast agents, based on the real part, δ, rather than on the imaginary part, β, of the refractive index.

Phase contrast imaging has been shown to have considerable potential in several biomedical applications (Bravin et al., 2013). In general, sensitivity to a new source of contrast can be helpful in highlighting features that have poor visibility in conventional imaging. Moreover, specific characteristics of phase contrast imaging can be exploited in particular applications. For instance, the edge-enhancement effect has been used in mammography to evaluate the borders of masses, in order to discriminate between benign and malignant lesions (Castelli et al., 2011), while sensitivity to small and ultra-small angle scattering has allowed unprecedented visualisation of the airways and important insight in lung imaging (Kitchen et al., 2008). Finally, the ability to obtain parametric images (such as apparent-absorption and refraction images, as shown in Figure 15.2) has been used

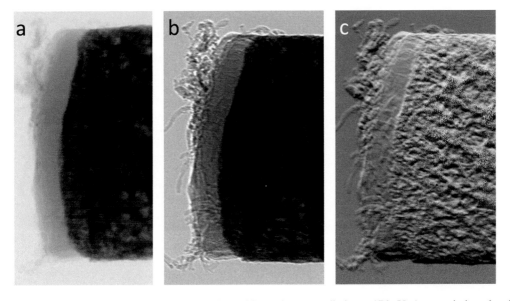

FIGURE 15.2 Femoral head core cut acquired with synchrotron radiation at 17 keV: a) transmission, showing no phase effects; b) and c) parametric images obtained with the analyser-based imaging technique, namely: b) apparent absorption, allowing a clear visualisation of the cartilage and of its internal structure; c) refraction, simultaneously depicting the articular cartilage and the underlying bone structure (Majumdar et al., 2004).

to simultaneously visualize articular cartilage, underlying subchondral bone and trabecular bone, facilitating earlier diagnosis (Majumdar et al., 2004; Muehleman et al., 2004).

15.3 RADIOMICS

15.3.1 Definition of Radiomics

Radiomics is an emerging area aimed at associating features extracted from large-scale radiological image analysis with biological or clinical endpoints, in order to develop diagnostic, predictive or prognostic models. The fundamental idea of radiomics is that the extraction of a large number of quantitative imaging features, also called 'radiomic features', can provide information that is distinct from or complementary to that provided by biopsy which, besides being invasive, does not always represent the entire volume of a tumour. The term 'radiomics' originates from the words 'radio' which refers to radiology, the science of medical imaging, and the suffix 'omics', first used in the term genomics to refer to mapping of the human genome (Avanzo et al. 2017, Gillies et al. 2016).

15.3.2 The Radiomics Framework

The goal of radiomics is to develop a function or mathematical model to classify patients according to their predicted outcome, such as complete response at the end of therapy, or to determine disease phenotype, by means of radiomic features.

The Image Biomarker Standardisation Initiative (IBSI) (Zwanenburg et al. 2016) provides standard definitions and nomenclature for radiomic features, reports guidelines for performing radiomic analysis and provides benchmark datasets and values to verify radiomic feature calculations.

The process of building a prognostic/predictive model, or radiomic signature, has four main stages (see Figure 15.3). The first step involves acquisition of images using different modalities (e.g.

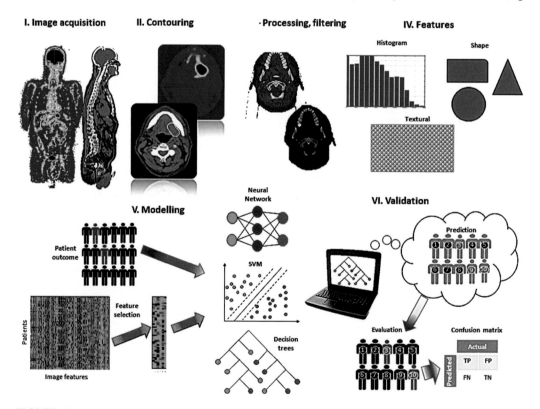

FIGURE 15.3 Model building.

CT, PET, ultrasound, MRI, dynamic contrast-enhanced (DCE) MRI, cone beam CT). Commercial and open source software packages are available to the scientific community that allow radiomic analysis to be performed, helping to standardise both feature definitions and computation methods (Castiglioni et al. 2019).

15.3.2.1 Segmentation of Region-of-Interest

The calculation of radiomic features is typically applied to a specific region of interest (ROI), such as the gross tumour, which first needs to be segmented manually on the images by an experienced radiologist or oncologist. However, a number of algorithms have been developed for automatic or semi-automatic segmentation, such as region growing, the relative threshold method in PET or variational approaches also in PET (Avanzo et al. 2017).

15.3.2.2 Extraction of Radiomic Features

Prior to feature extraction, the preprocessing steps of resampling and discretisation need to be applied to the images. These aim to reduce image noise and so enable reproducible and comparable analysis.

A large number of features, typically in the range of a few hundred to thousands, then can be extracted from each patient image. These variables can be calculated over the entire ROI or in a (n x n) pixel neighborhood of each voxel in the input image in order to map local properties.

The main categories of radiomic features are as follows (Zwanenburg et al. 2016).

Morphological features, such as volume, surface area, 2D and 3D maximal diameter, effective diameter, compactness, eccentricity, sphericity. These describe the size and shape of the ROI.

First-order or histogram statistics describe the distribution of the intensities of voxels within the ROI, ignoring the spatial interactions between them, and can be calculated from histogram analysis, as mean, minimum, maximum and standard deviation. Skewness and kurtosis measure the degree of histogram asymmetry and sharpness, respectively. The uniformity and entropy of the image histogram reflect inhomogeneity.

Second-order histogram or textural features describe the spatial distribution of voxel intensity levels. The term image texture refers to the perceived or measured spatial variation in intensity levels, which can be visualised as a grey level scale. Textural features can be calculated from the grey-level co-occurrence matrix (GLCM). This is a matrix with rows and columns representing grey level values. Cells at the intersection of each row and column indicate the number of times that pixels with the corresponding grey values in a certain spatial relationship with each other (such as a specific angle or distance) occur in the image (see Figure 15.4). GLCM-based features include entropy, energy and contrast. Other types of matrices describing image texture are the grey level run-length matrix (GLRLM) and grey level size zone matrix (GLSZM).

15.3.2.3 Machine Learning for Building Radiomic Models

In the language of machine-learning, the task of radiomics is equivalent to building an algorithm, or 'classifier', to correctly label images by analysing training data consisting of images of patients

Test image				GLCM 0°				GLCM 90°				GLCM 135°			
0	0	1	1	4	2	1	0	6	0	2	0	2	1	3	0
0	0	1	1	2	4	0	0	0	4	2	0	1	2	1	0
0	2	2	2	1	0	6	1	2	2	2	2	3	1	0	2
2	2	3	3	0	0	1	2	0	0	2	0	0	0	2	0

FIGURE 15.4   Example of calculation of GLCM.

(Adapted from Haralick et al. 1973.)

with known outcomes and their radiomic features (Avanzo et al. 2017). The building of a classifier has two steps, as follows.

> **Feature selection.** By feature selection, we mean an algorithm that is used to select features that are relevant to explaining a given output and are non-redundant. The simplest feature selection method is to use a scoring criterion for the variables, based on the degree of stability or correlation of variables, and remove those with worst ranking. Other feature selection methods used in radiomics are minimum redundancy maximum relevance (mRMR) and RELIEF (RELevance In Estimating Features).

> **Model building.** After the feature subset is obtained, various machine learning algorithms can be applied. In supervised approaches, the training data consists of a set of examples, each example representing a pair of an input vector made up of radiomic features from a patient with known outcome and a binary or continuous number describing that outcome. Among the most popular supervised classifiers in radiomics are logistic regression, decision tree-based classifiers, support vector machines (SVM) and artificial neural networks (ANNs) (Avanzo et al. 2019). Unsupervised models have no prior knowledge of the outcome and their goal is, for example, to identify cancer subtypes from a database of patients with similar characteristics.

15.3.2.4 Deep Learning

With the development of deep learning technologies based on multi-layer neural networks, especially convolutional neural networks (CNN), instead of training the machine to predict outcomes from pre-determined features we allow computers to learn for themselves which features optimally represent the data. A typical scenario is to build up a representation layer by layer. For example, the first level of the CNN may represent edges in an image oriented in a particular direction, the second may detect motifs in the observed edges, the third could recognise objects from ensembles of motifs (Le Cun et al. 2015).

15.3.2.5 Validation

Supervised feature selection methods are prone to overfitting, with the model reflecting noise in the image more than the original data. Once models are developed using the selected predictors, quantifying the predictive ability of the models (validation) is necessary. Based on the TRIPOD criteria, there are four types of validation of increasing strength. The weakest of these is developing and validating a model using the same data, which gives an optimistic estimation of the true performance. The strongest approach is to develop the model using one data set and validate using separate data (Collins et al. 2015). Classifiers are evaluated using either the confusion matrix or a receiver operating characteristic curve (ROC), which is a two-dimensional plot with the true positive rate on the y-axis and the false positive rate on the x-axis, the area under the resulting ROC curve (AUC) being a measure of performance.

15.3.3 OVERVIEW OF RESULTS OF RADIOMICS STUDIES IN ONCOLOGY

The radiomics approach, combining extraction of radiomic features with machine learning, can be used to detect/diagnose cancer, to automatically contour a tumour, and to predict tumour characteristics such as histology and genetic footprint. Models predicting response to therapy, recurrence, occurrence of lymph node or distant metastases, and survival, have been successfully implemented for a variety of pathologies. In radiotherapy, radiomics has also been able to predict side effects to organs at risk, such as radiation-induced lung injury or radiation pneumonitis. As well as applications in radiotherapy, radiomics has been applied to predict patient outcome in immunotherapy, molecular targeted therapy and laser therapy (Avanzo et al. 2017).

15.3.4 Reproducibility

Both digital and physical (Mackin et al. 2015) phantoms have been proposed for testing the entire radiomics chain from image acquisition to determination of prognostic output. When CT is the imaging modality used, the choice of scanner, methods of reconstruction, smoothing of the image and slice thicknesses are found to affect radiomic features. In PET imaging, textural features are sensitive to different acquisition modes, reconstruction algorithms, and user-defined parameters such as number of iterations, post-filtering level, input data noise, matrix size, and discretisation bin size (Leijenaar et al. 2013). Radiomic features extracted from MRI scans depend on the field of view, field strength, reconstruction algorithm and slice thickness. Segmentation affects the radiomics workflow, regardless of the imaging technique, because many extracted features depend on the segmented region. Semiautomatic segmentation algorithms may improve the stability of radiomic features, and recently available fully automatic segmentation tools may be as accurate as manual segmentation by medical experts.

15.3.5 Conclusions

The field of radiomics is constantly expanding and is an exciting opportunity for the medical physics community to participate in novel research in quantitative imaging. Machine and deep learning-based models have the potential to provide physicists and clinicians with new methods to improve diagnosis, select treatment and assess response in oncology and many other fields.

15.4 ULTRA-HIGH DOSE RATE RADIOTHERAPY – FLASH-RT

15.4.1 Introduction

As outlined in Chapter 9, radiotherapy is a major treatment option in the management of patients with cancer. Treatment is delivered with both curative and palliative intent with more than half of all cancer patients likely to receive radiotherapy during the management of their disease. The basis of radiotherapy relies on the observation that normal tissues can recover from the harmful effects of ionising radiation more effectively than tumours. However, the dose that can be delivered to the tumour is ultimately limited by the maximum dose that these surrounding normal tissues can withstand and recover from. In current conventional radiotherapy (CONV-RT), this differential effect is taken advantage of by fractionating the total dose administered with typical doses of 2 Gy per treatment fraction providing good protection of normal tissues and enabling them to recover more effectively between fractions than the tumour cells. In addition, technological developments have improved the accuracy with which this dose is delivered to the treatment volume, reducing further any collateral normal tissue damage. These two factors have enabled the dose to the tumour to be increased over the years, while maintaining or reducing side effects, improving the therapeutic benefit of radiotherapy regimes. State of the art CONV-RT is delivered today using intensity-modulated radiotherapy (IMRT), image-guided RT (IGRT), stereotactic ablative radiotherapy (SABR) and proton therapy.

Initial pre-clinical studies have shown that irradiation at dose rates far exceeding those currently used in CONV-RT contexts reduces radiation-induced toxicities even further while maintaining an equivalent tumour response. This irradiation at ultra-high dose rate is known as the FLASH effect with FLASH radiotherapy (FLASH-RT) emerging as a further potential method to increase normal tissue tolerances. This could enable further escalation of curative tumour doses and provide a new method to overcome the radiation resistance of tumours. However, much work needs to be done to understand the biology of the FLASH effect and develop the equipment and processes that would safely and effectively administer ultra-high radiation dose rates in a routine clinical setting.

15.4.2 What Is the FLASH Effect?

FLASH-RT involves the ultra-fast delivery of radiotherapy at dose rates several thousand times greater than those currently used in CONV-RT. In terms of mean dose rates, this is typically ≥ 40Gy/s for FLASH-RT versus ≥ 0.01Gy/s for CONV-RT (Bourhis et al. 2019a). A more complete definition of FLASH-RT parameters involves several inter-dependent physical factors. Due to the very high power required to generate radiation on a linear accelerator, it cannot be run in a continuous mode and instead is run in a pulsed mode. The generated radiation is therefore pulsed and for FLASH-RT the width and repetition rate of these pulses along with the overall duration of exposure are emerging as important factors. The dose rate delivered over one pulse will be significantly higher than the overall mean dose rate and this will depend in turn on the number and width of the generated pulses. For example, the FLASH-RT effect has been observed to be reproducible (Bourhis et al. 2019a) with 1-10 pulses of 1.8-2 μs duration with an overall treatment time of 200 ms and dose-rate within each pulse above 1.8×10^5 Gy/s. In all of these studies, performed on a range of animal models, the dose was delivered in a single treatment fraction. The remarkable observation made after exposure of these biological tissues was the relative protection of normal tissues compared to CONV-RT.

The increased viability of non-cancerous mammalian cells irradiated at ultra-high dose rates was initially reported in the 1960s. The reduction in normal tissue toxicity at ultra-high dose rates was described in the 1970s with further work reported in the early 1980s which demonstrated reduced mouse tail necrosis when irradiated using a 10MeV electron beam running at 50 pulses per second with a dose per pulse of 10^5 Gy/s when compared to more conventional lower dose rates (Hendry et al. 1982). It took several decades before this effect was 'rediscovered' in 2014 and the name FLASH introduced (Favaudon et al. 2014). In this reported work, an increased protection of normal tissue was observed combined with a similar anti-tumour effect for FLASH-RT compared with CONV-RT. These effects have been further investigated and largely corroborated by other groups. The results of the first patient treated with FLASH radiotherapy have been reported (Bourhis et al. 2019b). The treatment was for subcutaneous T-cell lymphoma and resulted in complete response with minimal toxicities.

15.4.3 What Causes the FLASH Effect?

The biological mechanism responsible for the reduction in normal tissue toxicities following FLASH-RT is not currently understood but several hypotheses have been proposed. One suggestion is that the differential response between FLASH-RT and CONV-RT is due to the radiochemical depletion of oxygen which occurs at ultra-high dose rates and the resultant radio-resistance conferred on the irradiated tissue (Wilson et al. 2020). It is generally accepted that hypoxic (oxygen depleted) tissues are more radio-resistant than well oxygenated tissues, as the lack of oxygen in the immediate vicinity of a cell limits the extent of radiation-induced DNA damage. Given that molecular oxygen is depleted as it reacts with free radicals generated from the radiolysis of water, irradiation at ultra-high dose rates is able to significantly deplete oxygen before it has time to replenish. This gives rise to a small window of radiobiological hypoxia. This effect is not observed with CONV-RT delivered at much lower doses per pulse and over longer periods, which enables diffusion of oxygen back into any regions where the levels may have been momentarily depleted. Although oxygen depletion seems to explain the reduced toxicity of FLASH-RT on normal tissue, it does not readily explain how FLASH-RT can maintain similar tumour response relative to CONV-RT. Although tumours are more hypoxic compared to normal tissue, most are not completely devoid of oxygen, so following FLASH-RT it would be expected that there is also radiochemical depletion of oxygen within the tumour, resulting in some additional radio- resistance and therefore reduced tumour control. However, experimental data does not support this with similar tumour control being observed in FLASH-RT as for CONV-RT. Work is currently underway to determine the possible reasons for this.

Another effect of ultra-high dose rates appears to be a reduction in cytokine activation. Cytokines are small proteins secreted from cells which have a specific role in the interaction and communication with other cells. One reported example is the release of transforming growth factor beta, which was significantly reduced in normal lung following FLASH-RT in mice when compared to the same dose delivered with CONV-RT (Rube et al. 2000). This particular cytokine is a well-documented molecular marker of radiation-induced pulmonary fibrosis. Pulmonary fibrosis is one example of normal tissue damage (following lung radiotherapy) that leads to scarring and thickening of the lung tissue, inhibiting normal function.

A modified immune response following FLASH-RT, relative to CONV-RT, has also been proposed as a possible mechanism contributing to the FLASH effect (Bourhis et al. 2019a). Fractionated treatment with CONV-RT results in the irradiation of a large proportion of lymphocytes (white blood cells comprising part of the immune system) compared to the total dose delivered in a single fraction. Following a standard CONV-RT regime of 30 fractions of 2 Gy, over 98% of the blood pool was reported as being exposed to more than 0.5 Gy (Yovino et al. 2013). In addition, the induction of chromosomal aberrations in the circulating blood is dependent on the total volume of blood pool irradiated. Therefore, it would be expected that the short irradiation time of FLASH-RT would significantly reduce the volume of blood irradiated, albeit to a higher dose, and therefore decrease the overall number of chromosomal aberrations so reducing negative effects on the immune response. It should be noted that it is unclear whether any effect linking immunity changes to the FLASH effect is as a consequence of it or contributes to it. The FLASH effect is observed in-vitro in bacteria and cell cultures which have no immune system and so effects on the immune response would only be a part of any underlying mechanism.

15.4.4 Clinical Applications of FLASH-RT

Translation of FLASH-RT into clinical practice would enable dose escalation for the treatment of radiation resistant tumours associated with poorer outcomes. The assumption in such cases is that an increased tumour dose could be delivered without inducing severe toxicities to surrounding tissues, as would be observed with CONV-RT. Also, FLASH-RT could be used instead of CONV-RT where good tumour control is currently achieved but with severe normal tissue toxicity. The assumption in this case is that the same tumour dose delivered with FLASH-RT will result in reduced toxicity to the normal tissues. Despite this exciting potential, the extent to which FLASH-RT could be used in clinical practice is still open to debate.

Initial results report an improved dose modifying factor of 20–40% when comparing FLASH-RT with CONV-RT but that this is only exhibited at total doses of 10 Gy or more (Bourhis et al. 2019a). A dose modification factor of 30%, for example, suggests that a dose of 10 Gy delivered using FLASH-RT would exhibit the same normal tissue damage as a dose of only 7 Gy delivered with CONV-RT. This dose may still be considered too large in many clinical scenarios, for example, in the treatment of larger, locally advanced tumours. However, the use of hypo-fractionated treatments (high doses per fraction) is becoming more widely used in the clinic for a range of treatment sites and could be proven even more beneficial with FLASH-RT and its potential for lower levels of normal tissue toxicity. It is clear that a lot of work is required to build clinical evidence to support the introduction of FLASH-RT in the clinic. Appropriately designed clinical trials need to be undertaken to demonstrate whether FLASH-RT has superior clinical outcomes over CONV-RT.

15.4.5 Technical Challenges

Most studies demonstrating a FLASH effect have been performed on dedicated linear accelerators generating electrons as the source of radiation. However, electrons only penetrate to a depth of a few centimetres in tissue and so new equipment needs to be developed in order to penetrate to greater depths. One solution may be the use of very high energy electron beams (100–250 MeV) which have improved penetration, sharp beam penumbra and are less sensitive to tissue heterogeneity than conventional x-rays. The generation of megavoltage x-ray energies at the required dose rates is more challenging on conventional linacs due to the high input power requirements which are many times more than those for generating electron beams of similar dose rates. Other ways of producing 6–10 MV FLASH x-ray beams are being explored to meet this challenge (Maxim et al. 2019). The mechanism of dose deposition in tissue and therefore possible FLASH effect needs to be confirmed for any new delivery method and type of radiation (i.e. electrons, photons or protons). Manufacturers of commercial proton radiotherapy devices are actively developing ways of delivering proton FLASH-RT. In fact, the very first in-human clinical trial evaluating the feasibility and safety of FLASH radiotherapy has recently been launched. The FAST-01 trial is investigating the treatment of patients with painful bone metastases of the extremities (hands, feet and wrist) delivered with a 250 MeV proton beam at FLASH dose rates. This is achieved by modifying a ProBeam proton beam therapy unit (Varian) to enable the FLASH dose rates.

There are potential risks associated with the delivery of FLASH-RT and these need to be carefully considered before clinical use. An ultra-high dose of radiation is being delivered in a very short time with a limited number of pulses (≤ 10) and so the beam monitoring system needs to be able to accurately monitor the dose on a pulse-by-pulse basis. Suitable high-speed detectors with rapid signal acquisition and processing are required to ensure accurate monitoring and beam termination to deliver the required radiation dose. The technology developed for use on large high-energy particle accelerators seems particularly promising for this purpose (Bourhis et al. 2019a).

15.4.6 OTHER BENEFITS OF FLASH-RT

Other factors could also increase clinical interest in FLASH-RT. The very short treatment duration of FLASH-RT minimises treatment delivery uncertainty caused by movement of the patient, effectively freezing physiological motion. This could enable a reduction in treatment margins and hence the irradiation of normal tissues in close proximity to the tumour. The technical challenge is to ensure the internal anatomy of the patient is exactly as intended at the point of beam delivery.

FLASH-RT delivered very quickly in only a small number of fractions compared to CONV-RT could also confer economic and logistical benefit to a radiotherapy department by increasing throughput and quality of life for patients who will not be required to attend for as many treatment sessions.

15.4.7 SUMMARY

FLASH-RT has been shown to markedly improve the differential effect between tumours and normal tissue compared to CONV-RT, with reported dose modifying factors of 20-40%. This results in reduced normal tissue damage for the same delivered tumour dose, or the possibility to increase the tumour dose while retaining comparable normal tissue damage. The consistency of the effect across a range of pre-clinical and animal studies justify its clinical translation, offering the ability to improve treatment, particularly for tumours that are resistant to radiation. There are many technical and clinical challenges ahead both to fully understand the FLASH effect and to effectively and safely deliver the required dose rates at depth within the patient. If these challenges are met, then FLASH-RT could become part of radiotherapy practice in the next 5–10 years.

15.5 EMERGING TECHNIQUES: CONCLUSION

The importance of medical physics in healthcare can be illustrated by considering the increasing number of procedures undertaken in the fields of medical imaging and radiotherapy. For example, in England, there were around 21 million general X-ray examinations in the period 2012–2013, while in the period 2017–2018, there were about 23 million. Over the same period, the number of CT scans increased from about 3.3 million to about 5.1 million.

The 2008 report of the United Nations Scientific Committee on the Effects of Atomic Radiation (UNSCEAR) shows that during the first decade of the twenty-first century there were 3600 million X-ray examinations, 37 million nuclear medicine examinations and 7.5 million radiotherapy treatments globally (UNSCEAR, 2008). Preliminary figures from the latest UNSCEAR report (yet to be published) show significant increases in these figures in the following decade.

This trend requires a significant increase in the number of medical physicists globally. The report of the Global Task Force on Radiotherapy for Cancer from 2015 (Rifat Atun et al., 2015) estimates that, for radiotherapy alone, by 2035 there will be a need for 17,200 newly qualified physicists in high-income countries, 12,500 in upper-middle-income countries, 7,200 in lower-middle-income countries and 2,400 in low-income counties. Adding to this medical physicists contributing to medical imaging and radiation safety, there is a need for approximately 60,000 medical physicists globally by 2035 (Tabakov, 2016). In order to meet this challenge and continue to support global healthcare provision effectively, there will need to be a significant increase in the number of trained medical physics specialists. This requires new educational courses and more students entering medical physics. We believe that this textbook demonstrates the breadth and potential of medical physics and that it will support the development of many new educational courses.

REFERENCES

Avanzo M, Pirrone G, Mileto M, Massarut S, Stancanello J, Baradaran-Ghahfarokhi M, et al. Prediction of skin dose in low-kV intraoperative radiotherapy using machine learning models trained on results of in vivo dosimetry. *Med. Phys.* 2019; 46: 1447–1454.

Avanzo M, Stancanello J, El Naqa I Beyond imaging: The promise of radiomics. *Phys. Med.* 2017; 38: 122–139.

Bourhis J, Montay-Gruel P, Goncalves Jorge P, Bailat C, Petit B, Ollivier J, et al. Clinical translation of FLASH radiotherapy: why and how? *Radiother Oncol.* (2019a); 139: 11–17. doi: 10.1016/j.radonc.2019.04.008

Bourhis J, Sozzi WJ, Jorge PG, Gaide O, Bailat C, Duclos F, et al. Treatment of a first patient with FLASH-radiotherapy. *Radiother Oncol.* (2019b); 139: 18–22.

Bravin, A, et al. X-ray phase-contrast imaging: From pre-clinical applications towards clinics. *Phys. Med. Biol.* (2013); 58: R1–R35.

Carroll, FE, et al. Pulsed tunable monochromatic X-ray beams from a compact source: New opportunities. *Am. J. Roentgenol.* (2003); 181: 1197–1202.

Castelli, E, et al. Mammography with synchrotron radiation: First clinical experience with phase-detection technique. *Radiology* (2011); 259: 684–694.

Castiglioni I, Gallivanone F, Soda P, Avanzo M, Stancanello J, Aiello M, et al. AI-based applications in hybrid imaging: How to build smart and truly multi-parametric decision models for radiomics. *Eur. J. Nucl. Med. Mol. Imaging* 46(13):2673–2699. 2019.

Collins GS, Reitsma JB, Altman DG, Moons KG Transparent reporting of a multivariable prediction model for individual prognosis or diagnosis (TRIPOD): The TRIPOD Statement. *BMC Med.* 2015; 13: 1-014–0241-z.

Diagnostic Imaging Dataset Annual Statistical Release 2017/18, *NHS England*, available free at https://www.england.nhs.uk/statistics/wp-content/uploads/sites/2/2018/11/Annual-Statistical-Release-2017-18-PDF-1.6MB-1.pdf

Diemoz, PC (2018) Non-Interferometric Techniques for X-ray Phase-Contrast Biomedical Imaging. In: Russo, P (Editor) *Handbook of X-ray Imaging - Physics and Technology*, pp. 999–1024. Boca Raton, FL, USA: CRC Press, Taylor & Francis Group

Favaudon V, Caplier L, Monceau V, Pouzoulet F, Sayarath M, Fouillade C, et al. Ultrahigh dose-rate FLASH irradiation increases the differential response between normal and tumor tissue in mice. *Sci. Transl. Med.* (2014).

Gillies RJ, Kinahan PE, Hricak H Radiomics: Images are more than pictures, they are data. *Radiology* 2016; 278: 563-577.

Haralick RM, Shanmugam K, Dinstein I Textural features for image classification *IEEE Trans. Syst. Man Cybern.* 1973; 3: 610–621.

Hendry JH, Moore JV, Hodgson BW, Keene JP The constant low oxygen concentration in all the target cells for mouse tail radionecrosis. *Radiat. Res.* (1982); 92: 172–181. doi:10.2307/3575852

Hertz, HM, et al. (2014) Electron-Impact Liquid-Metal-Jet Hard x-Ray Sources. In: Brahme A. (Editor) *Comprehensive Biomedical Physics*, vol. 8, pp. 91–110. Amsterdam: Elsevier.

Kitchen, MJ, et al. Dynamic measures of regional lung air volume using phase contrast x-ray imaging. *Phys. Med. Biol.* (2008); 53: 6065–6077.

LeCun Y, Bengio Y, Hinton G. Deep learning. *Nature* 2015; 521: 436–444.

Leijenaar RT, Carvalho S, Velazquez ER, van Elmpt WJ, Parmar C, Hoekstra OS, et al. Stability of FDG-PET Radiomics features: An integrated analysis of test-retest and inter-observer variability. *Acta Oncol.* 2013; 52: 1391–1397.

Lewis, RA Medical phase contrast x-ray imaging: Current status and future prospects. *Phys. Med. Biol.* (2004); 49: 3573–3583.

Mackin D, Fave X, Zhang L, Fried D, Yang J, Taylor B, et al. Measuring computed tomography scanner variability of radiomics features. *Invest. Radiol.* 2015; 50: 757–765.

Majumdar, S, et al. Diffraction enhanced imaging of articular cartilage and comparison with micro computed tomography of the underlying bone structure. *Eur. Radiol.* (2004); 14: 1440–1448.

Maxim PG, Tantawi SG, Loo BW PHASER: A platform for clinical translation of FLASH cancer radiotherapy. *Radiother Oncol.* (2019); 139: 28–33.

Muehleman, C, et al. X-ray detection of structural orientation in human articular cartilage. *Osteoarthritis Cartilage* (2004); 12: 97–105

Rifat Atun, DA, Jaffray, MB, Barton, F, Bray, M, Baumann, B, Vikram, TP, Hanna, FM, Knaul, Y, Lievens, TYM, Lui, M, Milosevic, BO, Sullivan, DL, Rodin, E, Rosenblatt, J, Van Dyk, ML, Yap, E, Zubizarreta, M Gospodarowicz Expanding global access to radiotherapy. *Lancet Oncol.* 2015; 16: 1153–1186.

Rigon, L (2014) X-ray Imaging with Coherent Sources. In: Brahme A. (Editor) *Comprehensive Biomedical Physics*, vol. 2, pp. 193–220. Amsterdam: Elsevier.

Rube CE, Uthe D, Schmid KW, Richter KD, Wessel J, Schuck A, et al. Dose dependent induction of transforming growth factor beta (TGF-beta) in the lung tissue of fibrosis-prone mice after thoracic irradiation. *Int. J. Radiat. Oncol.* (2000); 47: 1033–1042.

Smith P, Nusslin F Benefits to medical physics from the recent inclusion of medical physicists in the International Classification of Standard Occupations (ICSO-08). *Med. Phys. Inter.* (2013); 1(1): 10–14, available free at http://www.mpijournal.org/pdf/2013-01/MPI-2013-01-p010.pdf

Tabakov, S Global number of medical physicists and its growth 1965–2015. *Med. Phys. Inter.* (2016); 4(1): 78–81, available free from http://www.mpijournal.org/pdf/2016-02/MPI-2016-02-p078.pdf

United Nations Scientific Committee on the Effects of Atomic Radiation 2008 Report Volume I: General Assembly, Scientific Annexes UNSCEAR(2008).

Wilson JD, Hammond EM, Higgins GS, Petersson K Ultra-high dose rate (FLASH) radiotherapy: Silver bullet or fool's gold? *Front. Oncol.* (2020); 9: 1563. doi: 10.3389/fonc.2019.01563

Yovino S, Kleinberg L, Grossman SA, Narayanan M, Ford E The etiology of treatment related lymphopenia in patients with malignant gliomas: Modeling radiation dose to circulating lymphocytes explains clinical observations and suggests methods of modifying the impact of radiation on immune cells. *Cancer Invest.* (2013); 31: 140–144. doi: 10.3109/07357907.2012.762780

Zdora, M-C State of the art of x-ray speckle-based phase-contrast and dark-field imaging. *J. Imaging* (2018); **4**(5): 60.

Zwanenburg A, Leger S, Vallieres M, Lock S The, Image Biomarker Standardisation Initiative for. Image biomarker standardisation initiative. arXiv e-prints 2016; ar:1612.07003.

Index

Page numbers in *Italics* refer to figures; **bold** refer to tables and page numbers followed by 'n' refer to notes numbers